# System Innovation for Sustainability 1

# System Innovation for Sustainability 1

## PERSPECTIVES ON RADICAL CHANGES TO SUSTAINABLE CONSUMPTION AND PRODUCTION

EDITED BY ARNOLD TUKKER, MARTIN CHARTER,
CARLO VEZZOLI, EIVIND STØ AND MAJ MUNCH ANDERSEN

Greenleaf
PUBLISHING

2 0 0 8

ARCHBISHOP ALEMANY LIBRARY
DOMINICAN UNIVERSITY
SAN RAFAEL, CALIFORNIA 94901

© 2008 Greenleaf Publishing Ltd

*Published by Greenleaf Publishing Limited*
*Aizlewood's Mill*
*Nursery Street*
*Sheffield S3 8GG*
*UK*
*www.greenleaf-publishing.com*

Printed and bound by CPI Group (UK) Ltd, Croydon, CR0 4YY.

Cover by LaliAbril.com.

*All rights reserved. No part of this publication may be reproduced, stored in a retrieval system, or transmitted, in any form or by any means, electronic, mechanical, photocopying, recording or otherwise, without the prior permission in writing of the publishers.*

British Library Cataloguing in Publication Data:
    Perspectives on radical changes to sustainable consumption
        and production. - (System innovation for sustainability ;
        1)
    1. Sustainable development 2. Agricultural innovations
    I. Tukker, Arnold
    338.9'27

ISBN-13: 9781906093037

# Contents

Preface .................................................................................................................. ix

## Part 1: The context of this book ................................................................... 1

**1** Introduction ..................................................................................................... 2
*Arnold Tukker, Sophie Emmert, Martin Charter, Carlo Vezzoli, Eivind Stø,
Maj Munch Andersen, Theo Geerken, Ursula Tischner and Saadi Lahlou*

**2** Sustainability: a multi-interpretable notion: the book's normative stance .......... 14
*Arnold Tukker*

## Part 2: Business perspective ........................................................................ 45

**3** Review: the role of business in realising sustainable
consumption and production ........................................................................... 46
*Martin Charter, Casper Gray, Tom Clark and Tim Woolman*

**4** Business models for sustainable energy ........................................................... 70
*Rolf Wüstenhagen and Jasper Boehnke*

**5** Alternative business models for a sustainable automotive industry .................. 80
*Peter Wells*

**6** Sustainability-related innovation and the Porter Hypothesis:
how to innovate for energy-efficient consumption and production .................. 99
*Marcus Wagner*

**7** Marketing in the age of sustainable development .................................................. 116
*Frank-Martin Belz*

## Part 3: Design perspective ................................................................................. 137

**8** Review: design for sustainable consumption and production systems .............. 138
*Carlo Vezzoli and Ezio Manzini*

**9** Design for (social) sustainability and radical change ........................................ 159
*Ursula Tischner*

**10** Social innovation and design of promising solutions towards sustainability: emerging demand for sustainable solutions (EMUDE) ...................................... 178
*François Jégou*

**11** Eco-Innovative Cities Australia: a pilot project for the ecodesign of services in eight local councils .................................................................... 197
*Chris Ryan*

**12** Is a radical systemic shift toward sustainability possible in China? .................. 214
*Benny C.H. Leong*

## Part 4: Consumer perspective ........................................................................... 233

**13** Review: a multi-dimensional approach to the study of consumption in modern societies and the potential for radical sustainable changes ................. 234
*Eivind Stø, Harald Throne-Holst, Pål Strandbakken and Gunnar Vittersø*

**14** Product-service systems: taking a functional and a symbolic perspective on usership ............................................................................................................. 255
*Gerd Scholl*

**15** Social capital, lifestyles and consumption patterns ......................................... 271
*Dario Padovan*

**16** Linking sustainable consumption to everyday life: a social-ecological approach to consumption research ........................................ 288
*Irmgard Schultz and Immanuel Stieß*

**17** Emerging sustainable consumption patterns in Central Eastern Europe, with a specific focus on Hungary ...................................................................... 301
*Edina Vadovics*

## Part 5: System innovation policy perspective ... 319

**18** Review: system transition processes for realising sustainable consumption and production ... 320
*Maj Munch Andersen*

**19** System innovations in innovation systems: conceptual foundations and experiences with Adaptive Foresight in Austria ... 345
*K. Matthias Weber, Klaus Kubeczko and Harald Rohracher*

**20** Transition management for sustainable consumption and production ... 369
*René Kemp*

**21** Systemic changes and sustainable consumption and production: cases from product-service systems ... 391
*Oksana Mont and Tareq Emtairah*

## Part 6: Conclusions and integration ... 405

**22** Conclusions: change management for sustainable consumption and production ... 406
*Arnold Tukker*

Abbreviations ... 444
About the contributors ... 448
Index ... 455

# Preface

SCORE! (Sustainable Consumption Research Exchanges) is an EU-funded network project that supports the UN's 10 Year Framework of Programmes on Sustainable Consumption and Production (SCP). It runs between 2005 and 2008, consists of 29 institutions, and aims to involve and coordinate a larger community of several hundred professionals in this field, in the EU and beyond.

The SCORE! philosophy assumes that sustainable consumption and production structures can be realised only if experts that understand **business development, (sustainable) solution design, consumer behaviour and system innovation policy** work together in shaping them. Furthermore, this should be linked with experiences of actors (industry, consumer groups, eco-labelling organisations) in real-life consumption areas: **Mobility, Agro-food** and **Activities around the home** (Housing/Energy use). These areas are responsible for 70% of the life-cycle environmental impacts of Western societies. Consequently, the work in the project was organised as follows:

- The *first phase* of the project (marked by a workshop co-organised with the European Environment Agency [EEA] in Copenhagen, April 2006) aimed to convene a positive conjunction of conceptual insights, developed in the four aforementioned science communities, of how 'radical' change in SCP can be governed and realised

- The *second phase* put the three consumption areas centre stage. SCORE! work package leaders inventoried cases 'that work' with examples of successful switches to SCP in their field

The results of the SCORE! project will form the first four volumes of a new series on 'System Innovation for Sustainability', published by Greenleaf Publishing, which aims to bundle insights from the science community in the field of SCP for policy-making. This first book is the fruit of the first SCORE! workshop, organised by RISØ (a national laboratory at the Technical University of Denmark) and TNO (Netherlands Organisation for Applied Scientific Research) with support from the EEA and the European Topic Centre on Waste and Resource Management (ETC-WRM), and which took place on Thursday 20 and Friday 21 April 2006 in Copenhagen, Denmark. Selected, edited papers by the delegates have been complemented by review papers by the SCORE! team on each science domain. Additional chapters on the concept of SCP, and an integrated view of the governance of change in SCP, makes this one of the first comprehensive books discussing systemic change towards SCP. The three follow-up books will discuss change in SCP in the domains of food, mobility and housing.

The output of the SCORE! network has been much more extensive than we ever could have dreamed of at the start of the project. This reflects the fact that so many more people than expected made active contributions to the SCORE! events. The breadth of contributors and the excellence of their contributions illustrate this fact well, and present a clear indication that SCORE! is a viable network above and beyond the EU's period of financial involvement.

*The SCORE! Coordination Team, January 2008*

# Part 1
# The context of this book

# 1
# Introduction

**Arnold Tukker**
TNO, The Netherlands

**Sophie Emmert**
TNO, The Netherlands

**Martin Charter**
Centre for Sustainable Design, UK

**Carlo Vezzoli**
Politecnico di Milano, Italy

**Eivind Stø**
SIFO, Norway

**Maj Munch Andersen**
RISØ, Denmark

**Theo Geerken**
VITO, Belgium

**Ursula Tischner**
econcept, Germany

**Saadi Lahlou**
EDF, France

## 1.1 The relevance of sustainable consumption and production

The future course of the world depends on humanity's ability to provide a high quality of life for a prospective nine billion people without exhausting the Earth's resources or irreparably damaging its natural systems. It was on the basis of this recognition that the World Summit on Sustainable Development (WSSD) in Johannesburg in 2002 called on the international community to work toward improving global living conditions and to 'encourage and promote the development of a ten-year framework of programmes on sustainable consumption and production (SCP) in support of regional and national initiatives to accelerate the shift towards SCP'. Sustainable consumption focuses on formulating equitable strategies that foster the highest quality of life, the efficient use of natural resources and the effective satisfaction of human needs while simultaneously promoting equitable social development, economic competitiveness and technological innovation.

This quote is taken from the Oslo Declaration on Sustainable Consumption, a manifesto developed in February 2005 by about 50 scientists from around the world.¹ It is one of the (many) documents that suggest that significant efforts have to be made to make our production and consumption systems more sustainable to avoid a global crisis with regard to environmental and other resources.² In this context, at various levels the need for policies with regard to SCP is increasingly recognised as a priority. The most important policy statements include:

- The above-cited statement adopted at the WSSD, and the subsequent development of this 10-year framework under leadership of the UN Environment Programme (UNEP) and the UN Department of Environmental and Social Affairs (UN DESA) through the Marrakech process³
- The clear place for SCP in the sustainable development strategy adopted by the European Council in June 2006,⁴ which includes the task to develop an action plan for sustainable production and consumption in Europe by 2007

So, the change to sustainable production and consumption patterns has an important place on various policy agendas, but is it clear how to govern this change, and what such changes imply? The editors of this book, all affiliated to a major EU-funded project called Sustainable Consumption Research Exchanges (SCORE!), think that this is a hard and far from trivial nut to crack. They hence decided to dedicate the first of the two books that will be produced in the project to the question of how radical changes to SCP can be governed. Follow-up books will apply the lessons learned in the field of food, mobility, and energy and housing. This first chapter:

- Discusses the conceptual approach followed in SCORE! and the development of this book
- Introduces the structure of the book

---

1 See www.oslodeclaration.org [accessed 25 August 2007].
2 Compare as well the 'ecological footprint' calculations published by WWF and the related concept of 'World Overshoot Day' coined by the Global Footprint Network (GFN). As put by GFN director Mathis Wackernagel, in 2006 on 9 October humanity started 'to live from its ecological credit card', having used the resources that the Earth could regenerate in that year. We refer, for example, to org.eea.europa.eu/news/Ann1132753060 and www.neweconomics.org/gen/ecologicaldebt091006.aspx [accessed 25 August 2007].
3 Named after the city where the first major follow-up meeting on the 10-year framework after the WSSD was held. Under the umbrella of the 10-year framework, task forces—initiated by national governments with participation from all regions in the world—have been established to develop and test policies on sustainable procurement, sustainable products, sustainable tourism, sustainable building and construction, education for SCP, cooperation with Africa, sustainable lifestyles, and education for sustainable consumption (see www.uneptie.org/pc/sustain/10year/taskforce.htm [accessed 25 August 2007]).
4 ec.europa.eu/environment/eussd [accessed 25 August 2007]

## 1.2 Sustainable consumption research exchanges: conceptual approach

### 1.2.1 Introduction and project structure

SCORE! was set up in 2005 as an EU-funded network project in support of the UN 10-year framework of programmes on sustainable consumption and production. The project included 8 coordinators and 20 members (see Box 1.1), with the endorsement and involvement of a much larger group of SCP practitioners who participated in project events and who also contributed to this book. Three convictions, covered in more detail below, led to the project structure discussed in this section:

- The SCP challenge differs markedly per type of economy
- The SCP challenge can be solved only via an interdisciplinary and intradisciplinary approach
- The SCP challenge is of prime importance for the consumption domains of food, mobility, and housing and energy use

### 1.2.2 Sustainable consumption and production: challenges by type of country

The challenge of how to realise SCP patterns will probably differ greatly per country and world region (Hart and Milstein 1999; Tukker 2005). Even neighbouring countries with very similar economic structures can have different cultures and hence different success—and failure—factors with regard to governance approaches (compare Hofstede 1975; Jasanoff 1986). But such differences pale in comparison to the following division in countries (Hart and Milstein 1999; see also Table 1.1):

- Consumer economies (Western Europe, the US, Japan) with a high wealth per capita level and where poverty is all but eradicated (some 1 billion citizens)
- Emerging economies (e.g. China) that are rapidly changing and developing quickly into modern consumer economies (some 1–2 billion citizens)
- Base-of-the-pyramid (BOP) economies,[5] where the vast majority of people survive on a few dollars per day and which have consumer markets that are of relatively low importance to others in the global system (some 3–4 billion citizens)

This division is somewhat rough, since some economies may in fact fall in part in different classes.[6] Yet Hart and Millstein (1999) argue that these different economies face fundamentally different sustainability challenges:

---

5 These are also referred to as 'survival economies' (Hart and Millstein 1999). Prahalad (2004) suggested later the term 'bottom of the pyramid', more recently replaced by 'base of the pyramid'.
6 For instance, a large part of the population in India may be regarded as living in BOP economies. At the same time, in India a middle class is also rapidly developing, which would classify it as an emerging economy (compare Myers and Kent 2004).

**Box 1.1 Direct participants in the SCORE! project (as of May 2007)**

**Coordination team:**
- Arnold Tukker, TNO, Delft, The Netherlands (project manager)
- Sophie Emmert, TNO, Delft, The Netherlands
- Maj Munch Andersen, RISØ, Roskilde, Denmark (chair of the working group on system innovation policy)
- Martin Charter, The Centre For Sustainable Design, Farnham, UK (chair of the working group on business development)
- Carlo Vezzoli, Politecnico di Milano, Indaco, Milan, Italy (chair of the working group on design)
- Eivind Stø, SIFO, Oslo, Norway (chair of the working group on consumer research)
- Theo Geerken, Vito, Mol, Belgium (chair of the working group on mobility)
- Ursula Tischner, econcept, Cologne, Germany (chair of the working group on food)
- Saadi Lahlou, Electricité de France, Clamart, France (chair of the working group on energy use/housing)

**Members:**
- Robert Wimmer, GRAT, Austria
- Matthias Weber, ARC Seibersdorf, Austria
- François Jégou, SDS/Dalt, Belgium
- John Torgersen, Aarhus School of Business, Denmark
- Margit Keller, University of Tartu, Estonia
- Arouna. Ouèdraogo, INRA, France
- Frank Belz, Technical University of München, Germany
- Gerd Scholl, IÔW, Germany
- Michael Kuhndt, UNEP and the Wuppertal Institute Centre on Sustainable Consumption and Production, Germany
- Sylvia Lorek, SERI, Germany and Austria
- Wynand Bodewes, Erasmus University, The Netherlands
- Han Brezet, Technical University Delft, Industrial Design Department, The Netherlands
- Henk Mol, University of Groningen, The Netherlands
- René Kemp, MERIT, The Netherlands
- Edgar Hertwich, NTNU, Norway
- Cristina Rocha, INETI/CENDES, Portugal
- Oksana Mont, IIIEE, University of Lund, Sweden
- Rolf Wüstenhagen, University of St Gallen, Switzerland
- Sean Blair, Spirit Of Creation, UK
- Tim Cooper, Sheffield Hallam University, UK
- Tim Jackson, Surrey University/RESOLVE, UK

TABLE 1.1 **Sustainability challenges by type of economy**
Source: adapted from Hart and Milstein 1999

| Type of economy | Example countries | Main sustainability challenge |
| --- | --- | --- |
| Consumer | US, Japan, Western Europe | Dramatically lowering resource use while maintaining economic output ('Factor 10') |
| Emerging | China, South-East Asia, South America | Leapfrogging to sustainable structures of consumption and production without copying Western examples first |
| Base of the pyramid | Many countries in Africa | Developing dedicated solutions for the 'base of the pyramid'; providing a basis for sustainable growth |

- Consumer economies such as Western Europe and the US need to focus on reducing material use per consumption unit
- Emerging economies can look at how they can 'leapfrog' directly towards sustainable consumption and production structures
- In BOP economies, consumption and production structures need to be implemented that allow for basic needs to be covered and for subsequent sustainable growth

In fact, in each of these types of economies radical changes are required. But since the *goal* of change and the *nature* of the economy differ so tremendously, inevitably the type of governance will differ as well (Tukker 2005). With SCORE! being an EU-based project, the messages in this book will inevitably be biased towards consumer economies.

### 1.2.3 Sustainable consumption and production: contributions by the science field

As will be explained in more detail in this book, changes towards SCP are not usually simple. They often cannot be understood from a single perspective. With regard to consumption patterns, numerous studies have pointed at the fact that influencing the individual *willingness* or *need* to address environmental problems is not sufficient for change. The NOA model (Vlek *et al.* 1999) states that such a change is a function of a **need**, an **opportunity** (availability of means), and **ability** (access to means).[7] Experts on innovation convey, from a different point of departure, similar complexities with regard to sociotechnical change. They have come to the understanding that technology and technological development cannot be understood in isolation. A technology, the regulations built around it, related user practices, its symbolic meaning, the supporting

---

7 A simple example: 40–50 years ago it was the norm in Europe to take a bath only once a week. If you were to apply this (sustainable) habit now you would probably put your job and social life in peril, and hence it is not an option. Montalvo Corral (2002), on the basis of extensive research, arrives at somewhat different determinants for willingness to change, such as personal attitude, behavioural control and societal pressure.

infrastructure and the related maintenance and supply networks form a 'seamless web' or 'sociotechnical configuration' (Geels 2002; Hughes 1986). This and other work (EEA 2005; Jackson *et al.* 2004; OECD 2002; Segal 1999; Shove 2003) points out that one cannot ignore the systemic context of production and consumption if one is to understand processes of change.

The SCORE! philosophy therefore assumed that SCP patterns can be realised only if experts that understand business development, (sustainable) solution design, consumer behaviour and system innovation policy work together in shaping them (see Fig. 1.1). New sustainable solutions are often seen as key to changing production and consumption patterns—calling for a clear role of the design community in this process (Charter and Tischner 2001; Manzini *et al.* 2004). Yet such solutions will be implemented only when consumers accept and businesses can profitably produce them—calling for a clear role for experts with a business management background and for consumer scientists, who understand under which conditions consumers can be drivers for change.[8] And, last but not least, expertise is needed to understand innovation from a systemic perspective, particularly since many sustainability problems seem to be unsolvable by actors in the production–consumption value chain in an existing market (Elzen *et al.* 2004; Tukker and Tischner 2006).

**FIGURE 1.1** Knowledge communities involved in the SCORE! Project

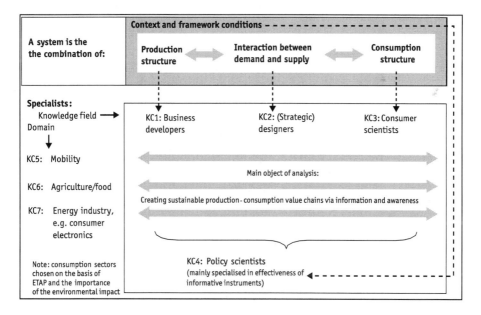

---

8 Indeed, contributors to this book have been surprised to learn that many studies in the field of sustainable product services, basically a function-oriented business model, were driven by normative desires and did not pay much attention to business management literature (Tukker and Tischner 2006).

8  SYSTEM INNOVATION FOR SUSTAINABILITY 1

### 1.2.4 Sustainable consumption and production: priority areas

Finally, there is the question: which are the priority areas in initiating the change to more sustainable production and consumption patterns? Of course, this question in part is determined by what one understands as SCP. Yet, with regard to the environmental dimension, this question can be answered unambiguously. In the period 2000–2006 a series of comprehensive studies and related reviews were performed of the life-cycle environmental impacts of final consumption expenditure in a great variety of countries (e.g. Belgium, The Netherlands, Norway, the EU-25 and the US (e.g. see Hertwich 2005; Suh 2004; Tukker 2006; Tukker *et al.* 2006; Weidema *et al.* 2005). We decided to include only illustrative results of this body of work in this book (see Table 1.2). For paper-length reviews and analyses we refer the reader to Hertwich 2005 or Tukker and Jansen 2006.

The variety of approaches in these studies for analysing the impacts of products was immense. Studies focused on different geographical areas, clustered products in different ways into groups, used fundamentally different data inventory methods (bottom-up life-cycle assessment or top-down input–output) and used different impact assessment methods. Still, the main priorities that all studies identified are crystal-clear. Taken together, mobility (car and air transport, including for holidays), food (meat and dairy followed by other types of food) and energy use in and around the home (heating, cooling and energy-using products) plus house-building and demolition cause, in most environmental impact categories, 70–80% of all life-cycle environmental impacts in society. The conclusion is obvious. If priorities (from an environmental perspective) have to be set, SCP programmes should focus on mobility (including tourism), food, and energy use and housing.

## 1.3 Implications for the structure of SCORE! and this book

The deliberations outlined above would in principle lead to a structure for approaching the change to SCP patterns with three dimensions:

- The type of economy: consumer, emerging or BOP
- The experts and expertise involved: business specialists, designers, consumer scientists and innovation policy experts
- The domains: with food, mobility, and energy and housing having top priority

With SCORE! largely confined to Europe for institutional reasons, the project could not cover the first dimension extensively.[9] The other two dimensions were included (see Fig. 1.1, page 7), in the following project structure (see Fig. 1.2):

---

9 In principle, the EU framework programmes under which the project was funded allow only EU and affiliated states to participate in the framework programmes. Exceptions exist but proved to be unavailable for SCORE!

**TABLE 1.2** Life-cycle environmental impacts of consumption, measured by energy-related indicators, per final consumption expenditure category[a]

Source: adapted from Tukker and Jansen 2006

| COICOP expenditure category[b] | Study | % of total expenditure in EU-25 (Tukker et al. 2006) | Coltins et al. 2006 | Dall et al. 2002 | Moll and Acosta 2006 | Nijdam et al. 2005 | Palm et al. 2006[c] | Peters and Hertwich 2006 | Huppes et al. 2006 |
|---|---|---|---|---|---|---|---|---|---|
| | Geographical focus | EU-25 | Cardiff | Denmark | Germany | Netherlands | Sweden | Norway | EU25 |
| | Indicator | | Ec. Footprint | Energy | Energy | GWP | $CO_2$ | $CO_2$ | GWP |
| | Main approach | | Top-down/ hybrid | Bottom-up | Top-down | Top-down | Top-down | Top-down | Top-down |
| CP01-02 | Food | 19.3% | 21.0% | 26.2% | 13.0% | 22.1% | 7.7% | 12.2% | 31.0% |
| CP03 | Clothing | 3.1% | 0.8% | 1.3% | 2.2% | 6.5% | 0.7% | 10.3% | 2.4% |
| CP04-05 | Housing | 25.1% | 30.8% | 40.8% | 54.3% | 33.4% | 29.1% | 23.0% | 23.6% |
| CP06 | Health | 3.9% | 0.3% | n/a | 1.8% | 0.3% | 1.0% | 1.1% | 1.6% |
| CP07 | Transport | 14.1% | 22.4% | 19.5% | 18.3% | 17.3% | 15.5% | 35.9% | 18.5% |
| CP08 | Communication | 4.0% | 0.5% | | | 0.0% | 1.7% | 2.1% | 2.1% |
| CP09 | Recreation | 9.1% | 8.3% | 7.2% | 8.1% | 15.1% | 0.5% | 0.5% | 6.0% |
| CP10 | Education | 1.4% | 0.3% | n/a | 1.8% | 0.7% | 0.3% | 0.1% | 0.5% |
| CP11 | Restaurants | 9.6% | 11.0% | n/a | n/a | 2.8% | 1.8% | 1.3% | 9.1% |
| CP12 | Miscellaneous | 10.3% | 4.5% | 5.1% | 0.4% | 1.8% | 6.6% | 13.1% | 5.2% |
| Other | Direct purchase of energy by households | | | | | | 35.0% | | |
| | TOTAL | 100% | 100% | 100% | 100% | 100% | 100% | 100% | 100% |

a Hence it is not concerned solely with the impacts in geographical regions. It concerns all environmental impacts generated in the often global production chains, caused by, for instance, the consumption of food in Cardiff.
b COICOP: 'Classification of Individual Consumption According to Purpose' (a UN statistical classification).
c Palm et al. discern a separate category that covers $CO_2$ emissions from direct energy purchases by households. This concerns mainly oil and gas for heating houses (CP04-05) and petrol for cars (CP07).

The top 3 in each study are highlighted in grey.

FIGURE 1.2 **The structure and timetable of SCORE!**

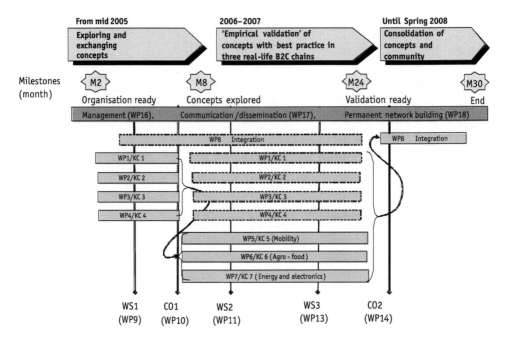

- The first phase of the project aimed to convene a positive confrontation of conceptual insights, developed in the four aforementioned science communities, of how 'radical' change to SCP can be governed and realised (resulting in the present book providing integrated conceptual insights into change to SCP)

- The second phase puts the three consumption areas centre stage. SCORE! workpackage leaders will inventory cases 'that work' with examples of successful switches to SCP in their field. In a series of conferences and workshops, cases will be analysed on 'implementability', adapted where needed, and policy 'prescriptions' will be worked out that can support implementation (resulting in a second book on practical guidance for change to SCP in the three areas of consumption)

This first book is the fruit of the 'positive confrontation' between 75 scientists from the four different communities during a workshop in April 2006 hosted in Copenhagen by the European Environment Agency (EEA). Not totally without coincidence, the four science fields usually have little interaction, each having their own scientific journals, professional societies, and so on. And, almost consequentially, each science field tends to have its own views with regard to how change to SCP should be realised. At the risk of creating caricatures, the positions of practitioners in these fields tend to be characterised as follows. Designers, who are action-oriented, simply start with all their great creativity to work on new sustainable solutions, only to be caught by the unpleasant surprise that the world for some reason does not implement many of their beautiful ideas.[10] An often-heard slogan on the business side is 'sustainability through the market': once perverse subsidies are removed and the environmental externalities are included in the prices, the market will solve it all (cf. Holliday and Pepper 2001). On the consumer side, the following story can often be heard: if only consumers were informed which products are 'sustainable', and take up their responsibility as citizens by buying 'green', the sustainable nirvana would be realised overnight.[11] Finally, system innovation specialists preach the need to perform analyses of complex systems, visioning exercises, and 'learning by doing' and 'doing by learning' experiments to understand how change should be fostered (Elzen et al. 2004)—an approach that would probably make the average sustainable designer or non-governmental organisation (NGO) rather impatient and, indeed, could provide a pretext to postpone virtually self-evident choices, since one can always learn more.

But, obviously, many smart scientists in these fields have developed very valuable insights that go well beyond these caricatures, whatever inherent limitation the field may still pose. What this book tries to do is to:

- Review the state of the art on the 'governance of change to SCP' from the perspective of the particular science field (Parts 2–5). This is done by a comprehensive review of the science field by one of the editors, and illustrated by three or four other (edited) papers from selected presenters during the workshop[12]

- Analyse the strengths and weaknesses of each perspective and to develop an integrated vision on how to govern change to SCP (the concluding chapter of this book, in Part 6)

First, in Chapter 2, we provide an analysis of the concept of SCP. After all, governance of change to SCP suggests a normative direction. Despite the problems involved, this book hence must discuss what SCP is, is not, or could be.

So, how should this book be read? We think that this depends on who you are. Our suggestions are as follows:

- If you have little time and want to understand only our key messages and the key dos and don'ts when stimulating SCP, read just the concluding chapter (Chapter 22)

- If you have bit more time and want to understand in more depth the contribution and limitations of business, design, consumers and system innovation policy in the change to SCP, read also the review chapters (Chapters 3, 8, 13 and 18) and Chapter 2 on SCP concepts

- If you are thrilled by the subject of SCP, read the book in full. We are convinced it is worth it

10 Compare Kazazian (2003) who did one of the more laudable attempts to overcome this implementation gap by checking his ideas with real-life companies. Unfortunately, only in exceptional cases did companies take on his sustainable business suggestions.
11 Indeed, this rather naïve philosophy seemed to be behind the work programme under which SCORE! was funded—the requirement being a project focused on informative instruments. We decided to take a much broader approach since such instruments alone won't do the trick.
12 Note that during the workshop itself all papers were thoroughly discussed. Usually, the discussant represented a different science field from that of the presenter.

# References

Charter, M., and U. Tischner (eds.) (2001) *Sustainable Solutions: Developing Products and Services for the Future* (Sheffield, UK: Greenleaf Publishing).

Collins, A., A Flynn, T. Wiedmann and J. Barrett (2006) 'The Environmental Impacts of Consumption at a Sub-national Level: The Ecological Footprint of Cardiff', *Journal of Industrial Ecology* 10.3: 9-24.

Dall, O., J. Toft and T.T. Andersen (2002) 'Danske husholdningers miljøbelastning' (working paper 13; Copenhagen: Miljøstyrelsen; www.mst.dk/udgiv/Publikationer/2002/87-7972-094-3/pdf/87-7972-095-1.PDF [accessed 25 August 2007]).

EEA (European Environment Agency) (2005) *Household Consumption and the Environment* (Report 11/2005; Copenhagen: EEA).

Elzen, B., F.W. Geels and K. Green (2004) *System Innovation and the Transition to Sustainability* (Cheltenham, UK: Edward Elgar).

Geels, F. (2002) 'Technological Transitions as Evolutionary Reconfigurations Processes: A Multilevel Perspective and a Case Study', *Research Policy* 31: 1,257-74.

Hart, S., and M.B. Milstein (1999) 'Global Sustainability and the Creative Destruction of Industries', *Sloan Management Review*, Autumn 1999: 23.

Hertwich, E. (2005) 'Life-cycle Approaches to Sustainable Consumption: A Critical Review', *Environmental Science and Technology* 39.13: 4,673-84.

Hofstede, G. (1975) *Allemaal Andersdenkenden: Omgaan Met Cultuurverschillen* (Amsterdam: Contact Amsterdam).

Holliday, C., and J. Pepper (2001) *Sustainability through the Market: Seven Keys to Success* (Geneva: World Business Council for Sustainable Development, 2nd edn; www.wbcsd.ch/DocRoot/xs6OhpvANJi0GJPFEkBH/stm.pdf [accessed 25 August 2007]).

Hughes, T.P. (1986) 'The Seamless Web: Technology, Science, Etcetera, Etcetera', *Social Studies of Science* 16.2: 281-92.

Huppes, G., A. de Koning, S. Suh, R. Heijungs, L. van Oers, P. Nielsen and J.B. Guinée (2006) 'Environmental Impacts of Consumption in the European Union: High-Resolution Input–Output Tables with Detailed Environmental Extensions', *Journal of Industrial Ecology* 10.3: 129-46.

Jackson, T., W. Jager and S. Stagl (2004) 'Beyond Insatiability: Needs Theory, Consumption and Sustainability', in L. Reisch and I. Røpke (eds.), *The Ecological Economics of Consumption* (Cheltenham, UK: Edward Elgar).

Jasanoff, S. (1986) *Risk Management and Political Culture: A Comparative Study of Science in the Policy Context* (New York: Russell Sage Foundation).

Manzini, E., L. Collina and E. Evans (eds.) (2004) *Solution-Oriented Partnership: How to Design Industrialised Sustainable Solutions* (Cranfield, UK: Cranfield University).

Moll, S., and J. Acosta (2006) 'Environmental Implications of Resource Use: Environmental Input–Output Analyses for Germany', *Journal of Industrial Ecology* 10.3: 25-40.

Montalvo Corral, C. (2002) *Environmental Policy and Technological Innovation. Why do Firms Reject or Adopt New Technologies?* (Cheltenham, UK/Northampton, MA: Edward Elgar).

Myers, N., and J. Kent (2004) *New Consumers: The Influence of Affluence on the Environment* (Washington, DC: Island Press).

Nijdam, D.S., H.C. Wilting, M.J. Goedkoop and J. Madsen (2005) 'Environmental Load from Dutch Private Consumption: How Much Damage Takes Place Abroad?', *Journal of Industrial Ecology* 9.1–2: 147-68.

OECD (Organisation for Economic Cooperation and Development) (2002) *Towards Sustainable Household Consumption? Trends and Policies in OECD Countries* (Paris: OECD).

Palm, V., A. Wadeskog and G. Finnveden (2006) 'Swedish Experience Using Environmental Accounts Data for Integrated Product Policy Issues', *Journal of Industrial Ecology* 10.3: 57-72.

Peters, G.P., and E.G. Hertwich (2006) 'The Importance of Imports for Household Environmental Impacts', *Journal of Industrial Ecology* 10.3: 89-109.

Prahalad, C.K. (2004) *The Fortune at the Bottom of the Pyramid: Eradicating Poverty through Profits* (Upper Saddle River, NJ/Philadelphia, PA: Wharton School Publishing).

Segal, J.M. (1999) *Graceful Simplicity: Towards a Philosophy and Politics of Simple Living* (New York: Holt).
Shove, E. (2003) *Comfort, Cleanliness and Convenience: The Social Organisation of Normality* (Oxford: Berg).
Suh, S. (2004) *Materials and Energy Flows in Industry and Ecosystem Networks: Life-cycle Assessment, Input–Output Analysis, Material Flow Analysis, Ecological Network Flow Analysis, and their Combinations for Industrial Ecology* (PhD Thesis, Leiden: CML, Leiden University).
Tukker, A. (2005) 'Leapfrogging into the Future: Developing for Sustainability', *International Journal for Innovation and Sustainable Development* 1.1: 65-84.
—— (ed.) (2006) 'Special Issue on Priorities for Environmental Product Policy', *Journal of Industrial Ecology* 10.3.
—— and B. Jansen (2006) 'Environmental Impacts of Products: A Detailed Review of Studies', *Journal of Industrial Ecology* 10.3: 159-82.
—— and U. Tischner (eds.) (2006) *New Business for Old Europe: Product-Service Development, Competitiveness and Sustainability* (Sheffield, UK: Greenleaf Publishing).
——, G. Huppes, S. Suh, R. Heijungs, J. Guinée, A. de Koning, T. Geerken, B. Jansen, M. van Holderbeke and P. Nielsen (2006) *Environmental Impacts of Products* (Seville, Spain: ESTO/IPTS).
Vlek, C.A.J., A.J. Rooijers and E.M. Steg (1999) *Duurzamer Consumeren: Meer Kwaliteit van Leven met Minder Materiaal?* (*More Sustainable Consumption: More Quality of Life with Less Material?*) (research report for the Dutch Ministry of Environment; Groningen: COV, Groningen University).
Weidema, B.P., A.M. Nielsen, K. Christiansen, G. Norris, P. Notten, S. Suh and J. Madsen (2005) *Prioritisation within the Integrated Product Policy*, 2.-0 (Copenhagen: LCA Consultants for Danish EPA).

# 2
# Sustainability: a multi-interpretable notion
The book's normative stance

**Arnold Tukker**
TNO, The Netherlands

## 2.1 Introduction

The goal is to realise more sustainable production and consumption patterns, but is it clear what this means? As with many concepts, there is agreement as long as it is expressed as a general notion, but there is interpretative conflict when the notion is specified. The same is the case with the notions of sustainable development and sustainable consumption and production (SCP). This chapter presents some views on how sustainable development and SCP can be interpreted. In order to do this, we first review and discuss some principles of sustainable development and SCP included in policy documents and publications of authoritative thinkers. We then translate this into notions of what sustainable societies may look like. We end by stating what we see as the minimum requirements for a sustainable society.

## 2.2 Sustainability and sustainable consumption and production

### 2.2.1 Official policy concepts: principles and related indicator systems

The concept of sustainable development originates in the 1980s. One of the early definitions was given in the World Conservation Strategy, developed by the United Nations Environment Programme (UNEP), the WWF and the International Union for Conservation of Nature and Natural Resources (IUCNNR). However, the concept made its true breakthrough when the Brundtland Report was published (WCED 1987). Over-consumption and grinding poverty had to be addressed by sustainable development: 'development that meets the needs of the present without compromising the ability of future generations to meet their own needs' (WCED 1987). At a global level, the sustainable development concept and related goals have been refined over time during the United Nations Conference on Environment and Development (UNCED) at Rio de Janeiro in 1992 and the World Summit on Sustainable Development (WSSD) in Johannesburg in 2002 (see Box 2.1).

The definitions in general have a clear anthropocentric character: human development is the key point. The definitions and concepts further often mention intergenerational and intragenerational equity and the eradication of poverty as key goals. Although most of the concepts suggest an environmental, economic and social dimension, it seems that the environmental dimension is 'more equal than the others'. Social and economic development should take place within the Earth's 'carrying capacity', suggesting that this environmental dimension forms a precondition for the other dimensions. Several indicator systems have been proposed to measure progress with regard to sustainable development.

An interesting point is that the Johannesburg Plan of Action and the EU Sustainable Development Strategy (SDS) of 2006 specify the concept of SCP within the context of sustainable development. They do so in slightly different ways. The Johannesburg Plan of Action works out SCP mainly as an issue of environmental protection and resource efficiency: 'de-linking economic growth and environmental degradation through improving efficiency and sustainability in the use of resources and production processes and reducing resource degradation, pollution and waste'. Actions mentioned in this Plan deal surprisingly little with consumption patterns, being more concerned with cleaner production and energy efficiency—in brief, with technologies (applied in processes and, at a distance, in products) to prevent environmental degradation.[1] The consumer *is* mentioned, but mainly in relation to awareness-raising programmes and information tools. The EU SDS places a very clear emphasis on the social element of sustainable development, although here also the true consumption side is under-represented. Here, the need to make products and processes environmentally and socially more sustainable and to support their markets seems to be the way forward (European Commission

---

1 The issue of social performance is mentioned in Point 18 of the Johannesburg Plan of Implementation, under the heading 'enhancing corporate environmental and social responsibility and accountability'.

Box 2.1 Some definitions and objectives of sustainable development

### International Union for Conservation of Nature and Natural Resources, the WWF and the United Nations Environment Programme

According to the International Union for Conservation of Nature and Natural Resources, the WWF and the United Nations Environment Programme (IUCNNR/WWF/UNEP 1980), 'for development to be sustainable, it must take account of social and ecological factors, as well as economic ones: of the living and non-living resource base, and of the long-term and short-term advantages and disadvantages of actions'.

### World Commission on Environment and Development

The World Commission on Environment and Development states that sustainable development is (WCED 1987): 'development that meets the needs of the present without compromising the ability of future generations to meet their own needs'.

### The Rio Declaration

The 27 principles of the Rio Declaration (UNCED 1992), agreed to support sustainable development, include the following statements:

- 'Human beings are at the centre of concerns for sustainable development. They are entitled to a healthy and productive life' (Principle 1)
- 'The right to development must be fulfilled so as to equitably meet developmental and environmental needs of present and future generations' (Principle 3)
- 'Environmental production shall constitute an integral part' (Principle 4)
- 'Eradicating poverty [is] an indispensable requirement for sustainable development' (Principle 5)
- 'States should reduce unsustainable patterns of production and consumption' (Principle 8)

### Johannesburg Plan of Implementation

The Johannesburg Plan of Implementation (UN 2002) states that the Plan 'strongly reaffirm[s] commitment to the Rio principles', and suggests a need to promote the integration of the three components of sustainable development—economic development, social development and environmental protection—as interdependent and mutually reinforcing pillars.

## UN Millennium Goals

The UN Millennium Goals (UN 2000), although of a more generic character, clearly embrace sustainability principles:

1. Eradicate extreme poverty and hunger
2. Achieve universal primary education
3. Promote gender equality and empower women
4. Reduce child mortality
5. Improve maternal health
6. Combat HIV/AIDS, malaria and other diseases
7. Ensure environmental sustainability
9. Develop a global partnership for development

## European Union

The EU (2006) sets in its Sustainable Development Strategy the following key objectives:

- Environmental protection: safeguard the Earth's capacity to support life in all its diversity, respect the limits of the planet's natural resources and ensure a high level of protection and improvement of the quality of the environment. Prevent and reduce environmental pollution and promote sustainable production and consumption to break the link between economic growth and environmental degradation
- Social equity and cohesion: promote a democratic, socially inclusive, cohesive, healthy, safe and just society with respect for fundamental rights and cultural diversity that creates equal opportunities and combats discrimination in all its forms
- Economic prosperity: promote a prosperous, innovative, knowledge-rich, competitive and eco-efficient economy that provides high living standards and full and high-quality employment throughout the EU
- Meeting our international responsibilities: encourage the establishment and defend the stability of democratic institutions across the world, based on peace, security and freedom. Actively promote sustainable development worldwide and ensure that the European Union's internal and external policies are consistent with global sustainable development and its international commitments

Box 2.2 **Example definitions and objectives of sustainable consumption and production**

### Soria Moria Symposium on Sustainable Consumption and Production, Oslo, 19–20 January 1994

According to the Soria Moria Symposium on Sustainable Consumption and Production, a follow-up meeting of the Rio Conference dedicated to SCP: SCP is 'the use of goods and services that respond to basic needs and bring a better quality of life, while minimising the use of natural resources, toxic materials and emissions of waste and pollutants over the life-cycle, so as not to jeopardise the needs of future generations' (Norwegian Ministry of Environment 1994).

### Johannesburg Plan of Implementation

According to the Johannesburg Plan of Implementation (UN 2002), the aim is to encourage and promote the development of a 10-year framework of programmes in support of regional and national initiatives to accelerate the shift towards sustainable consumption and production to promote social and economic development within the carrying capacity of ecosystems by addressing and, where appropriate, de-linking economic growth and environmental degradation through improving efficiency and sustainability in the use of resources and production processes and reducing resource degradation, pollution and waste.

The Plan then mentions tools such as life-cycle assessment, suggests the need to improve products and services while reducing health and environmental impacts, and to increase eco-efficiency and to develop and adopt awareness-raising programmes and consumer information tools relating to SCP.

### European Union

The EU (2006), in the SCP chapter of the Sustainable Development Strategy, mentions as targets the promotion of 'SCP by addressing social and economic development within the carrying capacity of ecosystems and decoupling economic growth from environmental degradation' and the improvement of 'the environmental and social performance [of] products and processes'; it further mentions green public procurement, labelling and target-setting for environmental and social performance, and information campaigns.

2006). Box 2.2 reflects on some definitions of SCP developed in international policy circles and documents. Indicator systems are typically subsets or complementary to those developed to monitor sustainable development and are focused more on final (household) consumption (e.g. OECD 1999; UNCSD 2001c).

### 2.2.2 Attempts to operationalise the notion of sustainable development

Particularly after the UNCED in Rio, various attempts have been undertaken to operationalise the notion of sustainable development in terms of clear conditions for economic and social development and indicators for monitoring progress. This body of literature is vast, and we will review here only the most authoritative attempts.

A first group of authors tried to come up with holistic **design rules** for economic processes. Examples include The Natural Step (Nattrass and Altomare 1999), the Hannover Principles (McDonough and Braungart 1998) and Datschefski's (2001) sustainability checks (Box 2.3). All these design rules have their roots in a holistic view of economy and ecology and often translate ecological development principles into criteria that should also apply to industrial systems (see e.g. the concept of industrial ecology [Frosch and Gallopoulos 1989; see also Ayres and Kneese 1969]). They tend to be source-oriented (i.e. posing direct demands on processes of production and consumption) rather than effect-oriented (i.e. estimating what nature can tolerate and then setting standards for emission and resource use). McDonough and Braungart prescribe the use of solar income, that waste should equal food, and that diversity should be stimulated. Following such philosophies, the Swedish Chemicals Inspectorate (KemI) suggested that persistent chemicals alien to nature should, in principle, be eliminated; for persistent chemicals that occur in nature, emission targets should be linked to natural background flows and concentrations in the environment; and, for other pollutants, emission targets should be set on the basis of known effects (KemI 1991). It goes without saying that such design rules end up being quite restrictive in comparison with the current way of organising production and consumption patterns.

Other studies aim to determine more or less objective boundaries for economic and social development. Typically, such studies have proposed more or less scientific criteria for values, objects or resources in the environment that need to be protected which policy-makers then could use to select a protection goal. This would then give insight into the maximum impact that is to be tolerated, and, by calculating backwards via, for example, the emission-effect chain, a maximum emission or resource-use level may be established. Depending on the geographical scale of the environmental problem, this then could serve as a basis for emission quota or targets for the world, regions, countries or sectors. Examples include many national and international goals for emission reductions, including the Kyoto Protocol (though inevitably targets were often relaxed in the political bargaining process; e.g. see Box 2.4). The ecological footprint can be seen as an

## Box 2.3 Principles for organising consumption and production systems

### The Natural Step (Nattrass and Altomare 1999)
- 'Substances from the Earth's crust must not systematically increase in the ecosphere'. Hence, waste must not systematically accumulate. In practical terms this implies a shift to the use of recycled materials and renewable resources
- 'Substances produced by society must not systematically increase in nature'. Therefore, synthetic substances should not be released into the environment more quickly than they can be safely broken down in nature, and emissions of persistent substances should be avoided altogether
- 'The physical basis for the productivity and diversity of nature must not systematically deteriorate'. Plant life is the only natural factor that uses solar energy to produce biomaterials out of less structured matter; this ensures that life on Earth does not slowly fade out as a consequence of the second law of thermodynamics
- 'Basic human needs must be met with fair and efficient use of energy and other resources'. Only under these conditions is cooperation possible in order to realise the above three conditions

### The Hannover Principles (McDonough and Braungart 2003)
- Insist that the rights of humanity and nature co-exist in a healthy, supportive, diverse and sustainable condition
- Recognise interdependence. The elements of human design interact with and depend on the natural world, with broad and diverse implications at every scale. Expand design considerations to recognise even distant effects
- Respect relationships between spirit and matter. Consider all aspects of human settlement—including community, dwellings, industry and trade—in terms of existing and evolving connections between spiritual and material consciousness
- Accept responsibility for the consequences of design decisions on human well-being, the viability of natural systems and the right of these to co-exist
- Create safe objects of long-term value. Do not burden future generations with the need to maintain or administer vigilantly any potential danger arising from the careless creation of products and processes or the use of inappropriate standards
- Eliminate the concept of waste. Evaluate and optimise the full life-cycle of products and processes so that they approach the state of natural systems, in which there is no waste
- Rely on natural energy flows. Human designs should, as in the living world, derive their creative forces from perpetual solar income. Incorporate this energy-efficiently and safely for responsible use
- Understand the limitations of design. No human creation lasts forever and design does not solve all problems. Those who create and plan should practise humility

in the face of nature. Treat nature as a model and mentor, not as an inconvenience to be evaded or controlled
- Seek constant improvement by the sharing of knowledge. Encourage direct and open communication between colleagues, patrons, manufacturers and users to link long-term sustainable considerations with ethical responsibility, and re-establish the integral relationship between natural processes and human activity

**Sustainability checks** (Datschefski 2001)
- Is it cyclic (i.e. made with organic materials or resources that are in closed loops)?
- Is it solar (i.e. using renewable energy during production and use)?
- Is it safe (i.e. are emissions to water, air and soil 'food' for other systems)?
- Is it efficient (i.e. can the same functionality be realised with less impact)?
- Is it social (i.e. does manufacture support basic social rights and natural justice)?

---

approach belonging to this category (Wackernagel and Rees 1996)[2] as are the following criteria developed by Daly (1991: 44):[3]

- Harvesting of renewable resources should not exceed regeneration rates
- Waste emissions should not exceed renewable assimilative capacity
- Non-renewable resources should be exploited, but at a rate equal to the creation of renewable substitutes

---

2 In essence, this method calculates how much bioproductive land is needed to produce the resources society needs and to absorb the emissions society creates (e.g. the amount of biomass that has to be produced to compensate for $CO_2$ emissions). By comparing the footprint with available bioproductive land it can be assessed if the world (or an individual country) possesses ecological reserves and/or has ecological deficits. Most industrialised countries have a clear deficit, and in 2004 the global footprint was 1.23: that is, humanity had overshot the carrying capacity of the Earth by 23% (WWF 2004). The ecological footprint method has been praised for its ability to reduce a complex issue into an easily understandable and appealing concept, but is has also been criticised for its value-laden and subjective choices (Van Kooten and Bulte 2000; van den Bergh and Verbruggen 1999). In similar vein, scholars such as Vitousek et al. (1986) and Imhoff et al. (2004) analysed how much organic material equivalent of the present net primary production in terrestrial ecosystems is being appropriated by humans; they concluded it is many tens of per cent each year (referred to as the 'human appropriation of net primary production' [HANPP]).
3 Daly further mentions that human scale (i.e. throughput in the economy) must be within the carrying capacity of the Earth, which requires choices to be made regarding population and consumption limits (and hence with regard to efficiency). The long-run marginal costs of expansion should be equal to the long-run benefits of expansion. Furthermore, technical progress for sustainable development should increase efficiency rather than throughput.

## Box 2.4 Target-setting for acidifying emissions: effect-based standard setting with regard to acidification in The Netherlands

In the 1980s, considerable concern over the effects of emissions of $SO_2$, $NH_3$ and $NO_x$ arose: it became clear that these substances, once deposited, would have an acidifying effect on soil and water, with clear effects on flora and fauna. The doomsday scenario of a massive *Waldsterben* (literally: forest-dying) was born. Studies analysed which 'critical load' deposition levels would be related to which types of effect in different ecosystems (e.g. conversion of moor into grassland at 700–1,100 mol H+ per hectare per year, or the dying of fish in surface water at 4,400 mol H+ per hectare per year). Also, given the drastic reductions in emissions needed to reach 700 mol H+ deposition per hectare, taking into account background concentrations caused by transboundary air pollution, the Dutch National Environmental Policy Plan went for a long-term emission target based on 2,400 mol H+ per hectare per year (RIVM 1991: 184-213).

A third group of authors operationalised the concept of sustainability as a number of interrelated capital stocks. Ekins (1992) proposed a distinction into manufactured, human, social and ecological capital. Wealth is created by the production of a flow of products and services making use of this (total) capital. The productive value of this capital (determined by its quality and quantity) must be kept intact (or, better, increased) to ensure the intergenerational sustainability mentioned in the Brundtland Report (compare with Solow 1992). The famous 'sustainability triangle', portraying sustainability as interrelated development along economic, social and environmental axes, reflects this approach (see Fig. 2.1).[4]

Closely related are conceptualisations of authors that base themselves on (complex) systems theory (e.g. Gallopín 2003). This theory sees development as a co-evolutionary process of interacting, self-organising (sub)systems (see e.g. Fig. 2.2). The human subsystem is not viable without the others, and the total system is capable of more than its parts (O'Connor and McDermott 1997). According to Bossel (1999), sustainability indicators and related criteria must describe therefore the performance of the individual subsystems, and the contribution to performance of other subsystems in terms of the satisfaction of the fundamental 'interests' of each subsystem. He claims these to be existence, effectiveness, freedom of action, security, adaptability, co-existence and psychological needs (for humans, and for systems with humans as components). Changes imposed on a subsystem by the other subsystems must not exceed the adaptive capacity of that subsystem. Such principles allow for a structured search of indicators that avoids redundancy or ad hoc expert judgement. Bossel seems keenly aware that his mix-

---

4 The capital approach also led to attempts to correct the most common indicator for 'progress', the gross domestic product, for externalities and other flaws: that is, for social and environmental capital losses, for the production of goods intended to compensate for such losses, or for 'inflation' of the market economy by inclusion of goods and services that used to be available for free. Such indicators have been published under names such as genuine progress indicator (Venetoulis and Comb 2004), measure of domestic progress (Jackson 2004) and index of sustainable economic welfare (Daly and Cobb 1989). Compare also Lawn 2003, Costanza *et al.* 1997 and our own contribution in Tukker and Tischner 2006.

FIGURE 2.1 The sustainability triangle

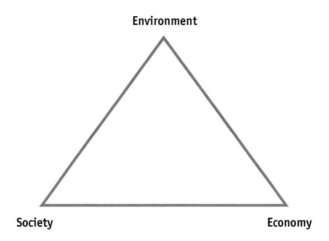

FIGURE 2.2 Subsystems in the anthroposphere

Source: adapted from Bossel 1999: 18

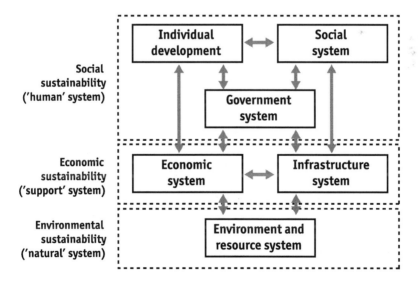

ing of principles of logic and judgement is essential. It is an illusion that all relations in the (sub)systems can be mapped and that the search for stock and flow indicators that matter has to be done through an interactive process. Such 'subjective qualitative assessment of viability and performance . . . suffices in many cases, but if results are to be reproducible, or assessments are to be compared, a more formalised quantitative procedure is required' (Bossel 2000: 9).[5]

### 2.2.3 Discussion

In sum, the definition of sustainability is far from unambiguous. A first discussion is often centred on uncertainty and ignorance. It is nice to put forward principles such as 'waste emissions should not exceed renewable assimilative capacity', but often it is impossible to assess within reasonable limits this assimilative capacity. The forceful warnings of Gore (2006), Stern (2006) and others (e.g. Lynas 2004) that global warming jeopardises the very existence of humanity are opposed by, for example, Lomborg (2001) and Livermore (2007), who see it as an exaggerated or solvable problem.[6]

Further discussion surrounds the substitutability of different types of 'capital' or 'stock'. This is far from trivial: social scientists tend to stress adaptability and the preservation of social and cultural systems, natural scientists tend to define sustainability in terms of the resilience of biogeophysical systems, and economists tend to relate sustainability to the preservation of productive capital stock (Munasinghe and McNeely 1995). The so-called 'weak sustainability' view accepts a full substitutability of stocks as long as the total productivity and output can be maintained or enlarged. In the words of Ekins et al. (2003): 'it is clear that sustainability may be consistent with the decline of one type of capital stock as long as another type of capital is increasing sufficiently to compensate for this decline'. Thus, from this perspective, pharmaceuticals can replace medicinal plants and livestock wild animals while biodiversity is maintained—so, in this view, there is no problem if African savannas are turned into agricultural land (cf. Munasinghe and McNeely 1995). In the 'strong sustainability' view crucial ecological capital stocks must be maintained at current levels: products of the ecosystem can be used at a rate only within that ecosystem's rate of renewal. Thus, in the aforementioned example, the African savannas should probably be protected—leading to cheerful comments from opponents of 'strong sustainability' that the present 'natural' landscape dominated by acacias was created only about a century ago after the human introduction of a pathogen (Munasinghe and McNeely 1995)! An intermediate position, dubbed 'sensible sustainability', accepts that ecological stocks may decline, but defines a criti-

---

5 Rotmans (2003) has proposed a similar approach using the sustainability triangle but, compared with Bossel, uses as guidance fewer criteria derived from systems theory.
6 As shown by, among others, Shackly and Wynne (1995) and van der Sluijs (1997), the rather narrow bandwidth of projections of global temperature rise put forward by the Intergovernmental Panel on Climate Change (IPCC) rests in large part on a converging social construction process within that body. Daily and Ehrlich (1992) comment that 'the complexity of these interactions makes it unlikely that they will be sufficiently well evaluated in the next several decades to allow firm calculations of any carrying capacity'.

cal level of certain stocks that must be maintained.⁷ The idea is that systems can become totally disrupted if a specific stock totally disappears and that, as a precaution, stocks should be allowed to build to higher levels if deemed necessary (see e.g. Munasinghe and Shearer 1995; Seragelding and Sheer 1994; Vuuren and de Kruijf 1998).⁸

Last, a final discussion questions whether 'sustainable development' is not a contradiction in itself. The great achievement of Brundtland was that her commission developed a concept that reconciled the positions of those pursuing economic growth and those pursuing environmental protection (*Economist* 2002; WCED 1987). However, doubt persists regarding whether this position is not too optimistic: 'the economy is an open subsystem of the Earth ecosystem, which is finite, non-growing, and materially closed. As the economic subsystem grows it incorporates an ever greater proportion of the total ecosystem into itself and must reach a limit at 100 per cent, if not before' (Daly 1993). This is at odds with the 'sustainable' (meaning 'perpetual') and 'development' (often set equal to 'growth') elements in the Brundtland Report definition. This is no minor problem in a global economic system that seems to need the inherent growth of a few per cent per year to stay stable (Jespersen 2004).⁹

All these discussions reflect that the complexity of the Earth's system is enormous— and it has to be noted that the most thorough discussions and analyses have focused almost entirely on the environmental dimension and have tended to neglect the social and equity side of sustainability. Questions on what sustainability means inevitably have a 'trans-scientific' character: 'questions that can be asked of science but that cannot be answered by science' (Weinberg 1972). Approaches such as The Natural Step reduce complexity by giving simple ground rules (rules that are too simple?), whereas authors taking a comprehensive systems view end up recognising that criteria and indicators cannot be but a mix of interactive judgement and logic. Many authors hence have concluded that the discussion on sustainability reflects a 'plethora of paradigms', an issue explored further in the next section (Fischer-Kowalski *et al.* 1994; Gallopín 2003; WRR 1994).

---

7 This 'sensible sustainability' position was developed by the World Bank, and probably, not entirely coincidentally, fits rather well with the Cobbs–Douglas production function, a well-known economic formula describing the substitutability of assets for production. It describes the output as a product of input of human and technical capital. As a consequence, if one of these capital stocks is zero, output is zero. There must thus be a minimum input. The concept of 'environmental utilisation space' can be seen as an elaboration of 'sensible sustainability' (Weterings and Opschoor 1994).

8 Various tales exist about populations ending up in crisis after the destruction of their natural capital by overpopulation and overuse. See Klein 1968 with regard to how a population of 29 reindeer introduced in St Matthew Island, Alaska, grew to 6,000 and then collapsed to 50, and see Young 2006 for a classic tale on the Easter Island crisis.

9 This may not look much but it implies a duplication of the economy in 20–25 years. In theory, the contradiction can be alleviated by pursuing development in terms of 'immaterial' growth, but this suggestion has been received sceptically. Most growth is needed in the Southern Hemisphere to ensure that the basic needs of the poor are met: needs that can be fulfilled only by material products such as houses, clothes and food. And even in Western economies it transpires that seemingly 'immaterial' services are produced by a very material supply chain. In practice, a massive shift from high-impact to low-impact expenditure, if indeed possible, would result in probably only a factor of 2 or less impact (Tukker *et al.* 2006). And, if Vitousek *et al.* (1986) were correct that humanity has already appropriated 40% of the ecologically available organic material equivalent in the early 1980s, Daly's inherent borders may be reached sooner rather than later.

## 2.3 Sustainability as a paradigmatic concept

### 2.3.1 Introduction

Unfortunately, sustainability paradigms are not neutral. Source-oriented paradigms such as The Natural Step pose much more stringent rules on production and consumption than do emission platforms based on 'sensible sustainability'. Such paradigms easily become part of a messy struggle for discursive and political hegemony between actor coalitions, with beliefs and interests usually hopelessly mixed up (Jasanoff 1990; Hajer 1995; Tukker 1999).[10] Science cannot play its desired role of arbiter. It can usually find a single and undeniable answer if one is dealing with relatively 'simple', not so open-ended, questions that by the vast majority of humans are conceptualised in a single way

**Box 2.5 Cultural theory and the management of toxic substances**

> Cultural theory postulates two dimensions of sociality by which an individual's involvement in social life can be described: group and grid (Thompson *et al.* 1990). The term 'group' refers to the extent to which an individual is bound in a unit. The term 'grid' refers to the degree to which an individual's life is determined by external prescriptions. By varying the group and grid position, one can see four forms of social organisation emerging: individualist, hierarchist, egalitarian and fatalist. Strong group boundaries with minimal prescriptions produce egalitarian social relations. Hierarchical relations are the result of strong group boundaries and strong binding prescriptions. For individualists there is neither group incorporation nor a prescribed role. Fatalists are subject to binding prescriptions and are excluded from group membership, being passive spectators of discussions between the other groups. Those in each of the three active ways of life (individualists, hierarchists and egalitarians) develop their own biases with regard to their 'myth of nature', how they prefer to match needs with resources and their preferences in dealing with blame, risk, economic growth and so on (see Table 2.1).
> 
> Tukker (1999) found these frames to be the driving factors behind controversies over the sustainable management of persistent, toxic, substances. Simply put, the egalitarians distrust knowledge and technology. The avoidance of the production and use of such substances is then perfectly logical, particularly if the change to alternatives is not costly to society as a whole. Individualists trust knowledge, technology and the resilience of nature. The setting of emission standards based on an assessment of effects is then perfectly logical: phase-outs would be an intolerable breach of entrepreneurial freedom and would leave individual companies as losers. Hierarchists choose the middle ground, opting for strict control of emissions but leaving production structures intact (see Fig. 2.3). This source-versus-effect management approach can be found in many of the principles discussed in Section 2.

---

10 The following is an illustrative quote from Kristalina Georgieva, Director for the Environment at the World Bank: 'I've never seen a real win–win [situation] in my life. There's always somebody, usually an elite group grabbing rents, that loses. And we've learned in the past decade that those losers fight hard to make sure that technically elegant win–win policies do not get very far' (*Economist* 2002).

## TABLE 2.1 Key characteristics of positions in cultural theory

Source: Thompson et al. 1990

|  | Individualist | Hierarchist | Egalitarian |
|---|---|---|---|
| **World-view** | | | |
| Myth of nature | Nature is benign | Nature is perverse/tolerant | Nature is ephemeral |
| Concept of human nature | People are self-seeking | People are sinful | People are born good |
| Needs and resources | Can manage both | Can manage resources | We can manage needs |
| **Management style** | | | |
| General approach | Adaptive | Controlling | Preventative |
| Attitude to nature | *Laissez-faire* | Regulatory | Attentive |
| Attitude towards humans | Channel rather than change them | Restrict their behaviour | Egality first |
| Attitude to needs and resources | Seek the limits of each | Increase resources | Reduce needs |
| Economic growth | For personal wealth | For collective wealth | Not preferred |
| Risk | Risk-seeking | Risk-accepting | Risk-averse |

## FIGURE 2.3 Frames for the sustainable management of persistent substances

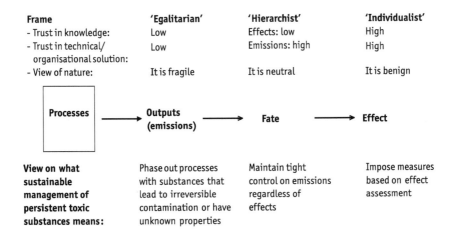

| Frame | 'Egalitarian' | 'Hierarchist' | 'Individualist' |
|---|---|---|---|
| - Trust in knowledge: | Low | Effects: low | High |
| - Trust in technical/organisational solution: | Low | Emissions: high | High |
| - View of nature: | It is fragile | It is neutral | It is benign |

Processes → Outputs (emissions) → Fate → Effect

| View on what sustainable management of persistent toxic substances means: | Phase out processes with substances that lead to irreversible contamination or have unknown properties | Maintain tight control on emissions regardless of effects | Impose measures based on effect assessment |

(cf. Knorr-Cetina 1993; Kuhn 1962, 1991). But for complex, 'trans-scientific', questions this situation is different. As a result of bounded rationalities (Simon 1957), differences in belief systems (Sabatier 1987) and so on, different actor coalitions come to answers that have a well-thought-out internal logic but which nevertheless are coloured by the cognitive lenses that each actor coalition (unknowingly) wears (Schwarz and Thompson 1990; Thompson et al. 1990; WRR 1994). Below, we will give three archetypical, and in part opposing, paradigms for sustainable development: one following roughly the 'sensible sustainability' approach, setting boundary conditions in which a free-market economy can flourish without too many binding rules; one applying basic principles for (presumed) healthy social and environmental development within the economic system; and one based on top-down management. The positions loosely relate to the 'individualist', 'egalitarian' and 'hierarchic' paradigms postulated by cultural theory that have been frequently recognised in practical case studies (see e.g. Thompson 1984; Tukker and Butter 2007; WRR 1994; see also Box 2.5).

### 2.3.2 An individualist view: sustainability through the market

The individualist has a basically optimistic view of the world and human nature and believes that individual ingenuity will in the end bring the solution to problems. Hence, the realisation of sustainable development becomes a matter of channelling individual incentives along the right direction but leaving a significant degree of freedom for individuals and groups to develop their own, preferred, solutions. The position reflects an entrepreneurial spirit and often coincides fairly well with the views of organisations representing industry in debates. Debates on the question of how to realise sustainability are no exception.

The World Business Council for Sustainable Development (WBCSD), probably the most influential industry think-tank on sustainability, even has a slogan that reflects this: 'Sustainability through the market' (Holliday and Pepper 2001). Their 'seven keys to success' leave no doubt that the market should be the primary means to reach a sustainable world (Holliday and Pepper 2001: 4):

> The rule of law, freedom of competition, transparent accounting standards, and a safe social context all contribute to the ability of business to create wealth . . . Proper valuation will help us maintain a diversity of species, habitats and ecosystems; conserve natural resources; preserve the integrity of natural cycles, and prevent the build-up of toxic substances in the environment. Realistic pricing is the optimal means of supporting change in the market.

The idea is that, once perverse subsidies are abolished, and the undesirable side-effects of our current production and consumption systems are internalised into market prices, market mechanisms will direct innovations along the right, sustainable direction, endorsed by informed consumer choice for sustainable products. Sustainable development then becomes a matter of time. And, indeed, there are various examples that show the power of market-based instruments in fostering change. Temporary tax measures paved the way for lead-free petrol and for cars with catalysts in various countries in the EU, and a tax exemption gave a massive boost to the introduction of 'green' electricity in The Netherlands.

In this view, free-market principles would also contribute to the solution of extreme poverty in the developing world: the so-called 'base of the pyramid' (BOP). Prahalad (2004) argues that innovative business models tailored to these BOP markets can work very well, be profitable and contribute to rising wealth by slashing transaction costs and other inefficiencies and by supporting entrepreneurship. The examples given by, for example, Prahalad (2004), Hart and Milstein (1999) and Holliday and Pepper (2001) include systems for providing micro-loans, a leasing system for mobile phones and so on.

This paradigm clearly does not question growth, an optimal economic scale or technological optimism and does not leave much room for ideas that economic power should be distributed, that the economy should be centred on locally oriented networks and so on. If in a market where externalities are included in the prices it is cheaper to fly shrimps from Norway to China to be peeled and then back to be consumed than peeling them in Norway: go ahead. If the market results in major differences in income, the global relocation of industries and hence in significant migration flows: no problem. Traditional development aid is sometimes even seen as being counterproductive and inefficient (Easterly 2006). 'Fair trade' premiums paid to farmers in developing countries may be seen as a distorting subsidy if, across the board, the sector meets basic environmental standards and working conditions.

The critique of this position is that the attainment of free markets is seen almost as goal in itself rather than as a means of reaching a high quality of life for the many. A division between sustainability-supportive and sustainability-adverse competition is not made (Scherhorn 2005). This position further neglects the powerful critiques within neoclassical economics that markets are not, in practice, perfect—for example, because bounded rationalities (Simon 1957), transaction costs (Williamson 1979), externalities and the evolutionary character of economic development result in unwanted 'lock-ins' (Dosi 1982). If these problems result in strong impediments to change that cannot be overcome by financial incentives, a market-based approach to SCP will not be effective.

### 2.3.3 An egalitarian view: distributed, sustainable economies via creative communities[11]

The egalitarian has a risk-averse, precautionary, attitude. The egalitarian society is characterised by operating in social groups without too many binding rules and with a high level of equality. Thinkers from this strand hence put much emphasis on the question of whether 'small' people can organise their own lives in the way in which they like: people should be able to determine their own destinies and to learn and grow from work and other experiences rather than being the passive beneficiaries of structures that take care of them—or, worse, that exploit them (compare Illich 1978; Schumacher 1973; see

---

11 This section relies heavily on a contribution by Carlo Vezzoli and Ezio Manzini to the April 2006 Copenhagen workshop of the SCORE! network (see also Chapter 8).

also Sen 1999).[12] Centralised systems of provision (e.g. energy provision) lead to centralisation of power and reduced opportunities for access to resources, which is a key factor in perpetuating poverty in the world (Rifkin 2002).[13] Implementation of alternative technologies for the production of energy based on, for example, the sun and hydrogen would not only reduce the global warming problem but also lead to a 'distributed economy' consisting of microplants set up close to the end-user, a democratisation of resources and energy, enabling individuals, communities and nations to reclaim their independence (Rifkin 2002). Such localised social networks of stakeholders tend to pay significant attention to preserving (resource) renewability (Sachs 2002). The prosperity of a community depends on the short-term and long-term availability of these resources; also, in this situation, control is easier and feedback better. Both externalities as free-rider behaviour can be better traced at the local scale.

The hope of these authors is that bottom-up initiatives by 'creative communities' will become drivers for change. Examples include communities that set up new participatory social services for the elderly and for parents, and new food networks to foster the consumption and production of food of quality (e.g. organically produced) and with particular characteristics (e.g. the Slow Food movement, solidarity purchasing groups and fair trade organisations).[14] The (alleged) positive environmental, sociocultural and political implications include the following:

---

12 Schumacher and Illich, important writers in the 1970s, had rather similar philosophies. Schumacher's book *Small is Beautiful* proposed the return of the dimensions of production to a 'human scale', so that individuals again could become the masters of technology rather than the reverse. This would allow people to learn and grow again rather than becoming passive, on a treadmill. 'Good work' would 'give a man a chance to utilise and develop his faculties; to enable him to overcome his egocentredness by joining with other people in a common task; and to bring forth the goods and services needed for a becoming existence' (Schumacher 1973). Illich stressed the danger that specialists (doctors, teachers, language specialists developing dictionaries and grammar, etc.) would start to determine elements of life that previously were 'owned' and 'developed' by individuals, resulting in lives with ever-less self-determination, capacities and development.
13 Many have observed that the rise in oil prices during the 1970s and 1980s was the main cause of increases in debt in the third world (e.g. Stiglitz 2002). Today, in many of the world's poorest countries, the cost of paying interest and settling debts is greater than the amount needed to provide essential services for the population of those countries.
14 For example, see the 70 examples of promising cases in the repository of the EU-funded EMUDE project: www.sustainable-everyday.net/EMUDE [accessed 25 August 2007]. Manzini and Jégou (2003) defined these people as 'creative' because they form groups of innovative citizens who are creative enough to invent new ways to solve a problem and/or to open up new possibilities. The British Design Council uses the expression **open model**: an economic system where consumers effectively become co-producers of products and services (Cottam 2004). Though different groups are involved, given the dynamism that characterises these creative communities one can connect the discussion relating to these communities to the older debate on **active minorities** (Moscovici 1979) and to more recent debates on the **creative class** (Florida 2002), the **cultural creatives** (Ray and Anderson 2000) and **creative cities** (Landry 2000).

- Socioeconomic implications: by bringing a large part of the value creation process to a local scale, distributed economies generate, and maintain, local wealth and local jobs. They intensify local activities and interactions (or **social fabric**) and prepare a favourable ground to optimise the use and regeneration of existing **social resources**

- Environmental implications: by reducing the scale of their individual elements, distributed systems permit the optimal use of local resources and facilitate forms of industrial symbiosis that reduce waste. By bringing production nearer to local resources and final users they permit a reduction in the average transport intensity of activities (and, therefore a reduction in congestion and pollution)

- Political implications: by bringing the power of decision nearer to final users and increasing the visibility of the systems on which decisions have to be taken distributed systems facilitate democratic discussion and choice. Since those advantages and problems that are related to a choice can be better compared, such systems facilitate individuals and communities in making responsible decisions

The critique of this vision from regular economists is obvious. They fear that such localised economies are highly inefficient, throwing away all the benefits to be gained from the recipe of specialisation, trade and enlarged scale of production that all but eradicated poverty in the Western world in the 20th century. Their question would be: 'if such communities are so efficient in delivering high-quality products and services, and if people really do want them, why are they still so marginal rather than a major business success?'

### 2.3.4 An hierarchic view: sustainable consumption and production by an Apollo project

The last view is probably more a strategy of how to realise the goal of SCP than a suggestion of goals in themselves. It relies heavily on a top-down approach in solving problems. In its extreme form an all-encompassing blueprint is developed and executed in an orderly, planned and stringently controlled fashion, all under the guidance of a central power. This type of approach to transition management is supported probably by those who call for a 'master plan' or 'Apollo programme' to save the environment. It should consist of an all-encompassing effort, with a lead role for (preferably a supranational) government in various areas of expertise to realise the necessary changes. In a more moderate form, the goals and planning are of a more indicative nature and the assessment of which means should be used to achieve the goals are more participatory, but there is still a powerful central actor that, when needed, can enforce progress on the process of change. Examples may be found in the US (Apollo) space programme launched in the 1960s by President Kennedy and in the (in)famous five-year plans in centrally planned economies. A moderate version of this approach would be to organise a number of binding treaties with strong environmental demands and targets.

This strategy for realising SCP be applied in the following cases. First, there must be a party in the system that has the power or legitimacy to apply such a hierarchic approach. Second, it must be fairly clear which ends should be reached, and what means are the most appropriate to do so.

The obvious weakness in this model is that it rests on a high level of centralised power, even at a global level. This is hardly a realistic proposition in a time where influential scholars have announced the advent of a 'network society' (Castells 2000), where power of the nation-state is perceived as declining in comparison with corporate organisations and when supranational organisations in the field of sustainability have only a limited influence (Fuchs and Lorek 2005).

## 2.4 Conclusions

### 2.4.1 Sustainable consumption and production: a debate on the meaning of life?

So, where are we now? Section 2.3 led probably to a bewildering range of visions of what sustainable development, and as a derivative, SCP could be. One can easily link these visions to programmes of political movements,[15] or, indeed, to religions. What should production and consumption deliver for humanity? Quality of life in a hedonic sense, and hence 'satisfaction', or 'happiness'? Or should it provide spiritual growth (i.e. with the creation of conditions that allow people to develop their capacities and potential and hence grow into 'more complete' and 'mentally richer' human beings)? Such questions have far from unambiguous answers and, indeed, touch on normative questions such as what one sees as the meaning of life.[16]

Although, on the one hand, we probably cannot avoid such questions in full, we are wary of portraying SCP as a box that contains everything. By keeping concepts vague one may have the advantage that many actors will support them.[17] However, any expectation that an SCP agenda to serves as a vehicle for solving all problems in the world, from environmental crises to world poverty, will simply backfire. The uptake of the SCP con-

---

15 Various authors have analysed this policy struggle over the sustainability agenda. Sneddon *et al.* (2006) suggest that social and environmental agendas draw the short straw because of developments such as: the relative decline in power of nation-states compared with 'agents of private capital'; the post-9/11 prioritisation of security from terrorism on international agendas; and the strengthening of intergovernmental institutions that champion economic growth and market liberalisation, most notably the World Trade Organisation (WTO)—where institutions related to the environmental and sociological dimensions are, at best, still in an embryonic state. According to some, this has concentrated power in the hands of parties that benefit from unsustainable growth and resource use (e.g. Woods 1999).
16 These examples were taken from the inaugural exhibition titled 'Happiness: A Survival Guide for Art and Life', held at the Mori Art Museum, Tokyo, from October 2003 to January 2004. The exhibition discerned four ideal types of happiness: a 'back-to-nature' Arcadia, a 'spiritual' Nirvana, 'hedonic' desire and, finally, harmony, in a way a blend of the first three types. See www.mori.art.museum/html/eng/exhibition/index.html [accessed 25 August 2007].
17 See, for example, our earlier comment on the acceptance of the term 'sustainable development' (see Section 2.2.3).

cept on policy agendas has been slow, and we firmly believe that this has had a lot to do with the vagueness with which SCP agendas are still clouded.[18] To solve this, as an editorial team, we must do two things:

- First, we must position the SCP agenda within the broader concept of sustainable development

- Second, we must make choices about our own goals that an SCP agenda should realise—thereby inevitably choosing sides in the partly subjective debate

### 2.4.2 The object of governance for sustainable consumption and production: sustainable consumption and production as part of sustainable development

One problem for the SCP agenda is that systems of consumption and production in principle cover almost the full 'human' (social) system and the (economic) support subsystem shown in Figure 2.2 (page 23). This is, apart from government institutions, the whole anthroposphere. In such a conception, 'sustainable development' and 'development for SCP' can hardly be separated: both deal with the full influence of humans on social and environmental systems.

In conformity with the Johannesburg Plan of Action and the EU Sustainable Development Strategy, we are, however, in favour of seeing the SCP agenda as a subset of a sustainable development agenda. An initial perspective on such a system, illustrated in Figure 2.4, helps us to find some loosely defined boundaries and to focus our attention:

- We are less interested in governance approaches for implementing change that have little effect on consumption and its interaction with production, such as end-of-pipe measures and (to a lesser extent) simple technical fixes of production processes, or policies that are dominated by the strengthening of institutions in, for example, developing countries

- We are interested in governance approaches that influence consumption patterns, the interaction between production and consumption, market dynamics, and the dynamics of change in sociotechnical systems; for example, we are interested in improved patterns of production, more intensive use of products, and consumption that is less material-intensive (either by shifting to immaterial consumption or through the creation of 'greener' products)

---

18 The realisation of tangible success in programmes dealing with a reduction of energy use per unit of use of a product is difficult enough for an operational policy-maker (e.g. a director within an environment ministry). Why should that person suddenly spend resources on a vague agenda with unclear goals that may even include poverty eradication, which has for decades been dealt with by very professional international organisations?

## FIGURE 2.4 The interaction between consumption and production

Source: adapted from Inaba 2004

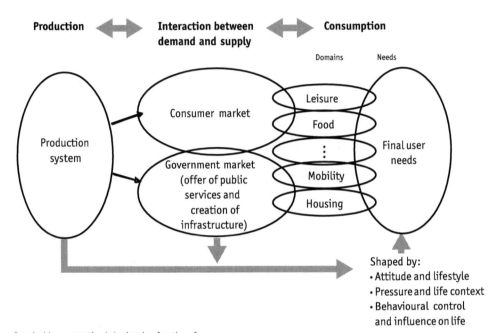

Sustainable consumption behaviour is a function of e.g.:
a  Needs, opportunities and abilities (Vlek *et al.*)
b  Attitude, social pressure and behavioural control (Montalvo Corral 2002)
   User awareness hence needs to be understood in the user's context.

### 2.4.3 The goals of governance for sustainable consumption and production: three fascinations of the editors

#### 2.4.3.1 Introduction

Within this focus, we now have to choose our sustainability goals. We opted for an approach that does not touch on the most value-laden elements in the sustainability discussion. Often, in paradigmatic debates, there are points that can be reasonably underpinned as 'hard' knowledge to which all paradigms (as reviewed in Section 2.3) must comply. Readers that may have expected a sweeping plea for the stimulation of distributed patterns of production and 'Buddhist' social development may be disappointed. We have, however a pragmatic reason for sticking to the basics. For an SCP agenda to be adopted and widely supported, a large number of stakeholders must be convinced of its sense of urgency or reasonableness—even if they do not like the consequences for themselves. Bringing a radical SCP agenda too early into a domain reigned over by opposing normative and policy views is killing it.

This position has a few implications. We will not question the 'growth' paradigm that underlies the prevailing market economies *a priori*, nor assume that the market is the source of all evil. The market economy has emerged from the 20th century as a system that best provides the conditions for freedom, initiative and (albeit in sustainability terms still dubious) prosperity, and we should not throw away the baby with the bathwater. To paraphrase Scherhorn (2005): we should, rather, learn to use the market to foster competition supportive of sustainability. We also will not embark on prescribing social and human development—where we are very well aware that some authors argue that the market economy may be experienced as an out-of-control treadmill forcing or luring consumers into using material goods, partly as a surrogate for immaterial needs (see e.g. Reisch and Røpke 2004; Segal 1999). There are in our view, however, with all the subjectivity that plagues the sustainability agenda, three points that either undeniably have to be acted on or are at least fascinating enough to pay attention to in SCP policy-making. We will discuss them in Sections 2.4.3.2–2.4.3.4.

### 2.4.3.2 The radical reduction of impact per unit of consumption

The first point that we hold as being undeniable is that a radical reduction of resource use and emissions per unit of consumption is essential for economic and social development to be viable over the next century. The world population ($P$) will rise from about 6 to about 9 million, and countries such as China and India are fast developing a wealthy consumer class, which will lead to a rise of affluence ($A$) and related consumption ($C$) with easily a factor 5 (Lutz *et al.* 2004; Myers and Kent 2004). If this occurs under current patterns of consumption and production ($T$), the famous Ehrlich and Holdren (1971) formula (Equation [2.1]) shows us that environmental impact ($I$) will rise tenfold or more:

$$I = PACT \qquad [2.1]$$

There may, of course, be dispute if current environmental limits are surpassed, but in our view it is inconceivable that such growth is possible without facing an ecological crisis. Across the board this implies that patterns of consumption and production must become radically more efficient. We must find ways to fulfil needs with a significant reduction of impacts per unit of consumption, a point from the SCP agenda already conveyed by many before us (Factor 10 Institute 1997; von Weizsäcker *et al.* 1997).[19]

---

19 In his book, *The Sceptical Environmentalist*, Björn Lomborg (2001) makes a passionate plea against what in his view is 'doomsday thinking'. He suggests that, for a great many impact categories, humanity can avoid crises. This may be true, but in his book he also states that technical progress and the substitution of resources will be needed to 'save the day'. The main difference in position between Lomborg and us may be that we think it will not do any harm to stress the sense of urgency for action—change will not come naturally. A more serious point is mentioned by Reijnders (1998): it may be necessary to adapt the Factor $X$ by impact type. This would lead us back to the difficult and often messy assessment of the Earth's carrying capacity (see Section 2.2.3 and footnote 6).

### 2.4.3.3 The eradication of poverty and the fostering of equity, in the South and the North

The second point that we hold as being obvious is that an SCP agenda must be supportive of poverty eradication and contribute to a reasonable level of equity in society: a world with billions of poor, deprived in most cases of basic facilities, does not live up to generally accepted standards of justice and ethics.[20] But what does this mean for an SCP agenda, particularly one looked at from the perspective of a project that is EU-based and hence biased towards the situation in the developed world? We feel it would overburden the SCP agenda if it were made the driving factor behind poverty eradication, and, besides, other institutions deal with this problem already.[21] We do see, however, two contributions SCP can make to the issue. It can do so:

- In terms of research and support projects: to analyse and see how development can be organised in such a way that countries 'leapfrog' to patterns of consumption and production that are viable in the long term, rather than those countries copying the problematic Western example first

- In terms of adjusting international production and trade patterns: to ensure that the externalities of resources and products that the developed countries import from others are reflected in prices. Alternatively, and better: that basic environmental and social and/or labour standards can be met and paid for.[22] Furthermore, it must be ensured that environmental measures have fair distributional effects

### 2.4.3.4 Enhancing the effectiveness of efforts to increase quality of life

The final point that we wish to raise is something with which we do not as yet know what to do, but it is highly fascinating to us and has a clear relation to SCP. Various sources (Hofstetter *et al.* 2005; Layard 2005; Shah and Marks 2004; Veenhoven undated) note that, despite massive economic growth over the past 30 years, in most Western countries the satisfaction with life, or quality of life, experienced did not rise. Apparently, our patterns of production and consumption have become highly inefficient in providing quality of life. This is illustrated dramatically in Figure 2.5, taken from the 'happy planet

---

20 Whereas this point, of course, primarily addresses problems in the Southern Hemisphere, it is also valid for the rich Northern Hemisphere. Extreme differences between rich and poor, in combination with structures that prevent the poor from lifting themselves out of poverty even with hard work, defies a reasonable sense of fairness. And, with environmental impacts largely proportional to consumption expenditure, it is the rich who are at the root of most sustainability problems.

21 However, one probably cannot but acknowledge that certain existing institutions and power structures may consolidate the poverty problem. To quote Archbishop Hélder Camara: 'When I give food to the poor, they call me a saint. When I ask why the poor have no food, they call me a communist'. en.wikipedia.org/wiki/Dom_Helder_Camara [accessed 25 August 2007].

22 This is a simple statement that, unfortunately, is difficult to implement in formal legislation. The WTO allows importing countries to set (non-discriminatory) standards in terms of the properties of product but it does not allow importing countries to set standards with regard to processes that create those products, making voluntary standards (e.g. Forest Stewardship Council [FSC] and corporate social responsibility [CSR] programmes) for the time being the main route for implementation. Another point for discussion will concern the extent to which minimum resource prices are required (as practiced by fair trade organisations), or the idea that price setting is acceptable once basic boundary conditions are met, allowing new entrants to out-compete incumbents.

FIGURE 2.5 Plot of 'happy life years' against ecological footprint, for 178 countries, by world region

Source: adapted from: Marks et al. 2006: 17

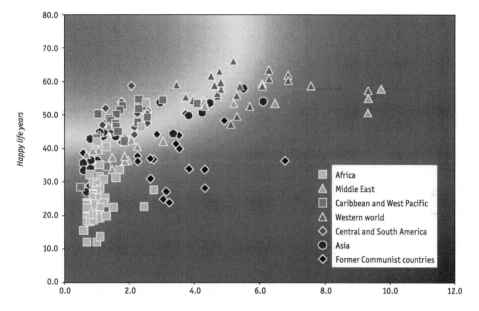

index' report published by the New Economics Foundation and Friends of the Earth. According to the authors (Marks et al. 2006): 'it measures the ecological efficiency with which, country by country, people achieve long and happy lives. In doing so, it strips our view of the economy back to its absolute basics: what goes in (natural resources, measured as the ecological footprint of consumption), and what comes out (human lives of different length and happiness)'. Figure 2.5 suggests strongly that large footprints (related to high consumption per capita and high GDPs) are no precondition for a high quality of life. One could even go as far as saying that having a very high GDP (and hence a large footprint) per capita is not a sign of progress but rather is a sign of inefficiency in providing what truly matters: countries with an equal quality of life and life years may differ by up to a factor 4 in terms of size of footprint. It may be too early to translate this directly into implications for policy, but these findings are fascinating enough to warrant further attention in any SCP agenda. Once the factors determining the shape of Figure 2.5 are better understood—and are shared—we will gain important guidance on how to structure patterns of consumption and production.[23] To be quite clear, this is not a simplistic plea to justify low levels of income simply because, even then, life can still

23 So, as the attentive reader will have noted, where initially we set out to avoid a position that is too value-laden with regard to how consumption and production patterns should develop, such a position may now be seen to be making a return via the back door. Marks et al. (2006) give, as an initial suggestion, the need to shift values, to encourage meaningful lives and to empower people—echoing somewhat the calls for a distributed economy and creative communities outlined in Section 2.3.3.

be pleasant. Rather, the research cited poses the question whether our current approaches to economic development do not, in fact, destroy the very (in)formal institutions, non-market goods and services, and the social fabric that may be essential for experienced quality of life.

### 2.4.4 In sum: our goals of governance for change to sustainable consumption and production

The deliberations in the above sections lead, in sum, to the following focus for our agenda for change to SCP. It concerns two elements: the object of governance, or a system focus (in order to separate it from a general sustainable development agenda), and the goals to be reached with governance:

With regard to the object of governance, we are interested mostly in approaches that influence consumption patterns, the interaction between production and consumption, market dynamics, and the dynamics of change in sociotechnical systems rather than end-of-pipe measures and (to a lesser extent) simple technical fixes of production processes, or policies that are dominated by institutional strengthening in, for example, developing countries.

With regard to the objectives of governance, we believe:

- A radical reduction of impact per consumption unit should be sought
- Support for poverty eradication and equity should be given; for example, financial compensation for compliance with basic environmental, labour and social standards in supply chains should be ensured
- The potential for 'leapfrogging' in developing economies should be investigated and tested
- The reasons for the apparently low efficiency of Western economies in providing high-quality life years should be investigated and understood, and its implications should be translated into guidelines for organising consumption and production patterns

# References

Ayres, R.U., and A.V. Kneese (1969) 'Production, Consumption and Externalities', *American Economic Review* 59.3: 282-96.
Bossel, H. (1999) *Indicators for Sustainable Development: Theory, Method, Applications* (Winnipeg, Canada: International Institute for Sustainable Development [IISD]).
Bossel, H (2000) 'Assessing Viability and Sustainability: A Systems-Based Approach for Deriving Comprehensive Indicator Sets', paper presented at the *Workshop on Integrated Natural Resource Management in the CGIAR*, Penang, Malaysia, 21–25 August 2000; www.inrm.cgiar.org/Workshop2000/docs/Bossel/bossel_main.pdf [accessed 25 August 2007].
Castells, M. (2002) *The Rise of the Network Society* (Malden, MA: Basil Blackwell).

Costanza, R., R. d'Arge, R.S. de Groot *et al.* (1997) 'The Total Value of the World's Ecosystem Services and Natural Capital', *Nature* 387: 253-60.
Cottam, H.L. (2004) *Open Welfare: Designs on the Public Good* (London: Design Council).
Daily, G.C., and P.R. Ehrlich (1992) 'Population, Sustainability, and Earth's Carrying Capacity: A Framework for Estimating Population Sizes and Lifestyles that could be Sustained without Undermining Future Generations', *BioScience* 42: 761-71.
Daly, H. (1991) 'Elements of Environmental Macro-economics', in R. Costanza (ed.), *Ecological Economics: The Science and Management of Sustainability* (New York: Columbia University Press): 32-46.
—— (1993) 'Sustainable Growth: An Impossibility Theorem', in H.E. Daly and K.N. Townsend (eds.), *Valuing the Earth: Economics, Ecology, Ethics* (Cambridge, MA: MIT Press): 267-73.
—— and J.B. Cobb Jr (1989) *For the Common Good: Redirecting the Economy Toward Community, the Environment, and a Sustainable Future* (Boston, MA: Beacon Press).
Datschefski, E. (2001) *The Total Beauty of Sustainable Products* (Mies, Switzerland: Rotovision).
Defra (UK Department of Environment, Food and Rural Affairs) (2005) *Sustainable Development Indicators in Your Pocket 2005 (SDIYP)* (London: Defra Publications).
Dosi, G. (1982) 'Technological Paradigms and Technological Trajectories', *Research Policy* 11: 147-62.
Easterly, W. (2006) *The White Man's Burden: Why the West's Efforts to Aid the Rest Have Done so Much Ill and so Little Good* (New York: Penguin Books).
*Economist* (2002) 'The Great Race', *The Economist*, 4 July 2002.
EEA (European Environment Agency) (2005) *EEA Core Set of Indicators: Guide* (Technical Report No. 1/2005; Copenhagen: EEA).
Ehrlich, P.R., and J.P. Holdren (1971) 'Impact of Population Growth', *Science* 171: 1,212-17.
Ekins, P. (1992) 'A Four-Capital Model of Wealth Creation', in P. Ekins and M. Max-Neef (eds.), *Real-Life Economics: Understanding Wealth Creation* (London/New York: Routledge): 147-55.
——, S. Simon, L. Deutsch, C. Folke and R. De Groot (2003) 'A Framework for the Practical Application of the Concepts of Critical Nature Capital and Strong Sustainability', *Ecological Economics* 44: 165-85.
EU (European Union) (2006) *Renewed Sustainable Development Strategy* (Publication 10117/06; Brussels: Council of the European Union; ec.europa.eu/environment/eussd [accessed 25 August 2007]).
Eurostat (2005a) *Complete List of Sustainable Development Indicators* (Luxembourg: Eurostat; epp.eurostat. ec.europa.eu/portal/page?_pageid=1998,47433161,1998_47437052&_dad=portal&_schema=PORTAL [accessed 25 August 2007]).
—— (2005b) 'Final Report of the Sustainable Development Indicators Task-force Theme 70', presented at the *57th Meeting of the Statistical Programme Committee*, CPS 2005/57/20/EN: epp.eurostat.ec. europa.eu/portal/page?_pageid=1998,47433161,1998_47437059&_dad=portal&_schema=PORTAL [accessed 25 August 2007].
Factor 10 Institute (1997) *Declaration of Carnoules* (Carnoules, France: Factor 10 Institute).
Fischer-Kowalski, M., H. Haberl and H. Payer (1994) 'A Plethora of Paradigms: Outlining an Information System on Physical Exchanges between the Economy and Nature', in R. Ayres and U.E. Simonis (eds.), *Industrial Metabolism: Restructuring for Sustainable Development* (Tokyo: United Nations University Press): ch. 14.
Florida, R. (2002) *The Rise of the Creative Class and How it is Transforming Work, Leisure, Community and Everyday Life* (New York: Basic Books).
Frosch, R.A., and N.E. Gallopoulos (1989) 'Strategies for Manufacturing', *Scientific American* 189.3: 152.
Fuchs, D., and S. Lorek (2005) 'Sustainable Consumption Governance: A History of Promises and Failures', *Journal of Consumer Policy* 28.3: 361-70.
Gallopín, G. (2003) 'A Systems Approach to Sustainability and Sustainable Development', paper presented at *Sustainability Assessment in Latin America and the Caribbean*, Santiago, Chile, March 2003, Medio Ambiente y Desarrollo 64, organised by the Sustainable Development and Human Settlements Division, ECLAC as part of the Government of The Netherlands Project NET/00/063.
Gore, A. (2006) *An Inconvenient truth: The Planetary Emergency of Global Warming and What We Can Do About It* (Emmaus, PA: Rodale Press).
GRI (Global Reporting Initiative) (2002) *Sustainability Reporting Guidelines* (Amsterdam: GRI).

Hajer, M.A. (1995) *The Politics of Environmental Discourse: Ecological Modernisation and the Policy Process* (Oxford, UK: Clarendon Press).
Hart, S., and M.B. Milstein (1999) 'Global Sustainability and the Creative Destruction of Industries', *Sloan Management Review*, Autumn 1999: 23.
Hofstetter, P., M. Madjar and T. Ozawa (2005) 'The Fallacy of Ceteris Paribus and Real Consumers: An Attempt to Quantify Rebound Effects', in E. Hertwich, T. Briceno, P. Hofstetter and A. Inaba (eds.), *Sustainable Consumption: The Contribution of Research* (Report 1/2005; Proceedings of an International Workshop, Gabels Hus, Oslo, 10–12 February 2005; Trondheim, Norway: NTNU [Norwegian University of Science and Technology], Industrial Ecology Programme [IndEcol], NO-7491): 48-64.
Holliday, C., and J. Pepper (2001) *Sustainability through the Market: Seven Keys to Success Brief* (Geneva: World Business Council for Sustainable Development, 2nd edn; www.wbcsd.ch/DocRoot/xs6OhpvANJi0GJPFEkBH/stm.pdf [accessed 25 August 2007]).
Illich, I. (1978) *Toward a History of Needs* (New York: Pantheon Books).
Imhoff, M.L., L. Bounoua, T.Ricketts, C. Loucks, R. Harriss and W.T. Lawrence (2004) 'Global Patterns in Human Consumption of Net Primary Production', *Nature* 429 (24 June 2004): 870-73.
Inaba, A. (2004). 'What is the Practical Way to Sustainable Consumption?', in K. Huback, A. Inaba and S. Stagl (eds.), *Proceedings of the International Workshop on Driving Forces of and Barriers to Sustainable Consumption*, University of Leeds, Leeds, UK, 5–6 March: 2-4.
IUCNNR (International Union for Conservation of Nature and Natural Resources)/WWF/UNEP (United Nations Environment Programme) (1980) *World Conservation Strategy: Living Resource Conservation for Sustainable Development* (Gland, Switzerland: WWF).
Jackson, T. (2004) *Chasing Progress: Beyond Measuring Economic Growth* (London: New Economics Foundation).
Jasanoff, S.S. (1990) *The Fifth Branch: Science Advisers as Policy-makers* (Cambridge, MA: Harvard University Press).
Jespersen, J. (2004) 'Macroeconomic Stability: Sustainable Development and Full Employment', in L. Reisch and I. Røpke (eds.), *The Ecological Economics of Sustainable Consumption* (Cheltenham, UK: Edward Elgar): 233-50.
KemI (Swedish Chemicals Inspectorate) (1991) *Risk Reduction of Chemicals: A Government Commissioned Report* (written in cooperation with the Swedish Environmental Protection Agency; Solna, Sweden: KemI).
Klein, D.R. (1968) 'The Introduction, Increase and Crash of Reindeer on St Matthew Island', *Journal of Wildlife Management* 32: 350-67.
Knorr-Cetina, K. (1993) 'Strong Constructivism: From a Sociologist's Point of View. A Personal Addendum to Sismondo's Paper', *Social Studies of Science* 23: 555-63.
Kuhn, T.S. (1962) *The Structure of Scientific Revolutions* (Chicago: University of Chicago Press).
—— (1991) 'The Road Since Structure', in A.M. Fine, M. Forbes and L. Wessels (eds.), *Proceedings of the 1990 Biennial Meeting of the Philosophy of Science Association* (East Lansing, MI: Philosophy of Science Association).
Landry, C. (2000) *The Creative City: A Toolkit for Urban Innovators* (London: Earthscan Publications).
Lawn, P.A. (2003) 'A Theoretical Foundation to Support the Index of Sustainable Economic Welfare (ISEW), Genuine Progress Indicator (GPI), and Other Related Indexes', *Ecological Economics* 44: 105-18.
Layard, R. (2005) *Happiness: Lessons from a New Science* (London: Allen Lane/Penguin Books).
Livermore, M (2007) 'All those scientists may still be wrong', *Daily Telegraph* (London), 1 March 2007; www.cps.org.uk/cpsfile.asp?id=684 [accessed 25 August 2007].
Lomborg, B. (2001) *The Sceptical Environmentalist: Measuring the Real State of the World* (Oxford, UK: Oxford University Press).
Lutz, W., W.C. Sanderson and S. Scherbov (eds.) (2004) *The End of World Population Growth in the 21st Century: New Challenges for Human Capital Formation and Sustainable Development* (London: Earthscan Publications).
Lynas, M. (2004) *High Tide: News from a Warming World* (London: Flamingo/HarperCollins).
Manzini, E., and F. Jégou (2003) *Sustainable Everyday: Scenarios of Urban Life* (Milan: Edizioni Ambiente).

Marks, N., S. Abdallah, A. Simms and S. Thompson (2006) *The (Un)Happy Planet: An Index of Human Well-being and Environmental Impact* (London: New Economics Foundation/Friends of the Earth).
McDonough, W., and M. Braungart (1998) 'The NEXT Industrial Revolution', *The Atlantic Monthly* 282.4: 82-92.
—— and M. Braungart (2003) 'From Principles to Practices', *MDMC Monthly Feature Story*, March 2003; www.mbdc.com/features/feature_july2003.htm [accessed 25 August 2007].
Montalvo Corral, C. (2002) *Environmental Policy and Technological Innovation: Why do Firms Reject or Adopt New Technologies?* (Cheltenham, UK/Northampton, MA: Edward Elgar).
Moscovici, S. (1979) *Psychologie des minorités actives* (Paris: PUF).
Munasinghe, M., and W. Shearer (1995) 'An Introduction to the Definition and Measurement of Biogeophysical Sustainability', in M. Munasinghe and W. Shearer (eds.), *Defining and Measuring Sustainability: The Biogeophysical Foundations* (Washington, DC: United Nations University/World Bank).
—— and J. McNeely (1995) 'Key Concepts and Terminology of Sustainable Development', in M. Munasinghe and W. Shearer (eds.), *Defining and Measuring Sustainability: The Biogeophysical Foundations* (Washington, DC: United Nations University/World Bank).
Myers, N., and J. Kent (2004) *The New Consumers: The Influence of Affluence on the Environment* (Washington, DC/Covelo, CA/London: Island Press).
Nattrass, B., and M. Altomare (1999) *The Natural Step for Business: Wealth, Ecology and the Evolutionary Corporation* (Gabriola Island, Canada: New Society Publishers).
Norwegian Ministry of the Environment (1994) *Proceedings of the Symposium on Sustainable Consumption* (Oslo: Norwegian Ministry of the Environment).
O'Connor, J., and I. McDermott (1997) *The Art of Systems Thinking: Essential Skills for Creativity and Problem-solving* (London: Thorsons/HarperCollins).
OECD (Organisation for Economic Cooperation and Development) (1999) *Towards More Sustainable Household Consumption Patterns: Indicators to Measure Progress* (Paris: OECD).
Prahalad, C.K. (2004) *The Fortune at the Bottom of the Pyramid: Eradicating Poverty through Profits* (Upper Saddle River, NJ/Philadelphia, PA: Wharton School Publishing).
Ray, P.H., S.R. Anderson (2000) *The Cultural Creatives: How 50 Million People are Changing the World* (New York: Three Rivers Press).
Reijnders, L. (1998) 'The Factor X Debate: Setting Targets for Eco-Efficiency', *Journal of Industrial Ecology* 2.1 (January 1998): 13-22.
Reisch, L., and I. Røpke (eds.) (2004) *The Ecological Economics of Consumption* (Cheltenham, UK: Edward Elgar).
Rifkin, J. (2002) *The Hydrogen Economy: The Creation of the Worldwide Energy Web and the Redistribution of Power on Earth* (New York: Tarcher/Putnam).
RIVM (Netherlands National Institute for Public Health and Environment) (1991) *Nationale Milieuverkenning 1990–2010 (National Environmental Forecast 1990–2010)* (Alphen aan den Rijn, Netherlands: Samsom H.D. Tjeenk Willink/RIVM).
Rotmans, J. (2003) *Transitiemanagement: Sleutel voor een duurzame samenleving (Transition Management: Key to a Sustainable Society)* (Assen, Netherlands: Van Gorcum).
Sabatier, P.A. (1987) 'Knowledge, Policy Oriented Learning and Policy Change: An Advocacy Coalition Approach', *Knowledge* 8.4: 649-92.
Sachs W., et al. (2002) *The Jo'burg Memo: Fairness in a Fragile World*. (Memorandum for the World Summit on Sustainable Development; Berlin: Heinrich Böll Foundation; www.boell.de).
Scherhorn, G. (2005) 'Sustainability, Consumer Sovereignty and the Concept of the Market', in K.G. Grunert and J. Thøgersen (eds.), *Consumers, Policy and the Environment: A Tribute to Folke Ölander* (New York: Springer): 301-25.
Schumacher, E.F. (1973) *Small is Beautiful* (New York: Harper & Row).
Schwarz, M., and M. Thompson (1990) *Divided We Stand: Redefining Politics, Technology, and Social Choice* (New York: Harvester Wheatsheaf).
Segal, J.M. (1999) *Graceful Simplicity: Towards a Philosophy and Politics of Simple Living* (New York: Holt).
Sen, A. (1999) *Development as Freedom* (Oxford, UK: Oxford University Press).
Serageldin, I., and A. Steer (eds.) (1994) *Making Development Sustainable: From Concept to Action* (Environmentally Sustainable Development Series, OP-2; Washington, DC: World Bank).

Shah, H., and N. Marks (2004) *A Well-being Manifesto for a Flourishing Society* (London: New Economics Foundation).

Shackley, S., and B. Wynne (1995) 'Global Climate Change: The Mutual Construction of an Emergent Science–Policy Domain', *Science and Public Policy* 22: 218-30.

Simon, H.A. (1957) 'Theories of Decision Making in Economics and Behavioural Science', *American Economic Review* 49: 253-83.

—— (1957) 'Administrative Behaviour', in, *Models of Man* (New York: Macmillan).

Sneddon, C., R.B. Howarth and R.B Norgaard (2006) 'Sustainable Development in a Post-Brundtland World', *Ecological Economics* 57: 253-68.

Solow, R. (1992) *An Almost Practical Step toward Sustainability* (Washington, DC: Resources for the Future).

Stern, N. (2006) *The Economics of Climate Change: The Stern Review* (Cambridge, UK: Cambridge University Press).

Stiglitz, J. (2002) *Globalization and its Discontents* (New York: W.W. Norton).

Thompson, M. (1984) 'Among the Energy Tribes: A Cultural Framework for the Analysis and Design for Energy Policy', *Policy Sciences* 17: 321-39.

——, R. Ellis and A. Wildavsky (1990) *Cultural Theory* (Boulder, CO: Westview Press).

Tukker, A. (1999) *Frames in the Toxicity Controversy* (Dordrecht/Boston/London: Kluwer Academic).

—— and U. Tischner (2006) *New Business for Old Europe: Product-Service Development, Competitiveness and Sustainability* (Sheffield, UK: Greenleaf Publishing).

—— and M. Butter (2007) 'Governance of Sustainable Transitions: About the 4(0) Ways to Change the World', *Journal of Cleaner Production* 15.1: 94-103.

——, P. Eder and S. Suh (2006) 'Environmental Impacts of Products: Policy Relevant Information and Data Challenges', *Journal of Industrial Ecology* 10.3 (Summer 2006): 183-98.

UN (United Nations) (1992) *Agenda 21* (Rio de Janeiro: UN Conference on Environment and Development).

—— (2000) *United Nations Millennium Declaration* (A/RES/55/2; New York: UN, 18 September 2000; www.un.org/millennium/declaration/ares552e.pdf [accessed 25 August 2007]).

—— (2002) *Plan of Implementation of the World Summit on Sustainable Development* (New York: UN).

UNCSD (United Nations Commission on Sustainable Development) (2001a) 'CSD Theme Indicator Framework', www.un.org/esa/sustdev/natlinfo/indicators/isdms2001/table_4.htm [accessed 25 August 2007].

—— (2001b) *Commission on Sustainable Development Work Programme on Indicators of Sustainable Development (Addendum)* (Ninth Session, E/CN.17/2001/4/Add.1; New York: UNCSD).

——(2001c) *Status Report on the Indicators of Consumption and Production Patterns* (New York: UNCSD).

Van den Bergh, J. (2005) 'BNP? Weg ermee!' ('GDP? Throw it away!'), *Economisch Statistische Berichten* 4475 (18 November 2005): 502-505.

—— and H. Verbruggen (1999) 'Spatial Sustainability, Trade and Indicators: An Evaluation of the Ecological Footprint', *Ecological Economics* 29: 61-72.

Van der Sluijs, J.P. (1997) *Anchoring amid Uncertainty; On the Management of Uncertainties in Risk Assessment of Anthropogenic Climate Change* (PhD thesis; Utrecht, Netherlands: Universiteit Utrecht).

Van Kooten, G.C., and E.H. Bulte (2000) 'The Ecological Footprint: Useful Science or Politics?', *Ecological Economics* 32: 385-89.

Van Vuuren, D.P., and H.A.M. de Kruijf (1998) *Compendium of Data and Indicators for Sustainable Development in Benin, Bhutan, Costa Rica and The Netherlands* (Report 807005001; Bilthoven, Netherlands: National Institute for Public Health and Environment [RIVM]).

Veenhoven, R. (undated) 'World Database on Happiness', www2.eur.nl/fsw/research/happiness [accessed 25 August 2007].

Venetoulis, J., and C. Comb (2004) *The Genuine Progress Indicator 1950–2002 (2004 Update): Measuring the Real State of the Economy* (Oakland, CA: Redefining Progress; www.rprogress.org/projects/gpi [accessed 25 August 2007]).

Vitousek, P., P. Ehrlich, A. Ehrlich and P. Matson (1986) 'Human Appropriation of the Products of Photosynthesis', *BioScience* 36.6 (June 1986): 368-73.

Von Weizsäcker, E., A. Lovins and L. Lovins (1997) *Factor Four: Doubling Wealth and Halving Resource Use* (London: Earthscan Publications).
Wackernagel, M., and W. Rees (1996) *Our Ecological Footprint: Reducing Human Impacts on the Earth* (The New Catalyst Bioregional series, 9; Gabriola Island, BC/Philadelphia, PA: New Society Publishers).
WCED (World Commission on Environment and Development) (1987) *Our Common Future* (The Brundtland Report; New York: United Nations).
Weinberg, A. (1972) 'Science and Transscience', *Minerva* 10: 209-22.
Weterings, R., and H. Opschoor (1994) *Towards Environmental Performance Indicators Based on the Notion of Environmental Space* (Publication 96; Rijswijk, Netherlands: Advisory Council for Research on Nature and Environment [RMNO]).
Williamson, O.E. (1979) 'Transaction Cost Economics: The Governance of Contractual Relations', *Journal of Law and Economics* 22: 233-61.
Woods, N. (1999) 'Order, Globalisation and Inequality in World Politics', in A. Hurrell and N. Woods (eds.), *Inequality, Globalisation, and World Politics* (Oxford, UK: Oxford University Press): 8-35.
World Bank (2005) *World Development Indicators 2005* (Washington, DC: World Bank; devdata. worldbank.org/wdi2005/index2.htm [accessed 25 August 2007]).
WRR (Wetenschappelijke Raad voor het Regeringsbeleid) (1994) *Duurzame Risico's, een blijvend gegeven (Sustainable Risks: A Lasting Issue)* (The Hague: Scientific Council for Government Policy).
WWF (2004) *Living Planet Report 2004* (Gland, Switzerland: WWF; www.panda.org/news_facts/ publications/key_publications/living_planet_report/index.cfm [accessed 25 August 2007]).
Young, E. (2006) 'Easter Island: A Monumental Collapse?', *New Scientist* 2,562 (31 July 2006): 30-34.

Part 2
**Business perspective**

# 3
# Review: the role of business in realising sustainable consumption and production

*Martin Charter, Casper Gray, Tom Clark and Tim Woolman*
The Centre for Sustainable Design, UK

## 3.1 Introduction

This chapter discusses the contribution of business to sustainable consumption and production (SCP) and the way in which this may be advanced through recognising the drivers and the experience of current approaches. It suggests opportunities for innovation and business models contributing to SCP.

The interpretation of SCP and its relationship to business varies according to the nature, aims and role of business; here, we take business to mean commercial firms involved in buying and selling. At a macro level, business is a powerful stakeholder in the national socioeconomic systems of consumption and production. Business can be viewed in consumption sectors, such as agriculture or electronics, by customer type; business to consumer (B2C), business to business (B2B) or business to government (B2G), or by function, such as retailing or manufacturing. At a micro level, business can take various forms, for example:

- Transnational corporations (TNCs): global subsidiaries
- Large, medium or small: exporters and non-exporters
- Micro enterprises
- Social enterprises
- Cooperatives and networks

How business can contribute to SCP is a complicated issue. In the following sections, we will address this question, first, by reviewing business drivers (Section 3.2), second, by analysing how business can contribute to change to SCP (Section 3.3), third, by synthesising insights into such patterns for change (Section 3.4) and, last, by listing limitations (Section 3.5). In Section 3.6 we summarise our conclusions.

## 3.2 Understanding business drivers

### 3.2.1 Introduction

Sustainable production is increasingly understood and considered feasible, while the shift to patterns of more sustainable consumption and SCP, in general, is an emerging area. This challenge is not minor: most of the 50 Factor 4 innovations deemed promising by von Weizsäcker *et al.* (1997) a decade ago are still confined to niche markets. The commercial movement towards SCP needs new business models that respond better to the drivers and opportunities for SCP and support policies that overcome barriers to change. Moving forward thus requires new innovative thinking from business, government and other stakeholders as well as leadership and smarter policy approaches. This is true in relation to what actually needs to be done now to make a difference (a radical step change) and what it is most practical to do immediately (incremental change, 'doing a bit better'). In this, it is key to understand the basic drivers that businesses face. We will discuss them first in general and then zoom in on drivers relevant to SCP.

### 3.2.2 Business drivers and strategies

Traditional internal drivers influencing the choices at company level include the need to generate profit, manage risk, reduce cost and meet internal ambitions. These internal ambitions are affected by the external drivers to increase shareholder value, deliver returns to shareholders and compete by satisfying customer demands within a developing legal and market framework; they also include issues such as compliance with general societal aspirations, norms and values, or the desire of business leaders to make a positive difference to society.

This business context is not stable. Global demand is increasing, particularly as the economies of China and India expand. The business context includes (mega) trends in industry and society: for example, towards globalisation, 'informatisation', networking, lean production and mass customisation. 'Time contraction' suggests that planning horizons may be reducing; the expression 'long term' may now imply perhaps only three years.

As a consequence of drives for cost reduction and market expansion, the outsourcing of non-core activities has been a trend in many industries. This means that design, assembly and manufacturing may be spread across different countries. Long supply chains and distributed networks spread their environmental and social impacts; production in 'the South' may be many thousands of miles away from the point of consumption in 'the North'.

A main strategy for business is to develop a fit between external opportunities and threats and internal strengths and weaknesses that cannot easily be copied by others and, at the same time, to remain flexible in order to adapt in a fast-changing environment. Porter (1985) suggests three generic strategies to respond to external competitive forces: cost leadership, product differentiation or a focus on one market segment. In this, Porter stresses the relevance of enhancing bargaining power *vis-à-vis* suppliers and customers and choosing strategies that fence off the market for competition and alternative products. An alternative strategy is to follow the unique internal capabilities of the firm. This resource-based strategy differentiates the business, enabling it to resist competition.[1]

A learning strategy stresses the importance of maintaining internal capabilities to make such changes in the business (Hamel and Prahalad 1994). Capabilities that enable the business to be reconfigured are particularly relevant: that is, strategic decision-making, knowledge creation and the transfer of capabilities to new products or services. Rather than seeking a fit between capabilities and business surroundings, a learning strategy enables the business to respond to new objectives.

Overlaying the need for business to be responsive to changes in surroundings with new technologies and markets emerging from niches is the view that industry goes through a life-cycle: from introduction or emergence, to growth, to maturity and to decline. The emphasis of innovation tends to shift from significantly improving product function and quality to competing on cost through process improvement. The decline stage can be countered only by 'strategic innovation' that involves redefining markets or activities, adding new but related products or services to avoid maturity and decline (Baden-Fuller and Stopford 1992).

### 3.2.3 Drivers relevant to sustainable consumption and production

#### 3.2.3.1 Crises related to environmental pressures

Over the past 20 years environmental crises have led to shifts in technological solutions and business practices. The wide-scale realisation of the causes and effects of the growing hole in the ozone layer in the late 1980s led to the removal of chlorofluorocarbons (CFCs) in products. Even before a consensus was reached and the stipulations of the Montreal Protocol, SC Johnson removed CFCs from products. This pioneering decision was consistent with its socially responsible ethos and helped it to secure its advantage ahead of competitors in CFC-free markets.

Today, scientific evidence of climate change is widely recognised. The summary of the fourth Intergovernmental Panel on Climate Change (IPCC) assessment report, revised by representatives from 113 governments, states that 'unequivocal' warming of the climate system is now evident, and it predicts a global temperature rise of 3°C by the end of the century, with an associated rise in sea level (IPCC 2007). The consequences for the economy, which provides the framework for business, have been estimated in the Stern Review (Stern 2006). For the UK, it has been shown that, by taking action now, the over-

---

1 Tukker and van den Berg (2006) summarise these theories for modelling change and deciding on, for example, whether to retain or acquire or outsource resources. One strategy is to move to a product-service system (PSS) that develops a closer, less imitable relationship with customers.

all costs of climate change could be reduced from the equivalent of 5–20% of global GDP each year to around 1%:

> If we take no action to control emissions, each tonne of $CO_2$ that we emit now is causing damage worth at least \$85—but these costs are not included when investors and consumers make decisions about how to spend their money. Emerging schemes that allow people to trade reductions in $CO_2$ have demonstrated that there are many opportunities to cut emissions for less than \$25 a tonne. In other words, reducing emissions will make us better off. According to one measure, the benefits over time of actions to shift the world onto a low-carbon path could be in the order of \$2.5 trillion each year (Stern 2006).

Stern called for support for the deployment of low-carbon technologies to increase by up to fivefold. Business is now recognising a growing competitive focus on low-carbon technologies as alternatives to the burning of fossil fuels, such as the use of hybrid cars, fuel cells and so on. This will contribute perhaps to our reaching a 'tipping point' for wider action by government and business to avoid the risk of climate change causing major disruption to economic and social activity.

### 3.2.3.2 Rising prices for energy and materials

Worldwide, transportation and industry are the major consumers of oil and natural gas (EIA 2006). Energy prices have risen as a result of concerns over the security of supply of fossil fuels because of conflicts (e.g. in Iraq), financial crises (e.g. in Asia in 1999) and environmental crises (e.g. Hurricane Katrina). Government taxation is also increasing the current and projected price of fossil fuels.

Material and production costs reflect these increases in energy costs. The need to reduce energy use and therefore carbon emissions through both energy efficiency and demand management is now an economic and, increasingly, a business priority.

### 3.2.3.3 The shift to an 'experience economy'

Pine and Gilmore (1999) argue that economic supplies have shifted from commodities, or materials, to products and, further, to services, towards, finally, experiences. Customer fulfilment may therefore become less material-intensive. Trends also show that there is a clearer link between happiness and environmental quality than there is between happiness and GDP (Kleanthous and Peck 2006). Such trends will support business models based on the creation of immaterial value (Tukker and Tischner 2006).

### 3.2.3.4 Societal expectations and owner aspirations

Societal expectations can be another driver for change. Well-known examples are the Brent Spar affair, expectations about meeting basic social and sustainability standards in supply chains and so on. The need to meet such standards forms a clear driver for firms in relation to protecting corporate or brand reputation, brand value and so on.

Interestingly, drivers are also coming from within firms. There is an interesting new group of entrepreneurs who have deliberately changed the course of their firms or who have invested heavily in sustainability. Reading Paul Hawken's book, *The Ecology of Com-*

*merce* (1994), is reputed to have resulted in Ray Anderson having his 'spear in the chest' moment that resulted in a decision to 'green' his company Interface. Anderson changed his direction, leading Interface to the goal of becoming 'sustainable' by 2020. As a result of the change, the company has developed many product and process innovations. Notable recent examples of entrepreneurs investing in 'sustainable technology' are:

- John Doerr (Google and Amazon): creating a $100 million fund for 'green technology'
- Bill Gates (Microsoft): providing $84 million finance for five ethanol bio-refineries
- Steve Case (AOL): making a $500 million investment in Revolution LLC, including Flexcar and Gaiam (Earth-friendly products)
- Vinod Khosla (Sun Microsystems): creating a fund to invest in clean-technology companies
- Paul Allen (Microsoft co-founder): starting up a company in Seattle to turn rape-seed (canola) oil into diesel fuel
- Richard Branson (Virgin Group): creating a $3 billion fund for renewable energy over the next ten years

The above examples offer unique cases of the potential effect of large investments in kick-starting developments, mostly in technologies related to sustainable energy, and are worthy of further investigation given their potential to change markets. These pioneering, successful and wealthy leaders may be less constrained by mainstream market forces and therefore able to make new forms of investment. Fulton (2002) suggests that if the primary driver of business survival is not fear and scarcity then people who have enjoyed a 20-year business boom may well start looking for 'meaning' and for ways to make a difference, perhaps explaining the some of the high-profile personal investments in 'clean tech'.

## 3.3 State of the art: insights into the contribution of business to changes to sustainable consumption and production

### 3.3.1 The role of business

The vital role of business in contributing to sustainable development is now well established as a principle both in governments and in leading business circles. It is clear that business has a pivotal role to play, in cooperation with governments and other stakeholders, in the necessary global transformation towards SCP. Business represents a significant proportion of the necessary resources, capabilities and mechanisms for the technological innovation that will be needed for this transformation. The role of business in this task has been highlighted as the power and influence of TNCs has increased relative to that of governments.

Business can contribute to SCP in a number of ways. Below, we discuss first the move towards sustainable production and the creation of sustainable products—actions that can be initiated largely within a firm (Section 3.3.2). We then move to look at sustainability approaches that deal with supply and downstream chains, often endorsed by labelling or certification (Section 3.3.3). Of particular interest for the SCP agenda is how business can contribute to sustainable consumption and radical sustainable innovation. Although in part overlapping with the former points, we hence discuss these topics specifically in Sections 3.3.4 and 3.3.5.[2] In Section 3.3.6 we end by deliberating on the forces that support business contributions to SCP.

### 3.3.2 The creation of more sustainable products and production processes

Traditional approaches to how business can contribute to SCP are to make products and production processes cleaner and more efficient. In particular, cleaner production, waste prevention and emission prevention is now fairly well understood and is common practice in the more mature industrial economies. Usually, the application of such 'best available technologies' is backed up by voluntary industrial standards, voluntary agreements and regulations (e.g. the EU IPPC Directive [EC 1996]). However, there are interesting examples of major firms going well beyond this. Lee Scott's mission to turn Wal-Mart 'green' came after his acceptance of the impact of Wal-Mart's footprint on the global environment and after considering the world in which his granddaughter would grow up:

> We are focused on three top-line goals: to be supplied with 100% renewable energy, to create zero waste and to sell products that sustain our resources and our environment. Those are ambitious goals, but we never think small at our company (Lee Scott, President and Chief Executive Officer, Wal-Mart [Wal-Mart 2006a])

This process has created change in the firm. Wal-Mart has established so-called sustainable value networks (SVNs) that have responsibility to share with others lessons in Wal-Mart's efforts to improve its carbon footprint (GreenBiz 2006). The aim of reducing solid waste by 25% in just two years gives an indication of the company's historically high level of waste. Wal-Mart's ambitious targets to reduce its environmental footprint should be commended given its potential impact as the world's largest retailer, serving 138 million customers weekly (Wal-Mart 2006b).

In a similar announcement, Marks & Spencer is to spend £200 million over five years on a wide-ranging 'eco plan', transforming the chain of 460 outlets into a carbon-neutral operation: banning group waste from landfill dumps, using unsold out-of-date food as a source of recyclable energy and making polyester clothing from recycled plastic bottles.

---

2 One could argue convincingly that companies should think about strategy (discussed in Section 3.3.5) first and then take the measures described in Sections 3.3.2 and 3.3.3 in that context.

Attention to the 'greening' of products via ecodesign is a more recent approach. It is, in some cases, driven by regulation (e.g. Tukker *et al.* 2001) but there are also firms pursuing this option for strategic reasons, bringing it close to the 'strategic sustainable innovation' to be discussed in Section 3.3.5. Depending on these reasons and the perceived reaction from a market segment, a business may or may not decide to communicate its strategy to its customers as contributing to sustainability.

Philips products compete globally through price and performance. However, ecodesign has been implemented by Philips to reduce cost and environmental impact. Philips products are not advertised as eco-products, but its Green Flagship range is projected to account for 30% of total revenues by 2012.[3]

On a smaller scale, Freitag produces fashionable shoulder bags made from discarded lorry canvas, a product that emerged from seeing a business opportunity rather than from a vision of sustainability. By creating a unique, competitive, fashionable and 'cool' product, Freitag does not have to rely on green ethics for customer attraction and the business is therefore not constrained to niche eco-markets.

### 3.3.3 Promoting sustainability in up- and downstream chains

#### 3.3.3.1 Corporate social responsibility

Corporate social responsibility (CSR) is defined as 'a concept whereby companies integrate social and environmental concerns in their business operations and in their interaction with their stakeholders on a voluntary basis' (EC 2006). CSR hence goes further than looking merely at upstream and downstream chains, encompassing the firm itself and other relations, but we pragmatically place our discussion of it in this section.

CSR schemes are often driven by large and powerful companies. An example is 'choice editing' (the practice of 'editing out' certain products [e.g. less-sustainable products]), implemented by retailers to reassure customers and to promote moves towards SCP. However, the Race to the Top (RTTT) project by the International Institute for Environment and Development (IIED) revealed that some major supermarkets are concentrating on short-term consumer values rather than on long-term 'citizen' values that emphasise a more responsible outlook. The effect of concentrating on the consumer is to absolve the supermarkets of the responsibility of moving towards SCP unless the consumer specifically demands it. According to two RTTT participants, surveyed by Fox and Vorley (2004), 'the consumer and the citizen are generally not the same person, and supermarket companies listen to the former first and the latter a long way second', and 'sustainability and CSR are often marginalised and not integrated into core business operations'.

This RTTT experience suggests that CSR schemes do not reside comfortably within a mainstream business culture that is based on short-term profits. In a consumer-oriented industry such as retailing, self-regulation and voluntary initiatives are likely to be effective only for issues that fall in line with immediate consumer interests or concerns. Creating incentives for retailers to drive positive contributions to other aspects of sustainability implies a more robust role for government (Fox and Vorley 2004).

3 www.philips.com/sustainability

### 3.3.3.2 Procurement

Sustainable procurement through the business supply chain contributes both to improving a business's sustainability performance and to providing incentives to lower-tier suppliers.

Most organisations currently concentrate their efforts on environmental supply-chain management through responsible purchasing driven by peer pressure, corporate or brand reputation and compliance (i.e. as a risk minimisation strategy [Collignon et al. 2006]). CSR approaches applied to procurement appear to be most common in the retail sector, where some companies are taking a significant lead. For example, Wal-Mart's move towards sustainable procurement will affect more than 60,000 suppliers in 70 countries—Wal-Mart imports more than $15 billion worth of goods from China annually (Wal-Mart 2006b).

### 3.3.3.3 Product certification and eco-labelling

Certification of products highlights another way for companies to 'green' their supply as well as downstream chains. Some certification schemes are initiated by business, whereas others are driven by government-related institutions or regulations. We will discuss here the Forest Stewardship Council (FSC) and Marine Stewardship Council (MSC) as examples of business-led initiatives, as well as some other labels.

Since its launch in 1993, global sales of FSC products have grown by 20% per annum, reaching $9 billion in 2004 (Sustainable Consumption Roundtable 2006). Two key factors have enabled the success of FSC: (a) retailers' policies of working with suppliers to establish the label and (b) FSC product availability in mainstream retail outlets. Consumer concern, media coverage and pressure from non-governmental organisations (NGOs), coupled with support from major retailers, such as B&Q, through 'choice editing', have also been important to the growth in sales of FSC products. However, recognition of the FSC logo and products remains very low—less than 5% of consumers surveyed in the UK recognised the logo. Lack of awareness is now being seen as a major barrier to encouraging more retailers to stock FSC products. Several years of high-profile promotion and funding in The Netherlands has achieved 63% recognition of the FSC logo, with 26% of the Dutch population claiming to actively seek FSC products (Sustainable Consumption Roundtable 2006).

Like the FSC, the MSC relies for its success on gaining 'critical mass' through consumer recognition. The MSC was initiated in 1995 through a partnership between Unilever and the WWF as a mechanism to manage the world's fish stocks sustainably. Wal-Mart's commitment to purchasing all of its wild fresh and frozen fish for the US market from MSC-certified sources (Wal-Mart 2006a) should boost the label's recognition and credibility in the USA. However, MSC has received considerable but not universal support (Porritt and Goodman 2005). The key to furthering the acceptance of MSC by all stakeholders is to increase the number of accredited fish types. In mid-2005 just over 4% of the world's marine wild fish supply was accredited. How Wal-Mart will be able to satisfy its commitment, and over what timescale given the relatively small proportion of MSC-certified fish stocks, is uncertain.

Government-led labels include the Energy Star®, introduced in the USA in 1992, and the A–G energy performance labels for domestic appliances in the EU. The Energy Star® covers more than 35 energy-consuming devices, including computers and printers.

Action on the part of regulators, retailers and manufacturers, influenced by such consumption labelling of domestic 'white' goods, has resulted in rapid efficiency gains. For example, even the least-efficient fridge freezers on sale in 2005 consumed only half as much energy as the least-efficient products on the market eight years before (Sustainable Consumption Roundtable 2006). Type III labels, which offer good quantitative information and some explanation of environmental issues, can also be used by retailers to 'choice edit' for sustainability (Frankl *et al.* 2005). Consumers may then select goods on the crucial factors of price and performance in the knowledge that energy-consumption issues have been managed by retailers.

However, there is an ongoing debate on the effectiveness of eco-label schemes (Porritt and Goodman 2005). Awareness of labels is often low (Forstater and Oelschaegel 2006) and information may not be easily processed rationally (Sanstad and Howarth 1994). Labels such as the FSC logo that offer one clear, concise, piece of information work best (Frankl *et al.* 2005)! Despite formal certification, labelling and award schemes, credibility can still be an issue (Forstater and Oelschaegel 2006). Partnerships between business and NGOs can help to overcome distrust or lack of credibility. Consumers appear more willing to trust brands or NGO-backed labels than government schemes, as seen by the success of Fair Trade labels compared with the limited success of the European eco-label (Sustainable Consumption Roundtable 2006).

The most important issue, however, is that, at best, eco-labels lead to incremental improvements, limited by a lack of measures to tackle levels of consumption. As will be discussed in more detail in Part 4 of this book, on consumer science, information alone cannot change the context in which a consumer operates. In such cases, 'information provision, whether through advertisements, leaflets or labelling, must be backed up by other approaches' (Collins *et al.* 2003).

### 3.3.4 The promotion of sustainable consumption

#### 3.3.4.1 Introduction

In many mainstream consumer markets, convenience, habit and budgets are currently more important influences on the purchasing phase of consumption than are environmental concerns. In all phases of consumption—planning, purchasing, use and disposal—there remains a large gap between environmental values and consumer behaviour. Rather than relying on weak, existing consumer signals there is a chance to create new opportunities through innovation. For example, the Apple iPod was not invented because consumers requested it, and Fair Trade created a market that previously did not exist.

Products and services satisfy not only basic needs for food, housing and transport but also needs for status, social identity and meaning. Consumption is influenced as much by peers and by habits as by price and values. Factors influencing consumption include:

- Rational choice. This involves the weighing up of costs and benefits, although there is a large gap between expressed values and concerns and actual purchasing choices; there is also a degree of confusion (e.g. sustainable products and services are often perceived as being costlier)

- Social norms. This relates to the modelling of behaviour on others (e.g. it concerns the status, identity and culture associated with goods and services)
- Convenience, habit and routines. People are unwilling to change these, particularly given the way in which options are structured (e.g. people typically overestimate the inconvenience of consuming sustainably)
- Trust. People may lack trust in brands or in information providers

Influencing the B2B market towards sustainable consumption has similar characteristics, and the conveyance of sustainable values can be a meaningful differentiator for B2B customers (CSR Europe 2007). Open information, quality assurance, labelling and certification are seen as key in facilitating better decision-making towards sustainable procurement and use in B2C, B2B and B2G markets (CSCP 2006).

There is currently weak overall demand for sustainable products outside particular market segments, but there is still 'deselection' activity (e.g. boycotts of companies, products and brands) by some consumers. The 2004 UK Ethical Purchasing Index (Cooperative Bank 2004) showed the small market share for ethical products growing by almost 40% in the previous five years, with the most significant growth in 'green' goods. However, the impact on market share for most products is not enough to create the 'demand pull' necessary to signal a radical and widespread transformation in the supply of sustainable products and services.

Business has the power to influence both the level and the nature of consumption through shaping perception, wants and needs. Businesses, especially larger corporations, can actively influence consumption through 'choice editing', marketing, design, 'demand-side' management and by participating in wider changes to the framework, for example by contributing to policy-making through lobbying. Whether business will do this to move customers towards the new generations of sustainable products and services and whether this will contribute significantly to SCP is an emerging question.

### 3.3.4.2 'Choice editing'

Probably one of the simplest ways business (particularly retailers) can contribute to making consumption more sustainable is through 'choice editing'. 'Choice editing' for sustainability is about shifting the field of choice for mainstream consumers, cutting out unnecessarily damaging products and 'getting real sustainable choices on the shelves' that consumers see as equally good or better (Sustainable Consumption Roundtable 2006). This approach does not aim to change levels or patterns of consumption, as in the approaches discussed below, but helps consumers 'automatically' choose sustainable products without the need to change habits or routines. The reasons for adopting 'choice editing' may be altruistic or for business interests such as brand building and risk reduction, or a combination of these.

### 3.3.4.3 Marketing and sustainability

Marketing is a business process related to the identification, understanding and management of customer needs. Kleanthous and Peck's (2006) survey of marketing professionals in the UK showed 48% believed there was unmet latent demand for more responsible mainstream brands, and 45% believed customers prefer brands that care about environmental and social impacts. Once a business case for sustainability is made, the marketing challenge is to make sustainable consumption innovative, exciting, cool and sexy—influencing the social norms and lifestyle aspirations that drive consumption. Ottman (1998) suggests marketeers need to:

- Understand the full range of current and future environmental, social and economic issues that affect the customer and product or service
- Help customers understand the sustainability-related issues in familiar terms and the benefits of the more sustainable product or service (i.e. the purpose of the product or service proposition in relation to the issues)
- Confirm what they can do to 'make a difference' while maintaining expectations of performance
- Establish credibility

Philips learned from its first attempt at marketing a compact fluorescent light bulb (CFL). The name 'Earth Light' confused people, the appearance was unfamiliar and the shape did not suit many lamps. At 20 times the price of an incandescent bulb, Earth Light could not climb out of a 'green' niche. The renamed, relaunched 'Marathon' solved several problems. The 'super long life' positioning appealed to the convenience-oriented mainstream, while the promise of saving 175% of the cost of the light bulb in energy costs over its lifetime extended the appeal to thrifty consumers. The US EPA's Energy Star® label added credibility, and sales in 2001 rose by 12% in a flat market. For further deliberations on marketing and sustainability we refer to Belz's contribution in Chapter 7 of this book, and also to the Sustainable Marketing Knowledge Network (www.cfsd.org.uk/smart-know-net).

#### 3.3.4.4 Experience-centred and use-centred design

Rather than 'manufacturing' customers' aspirations, a business may design its products or services based on a bottom-up approach, involving users and consumers, in order to address their needs and simultaneously influence their behaviour: for example, through 'user-centred design' or 'socially inclusive design'.

User-centred design is a bottom-up process that takes a systems view of sustainable consumption and lifestyles as a starting point for the design process, pioneered through participatory design of public services and infrastructure (e.g. by Christopher Day in architecture [Richardson et al. 2005]).

Design is more typically a multi-stakeholder activity in which engineers, marketing and brand experts, business strategists and research and development (R&D) play a more significant role than do designers. Sustainability issues are best considered upstream in the business and product development process, but designers often have little strategic influence at this stage. Sustainability and the opportunity to influence consumption may be seen as a minor part of the brief, pursued at the expense of other criteria such as revenue, quality and repeat business.

Business can use design to challenge consumer aspirations or empower customers to participate in creating new demand-led market opportunities. Design-led consumption offers the opportunity to increase the market demand for sustainable products, potentially creating a feedback loop between consumption and production.

### 3.3.4.5 'Demand-side' management

'Demand-side' management (DSM) is the planning, implementation and monitoring of activities designed to encourage consumers to modify consumption patterns, typically in terms of the timing and levels of utility usage, towards strategic conservation (REEEP 2004). Energy suppliers use demand–response programmes, for example, to reduce the energy consumed by customers when suppliers face premium prices, experience events compromising system reliability or need to reserve capacity. The reduction can be achieved by curtailment (e.g. by switching off) or turning on self-generation (e.g. backup generators), for which participating customers receive payments—a win–win opportunity for utilities and users to balance demand and supply, limiting the need for new sources of supply to provide capacity beyond managed peaks.

Similarly, transportation demand management has been applied in the USA to reduce overdependence on the automobile, although measures such as incentives to use public transport have had limited effect (World Energy Council 2002). Water utilities apply the following to influence demand:

- Education and advertising campaigns
- Revision of building codes
- Pricing instruments and other financial incentives
- Rationing

Renwick and Green (2000) conclude that small (5–15%) reductions in demand can be achieved through modest price increases and through voluntary measures such as information campaigns.

There are therefore price measures and non-price measures that businesses can apply to manage demand. Providing customers with clear information on the unsustainable consequences of consumption as well as collecting information from customers on their consumption patterns in response to market and non-market signals can help business influence customers towards more sustainable consumption.

## 3.3.5 Opportunities for strategic sustainable innovation

### 3.3.5.1 Introduction

At the interface between 'sustainable consumption' and 'sustainable production' is eco-innovation, or sustainable innovation—an emerging area that responds to those proposing radical change to achieve sustainability: for example, by Factor 10 and Factor $X$ approaches (Bartelmus 2002). This type of innovation goes well beyond the product and process changes described in Section 3.3.2 and concerns far-reaching 'competing for the future' innovations to meet sustainability demands (compare Hamel and Prahalad 1994; Christensen 1997). The European Union (EC 2006) described such eco-innovation as 'innovation aiming at demonstrable, radical (or significant incremental) progress towards the goal of sustainable development, through reducing impacts on the environment, or achieving a more efficient and responsible use of natural resources, including energy'.

Business can practise eco-innovation by pursuing the creation of novel and competitively priced goods, processes, systems, services and procedures, with minimal use of natural resources and pollution while satisfying legal standards, norms and stakeholder requirements. Eco-innovation therefore provides a link between environmentally sustainable consumption and production. Beyond eco-innovation, sustainable innovation, addressing environmental, social and economic objectives towards SCP, is a new area for most businesses. Sustainable innovation is process, the sustainability considerations of which (i.e. environmental, financial and social) are integrated into company systems from idea generation through R&D and commercialisation. This applies to products, services and technologies, as well as new business and organisation models (Charter in Charter and Clark 2007).

It can be argued that there are four tiers to eco-innovation and sustainable innovation: technology, products and services; process; organisational issues; and business issues. These different types and levels of eco-innovation and sustainable innovation need to be recognised in order—the stimulation of each level and type will need different approaches to be taken.

Sustainability is now increasingly being recognised as an area of business opportunity. A survey by Arthur D. Little (2005) of 40 technology firms across Europe, the USA and Japan suggests great potential. Some 95% of the firms believed 'sustainability-driven innovation' has the potential to bring business value, almost 25% believing it will definitely deliver business value. To illustrate this: General Electric's 'ecomagination' initiative, launched in 2005, intends to 'imagine and build innovative solutions that benefit [its] customers and society at large' (General Electric Company 2006). According to Jeff Immelt, Chairman of the Board and CEO of General Electric, 'we are committed to being the leader in environmental technology solutions. It is an area that fits our current product offerings and where we see significant growth and profitability' (General Electric Company 2006).

The Arthur D. Little survey clearly suggests that a strategic reconsideration of business models is at heart of the challenge: ' "Sustainability-driven" innovation means the creation of new market space, products and services or processes driven by social, environmental or sustainability issues' (Arthur D. Little 2005).

The issue of business models is discussed in more detail below. Since sustainability-driven innovation is usually rather radical and often starts as a business niche, the question of how to reach the mainstream market is also addressed (see Section 3.3.5.3).

### 3.3.5.2 Sustainable business models

A business model describes the planned or existing form of value creation and market orientation, integrating elements of the resource-based and market-based view of the firm. Hamel (2000) defines business models in terms of core strategy, strategic resources, customer interface and value network. By designing the elements of value proposition, value creation and revenue delivery appropriately a firm can tune its offering, although the challenge is to develop a business model that is environmentally, socially and economically sustainable. Developing sustainable (new) business models is an emerging area. There are three major areas of discussion:

- Organisational structures: that is, the impact of sustainability on organisational models. An example is the creation of 'hollow companies' such as *Metro* newspaper that exist only as brands, with manufacturing and other business functions outsourced (Freedman 2004)

- New offerings: that is, new, or reincarnated, product, service or product-service system (PSS) models that aim to reduce environmental impact. An example is Electrolux's washing solution that is based on payment for use of the service. Such a PSS may be product-oriented, use-oriented or result-oriented (Tukker and Tischner 2006)

- 'Bottom of the pyramid' (Prahalad 2005): that is, new products and solutions for the poor in the developing world. An example is Procter & Gamble's PUR® water treatment solution, in this case favouring a social-market model over a commercial model (WBCSD 2006)[4]

Many companies are keen to be followers and are not prepared to take the risk of making the first move (Sustainable Consumption Roundtable 2006). To get management 'buy-in' there is a need to achieve 'quick wins' and to show the value or payback. The development of sustainable (new) business models that are successful in implementation means involving, cooperating and forming partnerships with people, generating trust and 'buy-in' among the stakeholders. This may in turn necessitate new forms of partnership and cooperation, exemplified by voluntary agreements. Business will need to raise environmental and social awareness and learn the associated skills and how to apply them. For example, production-based businesses must extend knowledge from assembly to disassembly and to closing loops in material flows, perhaps adopting a cyclic remanufacturing model. The challenges can be identified as:

- The dominance among publicly owned companies of a system that is preoccupied with delivering short-term results to shareholders, potentially limiting decision-making freedom. To move towards sustainability, publicly listed companies may benefit from moving to or emulating models of private ownership or from selecting shareholders according to their desire to facilitate SCP

- The need to establish systems to identify and recognise emerging sustainable (new) business opportunities. This may require a change of tack from 'risk management' to 'opportunity capitalisation'

- Protecting brand, reputation and 'intangibles', increasingly important risk issues for companies, particularly in markets less tolerant of sustainable marketing

4 Note that PSS or 'hollow companies' are not more sustainable per se, since they simply may be an alternative organisational structure with the same products and processes—and hence the same environmental impacts. Only if the new organisational form leads to greater efficiencies, stimulate less use of artefacts, etc., will sustainability gains be expected.

A key issue is to ensure that appropriate organisational and business processes and models are in place to continuously integrate sustainability into existing processes. Few companies have fully integrated sustainability into their business or new product development (NPD) processes. However a few leading-edge companies already exploit it as an opportunity, for example, Philips with its Green Flagship programme and Electrolux with the Green Range. The potential for sustainability-driven innovation is recognised among many mainstream companies (Arthur D. Little 2005), particularly those sensitive to niche competitors—competitors who may grow with appropriate market expansion or investment.

### 3.3.5.3 Niche → mainstream

Niche markets are relevant to SCP in representing a testing ground for radical innovations which, if successful, can become mainstream.

Ethical companies successful in becoming established from a niche market may be acquired by market leaders: for example, The Body Shop by L'Oréal (2006), Ben & Jerry's by Unilever (2000) and Green & Black's by Cadbury Schweppes (2005). This raises questions over the implications both for the smaller company and for the new parent company—will the smaller company's ethical position become diluted by the parent company mission, or will the 'David' green the 'Goliath'? Considering the motivations of the smaller and larger company in bringing the niche player to a more mainstream position is part of understanding whether more sustainable (niche) practice is also brought to the mainstream.

To capitalise on the growth of ethical consumerism seen in recent years (Cooperative Bank 2004; Harrison 2006) companies have recently been exploring how they can build brand reputation by aligning with consumers' concerns and desires (Forstater and Oelschaegel 2006). Larger companies with a poor ethical reputation can have difficulty in launching their own ethical product ranges but can bypass these problems and address consumer concerns by acquiring ready-made ethical brands (Harrison 2006). The acquiring company gains a high-quality, ready-made and proven product range and also a 'laboratory' and 'beacon' for the rest of the organisation (Kleanthous and Peck 2006).

The opportunity to influence the new parent company and therefore the market sector may be an important element of the decision for those leading the niche player facing acquisition. Anita Roddick (The Body Shop) and Craig Sams (Green & Black's) both voiced their belief that an acquired ethical company can influence its new parent to improve its corporate behaviour, although it is unclear and perhaps unlikely if this was their prime objective in selling. Roddick's call to boycott all Nestlé-owned products, including those from The Body Shop, calls into question the degree to which The Body Shop has had a positive influence over its owners to date.

Ben & Jerry's hostile takeover by Unilever has been widely recognised to have been detrimental to the company's ethical values, with its latest social audit finding that only 45% of Ben & Jerry employees think the company is taking its social values seriously (Harrison 2006). Such a loss of mission appears typical in other acquisitions in the USA, where there is a wider trend in large companies taking over small 'ethical' companies (e.g. Seeds of Change, by Mars, and Stoneyfield Farm, by Danone). This loss of mission in the acquired company may result from a new primary focus on generating quarterly profits and increased shareholder value, which may constrain innovation in ethical product development (Harrison 2006). Other challenges that need to be tackled include organisational 'buy-in' to the ethical objectives from the parent company and maintaining the recognition and performance of human and capital resources during any integration.

### 3.3.6 Fostering change: where does the power lie?

As many of the examples above have shown, to initiate real change in an organisation it is necessary to gain board-level support. Moves towards SCP can be facilitated by other stakeholders in a firm, particularly those who connect companies with consumers. Marketing should know the business's customers best and has a key role in influencing consumption (see Section 3.3.5.1). However, some literature indicates that marketing does not understand the language of sustainability, SCP and CSR, not so much because of the language itself but because of the basic assumptions on which the language is based (Kleanthous and Peck 2006). From the perspective of some chief executives, marketing exists to increase revenue, whereas sustainability or CSR departments generally exist to minimise and manage risks. This sets up a tension at a departmental level, where sustainability or CSR managers are regarded with suspicion by marketeers and remain excluded from the planning process (Charter *et al.* 2004; Kleanthous and Peck 2006).

Finance departments and accountants often have the real power in firms by 'holding the purse strings', facilitating information flows and so on. Accountants' interpretation of data makes them key influencers and often 'decision-makers'. They can thereby support or hinder a company's movement towards sustainability but are not generally engaged in the SCP discussion (ICAEW 2004).

Apart from such internal blockages in a firm, widespread adoption of radical action may be halted by a lack of consensus among competitors.[5] This suggests a lack of conviction in leaders' ambitions to realise SCP or a fear of moving beyond conventions and 'comfort zones'. And, here, the mutually reinforcing action of business and government is inevitably needed to break the deadlock. In the words of Patrick Burrows, of Tesco plc (Sustainable Consumption Roundtable 2006), 'businesses need to be brave. They need to go out speculatively and push technology that makes sustainable development economically viable. But Government needs to help businesses to be brave—by incentivising and supporting innovation'.

## 3.4 Synthesis: models of change

In summary, the number of approaches that businesses can use to contribute to change towards SCP is surprisingly large and in most cases not overwhelmingly complex. It concerns undertaking actions such as:

---

5 Email interview with P. Maltby, 14 July 2006.

- Making products and production more sustainable by implementing classical measures such as cleaner production, planning for waste and emission prevention and developing greener products via ecodesign
- Making their supply chains and downstream chains more sustainable by ensuring that social and environmental standards are complied with. Such sustainable corporate procurement or downstream 'responsible care' usually goes hand in hand with CSR and supportive certification schemes, such as FSC, MSC and other labels
- Supporting sustainable consumption by applying 'choice editing' (particularly in retail), sustainability marketing, experience design and demand-side management, all aiming at reducing the use of material artefacts or promoting the use of more sustainable products
- Developing capabilities for 'strategic sustainability innovation': that is, thinking through the sustainability challenge for the longer term, deriving the opportunities and threats for the firm and translating these into novel business models and innovation opportunities (e.g. compare Chapter 4 by Wüstenhagen and Boehnke and Chapter 5 by Wells in this book)

Hence, identification of what firms can do is not the main problem. The main question is: when are they compelled to do it? In Section 3.2 we pointed at a number of simple reasons. First, the actions listed may enhance a firm's competitive position directly by the reduction of costs and risks, by placing a higher-valued product in the market and by creating more certainty over quality in the supply chains or fewer risks for brand and image. Second, the firm may operate in a sector where sustainability crises, fast-increasing resource prices or social expectations are likely to change the business context in future. This may result, for instance, not only in opportunities for strategic innovations but also for the need for defensive strategic moves. Last, inspired and proactive business leadership may make the difference. Business always has significant room to manoeuvre, but when focused on bottom lines and shareholder value a business may oversee or neglect interesting moves towards sustainability. Once such an essentially fear-driven and scarcity-driven attitude is overcome (e.g. by the certainty of having 'deep pockets'), behaviour driven by a quest for meaning can be found, such as that of Ray Anderson, who repositioned Interface to become fully sustainable by 2020, or that of business leaders such as Bill Gates, Richard Branson and Paul Allen, who all used their own capital to support heavy investments in 'sustainable technology'.

Yet the final quote, by Patrick Burrows, in Section 3.3 indicates that some government backup can make a considerable difference. Business has long been forced to change by legislation—for instance, when pursuing cleaner production—and market forces and conventional change strategies have emerged and evolved as a result. Regulations and financial instruments applied by government, via voluntary agreements or in the form of self-regulation by business, help to create 'a level playing field' and to ensure that minimum demands are met. This is particularly relevant for the first two actions listed at the start of this section. Yet strategic innovation and, to a lesser extent, support for sustainable consumption requires different support. Innovation towards SCP is not just about new concepts; it is also about investment in, and commercialisation of, technologies, products and services and entrepreneurship to create and capture the sources of sustainable value in the market. There is a need for governments to foster greener markets: for example, those that reward technologies, products and services that have superior environmental and social performance. The 'greening' of public procurement provides B2G opportunities for eco-innovation, as Japan has shown via its green purchasing law. The need to address market failures is a key issue, and governments need to understand how they can intervene earlier in B2C and B2B product life-cycles to reduce the environmental and social impacts of technologies, products and services. We, in fact, enter here the discussion on how eco-innovation can best be fostered, an issue discussed at length by Andersen in Chapter 18 of this book. We also refer to Wagner's contribution on the 'Porter Hypothesis' in Chapter 6 (i.e. the claim that the sustainability incentives of markets and regulators do not threaten but rather foster the competitive edge of an industry).

## 3.5 Limitations

### 3.5.1 Introduction

From the previous sections one can easily deduce a number of limitations in the role that business can play in the change to SCP. The essence is that business cannot neglect business fundamentals, and these do not necessarily reward changes towards SCP.

There are many well-known obstacles to change: resistance and inertia (organisational or individual) within businesses with vested interests in the status quo, market conservatism, risk aversion and a focus on short-term pressures according to the ownership and financial structure (e.g. reflecting the interests of institutional investors). A financial system that is focused on quarterly results to shareholders is obviously a major limitation to the development of SCP practices that require a longer-term vision to justify investment. Particularly in mature and/or capital-intensive businesses there is a fear of increased costs related to changing products and processes. Added to these are negative perceptions or lack of understanding of sustainable development and a lack of incentives for more sustainable markets.

Furthermore, sustainable consumption may be interpreted by business to be about limiting or reversing growth, which would reduce revenue and profits and therefore be viewed with hostility. Indeed, with notable exceptions,[6] growth of markets, profits and shareholder value seem to be the main goals of business. Inevitably, there will be businesses that pursue such growth by taking advantage of the fact that they can turn any external costs on third parties or that try to create a market out of activities that hitherto were delivered for free via social structures outside the market.[7] Of course, there is

---

6 Patagonia reputedly capped turnover and therefore growth. This company was prepared to think more radically about what is necessary to achieve sustainable consumption, but obviously this is the exception to the rule.

7 In their otherwise inspiring book, *The Experience Economy*, Pine and Gilmore (1999: 97) give in this respect a telling quote: 'the history of economic progress consists of asking money for something that once was free'. Such business expansion obviously may just have a questionable contribution to make to improved sustainability and quality of life in society.

also the accusation that businesses use their substantial marketing power to promote greed and the creation of new 'needs', leading to much higher material flows in society than necessary for real well-being (cf. Schumacher 1979). It also has been argued (Beder 1998; Korten 1995) that TNCs purporting to be leaders are often controlling and slow the agenda in order to buy time while trying to understand the emerging agenda and develop strategies and plans to minimise threats and maximise opportunities.

### 3.5.2 Limits to the role of business in radical eco-innovation

The limitations with regard to pursuing more radical innovations or eco-innovations are even larger. As will be discussed in more detail in Chapter 18 by Andersen, various conditions must be fulfilled simultaneously to get the 'innovation dynamo' started. It concerns, for example, the availability of lead markets, venture capital, entrepreneurship and stability in policy and institutional context and so on. But, as concluded during the Dutch presidency of the EC in 2004, such linkages needed to 'green' the innovation system are often lacking, as they are between traditional innovators, investors and entrepreneurs, on the one hand, and sustainability experts on the other hand (Charter and Adams 2005).[8]

On top of this, true systemic innovation is even more complex, requiring a greater number of stakeholders to be involved than an individual firm may be able to engage. For example, to deal with transport pollution and congestion in the planned eco-city, Dongtan, in China, a sustainable 'mobility subsystem' is being designed within the 'city system'. It will be essential to develop a holistic and interdisciplinary approach to design, bringing in city planners, architects, civil engineers, product designers and others, to achieve this aim effectively.

### 3.5.3 Conclusions

There are therefore few examples of radical sustainable innovation that have been introduced onto the market today other than in niche markets. This may be because of the obstacles that disruptive innovation faces by its very nature (Stefik and Stefik 2004). The disruptive nature of radical innovations and people's general resistance to change emphasises the need to protect emerging innovations in a niche and to make the benefits of the innovation clear and accessible—this is key to accelerating changes, as in the case of the internet. Therefore, important questions remain:

- Is it reasonable to assume that far-reaching economic and environmental or social win–win situations can become established in existing markets and with realistic changes to the business models?
- Are such radical innovations the way forward for SCP?

8 Traditional entrepreneurs often do not recognise 'green' opportunities, and the investment community is not proactively engaged: business concepts and technologies seem too risky, financial returns are not seen as sufficiently significant to justify investment, and trusted, successful entrepreneurs with track records in the area are scarce. Indeed, often, inventors and developers with sustainable 'start-up' technology, product or service may be driven by a personal ethos but often appear to lack solid business and marketing skills.

- If radical innovations are not contributing to SCP what can be done to make them contribute?
- How can radical eco-innovations and sustainable innovations be accelerated (e.g. what is the basis for the business case)?

The answer to these questions seems self-evident. As it was put by the Sustainable Consumption Roundtable (2006), 'Who genuinely believes that sustainable development is achievable without reductions in consumption, without changing the ways in which we generate material wealth, without developing new investment models, new accounting methods and tax systems, without profound changes in behaviour, values and worldview?'.

In summary, to stimulate the development of eco-innovation and sustainable innovation there is no option but to generate an ongoing building of public policy pressure on both the supply side and the demand side, working with business and other stakeholders, using long-term thinking to direct short-term policy.

## 3.6 Conclusions

Business, in its various forms, is a powerful stakeholder in the macro-level and micro-level systems of consumption and production. Progress towards SCP has been incremental rather than radical, and increases in aggregate consumption appear to be eliminating much of the progress towards levels and types of business activity that are environmentally and socially sustainable. The significant resources directed by business make it a key part of the debate on the level of technological efficiencies required to reach sustainability (e.g. Factor $X$). Given the economic as well as environmental and social impacts of climate change, and given increasing aspirations and affluence in developing countries, the vital importance of moving towards less material and cleaner patterns of sustainable consumption is becoming increasingly recognised.

Businesses likely to evolve and avoid decline are those adaptable in the face of external drivers such as environmental crises, rising energy and material prices and global trends—transforming their competence through rejuvenating entrepreneurship, experimentation, learning and innovating strategically. A 'third green wave' centred on responses to climate change may induce change, with associated cost-cutting measures, to reduce energy and material consumption on the demand side and incentives to stimulate the commercialisation of new energy technologies on the supply side. To sustain profits in the face of increasing input costs, business may shift focus from production to value-added service-based models

Emerging recognition of business models that can influence sustainable consumption can be coupled with the traditional focus on technology and production. However, concrete discussions on the realities of sustainable consumption are still in their early stages, and business does not yet fully understand the concept or its implications following a patchwork of initiatives, such as CSR and sustainable design, and certification schemes, such as FSC, MSC and other labels. The role of business in influencing sustainable consumption—for example, through 'choice editing', marketing or design—requires further investigation by government, academia and business itself. The preconditions required for fundamental change to bring products and services pioneered in a niche to the mainstream where there remains weak overall demand can be categorised as framework changes, crises and leadership.

Although sustainability is now seen as a focus for business opportunities, government policies and regulations can have an influence on moving business towards SCP where the business value is hidden or missing. The right framework, with appropriate incentives, needs to be created to enable the implementation of 'push' and 'pull' strategies towards eco-innovation or sustainable innovation, to encompassing technologies, products and services, processes, organisational models and their associated drivers. Legislation, such as the Green Purchasing Law in Japan, has proven to be effective in moving the supply side and could be extended to other countries.

Leadership at all levels is crucial to establishing SCP. Entrepreneurs and marketers are key to establishing new business models that embed the SCP agenda, and strong personalities also act as an example to others to overcome the 'I will if you will' barrier. However, the scope and timescales of their impact mean that sustainability is unlikely to be achieved alone by businesses, even if they are prepared to push. Governments, NGOs and the (non-SRI [socially responsible investment]) finance community must also buy into the agenda as key players in creating a framework that is conducive to change and in addressing market failures. Mechanisms such as training to help build sustainability into leaders' vision, and for facilitating and embedding that vision throughout organisations, need to be further investigated.

On a basic level, the real and necessary aims of the SCP agenda need to be more clearly defined and accepted by business, government and others; targets and goals need to be set to stimulate progress. In particular, the gaps between what needs to be done, what is realistic and what is easy need to be exposed and challenged. Eco-innovation and sustainable innovation is a promising route towards SCP but requires more catalysts, networks and learning to be applied (e.g. to support start-ups and help entrepreneurs recognise 'green' opportunities and harness SRI funds).

SCP policy has focused on decoupling the link between consumption of products and services and environmental impact. Business can go further by influencing the level and nature of sustainable consumption, potentially going beyond gains in the efficiency of consumption to address quality of life. By challenging what we mean by sustainable growth we can begin to understand what we aim to achieve by business—is it financial gain or is it what we perceive money can give us (i.e. happiness)? Considering other ways of meeting human needs and aspirations may result in radical new business models.

# References

Arthur D. Little (2005) 'Innovation High Ground Report: How Leading Companies are Using Sustainability-driven Innovation to Win Tomorrow's Customers', www.adl.com/insights/news/25hw6f93hfc3471h6e6hcnhv38/ADL_Innovation_High_Ground_Report_2005.pdf [accessed 7 March 2007].

Baden-Fuller, C., and J.M. Stopford (1992) *Rejuvenating the Mature Business: The Competitive Challenge* (Boston, MA: Harvard Business School Press; London: Routledge).

Bartelmus, P. (2002) *Unveiling Wealth on Money, Quality of Life, and Sustainability* (Dordrecht, Netherlands: Kluwer).

Beder, S. (1998) *Global Spin: The Corporate Assault on Environmentalism* (White River Junction, VT: Chelsea Green).

Charter, M., and G. Adams (2005) *Investigating the Sustainability Issues in the Information Technology (IT) Sector and Minimising the Related Impacts to the Supply Chain* (Farnham, UK: Centre for Sustainable Design).

—— and T. Clark (2007) *Sustainable Innovation* (Farnham, UK: The Centre for Sustainable Design).

——, L. Elvins and G. Adams (2004) *Sustainable Marketing: Understanding the Obstacles to and Opportunities for Involvement of Marketing Professionals in Sustainable Consumption* (Farnham, UK: Centre for Sustainable Design).

Christensen, C.M. (1997) *The Innovator's Dilemma: When New Technologies Cause Great Firms to Fail* (Cambridge, MA: Harvard Business School Press).

Collignon, M.X., C.J.L. Hogenhuis-Kouwenhoven and E.J. Stork (2006) 'Responsible Purchasing: A Practical Business Guide', in *SCORE! Proceedings: Sustainable Consumption and Production: Opportunities and Challenges, Wuppertal, 23–25 November 2006, Volume I* (Delft, Netherlands: TNO): 87-112.

Collins, J., G. Thomas, R. Willis and J. Wilsdon (2003) *Carrots, Sticks and Sermons: Influencing Public Behaviour for Environmental Goals* (Green Alliance report for the Department of Environment, Food and Rural Affairs [Defra]; London: Defra).

Cooperative Bank (2004) '2004 Ethical Consumerism Report', www.co-operativebank.co.uk/servlet/Satellite?c=Page&cid=1134717168931&pagename=CoopBank%2FPage%2FtplPageStandard [accessed 25 August 2007].

CSCP (Centre on Sustainable Consumption and Production) (2006) *Creating Solutions for Sustainable Consumption and Production* (Wuppertal, Germany: UNEP/Wuppertal Institute, in collaboration with CSCP).

CSR Europe (2007) *B to B Working Group on Sustainable Marketing* (Brussels: CSR Europe).

Defra (UK Department of Environment, Food and Rural Affairs) (2006) 'Whatever Next?', *SD Scene Newsletter*, May–June 2006; 216.239.59.104/search?q=cache:yWRxUPUYiuMJ:www.sustainable-development.gov.uk/what/newsletter/may-june.htm+%22To+demonstrate+the+Government%E2%80%99s+commitment+to+leading%22+UN&hl=en&ct=clnk&cd=1&gl=uk [accessed 25 August 2007]).

EC (European Commission) (1996) 'Council Directive 96/61/EC of 24 September 1996 Concerning Integrated Pollution Prevention and Control', *OJ* L 257 of 10 October 1996.

—— (2006) 'Corporate Social Responsibility', ec.europa.eu/enterprise/csr/index_en.htm [accessed 25 August 2007].

EIA (Energy Information Administration) (2006) *International Energy Outlook 2006* (Washington, DC: US Department of Energy; www.eia.doe.gov/oiaf/ieo/index.html [accessed 25 August 2007]).

Forstater, M., and J. Oelschaegel (2006) *What Assures Consumers?* (London: AccountAbility/National Consumer Council).

Fox, B., and B. Vorley (2004) 'Stakeholder Accountability in the UK Supermarket Sector', www.racetothetop.org/documents/RTTT_final_report_full.pdf [accessed 25 August 2007].

Frankl, P., L. Pietroni, E. Montcada and F. Rubik (2005) 'Conclusions', in F. Rubik and P. Frankl (eds.), *The Future of Eco-labelling: Making Environmental Product Information Systems Effective* (Sheffield, UK: Greenleaf Publishing): 236-90.

Freedman, P. (2004) 'The Age of the Hollow Company', *The Sunday Times*, 25 April 2004.

Fulton, K. (2002) 'The Possibility of a New Era of Social and Political Innovation', in E. Kelly and P. Leydon (eds.), *What's Next?* (Chichester, UK: John Wiley): 90-101.

General Electric Company (2006) 'Ecomagination', www.ge.com/company/citizenship/ecomagination/index.html [accessed 25 August 2007].

GreenBiz (2006) 'Wal-Mart Sustainability Meeting Focuses on Climate Change, Supply Chain', www.greenbiz.com.

Hamel, G. (2000) *Leading the Revolution* (Cambridge, MA: Harvard Business School Press).

—— and C.K. Prahalad (1994) *Competing for the Future* (Boston, MA: Harvard Business Review Press).
Harrison, R. (2006) 'Swallowed Up', *Ethical Consumer* 101: 31-33.
Hawken, P. (1994) *The Ecology of Commerce: How Business Can Save the Planet* (London: Weidenfeld & Nicolson).
ICAEW (Institute of Chartered Accountants in England and Wales) (2004) 'Sustainability: The Role of Accountants', www.icaew.com/index.cfm?route=127769 [accessed 25 August 2007].
IPCC (Intergovernmental Panel on Climate Change) (2007) *Climate Change 2007: The Physical Science Basis. Summary for Policy-makers*, www.ipcc.ch/SPM2feb07.pdf [accessed 25 August 2007].
Kleanthous, A., and J. Peck (2006) *Let Them Eat Cake* (Godalming, UK: WWF-UK).
Korten, D. (1995) *When Corporations Rule the World* (San Francisco: Berrett-Koehler).
Ottman, J.A. (1998) *Green Marketing: Opportunity for Innovation* (New York: NTC–McGraw-Hill).
Philips Electronics NV (2007) 'Improving Lives Delivering Value: Sustainability Report 2006', www.philips.com/assets/Downloadablefile//Sustainability-Annual-Report-2006(2)-16090.pdf [accessed 25 August 2007].
Pine II, B.J., and J.H. Gilmore (1999) *The Experience Economy* (Boston, MA: Harvard Business School Press).
Porritt, J., and J. Goodman (2005) *Fishing for Good* (London: Forum for the Future).
Porter, M.E. (1985) *Competitive Advantage: Creating and Sustaining Superior Performance* (New York: The Free Press).
Prahalad, C.K., (2005) *The Fortune at the Bottom of the Pyramid* (Upper Saddle River, NJ: Pearson Education).
REEEP (Renewable Energy and Energy Efficiency Partnership) (2004) 'Glossary of Terms in Sustainable Energy Regulation', www.reeep.org/media/downloadable_documents/9/0/SERN%20Glossary.doc [accessed 25 August 2007].
Renwick, M.E., and R.D. Green (2000) 'Do Residential Water Demand Side Management Policies Measure Up? An Analysis of Eight California Water Agencies', *Journal of Environmental Economics and Management* 40.1: 37-55.
Richardson, J., T. Irwin and C. Sherwin (2005) *Design and Sustainability: A Scoping Report for the Sustainable Design Forum* (London: London Design Council).
Sanstad, A.H., and R.B. Howarth (1994) ' "Normal" Markets, Market Imperfections and Energy Efficiency', *Energy Policy* 22.10: 811-18.
Schumacher, E.G. (1979) *Good Work* (New York: Harper & Row).
Stefik, M., and B. Stefik (2004) *Breakthrough: Stories and Strategies of Radical Innovation* (Cambridge, MA: MIT Press).
Stern, N. (2006) *The Economics of Climate Change: The Stern Review* (Cambridge, UK: Cambridge University Press).
Sustainable Consumption Roundtable (2006) *Sustainable Consumption Business Dialogue: A Report to the Sustainable Consumption Roundtable from the University of Cambridge Programme for Industry* (London: Sustainable Consumption Roundtable; www.ncc.org.uk/nccpdf/poldocs/NCC122rr_business_dialogue.pdf [accessed 25 August 2007]).
Tukker, A., and U. Tischner (2006) 'Conclusions', in A. Tukker and U. Tischner (eds.), *New Business for Old Europe: Product-Service Development, Competitiveness and Sustainability* (Sheffield, UK: Greenleaf Publishing): 350-74.
—— and C. van den Berg (2006) 'Product-Services and Competitiveness', in A. Tukker and U. Tischner (eds.), *New Business for Old Europe: Product-Service Development, Competitiveness and Sustainability* (Sheffield, UK: Greenleaf Publishing): 35-71.
——, P. Eder and E. Haag (2001) *Eco-design: European State of the Art. Part I. Comparative Analysis and Conclusions* (European Science Technology Observatory Report, EUR 19583 EN; Seville, Spain: DG European Commission Joint Research Centre Institute for Prospective Technology Studies; esto.jrc.es/reports_list.cfm [accessed 25 August 2007]).
Von Weizsäcker, E., A.B. Lovins and L.H. Lovins (1997) *Factor Four: Doubling Wealth, Halving Resource Use* (London: Earthscan Publications).
Wal-Mart (2006a) 'Wal-Mart greening could have huge impact', www.Wal-Martfacts.com/articles/3832.aspx [accessed 25 August 2007].

—— (2006b) 'Wal-Mart opens first experimental Supercenter', www.Wal-Martfacts.com/articles/2081.aspx [accessed 25 August 2007].

WBCSD (World Business Council for Sustainable Development) (2006) *Procter & Gamble: Treating Water at its Point of Use* (Geneva: WBCSD).

World Energy Council (2002) 'UN Commission on Sustainable Development Secretary-General's Note for the Multi-stakeholder Dialogue on Sustainable Energy and Transport: Addendum No. 1. Dialogue Paper by Business/Industry', www.worldenergy.org/wec-geis/publications/default/archives/other_documents/sust_dev/topic_4.asp [accessed 7 March 2007].

# 4
# Business models for sustainable energy*

*Rolf Wüstenhagen and Jasper Boehnke*
Institute for Economy and the Environment, Switzerland

## 4.1 Introduction

The energy sector is one of the largest sectors of the economy, accounting for annual sales of about $2,000 billion worldwide (SAM 2002). Total investment required for energy-supply infrastructure worldwide over the period 2001–2030 is estimated at $16 trillion by the International Energy Agency (IEA), which is a substantial increase compared with the prior 30-year period (IEA 2003). A set of environmental and security concerns, in conjunction with technological innovation, is currently leading to fundamental changes in the energy industry. For example, more than 80% of electricity worldwide is generated based either on fossil fuels, one of the main reasons for global warming, or on nuclear energy, which involves security concerns and issues regarding hazardous waste. If we look at all energy (including fuels for transportation and heating), the lack of sustainability becomes even more evident, with the combined share of oil, coal and gas in 2001 achieving 86% of global energy consumption, and nuclear power adding another 6.5% (www.eia.doe.gov). With increasing concentration of oil reserves in a few countries of the Middle East, as well as new, strong, demand coming from emerging countries such as China, concerns over the security of supply add to environmental drivers for change. Also, more than two-thirds of primary energy gets lost as a result of inefficiencies in the energy sector and on the demand side (UNDP/WEC/UNDESA 2000).

---

* This research has been supported by the Swiss National Science Foundation, NRP 54 'Sustainable Development of the Built Environment' (Project No. 405440-107239).

Therefore, the energy sector has been identified as a key target area for efforts to promote sustainable consumption and production (BMU 2005).

Sustainable energy technologies can be defined as those technologies providing energy services (such as light, heat or mobility) with a lower environmental impact than currently experienced, while maintaining economic efficiency (including external costs) and being socially acceptable. Many of the possible solutions to the sustainability challenge in the energy sector are marked by the need simultaneously to address consumption and production. For example, a shift towards renewable and distributed energy systems turns consumers into co-producers of heat and electricity (Sauter and Watson 2007; Watson 2004). Increasing demand-side energy efficiency, as promoted by governments on both sides of the Atlantic (EC 2005; US DOE/US EPA 2006), is also inextricably linked to consumer involvement, and so is the decision of the increasing number of both private and corporate energy users that seem to be willing to 'do their bit' and purchase green power (Bird *et al.* 2002).

Hence, in addition to ongoing policy efforts to facilitate a transition towards a sustainable energy future, there is also a case for investigating this important sector from a sustainable consumption and production (SCP) perspective: that is, focusing on the processes that influence consumer demand on the one hand, and factors that characterise successful firm strategies to match that demand on the other. Within an ongoing research project, we are taking an SCP perspective to investigate factors that facilitate or hamper the development of distributed energy systems, or micropower, in Switzerland (Wüstenhagen *et al.* 2006). The present chapter focuses on the sustainable production perspective of this broader picture by looking at business models for sustainable energy.

## 4.2 Challenges in commercialising sustainable energy technologies

### 4.2.1 Environmental externalities

Sustainable energy technologies, such as solar cells, solar thermal collectors, micro cogeneration plants, Stirling engines and heat pumps (Dunn 2000), use renewable or non-renewable fuels to generate electricity and/or heat with a lower environmental impact than conventional energy technologies such as large coal power plants or oil-based systems. This reduced environmental impact, however, does not necessarily translate into reduced private cost for the consumer, because the environmental externalities of conventional energy systems (such as the damage caused by $CO_2$ emissions, or the risk of a nuclear accident) are not fully internalised in market prices. Therefore, switching from conventional to 'green' electricity or from fossil fuels to renewable heating means a lower cost to society, but not necessarily lower cost for the consumer. The discrepancy between private and public benefit (and cost) is a serious barrier to both consumer and investor in making decisions regarding sustainable energy (Wüstenhagen and Teppo 2006).

At the same time, most products, including sustainable energy technologies, are characterised by a bundle of private and public benefits. Solar thermal collectors reduce emissions (public benefit) but also warm water and insure against oil price increases (private benefit). Generating electricity from biogas on a farm reduces dependence on non-renewable resources (public benefit) and also reduces the farmer's electricity bill and his or her waste-stream (private benefit). Therefore, while the full exploitation of the market potential for sustainable energy technologies requires policy intervention for internalisation of external cost, a (possibly substantial) share of the market is achievable even today if marketers find appropriate ways to emphasise in their communication efforts the private benefits that sustainable energy solutions provide (Villiger et al. 2000).

### 4.2.2 Capital intensity and long lead-times

Developing new energy technologies takes a lot of time and resources. Compared with many other industries, the energy industry is characterised by a high capital intensity and long lead-times. Developing a new energy conversion device, such as a microturbine or a fuel cell, will typically require investments in research and development (R&D) of several hundred million euros. Setting up a factory for manufacturing solar cells requires investments of at least some tens of millions of euros. And buying a new heating system for a residential house will typically require the consumer to incur upfront cost of several thousand euros, and probably twice that amount if he or she opts for renewable energy. Therefore, successfully marketing new energy technologies is a challenging task, and finding investors for a new venture in this business is equally challenging (Wüstenhagen and Teppo 2006).

### 4.2.3 The power of incumbents

Another consequence of the capital-intense nature of the energy business is that huge amounts of capital have been invested in existing infrastructure, and hence the companies that own these assets have a very strong position in the market and tend to be reluctant to do anything that would cannibalise their existing business. This poses serious challenges to new players who want to enter that market, for example, in order to introduce a new sustainable heating system.

## 4.3 Business models

### 4.3.1 Business models as a unit of analysis in management research

Over recent years the business model concept has become increasingly popular in management theory and practice (Magretta 2002; Osterwalder et al. 2005; Shafer et al. 2005). In general, a business model can be defined as a description of a planned or existing business and its specific characteristics with respect to value creation, on the one hand, and market-orientation on the other (Hedman and Kalling 2003; Osterwalder et al. 2005; Stähler 2001). The business model concept combines elements of the resourced-based and the market-based view of the firm and thus takes an integrated point of view (Kalling 2002; Morris et al. 2005).

Initially, research on business models emerged in the field of e-commerce (Afuah and Tucci 2002; Alt and Zimmermann 2001; Mahadevan 2000; Morris et al. 2005; Tapscott et al. 2000; Timmers 1998). More recently, the business model has been discussed in a broader set of management-related publications (e.g. Chesbrough and Rosenbloom 2002; Rentmeister and Klein 2003; Shafer et al. 2005). E-commerce activities seem no longer to be a prerequisite for applying business model analysis to explain value creation (Magretta 2002; Sillin 2004). Even more than academic research, management practice has emphasised the importance of business models (Morris et al. 2005). Although a considerable number of publications have tried to explain the business model concept, no generally accepted definition has evolved so far (Osterwalder et al. 2005; Porter 2001; Shafer et al. 2005). Most practitioners are not able to explain the concept either (Linder and Cantrell 2000). Researchers seem to agree that a business model describes how a business creates value and that it is an important new unit of analysis, highly relevant to both management theory and practice (Belz and Bieger 2004; Chesbrough and Rosenbloom 2002; Morris et al. 2005; Rentmeister and Klein 2003). Business model analysis can help to understand and communicate the key success factors of value creation. Furthermore, it can be used to measure, compare or even change the business logic (Morris et al. 2005; Osterwalder et al. 2005; Shafer et al. 2005).

### 4.3.2 Why have business models become a 'hot topic'?

The increasing popularity of business models in management research and in the popular business press can be explained by several factors. First, there is a clear link between the popularity of the term 'business model' and the emergence of e-commerce and the Internet in the 1990s. The New Economy enabled completely new forms of value creation, asking for explanations beyond the traditional bricks-and-mortar modes of production previously discussed in management literature. Second, the boom in entrepreneurship and venture capital funding that resulted from the new technological opportunities led to a creative search for new business models, relatively unconstrained by traditional ways of doing business that were embedded in large incumbent firms. Third, even within 'Old Economy' sectors such as the airline industry, computer hardware or furniture retailing, some companies boosted their profitability compared with industry standards by adopting new business models. Famous examples included:

- 'No frills' airlines such as EasyJet and Southwest Airlines, which revolutionised pricing, booking systems, route networks and service concepts

- Dell Computer, which introduced radically new ideas about logistics, distribution and financial management in the computer hardware retail business

- IKEA, which turned the traditional furniture retail business upside down by changing prevalent assumptions about logistics

### 4.3.3 Key elements of a business model

Timmers (1998) was among the first to put forward a prominent formal definition of a business model, viewing it as a description of the architecture of value generation (across firm boundaries), of the potential value generated for partners and final consumers, of the sources of revenue and of the marketing strategy. Another early definition was given by Hamel (2000), who proposed four components to the business model: core strategy (including the product or service), strategic resources (i.e. competences, assets and processes and their configuration), customer interface (by which benefits are delivered) and value network. In this chapter we will follow a definition of the business model as proposed by Stähler (2001), who defined a business model as a description of a planned or existing business including three key elements, namely:

- Value proposition
- Configuration of value creation
- Revenue model

The value proposition describes how the products and services offered by the business create value for the customer and other stakeholders. Regarding the configuration of value creation, the most important question concerns which steps of the value chain are to be performed by the business. The core processes must be defined, as well as the targeted customer groups of the business. Ideally, a business develops core competences by which it can be distinguished from its competitors. Also included in the analysis of the configuration are aspects such as cross-selling effects, complementary products and the design of processes that transcend the value step. Finally, the revenue model describes how the business generates its sales revenue (e.g. by selling the actual products or by selling leasing contracts) and ultimately profits.

### 4.3.4 The relevance of business models in the context of sustainable energy

Why is it relevant to look at business models when doing research on the commercialisation of sustainable energy technologies? By appropriately designing the three elements of a business model described in Section 4.3.3, a firm can tune its offerings to meet the three challenges of sustainable energy technologies as discussed in Section 4.2 and therefore achieve higher market penetration of these technologies as well as commercial success at the firm level.

#### 4.3.4.1 Value proposition

When it comes to the value proposition, as indicated in Section 4.2.1, sustainable energy technologies are typically characterised by a strong component of public benefit. As a result, they pose a tougher challenge to marketers who try to convince consumers to buy such products rather than products with a stronger component of private benefit, such as a new mobile telephone or a new car. While this has sometimes led to the conclusion that the promotion of sustainable energy should be left to governments, proponents of green marketing have argued that there is still room to manoeuvre for private marketers

(Ottman 2004). Public policy can certainly help, but by carefully focusing on an important customer problem that the product can solve—that is, by highlighting the private benefit on top of the obvious public benefit—a sustainable energy venture will be able to attract a broader set of customers beyond the 'eco-niche' (Villiger *et al.* 2000).

The case of the Swiss municipal utility EWZ (www.ewz.ch), based in Zurich, can help to illustrate this point. As other 'green' electricity marketers it initially struggled with the fact that the difference between the buying of its product and that of conventional electricity is not really tangible for the consumer. EWZ has overcome this problem by inviting its solar electricity customers to a party in the local neighbourhood whenever the contributions of an increasing number of customers allows the inauguration of a new solar power plant, thereby creating a tangible component of private value (Wüstenhagen 2000). Another example is the case of micro combined heat and power (CHP) generation. Traditionally, many suppliers of micro CHP were small engineering-driven firms that worked hard to maximise the electrical (and hence environmental) efficiency of their technology, assuming that this would cause consumers to switch from their conventional heating system to producing their own electricity and heat in-house. WhisperTech (www.whispergen.com), a New Zealand-based manufacturer of Stirling engines (the WhisperGen®) has changed this assumption and radically rethought the value proposition of its product. As a result, the first market that it addressed was sailing boats, where the ability of the WhisperGen® to generate electricity silently met with a high customer willingness-to-pay. After harnessing this particularly profitable market segment, the company went on to enter the more competitive stationary micro CHP market. There again, the company clearly took customer value as the starting point, and consequently positioned its products as a convenient, silent, home appliance that replaces a conventional heating boiler, while, among other things, providing environmental and cost savings.

### 4.3.4.2 Configuration of value creation

A properly designed configuration of value creation can help to address the second and third barriers described in Section 4.2: namely, capital intensity and the power of incumbents. In a globalised economy, firms now have the possibility to outsource large parts of the value chain and focus on only those components that are key to their competitive advantage. By doing so they can reduce capital intensity, as compared, for example, with the 'Old Economy' model of doing everything in-house, from R&D to manufacturing to distribution. Similarly, configuring value creation in a way that minimises head-on competition with incumbents can be a smart way for entrepreneurial firms to address the market power issues that are omnipresent in the energy sector. Cooperating with existing heating manufacturers in distribution may be a wise decision for a capital-constrained start-up firm developing a new co-generation system, rather than getting distracted from tackling the ramp-up of a new distribution system while at the same time dealing with technology development challenges.

A good example of a sustainable energy company pursuing an innovative approach to configuring its value creation is Pelamis Wave Power (www.pelamiswave.com, formerly Ocean Power Delivery), a Scottish manufacturer of wave energy converters. Pelamis Wave Power has designed a new 750 kW offshore device that converts the power of waves to electricity. Developing and manufacturing these 120 metre long steel devices requires substantial capital investment. Unlike some of its competitors in the marine energy field, and probably assisted by advice from its experienced venture capital investors, Pelamis Wave Power has adopted a modular design, where only the 'intelligent core' of the machines, the three power conversion modules, are manufactured in-house, whereas the more standardised parts are subcontracted to other firms.

#### 4.3.4.3 Revenue model

Finally, new innovative revenue models have been at the core of success for business model innovators in traditional industries such as EasyJet, and they can be equally important for sustainable energy ventures. Given the capital intensity of sustainable energy technologies, discussed above, reduction of the upfront cost for consumers is one of the key concerns in marketing innovation in this sector. Leasing or contracting can provide successful solutions to this problem. Also, many sustainable energy technology firms tend to be focused on selling hardware, while other industries have demonstrated that after-sales services are often much more relevant for long-term profitability. Finally, as many governments have set up incentive schemes for renewable energy, the intelligent management of available subsidies may also be seen as an important element of revenue models for sustainable energy.

In terms of examples from the sustainable energy sector, a particularly innovative example is the German company Schmack Biogas AG (www.schmack-biogas.de), which sells not just biogas power plants but also offers a comprehensive technical and biological service. The starting point for developing these highly profitable services was the observation that a major risk for the profitability of operating a biogas plant is the varying quality of the resource. Controlling the fermentation process and conducting microbiological analyses of the biogas before actually fuelling the plant can help to avoid expensive shutdowns of the plant. Therefore, Schmack Biogas's customers are willing to pay a premium for these 'biological maintenance contracts', because they will benefit from the greater reliability and higher output of their plant.

As for the leasing and contracting examples mentioned, an instructive case is the Austrian manufacturer of biogas power plants, Jenbacher (www.ge-energy.com/businesses/ge_jenbacher/en/index.htm). The company has long been known for the high technical quality of its products, but when it was acquired by General Electric in 2003 the combination of its hardware offerings with the full range of financing solutions offered by the new parent firm created new opportunities to address customers' capital constraints and generate new revenues.

## 4.4 Conclusions

Appropriately designed business models are an important opportunity to overcome some of the key barriers to the market diffusion of sustainable energy technologies (see Fig. 4.1). Consequently, focusing the value proposition on the aspects that create the highest (private) customer value, rather than primarily highlighting the public benefits of sustainable energy, is a means of addressing the challenges posed by environmental

FIGURE 4.1 How business model configuration addresses challenges in commercialising sustainable energy technologies

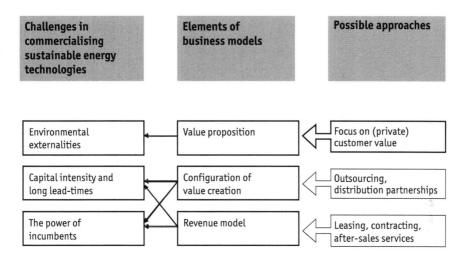

externalities. Configuring the value creation in a way that provides for efficient use of own capital, and making use of outsourcing opportunities, can mitigate the implications of the capital-intensive nature of the energy industry. By entering into distribution partnerships with established players, sustainable energy ventures can address both capital intensity (of building up distribution networks) and the power of incumbents. Innovative revenue models such as leasing or contracting, but also the targeted use of available subsidies, should be pursued to reduce the upfront capital cost for customers. If incumbent players, such as installers, are provided with opportunities for offering after-sales services, their resistance to innovation is likely to be reduced.

To summarise, this chapter has demonstrated the importance of new business models for enhancing sustainable production and consumption in the energy sector, particularly with regard to successful commercialisation of distributed energy systems.

# References

Afuah, A., and C.L. Tucci (2002) *Internet Business Models and Strategies* (New York: McGraw-Hill).
Alt, R., and H.-D. Zimmermann (2001) 'Preface: Introduction to Special Section. Business Models', *Electronic Markets* 11: 3-9.
Belz, C., and T. Bieger (2004) *Customer Value: Kundenvorteile schaffen Unternehmensvorteile* (St Gallen, Switzerland: Thexis).
Bird, L., R. Wüstenhagen and J. Aabakken (2002) 'A Review of International Green Power Markets: Recent Experience, Trends, and Market Drivers', *Renewable and Sustainable Energy Reviews* 6.6: 513-36.

BMU (Bundesministerium für Umwelt, Naturschutz und Reaktorsicherheit, Germany) (2005) *Sustainable Energy Consumption: Meeting Report of the European Conference under the Marrakech Process on Sustainable Consumption and Production (SCP)*, Berlin, 13–14 December 2005.

Chesbrough, H., and R.S. Rosenbloom (2002). 'The Role of the Business Model in Capturing Value from Innovation: Evidence from Xerox Corporation's Technology Spin-off Companies', *Industrial and Corporate Change* 11.3: 529-55.

Dunn, S. (2000) 'Making Way for Micropower', *Co-generation and On-site Power Production* 1.5.

EC (European Commission) (2005) *Doing More With Less: Green Paper on Energy Efficiency* (Brussels: EC).

Hamel, G. (2000) *Leading the Revolution* (Cambridge, MA: Harvard Business School Press).

Hedman, J., and T. Kalling (2003) 'The Business Model Concept: Theoretical Underpinnings and Empirical Illustrations', *European Journal of Information Systems* 12: 49-59.

IEA (International Energy Agency) (2003) *World Energy Investment Outlook* (Paris: IEA).

Kalling, T. (2002) *The Business Model and the Resource Management Model: A Tool for Strategic Management Analysis* (Working Report 2002/4; Paper and Packaging Research Programme; Lund, Sweden: Institute of Economic Research, Lund University).

Linder, J.C., and S. Cantrell (2000) *Changing Business Models: Surveying the Landscape* (Cambridge, MA: Accenture Institute for Strategic Change).

Magretta, J. (2002) 'Why Business Models Matter', *Harvard Business Review* 80.5: 86-92.

Mahadevan, B. (2000) 'Business Models for Internet-Based E-commerce', *California Management Review* 42: 55-69.

Morris, M., M. Schindehutte and J. Allen (2005) 'The Entrepreneur's Business Model: Toward a Unified Perspective', *Journal of Business Research* 58: 726-35.

Osterwalder, A., P. Pigneur and C.L. Tucci (2005) 'Clarifying Business Models: Origins, Present and Future of the Concept', *CAIS (Communications of the Association for Information Systems)* 15 (May 2005).

Ottman, J. (2004) *Green Marketing: Opportunity for Innovation* (Charleston, SC: BookSurge, 2nd edn).

Porter, M.E. (2001) 'Strategy and the Internet', *Harvard Business Review* 79: 63-78.

Rentmeister, J., and S. Klein (2003) 'Geschäftsmodelle: Ein Modebegriff auf der Waagschale', *Zeitschrift für Betriebswirtschaft* 1.73: 17-30.

SAM Sustainable Asset Management (2002) *Changing Climate in the Energy Sector: A New Wave of Sustainable Investment Opportunities Emerges* (Aurich: Zollikon, 2nd edn; www.sam-group.com).

Sauter, R., and J. Watson (2007) 'Strategies for the Deployment of Microgeneration: Implications for Social Acceptance', *Energy Policy* 35.5 (Special Issue on Social Acceptance of Renewable Energy Innovation): 2,770-79.

Shafer, S.M., H.J. Smith and J.C. Linder (2005) 'The Power of Business Models', *Business Horizons* 48: 199-207.

Sillin, J.O. (2004) 'The Electric Power Industry Business Model for the 21st Century', *The Electricity Journal*, April 2004: 42-51.

Stähler, P. (2001) *Geschäftsmodelle in der digitalen Ökonomie: Merkmale, Strategien und Auswirkungen (Business Model Innovation in the Digital Economy: Characteristics, Strategies and Repercussions)* (PhD thesis; Köln-Lohmar, Germany: Josef Eul Verlag/University of St Gallen).

Tapscott, D., D. Ticoll and A. Lowy (2000) *Digital Capital: Harnessing the Power of Business Webs* (Cambridge, MA: Harvard Business School Press).

Timmers, P. (1998) 'Business Models for Electronic Markets', *Electronic Markets* 8.2 (July 1998; Special Issue on Electronic Commerce in Europe; ed. Y. Gadient, B. Schmid and D. Selz; www.electronicmarkets.org/modules/pub/view.php/electronicmarkets-183 [accessed 25 August 2005]).

UNDP (United Nations Development Programme)/WEC (World Energy Council)/UNDESA (United Nations Department of Economic and Social Affairs) (2000) *World Energy Assessment* (New York: UNDP/WEC/UNDESA).

US DOE (Department of Energy)/US EPA (Environmental Protection Agency) (2006) *National Action Plan for Energy Efficiency* (Washington, DC: US DOE/US EPA, July 2006).

Villiger, A., R. Wüstenhagen and A. Meyer (2000) *Jenseits der Öko-Nische* (Basel, Switzerland: Birkhäuser).

Watson, J. (2004) 'Co-provision in Sustainable Energy Systems: The Case of Micro-generation', *Energy Policy* 32: 1981-90.

Wüstenhagen, R. (2000) *Ökostrom: Von der Nische zum Massenmarkt* (*Green Electricity: From Niche to Mass Market*) (Zurich: vdf).

—— and T. Teppo (2006) 'Do Venture Capitalists Really Invest in Good Industries? Risk–Return Perceptions and Path Dependence in the Emerging European Energy VC Market', *International Journal of Technology Management* 34.1–2: 63–87.

——, J. Boehnke and J. Känzig (2006) 'Micropower in Residential Buildings: An Analysis of Customer Preferences and Business Models', in International Association for Energy Economics (IAEE) (ed.), *Securing Energy in Insecure Times, Proceedings of the IAEE 2006 International Conference*, Potsdam, 7–10 June 2006, www.iwoe.unisg.ch/micropower [accessed 25 August 2007].

# 5
# Alternative business models for a sustainable automotive industry

*Peter Wells*
Centre for Business Relationships, Accountability, Sustainability and Society, UK

## 5.1 Introduction

With respect to many consumer goods, more efficient production or a more efficient product would not in itself constitute the attainment of sustainability because such improvements are still embedded within essentially linear value chains and are articulated by business models that emphasise 'fire and forget' production. Taking the case of the automotive industry, this chapter seeks to demonstrate that alternative business models have the potential to redefine the terms of competition and in so doing realign business with society at the local level while finding a solution to the wastefulness that is characteristic of the profusion society. The chapter draws on many years of research but it is informed specifically by a five-year programme conducted within the Centre for Business Relationships, Accountability, Sustainability and Society (BRASS) at Cardiff University investigating the concept of micro-factory retailing. It is concluded that there are many, but not necessarily insurmountable, structural barriers to the creation of sustainable patterns of production and consumption with respect to complex engineered products such as cars.

In so doing, this chapter draws together two lines of discussion. On the one hand, it is argued that sustainable technology with the characteristics to achieve 'Factor $X$' scale improvements needs a pathway in which to develop. Given that in contemporary society technologies are generally delivered by companies, acting in markets to reach customers, it is therefore pertinent that this pathway needs to be competitive with the prevailing market offerings and to be viable at low volumes—because the market will initially be small compared with that for established products. While much has been done in terms of understanding the ways in which strategic niches may form the stepping stones along the way to successful market penetration, there has been rather less consideration of the ways in which business may have to change in order to challenge the prevailing orthodoxy.

On the other hand, it is also argued that contemporary patterns of production and consumption are essentially unsustainable, whatever the technologies or materials involved. This is perhaps a more difficult case to make, but in the example explored here (that of the automotive industry) the explanation rests primarily on the structural conditions evident in the industry that result in a distinctive business model. This approach to business is by no means confined to the automotive industry and is characterised by high capital intensity leading to over-production and thence over-consumption. The symptoms of this trajectory include continued downward pressure on the price of new cars, high rates of product depreciation, baroque innovation and the arrival of economic obsolescence before technical obsolescence. Despite the widespread evidence for the lack of economic, social or environmental sustainability in the automotive industry, it remains characterised by high barriers to entry and exit, with considerable resilience to disruptive technological change.

The chapter begins with an introductory section that seeks to locate the case presented here within the wider debate on sustainable production and consumption, making the basic case that the assumption that the 'market' will allow the emergence of appropriate business models is somewhat optimistic. This is followed by a definition of what is meant by the term 'business model' in the context of this chapter, and how that concept can be related to the ways in which value is created by a company.

Thereafter, the discussion is more closely aligned to the specifics of the automotive industry, not least on the premise that the transition from generic understanding to specific solution requires embedded knowledge. Put simply, the solutions appropriate to the automotive industry may not necessarily be those appropriate to other aspects of sustainable production and consumption. Note also that the approach adopted here is not deterministic, in that new ways of making things (production technology) and new things themselves (product technology) are envisaged as enabling changes in, for example, the organisation of capital or modes of consumption but not necessarily requiring them. This theme is expanded on later, particularly in the context of the discussion of barriers to change at the conclusion of the chapter.

## 5.2 Sustainable consumption and production: the issue of business models

The need for sustainable consumption and production (SCP) is broadly understood. That is, with expected world population levels, existing patterns of material consumption are beyond the capacity of the planet if the majority of the world population were to achieve the patterns currently exhibited by the industrialised nations of North America, Europe and Japan (von Weizsäcker *et al.* 1997). In essence, there are two related problems here: to reduce the already unsustainable patterns of production and consumption evident in those industrialised nations, and to allow a level of material and quality-of-life prosperity to develop in the economically emergent nations in a manner that does not further undermine the carrying capacity of the planet for existing and future generations. Food consumption, domestic energy requirements and, significantly in this case, mobility are seen as the three major dimensions of modern life that account for the majority of the ecological impact of contemporary consumption patterns (Tukker et al. 2005).

The extant SCP literature is testimony to the scale of the problem and also to the potential availability of solutions. Enormous effort has gone into understanding how ameliorative technologies might develop, most notably by the 'industrial transformation' school of thought. As Tucker et al. (2005) rather dejectedly note, von Weizsäcker et al. (1997) actually listed some 50 examples of innovations that would substantially reduce environmental impact for a given 'service unit' made available to consumers, yet ten years on such items are confined to niche markets. The explanation pursued in this chapter is that, at least in part, it is because there has been insufficient consideration given to the need to have viable business models to deliver SCP. While attention has been given, deservedly, to significant social and cultural forces that underpin and reinforce contemporary patterns of consumption, the proponents of SCP have to date largely given their attention to the required changes in regulatory frameworks and to other government intervention that might provide the conditions for change towards SCP. Studies of innovation, particularly those grounded in ecological economics, tend to focus on the sociotechnical landscape—and the creation of niches within which innovations might form an initial presence—and on the sociotechnical regime that might be thought of as the ecosystem within which such niches can be nurtured (Geels 2002).

In the context of sustainable mobility, the approach adopted by Geels (2002) is illustrative both of the systemic coverage of the SCP concept and of the gap in analysis with respect to business models. While Geels admits the importance of industrial structure as one determinant of the sociotechnical regime, there seems to be scant translation of this to an understanding of the dynamics of competition and change by businesses within the mobility sector. Obviously, vehicle manufacturers and their suppliers are major actors for change, or non-change, in terms of how mobility aspirations are defined and met. There appears to be an implicit faith in the 'invisible hand' of the market, indeed in the essential power of capitalism, to respond with appropriate new businesses as required.

Transition management as a school of thought employs a broad definition of 'technology' that extends well beyond either the artefact or the process, such that particular technological configurations are seen as formed within and deeply embedded in socioeconomic, cultural and institutional systems and structures (Berkhout et al. 2003) while also serving to change those structures and systems. In the language of these theorists (Geels 2002; Rip and Kemp 1998; Schot 1998) the automotive industry can be understood as a regime with characteristic ways of both defining the problem and arriving at the solution (of mobility). This regime comprises accepted ways of working, of engineering practices, of product and process definitions and of related institutional practices, infrastructures, policy and industrial structure. The transition to SCP is thereby entailed as a meso-level problem whereby the (unsustainable) aspects of the regime must be changed. At this relatively abstract level, then, there is clearly a 'space' within which business per se is located. However, in practice, the major thrust of the analysis to date has been on creating the policy environment within which, for example, strategic niche management (Kemp et al. 1998) might nurture the seedlings of regime shift.

Meanwhile, and from a rather different perspective, in the context of emerging economies there are quite different macro-economic and social conditions within which demands for mobility are emerging. For example, as a crude characterisation, these economies are notable for a relative paucity of capital and surfeit of labour (often somewhat unskilled in terms of modern industrial requirements) and have large but dispersed rural populations with governments having a strong incentive to reduce the rural–urban drift that is powering the growth of new cities. In these conditions, the simple replication of business models from the established industrial nations may be inappropriate (Wells 2001) or inadequate to the task (Hart 2005).

To summarise, then, the proposal here is that one reason for the inability to achieve the transition to SCP is the failure to achieve innovative business models. Innovation forms a central feature of the SCP analysis, and again the concept is very broad, to include social, cultural and political dimensions of innovation. Once again, however, the analysis seems to stop at this point. While not seeking to discredit the overall project that is SCP, this chapter therefore seeks to establish both the significance of business for achieving SCP and to put some proposals forward as a means of recognising sustainable business when we see it.

There are many instances where observers may identify what they believe to be examples of sustainable mobility, or at least of mobility that is 'more sustainable' than that which preceded it. These instances may include technology innovations such as fuel cell or hybrid vehicles or social innovations such as car clubs. Whatever the comparative merit of these cases, the stance adopted here is that they are manifestly not examples of sustainable consumption *and* sustainable production. Similarly, there are many instances where observers may identify what they believe to be examples of 'more sustainable' production in the automotive industry, using fewer physical resources or creating lower levels of pollutants for example. Again, while this may be an instance of more sustainable production, it contributes little or nothing to the consumption issue. It is the simultaneous solution of the problem that will deliver something approaching genuinely sustainable mobility, and the stance adopted here is that finding the correct formula for the business model is the key to this solution.

In the first instance, the starting assumption is that business is the main vector of innovation; it bridges the gap between production and consumption. The second basic assumption is that, if these vector businesses are not competitive, they will go out of business and the regime will once again lapse back to the default position of the present incumbents: that is, the status quo will not change substantially. It is recognised that both assumptions could be questioned, and obviously there are many other sources of change or indeed non-change in society. The value of sociotechnical experimentation (Brown and Carbone 2004) is not denied. Neither, also, is the contribution from government or from entities such as social enterprises that offer alternative discourses to the mainstream mantra of compete and survive (Bristow and Wells 2005). Rather, this contribution can be seen as something of an antidote to the tendency of the transition management literature to treat business innovation as unproblematic and possibly trivial. Hence, the third basic assumption underpinning this chapter is the proposition that business (in this case the automotive industry) as it is currently structured and practised cannot produce sustainable mobility—but, ironically, in business terms at least it has proven itself to be remarkably sustainable.

Finally, in this section, it is worth considering the issue of time with respect to SCP. One of the elemental features of a deeply embedded regime such as that which characterises the automotive industry is that it is extremely slow to change and that apparently it changes more slowly than the scale of the sustainability problems that accumulate around it. Alternatively, it seems important that change happen rather quickly with respect to many issues of concern, or it will simply be too late: climate change may be one such example. Hence it might be contended that an alternative business model is more sustainable simply because it gets to the end-point more quickly than the incumbent business model. This significant point is not further developed in this chapter but implicitly underlies the discussion of economic scale and the pace of innovation diffusion, be it technical or social.

## 5.3 Defining business models

A starting point for business models is to understand the relationship between the product as an artefact, the production processes needed to make it and the structure of industry that emerges as a result. In this respect, the analysis is actually narrower than that adopted by the transition management school of thought, but in the search for breadth and comprehensiveness there is some danger that transition management has sacrificed important detail. In any case, the issue is partly about what element forms the centre for the analysis. In the account presented here, business has been put at the centre, in particular the pivotal business that emerges as the most influential through the linear value system, from supplier right through to customers. Hence, the business model undoubtedly influences how consumers think about the product, and the normative rules that shape expectations. This product–process–structure triumvirate is rarely entirely solid but is equally highly influential in mediating between production and consumption. Of course, it is possible to have an innovative business model while utilising existing product and production technology (and an example is outlined in the case studies below). Equally, it is possible to attach innovative technologies to existing business models and indeed existing businesses. Innovation is more fluid and 'fuzzy' than such simple prescriptions would allow. Nonetheless, and especially in the face of entrenched business practices, it is also the case that novel technologies can be considered as enabling new business models, thereby providing a new entrant with a survivable route to market expansion.

The concept of business models is rarely defined in the strategic literature, although, recently, greater attention has been given to it because of the attentions of leading scholars such as Hart (2005). Arguably, the concept came into popular use in the so-called 'dot.com' era, when many new businesses were searching for a profitable means of exploiting the possibilities of widespread Internet usage. As that era demonstrated, a great many such business models manifestly failed. What was interesting about the era, however, was the explicit recognition of two key features. First, that a significant technological change (the Internet) could create the conceptual space for new ways of intermediating between businesses and consumers; and, second, that by defining new business models those conceptual opportunities could be translated into competitive advantages against traditional companies or approaches to the market in question. Of course, it could also be argued that the dot.com era also illustrated an important third facet of relevance here: just how difficult it is to define and put into operation a genuinely successful new business model, whatever the sector under consideration.

Other areas of business activity have also attracted the redefinition of business models. One can consider privatisation as a case in point, with a transition from a social–governmental 'business' model to one with at least some of the characteristics associated with openly competitive markets. In a related manner, trade liberalisation and the general policy of opening up markets may at least in part be responsible for providing the conditions for new business models in some sectors, airlines being a potential case in point, with the so-called 'low-cost, no-frills' airlines emerging in the 1990s.

The business model can be thought of as being intermediary between an overarching business concept and the detail of a business plan. It is a combination of the structure of the business, the product-service offering and the way in which added value is provided to customers. The business model, furthermore, is a statement of the terms of competition between businesses; entrenched business models become part of the accepted norms of behaviour—accepted by regulators, consumers and others as well as competitor businesses. In this respect, business models can be extremely pervasive and enduring. More theoretically, the concept of the business model is about value creation frameworks. As Stabell and Fjeldstad (1998) have shown (derived from earlier analysis from Porter 1980, 1985), the traditional (and especially manufacturing) value creation framework has been linear. That is to say, materials are created and then transformed into products and are then dispatched to customers with appropriate after-sales support until the product expires or is disposed of. Meanwhile, value passes back down this linear chain, along with other 'backflows', principally information. It can be readily appreciated that this linear framework lends itself to unsustainable business operations, although this was not the point of the analysis from Stabell and Fjeldstad. That is, the model closely approximates to the 'dig it up, use it, throw it away' approach to the satisfaction of material wants and needs that lies at the heart of many of the problems associated with the reconciliation of production and consumption. This concept is distinct from 'value capture', which has more to say about how distinct companies compete to capture value from each other—though it is equally true that the pattern of value creation and the strategies of value capture are often closely related. Indeed, the critical point discussed below is that innovative business models are one means to redefine the terms of competition, and hence both a new way of creating value and a new way of capturing that value.

In part, the interest in value creation patterns and the characteristics of the business model have been stimulated by another environmental debate: that concerned with recovering waste-streams in production and post-consumer use (Wells and Seitz 2005). Closed-loop supply chains, embodying remanufacturing and reverse logistics, might be expected to be an important means to enable businesses to meet the growing demands of corporate social responsibility and to meet wider social goals to reduce the resource intensity of contemporary economic life (CEC 2000; Desai and Riddlestone 2002; Hart 1997; Steinhilper 1998; US EPA 1997). There is a clear resonance with the concepts of eco-efficiency (Schmidheiny 1992) and eco-modernism (Ayres *et al.* 1997): that closed loops offer opportunities to achieve the so-called 'triple bottom line' of social, business and environmental benefits (Hawken *et al.* 1999). Waste-streams—including mechanical products that can be made serviceable again—can provide useful value-added business opportunities (Ferrer and Whybark 2000). On the one hand, it becomes apparent in the case of reverse logistics and other features of closed-loop systems that in some respects the traditional linear value creation chain has been altered. On the other hand, one of the most significant impediments to the widespread adoption of closed-loop value systems is the empirical reality of the pre-existing linear value system. That is, in myriad hard engineering and soft attitudinal ways the primary linear value creation system is unable to accept returned products, components or materials. This suggests that SCP, which in itself would probably embody the concept of dematerialisation (or no new net consumption of raw materials), would perforce need non-linear value creation systems: alternative business models are fundamental to the achievement of sustainable production and consumption.

## 5.4 Disruptive technologies: the business dimension and the automotive industry

This section presents a brief analysis of business models in the automotive industry as they pertain today, along with a consideration of why disruptive technologies have thus far failed to materialise in the industry. Exactly what constitutes disruptive technology is of course somewhat elusive. There are perhaps two main categories of disruption: that which emanates from outside the sector (i.e. an alternative means of achieving the same or similar ends) or internally (i.e. a radical improvement of the existing means of achieving an end). In either case, the automotive industry appears to have been immune to such challenges. In this industry it is the vehicle manufacturers that may be regarded as the pivotal businesses, orchestrating component suppliers into the car production process and the distribution and retailing system that presents those cars to consumers. The discussion on business models is therefore focused on the vehicle manufacturers by virtue of their privileged position within the system. It is recognised that alternative business models may constitute a redefinition of these power relationships.

### 5.4.1 The traditional business model and the automotive industry

The traditional business model in the automotive industry has been remarkably enduring. In its basic form it dates back to a series of innovations, both technical and managerial, in the early years of the 20th century. Henry Ford is often cited as the 'father' of mass production in the automotive industry, with the introduction of the moving assembly line, the standardised product design, the fragmentation of work and the strategy of expanding the market through price reductions. In some important respects, this conceptualisation of the history of the industry is flawed, for it neglects the fact that Ford mass-produced a 'pre-industrial' design that lacked a vital element of the product–process–structure equation: the all-steel body developed by Budd (Nieuwenhuis and Wells 2007). General Motors (GM) then introduced some equally important but non-technical innovations, including the GM Acceptance Corporation (GMAC) to provide credit facilities to buyers, the idea of the annual model change and the idea of the multiple-brand group with a range of cars for 'every purse and every purpose'.

Many years of refinement then followed, leading to the characteristic business model of the automotive industry. Under this model the key determinant of the scale and scope of production is the requirement for economies of scale in manufacturing, a consideration shaped by the capital intensity of the production process, including tooling for that process. This has resulted in a pattern of centralised manufacturing facilities, linked to the market via long logistics lines and extensively distributed franchised dealerships. Under this framework, the widespread adoption of the Toyota Production System from the 1990s onward represented a further refinement of the business model, but not the radical change presented at the time (Womak *et al.* 1990). Vehicle manufacturers earn their revenue primarily through the sale of new vehicles and associated finance packages, not from vehicles in use (apart from the lucrative spare parts business). The main measure of competitive status is market share for new car sales. Given the very high fixed costs of production, the achievement of low prices per unit can be obtained only by high volumes of unit production. In other words, economies of scale are a classic 'double-edged sword' in that, with this business model, the vehicle manufacturers absolutely have to achieve high output or suffer financial loss. It is hardly surprising, then, that the industry as a whole suffers from endemic overcapacity, an ability to produce many more vehicles than the market can absorb, with consequential impacts on over-production and over-supply, leading to prevalent discounting of new cars in the market. Returns to capital are low, typically below 5%, and often negative, with periodic crises enmeshing the vehicle manufacturers. The vehicle manufacturers do not capture most of the profit streams generated by cars in use; these go to franchised dealerships, independent garages, insurance companies, fuel companies and many other parties (for a fuller account of the prevailing business model, see Wells and Nieuwenhuis 2001).

The high costs of production and of new model development leads to concerns regarding risk reduction, thereby embedding a conservative culture. With respect to existing technologies in vehicle production, given the very long experience they are extremely stable and predictable; all the performance parameters are understood. All alternatives are judged against the standards of the prevailing regime and are almost inevitably found wanting. This is particularly relevant for the introduction of disruptive technologies, which almost by definition would need to begin at low volumes and hence high per-unit prices. Note that these observations apply not only to what the industry terms 'power train' items but also to the basic design of vehicles, to the architecture of the body structure and to the materials used.

Now, this overall business model has perhaps three variants that are worthy of consideration. These are:

- High-volume producers
- Premium producers
- Specialist producers

The high-volume producers adopt the business model outlined above. The so-called premium producers have historically occupied a sort of middle ground, offering a degree of exclusivity and performance or a specification not provided by the mainstream, high-volume producers. The specialist producers that occupy niche markets tend to be characterised by products with extreme performance or luxury. This pattern has rather broken down in recent years, as mass-market high-volume producers have sought to capture some of the premium market and as premium producers have expanded into offering a fuller range of products at higher volume. Specialist producers are often now owned by larger groups. Still, in broad terms, all three categories adopted the same basic approach of deriving revenue from the sale of new cars.

Despite these observations, the automotive industry has not and is not entirely bereft of innovations with respect to the business model and the introduction of disruptive technologies. The following section discusses the extent to which the constituents of a radical business model might be observed within existing practices, particularly with respect to low-volume specialist producers.

### 5.4.2 The traditional niche business model and the automotive industry

As noted above, there has been an enduring niche business in the automotive industry, particularly that in Europe. It is interesting to speculate why Europe is such a fertile place in terms of nurturing these specialist producers when they were squeezed out of existence in the USA and never really appeared at all in Japan or South Korea. Whatever the reasons, these niche businesses exhibit some features that are worthy of attention with respect to the development of alternative business models and the achievement of SCP.

In the traditional niche business model the following characteristics are often present, though not necessarily all in one business at one time. It is useful to consider that in some cases the businesses concerned have demonstrated remarkable levels of sustainability in an economic and social sense, even if the products themselves have not shown particular environmental merit. For example, the UK niche vehicle manufacturer Morgan has the distinction of having made a profit every year of existence and of never having made any employees redundant. Very few businesses can make this claim.

The niche manufacturers could be said to include Rolls Royce, Bentley, TVR, Lotus, Bugatti, Ferrari and Maybach—all with production volumes under 10,000 units per annum. The characteristics of note include the following:

- Production capacity is pitched to meet the lowest point in the demand cycle, with excess demand managed by greater or lesser waiting times for products (often measured in years)
- The product is retained in production for a long time, with many minor iterations rather than radical redesigns
- The product usually serves a distinct niche demand and is therefore specifically designed for this purpose rather than being a 'general purpose' machine
- The product has a high value that does not depreciate rapidly with time, as is the case for mainstream products. As a consequence, the product has an impressive longevity
- Production itself is limited; in general, these manufacturers would more accurately be termed assemblers, with all major components bought in (especially engines and gearboxes)

- Actual production is often confined to the vehicle body, with alternatives to the all-steel Budd system adopted, offering low-volume viability and often performance advantages such as lower weight
- The assembly process is conducted without automation and allows a high degree of variety, thereby offering high levels of customisation
- Cars are built to order, not in anticipation of demand
- There is often high engagement with customers in an iterative process whereby product specification is refined. The engagement continues into the vehicle use phase
- The factory often serves as the point of vehicle service, especially for major product overhauls
- The business derives a high proportion of income from non-manufacturing activities, including 'heritage' parts businesses and motor racing

It is certainly the case that the vehicles produced are expensive relative to mainstream products. Therein lies a conundrum. From an environmental perspective, the high economic value is a useful attribute, because it provides the rationale for product longevity: these vehicles are worth keeping. Product longevity is an attribute of debatable merit in the context of contemporary cars but would certainly be a feature of sustainable mobility in the future. However, high unit prices obviously preclude many from the purchase of such cars: we cannot all afford to drive a Rolls Royce! Therefore, it might be reasonably proposed that these business models and their products are essentially irrelevant to the challenge of mass sustainable mobility. This is clearly the case, unless changes can be made to the business model that bring the advantages of niche manufacturing together with technological innovation and affordability.

### 5.4.3 Innovative business models and the automotive industry

It is interesting to speculate on the degree to which the prevailing business model is historically and spatially specific, bounded to time and place. Put another way, has the prevailing business model had its day? The prevailing business model is best suited to conditions of market expansion and indeed contributes significantly to such expansion through the reduction in (quality-adjusted) prices. In the stagnant and fragmenting markets evident through much of the industrialised world today it becomes much more difficult to reconcile the demands of the production system (for standardisation) with the demands of the market (for variety). In other words, regardless of any sustainability considerations, the structural conditions of the regime are such that the prevailing business model is under pressure. These pressures are manifest in the economic performance of the business model, with many major vehicle manufacturers and suppliers struggling to attain profitability. Of course, the vehicle manufacturers and the industry are responding with a range of strategies, including globalisation (finding new, expanding, markets), platform designs, modular supply, outsourcing to low-cost production regions, outsourcing of manufacturing to third parties and attempts to drive up the per-unit value of vehicles sold through the integration of many new features.

Still, the economic pressures are being compounded by growing concerns over sustainability. To date, the main interests have been with respect to emissions of toxic gases. More recently, for environmental and strategic reasons, the issue of fuel consumption (and of $CO_2$ emissions) has come to the fore, while in Europe, at least, specific demands are now made in terms of recycling so-called end-of-life vehicles. Other aspects of unsustainability remain neglected: principally, the extraction and use of raw materials by the automotive industry. Nonetheless, with these sustainability concerns comes an imperative for radical technological change in the power train and, probably, the body architecture.

Taken together, then, there is scope now for the introduction of new business models that can embrace both the economic and the environmental challenges faced by the industry. This leaves two basic questions: how can new business models emerge, and what happens to the existing businesses? The social cost of structural readjustment is already high, counted in terms of plant closures and redundancies, and is indicative of an industry with high barriers to entry and, crucially, high barriers to exit. In turn, this raises a third issue: will the impetus for change come from within the industry or from outsiders with no vested interest in the contemporary status quo? Evidently, vehicle manufacturers, for all their difficulties, are in many respects the most able to institute change. They have the technical, financial and human resources, and the experience and knowledge, to produce something as complex and safety-critical as a car and then bring it to market. They also have the strength of existing brands, an issue that should not be dismissed lightly.

Alternatively, outsiders (perhaps from the industry but no longer 'of' the industry) may see the potential for a business model redesign that existing vehicle manufacturers dismiss as unworkable. Moreover, with new technologies come new competences, new suppliers and sometimes an opportunity to change the rules. Lacking in historical baggage or the sunk costs of previous investments, unconstrained by considerations such as protecting the existing brand character, new entrants may offer innovative business models that find an echo in the need to change the way in which value is created and captured. In recent years there have been several attempts to do just that. Some have been documented elsewhere (e.g. see Wells and Orsato 2005); however, the following section provides some thumbnail sketches of examples to illustrate the main points.

The examples outlined below are concerned primarily with the interlinked aspects of design, production and use. It is the case, of course, that alternative business models have emerged both at the theoretical level and at the empirical level with respect to sustainable mobility. At the theoretical level, there has been considerable interest in so-called 'product-service systems' as they might apply to the automotive industry (Wells and Williams 2006; Williams 2006). In brief, by selling the service of mobility rather than vehicles the idea is that various environmental and other advantages could accrue both to consumers and to producers. Indeed, the closed-loop product-service system combines this with the idea of circularity in the production–consumption system. At a practical or empirical level there have been many car-sharing schemes developed over recent years that in various ways seek to separate the activity of mobility from personal ownership of vehicles. Still, it must be admitted that these schemes remain marginal compared with the mainstream practice of automobility.

## 5.5 Case studies

This section provides four brief case studies of innovative business models from a range of actors. The list is neither exhaustive nor particularly detailed but is intended to provide a 'flavour' of these alternative models. In part, the point is to illustrate that there is not one definitive alternative available and that the degree of deviation from the existing business model may be more or less marked. None of the alternatives discussed has come to pass: again an important consideration in the context of understanding why and how transitions may fail to emerge. Table 5.1 provides a summary.

TABLE 5.1 **Alternative business models in the automotive industry: an outline of the main features using four examples.**

| Feature | MDI air car | TH!NK | PetrolCo[a] | AUTOnomy |
| --- | --- | --- | --- | --- |
| Source | Ex-Formula 1 engineer | Industry outsiders; plastic injection moulding | Petroleum company | General Motors, established vehicle manufacturer |
| Technology | Compressed-air engine; steel frame plus plastic panels | Battery electric vehicle; PSA 205 running gear; plastic panels | Either normal new cars or bespoke design; standard technology | Fuel cell and drive-by-wire in skateboard chassis |
| Component supply | Bought in via central purchasing operation | Large modules for all key items | Not known | Large modules made in-house. Rest purchased via central operation |
| Manufacturing | Franchised plant at 2,000 units per annum; in-house production of panels | In-house production of panels; 5,000 units per annum | Not known | Large-scale production of chassis; localised assembly of body |
| Retail and servicing | Production and retail combined | Localised manufacturing; no dealerships; use of Internet | Pay-per-use concept with regular 'refresh' | Not known, but acknowledged to allow different approach |
| End-of-life vehicle | Not within business model | Not within business model (except battery replacement) | Continued re-use of vehicles | Not known |

MDI = Motor Development International

a  The name of this major petrol company cannot be given for reasons of confidentiality.

### 5.5.1 MDI air car

This particular business model provides a good illustration of an attempt to combine in one innovative package a new approach to vehicle design, vehicle production and exploitation of the market. Motor Development International (MDI) is the company formed to bring to market the ideas of the inventor of the compressed-air engine, Guy Negre (Wells 2002). The technical concept and the business plan have generated much controversy in the automotive industry, and doubts over both remain.

In this vehicle, compressed air is held in a suitable canister and is then fed into a cylinder and allowed to expand and in so doing the expansion provides the motive force to push a piston and hence turn the engine. There is no combustion so there are no emissions at the point of use other than air. A useful attribute of the technology is that any sort of dedicated infrastructure would not be technically difficult or expensive to install—air-refilling points could easily be added to existing petrol stations, for example. Simple air compressors could be run from domestic electricity and recharge the cylinders overnight.

The car was positioned in the market and performs rather like a battery electric vehicle but without the weight and cost penalty of high-performance batteries. Compared with contemporary petrol and diesel cars, the range, top speed and acceleration are limited. An interesting by-product of the technology is that the exhaust air is at $-15°C$, so air conditioning for the cabin is easy to obtain.

Of equal interest is the business plan developed by MDI. With many innovators, the core problem is usually lack of investment resources allied to the need to break the hold of the existing market leaders. MDI is no exception, but rather than seek to persuade an existing vehicle manufacturer to take up the technology MDI has tried a quite different approach. The core of the MDI approach is to grant licences to third parties which, in effect, take on an MDI franchise for a defined territory in return for the investment needed to create the factory to serve that territory. MDI has designed a standardised or modular factory and claims that 50 factories have already been allocated in various locations around the world. In addition, the standardised factory includes office space and a showroom, because in the MDI concept the point of manufacturing is also the point of retail, service and maintenance delivery. A prototype factory is claimed to exist in Nice. The factory therefore includes 4,200 $m^2$ of workshop space, 500 $m^2$ of offices and 300 $m^2$ of showroom space. On a single shift, with 70 workers, the factory is expected to produce about 2,000 vehicles per annum. In terms of operations, the factory would manufacture and assemble engines, car parts and the chassis and undertake final assembly. The large plastic body panels would be manufactured at the factory as well. Of course, in addition, the factory would undertake promotion and sales, distribution, the sale of spare parts, repairs and service within the zone allocated to it.

Despite initial interest, this project does not appear to have come to market—though it is difficult to know precisely why the concept has thus failed. Certainly, the designs themselves, and their styling, were controversial, while the inability to offer demonstration vehicles undermined confidence in the concept. Perhaps the biggest problem was one of credibility: potential investors just did not have sufficient faith in the company or the concept. In many respects, considering the huge sums squandered by complacent investors in the infamous 'dot.com' boom of the late 1990s, this is quite remarkable and goes to illustrate that fashion and prejudice can be important in determining outcomes on financial markets.

## 5.5.2 TH!NK

One version or approach was the TH!NK. The basic design concept was a two-seat city battery electric vehicle with a thermoplastic body, for use by urban commuters and utilities (Wells and Nieuwenhuis 1999). The TH!NK employed a lower frame constructed from 90% high-strength steel cut, folded and welded rather than pressed into shape—the design for which was developed in cooperation with the British Steel Automotive Engineering Group. Normal steel pressings would have required large investments in tooling. Mounted onto the lower frame was an upper frame constructed from aluminium extrusions, seam-welded at the joints—this time Norsk Hydro provided useful expertise. The thermoplastic body was moulded in one operation, with separate mouldings for the doors, roof and a few smaller parts, and was non-structural. The factory in Norway had a design capacity of 5,000 units per annum and was characterised by a highly modular assembly process: the production line was only ten stations long. The wider business model included the use of Internet sales and mobile service delivery to obviate the need for dealerships. Furthermore, the intention was to supply potential new markets such as California by locating a 'cloned' factory in the market.

Having initially been rescued from bankruptcy by Ford, and placed into a 'green' portfolio of vehicles, TH!NK was sold off after the ousting of CEO Jak Nasser from Ford. TH!NK officially went into bankruptcy in January 2006, having failed to find sufficient markets for the products. The project has not entirely failed as such; it is perhaps best described as not having yet succeeded!

## 5.5.3 PetrolCo

This example derives from a consultancy project undertaken on behalf of a major petroleum company; for reasons of confidentiality this company cannot be named. The project explored two variants of the same basic scheme. The concept in principle was simple: consumers would pay a deposit to have a vehicle 'on their driveway' for a contract period of three years. Thereafter, consumers would pay a per-mile fee, possibly with other items such as insurance bundled into the package in the same way. The vehicles would then be returned to the company and would be put through a 'refresh' process, including replacement interiors, before being returned to the consumer.

The business model was proposed both for existing new cars, purchased for the business from an established vehicle manufacturer, and for a purpose-designed new car. Indeed, some designs were commissioned from leading designers. Interestingly, if the target price for the cars were to be met, the business plan would work on the basis of the deposits alone, with significant profitability coming if the consumer were to travel more than 5,000 miles per annum.

From an SCP perspective, this business model has some affinity with other product–service system concepts. The consumer is faced with a more accurate reflection of the marginal cost of motoring and may therefore be expected to adjust their travel behaviour accordingly. The refresh process keeps cars in circulation for much longer, reducing the need to build new ones, while the business model enables the company to earn revenue from managing the fleet of cars in use rather than from production per se.

The proposal was not pursued in the end, mainly because it was felt to be too far away from the existing core business of the petroleum company and would potentially bring the company into competition with their (vehicle manufacturer) customers. Again, this is not a concept that has failed as such; perhaps it is an idea whose time has not yet come.

### 5.5.4 GM AUTOnomy

Even the vehicle manufacturers have shown potentially radical ideas—an example being the GM AUTOnomy. Designed by a small team within GM, the brief was essentially to reinvent the automobile in the light of the fuel cell and drive-by-wire. The vehicle is split, with a running chassis on which a separate body can be mounted. The chassis contains the fuel cell and all related power-train components as well as the physical and electronic docking points for the body. With drive-by-wire there is scope for redesign as demonstrated by the Hy-wire concept vehicle (GM 2003).

The interesting feature of this car is the manufacturing and market strategy potential it contains. In interviews, GM outlined the idea that the 'skateboard' chassis could be manufactured in very high volumes in a centralised facility and then distributed out to localised assembly points, where the rest of the car could be assembled, quite possibly by third parties. The ability to swap the upper body elements, including, for example, the seats, instruments and steering wheel, means that the vehicle could be changed in configuration during its useful working life.

Given the financial difficulties besetting GM, and the technical challenges of developing viable fuel cell vehicles, it is perhaps unsurprising that this concept has yet to reach the market. However, ideas such as this may be adopted, in part, in the future and in any case illustrate how new technologies can offer the potential to redesign the product itself and the manufacturing system. In many respects the project has been successful (Borroni-Bird 2006), having achieved the goals set out, and features developed for this vehicle may yet find their way into applications. The problem for the mainstream automotive industry is the huge time-lag involved.

### 5.5.5 Other successes and failures

It is worth considering in this section the wider context within which the case examples may be judged. In the first instance, it is important to recognise that the mainstream automotive industry is hardly an overwhelming success story even in terms of the most basic of indicators such as profitability. At the time of writing this chapter, in mid-2006, several major vehicle manufacturers are in financial crisis, including GM, Ford, Fiat, Mitsubishi, Saab and Jaguar. The once-dominant UK producer MG Rover went bankrupt in 2005. Leading suppliers, including Visteon, Delphi, Lear and others, are also bankrupt. Put simply, the prevailing or orthodox business model in the automotive industry is itself hardly sustainable. As a further consideration, business in general will (almost) always fail in the end. Even a cursory examination of the industrial history of a region, sector, product or process will show that the life-span of such things is short and often characterised by significant volatility. There are few companies in the world that are more than 100 years old, for example; equally, there are few locations that have made the same product for that period of time. Failure is the inevitable consequence and necessary condition of the market working.

Second, there are plenty of failures in terms of product introductions by the established vehicle manufacturers. Where these companies have sought to introduce new brands or product concepts, such as the Saturn (GM in North America) or the Smart car (DaimlerChrysler), even market success has not been matched by profitability. That is, the established industry hardly has a strong record of radical innovations or the introduction of new technologies or new business models. It is not the purpose of this chapter to explore the reasons why change does not happen in an established industry such as that which produces cars, though it must be suspected that there are cultural and institutional barriers to change. Rather, it is important to recognise that the industry itself need not necessarily be the source of radical innovation.

It might be contended that the automotive industry has achieved dramatic and significant change, and some have pointed to the introduction of hybrid vehicles (notably the first-generation and second-generation Toyota Prius) as exemplars of the way in which the industry itself can lead the way to sustainable mobility. Again, in some respects, the discussion of the Prius, or of hybrid technology generally, is one that deserves separate treatment. Here, however, it might be observed that (a) the fuel economy performance benefits of hybrid vehicles in use might be rather less in practice than is suggested in official test figures; (b) hybrid power trains add cost, complexity and weight to a vehicle; (c) hybrid power trains are increasingly being used not to improve fuel economy as such but to improve acceleration in heavy vehicles such as US 'light trucks'; (d) there are various compensating actions taken by consumers that will reduce the positive contribution made by hybrids to reduced fuel consumption; (e) hybrid engines address and slightly improve the aspect of $CO_2$ emissions only, which is only one element in the much broader concept of sustainable mobility; (f) hybrid engines are applied in otherwise standard vehicles and in many respects offer a safe incremental pathway for vehicle manufacturers in a manner that does not challenge the basic business. Indeed there is no fundamental change to the business model adopted by Toyota; the company simply sees hybrid engines as a way of selling more and more cars, increasing its market share.

These measures, therefore, along with items such as the promotion of biofuels, represent a slight amelioration of the existing way of business, not a dramatic change that seeks to unify sustainable consumption with sustainable production.

## 5.6 Overall conclusions

If nothing else, the above examples illustrate the point that the emergence of new businesses and new business models to usher in the era of SCP is far from unproblematic. At the same time, the pervasive contemporary business model in the automotive industry is itself a significant impediment to change and hence a significant reason why innovative technologies in the broadest sense have not yet emerged. There is no absolute reason why the current automotive industry cannot itself generate the novel technologies themselves and the novel approaches to the market that could presage the end of overproduction and over-consumption. The GM AUTOnomy case is at least illustrative that even the largest vehicle manufacturers can arrive at imaginative solutions.

The examples have also been selected to illustrate the diverse character of business model and new technology combinations that are possible. Indeed, perhaps the future is one in which there is no one single dominant business model but several co-existing models. Hence, the examples show different sorts of company seeking to redefine the automotive mobility 'space', ranging from entrepreneurial new entrants (MDI and TH!NK), a mainstream incumbent (GM) and a large company with a major interest in contemporary patterns of mobility (PetrolCo). The examples show varying degrees of technological innovation and, indeed, quite different solutions to the problem of sustainability in this sense, with fuel cell vehicles (GM), battery electric (TH!NK) and compressed air (MDI), while one (PetrolCo) shows no change in motive power technology at all. Again, three of the examples (GM, TH!NK and MDI) show various solutions to body engineering that are more suited to low-volume production. The examples also illustrate to varying degrees an attempt to redefine the nature of the value creation process, taking the 'vehicle manufacturer' beyond mere manufacturing into product service systems in which the car becomes a corporate asset to be supported through many years of use and several iterations of customer. These dimensions are important because they represent ways of breaking the over-production problem.

This chapter has not discussed greatly the question of value capture or the extent to which alternative business models may enable the terms of competition to be changed. At one level, the evidence so far is that none of these alternatives has made it to the market and therefore the logical conclusion is that they are not competitive. This is perhaps unfair on the alternative models discussed here, or of others that could have been raised in the chapter but were not. It is, however, suggestive of the two critical problems blocking the introduction of alternative business models: finance, and customer acceptance. The first problem is probably amenable to solution, providing the business can be structured correctly. The second is more challenging, because consumers have for years been 'told' to define motoring in the terms offered by the mainstream automotive industry. Expectations in this sense can be very powerful and can override the logical analysis of the merits of the product or the product-service offering.

As ever, more research is needed.

## References

Ayres, R., G. Ferrer and T. Van Leynseele (1997) 'Eco-efficiency, Asset Recovery and Remanufacturing', *European Management Journal* 15.5: 557-74.
Berkhout, F., A. Smith and A. Stirling (2003) *Sociotechnological Regimes and Transition Contexts* (ESRC Sustainable Technologies Working Paper 2003/3; Brighton, UK: Science Policy Research Unit, University of Sussex).
Borroni-Bird, C. (2006) 'The Reinvention of the Automobile', in P. Nieuwenhuis, P. Vergragt and P. Wells (eds.), *The Business of Sustainable Mobility: From Vision to Reality* (Sheffield, UK: Greenleaf Publishing): 209-22.
Bristow, G., and P. Wells (2005) 'Innovative Discourses for Sustainable Development', *International Journal of Innovation and Sustainable Development* 1.1–2: 168-79.
Brown, H.S., and C. Carbone (2004) 'Social Learning through Technological Innovations in Low-impact Individual Mobility: The Cases of Sparrow and Gizmo', *Greener Management International* 47: 77-88.
CEC (Commission of the European Communities) (2000) *Directive 2000/53/EC of the European Parliament and of the Council on End-of-life Vehicles* (Document 300L0053 [OJ 2000 L269/34]; September 2000; Brussels: Commission of the European Communities).

Desai, P., and S. Riddlestone (2002) *Bioregional Solutions for Living on One Planet* (Schumacher Briefing No. 8; Totnes, UK: Green Books).

Ferrer, G., and D.C. Whybark (2000) 'From Garbage to Goods: Successful Remanufacturing and System Skills', *Business Horizons* 43.6: 55-64.

Geels, F. (2002) 'Technological Transitions as Evolutionary Reconfiguration Processes: A Multi-level Perspective and Case Study', *Research Policy* 31: 1,257-74.

GM (General Motors) (2003) *Hy-wire and HydroGen3* (media information CD; Zurich: GM Europe).

Hart, S.L. (1997) 'Beyond Greening: Strategies for a Sustainable World', *Harvard Business Review*, January/February 1997: 66-76.

—— (2005) *Capitalism at the Crossroads: The Unlimited Opportunities in Solving the World's Most Difficult Problems* (Upper Saddle River, NJ: Wharton School Publishing).

Hawken, P., A. Lovins and L.H. Lovins (1999) *Natural Capitalism* (Snowmass, CO: Rocky Mountain Institute).

Kemp, R., J. Schot and R. Hoogma (1998) 'Regime Shifts to Sustainability through Processes of Niche Formation: The Approach of Strategic Niche Management', *Technology Analysis and Strategic Management* 10.2: 175-95.

Nieuwenhuis, P., and P. Wells (2007) 'Production Technology or the Organisation of Labour? The Foundations of the Mass Production Car Industry', *Industrial and Corporate Change* 16.2: 183-211.

Porter, M.E. (1980) *Competitive Advantage: Creating and Sustaining Superior Performance* (New York: The Free Press).

—— (1985) *Competitive Advantage: Creating and Sustaining Superior Performance* (New York: The Free Press, 2nd edn).

Rip, A., and R. Kemp (1998) 'Technological Change', in S. Rayner and E. Malone (eds.), *Human Choices and Climate Change, 2* (Columbus, OH: Columbia University Press).

Schmidheiny, S. (1992) *Changing Course: A Global Perspective on Development and the Environment* (Cambridge, MA: MIT Press).

Schot, J. (1998) 'Towards New Forms of Participatory Technology Development', *Technology Analysis and Strategic Management* 13.1: 39-52.

Stabell, C.B., and O.D. Fjeldstad (1998) 'Configuring Value for Competitive Advantage: On Chains, Shops, and Networks', *Strategic Management Journal* 19.5: 413-37.

Steinhilper, R. (1998) *Remanufacturing: The Ultimate Form of Recycling* (Stuttgart, Germany: Fraunhofer IRB Verlag).

Tukker, A., A. Collins, F. Hines and P. Wells (2005) *Reducing the Welsh Footprint: A Contribution to the UN's 10-year Programmes on Sustainable Consumption and Production* (Cardiff, UK: Centre for Business Relationships, Accountability, Sustainability and Society [BRASS], Cardiff University).

US EPA (Environmental Protection Agency) (1997) *Waste Wise: Remanufactured Products: Good as New* (Washington, DC: US EPA).

Von Weizsäcker, E., A.B. Lovins and L.H. Lovins (1997) *Factor Four: Doubling Wealth, Halving Resource Use* (London: Earthscan Publications).

Wells, P. (2001) 'Micro Factory Retailing: A Business Model and Development Trajectory for Emerging Economies', paper presented at the *Greening of Industry Network Conference: Sustainability at the Millennium: Globalisation, Competitiveness and the Public Trust*, Thailand, 22–25 January 2001 (published on CD-ROM; www.greeningofindustry.org).

—— (2002) 'The Air Car: Symbiosis Between Technology and Business Model', *Automotive Environment Analyst* 92: 25-27.

—— and P. Nieuwenhuis (1999) 'Th!nk: The Future of Electric Vehicles?', *Automotive Environment Analyst* 53: 20-22.

—— and P. Nieuwenhuis (2001) *The Automotive Industry: A Guide* (Cardiff, UK: CAIR).

—— and R. Orsato (2005) 'Redesigning the Industrial Ecology of the Automobile', *Journal of Industrial Ecology* 9.3: 1-16.

—— and M. Seitz (2005) 'Business Models and Closed Loop Supply Chains: A Typology', *Supply Chain Management: An International Journal* 10.4: 249-51.

—— and A. Williams (2006) 'Structure and Strategy for Sustainable Business Models: Closed Loop Product Service Systems and Sustainable Production and Consumption', paper presented at the *Greening of Industry Network Conference*, Cardiff, UK, 2–5 July 2006.

Williams, A. (2006) 'System Innovation in the Automotive Industry: Achieving Sustainability through Micro-factory Retailing', in P. Nieuwenhuis, P. Vergragt and P. Wells (eds.), *The Business of Sustainable Mobility: From Vision to Reality* (Sheffield, UK: Greenleaf Publishing): 80-91.

Womak, J., D.T. Jones and D. Roos (1990) *The Machine that Changed the World* (New York: Rawson Associates).

# 6
# Sustainability-related innovation and the Porter Hypothesis
## How to innovate for energy-efficient consumption and production

*Marcus Wagner*
BETA and TUM Business School, Germany

## 6.1 Introduction

In 1991 the American economist Michael E. Porter proposed that stringent environmental regulation (under the condition that it is efficient) can lead to win–win situations in which social welfare as well as the private net benefits of firms operating under such regulation can be increased (Porter 1991). It is obvious that production and consumption (in the case where improvements are product-related) becomes more sustainable if this is the case. Innovations play a pivotal role in this, since they are the mechanism that allows additional compliance to offset tightening environmental regulation. However, opponents of this Porter Hypothesis criticise the hidden assumption that firms systematically overlook opportunities for improving environmental quality that also increase their competitiveness or other private benefits. This debate is important for sustainable consumption and production (SCP) since it clarifies the scope of a business case for SCP.

An aspect of the Porter Hypothesis that is frequently criticised is the assumption that existing regulatory regimes represent stringent and at the same time efficient environmental regulation, and this critique indeed seems to have some merit. This is relevant to SCP because (e.g. environmental) regulation is suggested as a means to achieve the objectives of SCP. This is very closely linked to aspects of market structure and therefore this chapter analyses the Porter Hypothesis and its link to SCP with regard to two aspects: theoretical considerations and an empirical analysis of energy efficiency.

First, expanding on the relationship between the Porter Hypothesis and SCP, it will analyse the role of innovation and market structure based on theoretical reasoning behind the hypothesis by discussing and analysing the arguments brought forward in favour of and against the hypothesis based on different theoretical analyses and models proposed in the field. This will be done in Sections 6.2–6.4. These sections will provide insight into the conditions under which the hypothesis holds as well as important information for future policy-making on SCP, especially with regard to innovation and market structure.

Second, the chapter will expand on Sections 6.2–6.4 by applying Porter's ideas to one specific aspect of environmental quality: namely, energy efficiency. This is relevant for the consumption and production aspects of SCP. This second part of the chapter reveals opportunities for energy efficiency with respect to new (and more sustainable) business models and additional potential for innovation. This perspective will be discussed in Sections 6.5–6.7 with respect to the interaction of innovation and market structure aspects in the Porter Hypothesis and with regard to their relevance for innovation as well as SCP.

## 6.2 Linking the Porter Hypothesis to sustainable consumption and production

In 2002 the World Summit on Sustainable Development (WSSD) formulated a framework for sustainable consumption and production (United Nations 2002) motivated, among other things, by the fact that households, because of their sheer number, significantly contribute to climate change, waste production and land use as well as to water and air pollution (OECD 2002). The Oslo Declaration on Sustainable Consumption details this framework (e.g. in terms of a research programme on SCP) and points out that, to date, the focus of SCP has been largely on the production aspects (Tukker *et al.* 2006). The Porter Hypothesis can help considerably in achieving the goals of the WSSD framework, since the simultaneous triggering of innovations with economic, social and environmental benefits is at its heart. Being targeted at firms, its direct influence is primarily on the production side, but, as will be seen later, it can have important indirect effects on the consumption side as well.

When looking for areas of empirical analysis of the linkages between the Porter Hypothesis and SCP, energy stands out as a topic of relevance for production (i.e. for industry and commerce) and for households alike. The share of households in total energy consumption has increased in the past ten years in almost all the EU-15 countries, contributing a constant 10% of $CO_2$ emissions from 1992 to 2002 (EEA 2004), with heating accounting for 70% of household energy use (EEA 2005). In addition to this, the consumption of water and energy, as well as energy generation, represents 27% of the per capita consumption expenditure in the EU-15, and similar figures apply to the new member states (EEA 2005).

Recent models analysing the direct and indirect energy consumption in several of the EU-15 countries employ input–output and process analysis and find that direct energy consumption is growing less than proportionally to household spending and that indirect energy use is increasing proportionally to household spending and that the variation across countries is considerable (Moll *et al.* 2005). Linking the Porter Hypothesis and SCP hence seems to concern mainly two areas: innovation and the effect of different market structures. Analysing these in more detail is thus the focus of Sections 6.3 and 6.4.

## 6.3 The role of innovation

Porter and van der Linde (1995) note the particular role of innovation as a mechanism for putting the Porter Hypothesis to work. This is because the concept of innovation offsets (i.e. private benefits to firms from innovatory activity triggered by stringent, yet efficient, environmental regulation) is central to the Porter Hypothesis. Opponents to this view argue that the existence of profitable, or at least cost-effective, opportunities at the firm, industry or national level to reduce environmental pollution as proposed in the Porter Hypothesis are not likely in most industries. In the best case, it benefits firms in such industries to pursue emission reductions until they meet their industry's regulatory standards. Over-compliance in such industries is unlikely, since such over-compliance would be rational for firms only if it could be achieved through cost-effective pollution abatement, which, by definition, is not possible. Therefore, regulation beyond emission levels corresponding to the private optimum would increase production costs and in turn reduce profitability of those firms exposed to it (Romstad 1998).

One important assumption of this last view is, however, that firms pursue maximisation and not satisficing behaviour (Simon 1945). In the latter case, firms may not have explored specific areas of technology or innovation and may be triggered into action by the tightening of regulations or the introduction of stringent regulations. This could lead to innovation activities, aimed at achieving compliance with novel regulation, crowding out discretionary innovation, aimed at new product development. At the same time a compliance orientation can lead to the discovery of new areas of product development, to a search for processes in different fields of technology and to the acquisition of additional knowledge and capabilities by the firm (e.g. see Roediger-Schluga 2003). Given their pivotal role in the Porter Hypothesis, future research on the specific mechanisms of how environmental regulation influences innovation seems to have some merit. This could concern the relative influence of innovation drivers such as private investment or government expenditure in research and development (R&D), spending for improvement of education systems, regulation or price changes and whether innovation policy should focus, for example, on the correction of negative externalities, or on the elimination of subsidies.

Another aspect is that environmental regulation should, in general, and ideally, provide incentives to innovate. These characteristics apply mainly to environmental taxes and tradable emission permit systems, which should thus be used more often in practice if one wishes to create conditions conducive to the Porter Hypothesis (Hemmelskamp 1997; however, for a more detailed analysis of innovation incentives provided by different types of instruments see Montero 2002). As is well known from innovation theory and from regulation theory in general, innovation should not be focused on specific technologies and should equally take into account the rate and direction of innovation. In addition, regulation should be close to the end-user of the technology and consider voluntary agreements as well as standards and labels for environmentally more benign products (for discussions of specific aspects mentioned here, see Jaffe *et al.* 2002; Rennings *et al.* 2003).

Despite the limitations of the Porter Hypothesis, it provides additional arguments for preferring incentive-based regulations over command-and-control-type regulation, since the former are likely to reduce abatement costs. Incentive-based regulation, and in particular tradable emission permit systems, maintain incentives for firms in an industry to reduce emissions through innovation, provide cost-effective allocation and abatement solutions and are therefore likely to limit decreases in the profitability of firms. The economic efficiency of regulations is pivotal here, because low-cost regulatory approaches are most likely to reduce trade leakages, which have adverse effects on overall social welfare and can reduce barriers to international cooperation on issues of transboundary pollution (Romstad 1998). In doing so, efficient regulation can indirectly produce competitive advantages (or at least reduce competitive disadvantages) for firms competing internationally, since it reduces part of their regulatory costs while at the same time having the potential to trigger innovation offsets.

## 6.4 The influence of market structure

Next to the role of innovation, an analysis of the influence of the Porter Hypothesis on sustainability innovation has to take into account the findings of industrial economics and, in particular, of the theorising and empirical evidence concerning market structure. Market structure is defined through the demand side and the supply side of an industry.

In order to assess the supply-side aspects of market structure it is necessary to define a market. Once this has been achieved, an industry can then be defined as consisting of all firms that operate in that market. Markets can be defined based on physical features of the product or on buyer or producer characteristics (Moschandreas 1994). The supply side is one important aspect of the market structure that determines levels of innovation, with the demand side being another important element of this.

The intensity of competition in an industry is an important element of the supply side. The level of competition is usually measured by the degree of seller concentration in an industry (i.e. the number and size distribution of the firms selling in an industry). Another measure to assess the intensity of competition between producers is the cross-price elasticity of supply (Moschandreas 1994).[1] The more elastic the supply between competing products in an industry, or, equivalently, the higher the level of concentra-

---

1 The cross-price elasticity of supply measures the percentage change in the quantity supplied of one product in response to a change of one unit in the price of a competing product.

tion in an industry, the more difficulty firms experience in passing environmental costs on to consumers.

In the extreme of a monopoly it is possible to pass on most environmental costs to customers (as has been the case for a long time in electricity generation). In the case of a perfectly competitive industry, producers have difficulty passing on environmental costs, since they are price takers. The five-firm concentration ratio for the UK electricity generation industry in 1996 was estimated to be 74.9%,[2] implying a lower degree of concentration compared with that in Germany, where, in 2004, the concentration ratio was approximately 80%. However, compared with the situation prior to privatisation and deregulation of the electricity generation and supply industries from 1989 onwards, the level of concentration is now lower. This is also a good example of how different implementation approaches have resulted in very different evolutionary paths in the market structure, especially the supply side. Whereas in the UK the market was largely deregulated while at the same time establishing itself with the Office of Electricity Regulation (OFFER), a very powerful central regulator, in Germany the concentration process on the supply side has in many ways even increased the oligopolistic market structures that existed before. In the UK, deregulation of the electricity distribution market was finalised with the introduction of full competition for the household retail market.

Next to the supply side, the demand side is another core element of market structure. In the following the relevance for sustainability innovation of the demand and supply side is discussed, taking (product-related and process-related) energy efficiency in industry, households and commerce as an example. I analyse and discuss for this special aspect of environmental quality how the mechanisms underlying the Porter Hypothesis may bring about innovation that enables more sustainable production and consumption. To do so I first review the extant literature on the economic aspects of energy efficiency and, in particular, the question of market failure. Following this, I describe linkages to the Porter Hypothesis and discuss how changes in regulation and market structure jointly determine innovatory activity that increases energy efficiency.

## 6.5 The case of energy efficiency and the debate on the energy-efficiency gap

### 6.5.1 The issue of market barriers to energy efficiency

Among researchers and policy-makers in the field of energy production it is discussed, controversially, to what extent there is an energy-efficiency gap (i.e. whether there is a divergence between the socially optimal levels of investment in energy efficiency and the levels currently observed in practice). According to Huntington *et al.* (1994), some of the issues that have to be addressed to proceed towards a solution of this controversy are (see also Section 6.5.2):

---

2 Estimation by the author based on the net generation capacity and data published by the UK Electricity Association.

- The general relationship between energy efficiency and economic efficiency
- The identification and measurement of technical inefficiency in energy use and how it can be distinguished from neutral or even energy-saving inefficiency
- The degree of market inefficiency and possibly market failure in energy markets with regard to energy efficiency, and the nature, extent and severity of 'market barriers'
- The influence of discount rates and hidden costs for energy-efficiency investments not being carried out
- Bounded rationality in consumer energy decision-making as well as the role of information asymmetries and transaction costs as causes of inefficient behaviour
- If, and to what extent, there are shortcomings in the market mechanism in achieving an optimal diffusion of energy-efficient technology[3]

The re-regulation and liberalisation processes in the European electricity markets had considerable implications for energy-efficiency activities, especially on demand-side measures and other energy services. The issue of an energy-efficiency gap is therefore important for policy-makers. If energy markets express market failure as far as demand-side energy efficiency is concerned, then there would be a need for government intervention.[4] Those policies, if any, that should be applied to promote energy efficiency, demand-side measures and energy services will depend, of course, on the severity of market imperfections, on technical potentials and on political preferences.

For example, it is believed that the level of service in households and in the service sector (especially commerce) could be achieved with only 26% of today's electricity use and that the cost of changing appliances in order to achieve this is, on average, about 2.5 cents per kilowatt-hour (von Weizsäcker *et al.* 1995).

As energy-related emissions account for a large part of global greenhouse gas emissions, the level of energy efficiency, demand-side measures and energy services will impact on global climate change. Policy-makers therefore need to decide the scope of policies that respond to global warming that aim for demand-side energy efficiency. Several advocates who are in favour of the energy-efficiency gap hypothesis agree that neither the econometric concept (that the operative criterion for most energy policy should be that of economic efficiency) nor the technological concept (that a range of cost-effective energy-efficiency investments are inhibited by market barriers) of the economic potential for energy efficiency alone correctly represents society's best outcome (Huntington *et al.* 1994). In the following, some of the previously mentioned aspects surrounding the efficiency-gap debate will be examined in greater detail.

---

3 The last two points in this list are likely to be the most relevant to the Porter Hypothesis.
4 Only if this is the case would government intervention through regulation be justified and, consequently, the mechanisms proposed in the Porter Hypothesis may also be better enabled.

## 6.5.2 Arguments for and against the energy-efficiency gap hypothesis

### 6.5.2.1 The basic points of view and the general relationship between energy efficiency and economic efficiency

According to Sutherland (1996), proponents of the energy-efficiency gap hypothesis first used engineering-based economic analyses, where the present value of saved energy was estimated to exceed the initial capital investment to support energy conservation programmes, and then developed a justification for such projects based on conventional economic analysis of market failures, asserting that energy conservation programmes enhance economic efficiency by reducing market failures. It is claimed that the level of energy-efficiency investment undertaken in uncorrected markets is short of the truly cost-effective level and that cost-effective energy efficiency should be encouraged by government policy (e.g. requiring energy-efficiency labelling or mandatory standards of energy efficiency) and by the utilities themselves (e.g. through demand-side management [DSM] programmes) to close the energy-efficiency gap. Such policy intervention may trigger (as proposed in the Porter Hypothesis) innovations in energy technologies and services that simultaneously increase social and private benefits. These could also be organisational innovations such as restructuring or the unbundling of large vertically integrated energy suppliers into smaller or independent units.

A telling example is the change in US legislation in 1978 that enabled competition in energy supply. As a result of this the Public Utility Commission in many federal states based its decisions on least-cost planning (LCP; i.e. the choice of the cheapest option), which in many cases focused on energy services or DSM. Since small energy service companies were successfully offering these alongside large utilities, in 1992 a federal law was introduced that required all states to apply integrated resource planning in energy supply, in turn fundamentally changing electricity markets in the USA (von Weizsäcker et al. 1995).

Although energy efficiency is at the forefront of the debate on sustainability issues, according to Patterson (1996), it is surrounded by a number of critical methodological problems. According to Sutherland (1996), conservationists claim that if an energy-efficiency investment is cost-effective it necessarily contributes to economic efficiency. Sutherland criticises this point of view, stating that there is a conceptual difference between economic and energy efficiency, and he gives the example of implementing time-of-use rates. This is a DSM measure and may contribute to economic efficiency by matching retail prices and marginal costs, but there is the possibility that energy use will increase. Similarly, in the case of fuel substitution leading to reduced energy costs, energy consumption might increase or decrease, thus indicating that improvements in economic efficiency can have ambiguous effects on energy efficiency.

Generally, the figures put forward by proponents of the energy-efficiency gap hypothesis are based on engineering-based economic analyses that estimate the net benefits from energy-efficiency improvements as the difference between the net present value of energy saved and the initial investment costs, whereas the appropriate measure of benefits from the point of view of economic efficiency would be consumer surplus or willingness to pay. Therefore, it is claimed that if performance or appliance standards, as well as utility DSM programmes, are to pass the test of economic efficiency they have to equate the marginal costs and benefits of reducing the intensity of energy use.

### 6.5.2.2 The degree of market inefficiency and possible market failure in energy markets

To answer the questions surrounding the energy-efficiency gap hypothesis, the general relationship between energy efficiency and economic efficiency, and the identification and measurement of technical inefficiency in energy use, are necessary preconditions to decide on the degree of market inefficiency and possible market failure in energy markets.

Opponents to the concept of market barriers claim that such barriers (e.g. uncertainty in future fuel prices, limited access to capital, or low energy prices) are typically not market failures and are therefore not producing the economically inefficient market outcomes. Consequently, regulatory policies that attempt to encourage energy-efficiency investment are rarely based on the failure of private markets to make efficient decisions. Even if market failures are identified, the benefits of government intervention should exceed the costs, and intervention therefore has both to reduce market failure and pass a cost–benefit test (Sutherland 1996).

The response of conservationists to this criticism is that, although the concept of market barriers has rarely been developed in terms of well-established economic concepts, analyses that are sceptical of the idea of market barriers tend to be based on the simple proposition that 'normal' markets are efficient and that the intuition expressed in the market-barrier concept may be closer to ideas from the theoretical mainstream than from those who hold a more sceptical view (Sanstad and Howarth 1994). Levine *et al.* (1994), for example, assert that market imperfections (i.e. market barriers) preclude private decisions from attaining a level of energy efficiency consistent with economic efficiency. In a step further, it is claimed that the view that competitive markets produce optimal outcomes follows from libertarian political theory and that the most conservative 'economic' case can thus be seen as primarily political in origin (Sanstad and Howarth 1994).

Huntington (1994) takes a middle ground in this respect, making the point that minor modifications to market institutions might be preferable to widespread end-use (performance or appliance) standards or utility DSM programmes using rebates or subsidies. He further remarks that the conservationists' faith in widespread government intervention in the markets for energy and energy-efficient equipment may be unjustified, as government failure can create economic distortion just as can market failures. The considerations discussed in Sections 6.5.1 and 6.5.2 are of relevance to the role of the Porter Hypothesis since they indicate that demand-side and supply-side measures to increase energy efficiency are actually substitutes. In the next section I relate the insights of Section 6.5 to the Porter Hypothesis.

## 6.6 The role of the Porter Hypothesis for energy-efficiency improvements

The starting point of the investigation was the issue of how the role of innovation and market structure in the mechanisms proposed in the Porter Hypothesis can achieve more sustainable consumption and production processes. In order to assess this, the role of innovation and market structure assumed in the Porter Hypothesis (as well as the interaction between them) was analysed. Following this, the issue of energy efficiency in industry, commerce and households was discussed from various angles in order to evaluate the relative influence of market barriers, market failure and general anomalies in the behaviour of economic agents in bringing about the levels of energy efficiency that can be observed empirically. An important conclusion from the review of these two lines of thinking and debate is that, in many ways, if the mechanisms of the Porter Hypothesis are enabled then the issue of market failure with regard to energy efficiency becomes much less relevant.

The reason for this is that without the mechanisms of the Porter Hypothesis (i.e. innovation as a result of stringent, yet efficient, regulations that produce social as well as private benefits) government intervention would be efficient only from the point of view of social welfare, if energy inefficiencies are in fact a market failure. However, if strict regulation leads to innovatory activities with the effects proposed in the Porter Hypothesis (i.e. increased competitiveness of firms and improved environmental quality, in this case especially improved energy efficiency) then even in the absence of market failure government intervention in terms of stringent, yet efficient, regulation may be justified. However, as argued in Section 6.4, in the absence of market failure the Porter Hypothesis is most likely to hold if the market structure of the industry in which the firm operates is conducive in terms of supply-side aspects (such as the capital structure characteristics of the industry or the firm) or a favourable demand side and if reward structures, organisational inertia or inefficiency in firms' routines (Gabel and Sinclair-Desgagné 1999) are such that the result is organisational failure.

Examples from industry where such organisational failures exist and where industry-wide or firm-specific targets and self-regulation led to significant reduction of energy consumption are numerous. The International Technology Roadmap for Semiconductors points to the need to compensate, by greater efficiency, for the higher energy requirements resulting from increased wafer sizes that resulted from the industry transition to 300 mm wafers. It has been shown that the semiconductor industry could save nearly half a billion US dollars through improvements to tools and facility support systems, such as through the optimisation of air conditioners and reduced air velocity through clean rooms, and by other energy-efficiency measures (Singer 2006). Dow Chemicals realised an average return on investment of 204% (over 12 years), achieving a 470% return in 1989, whereas ST Microelectronics realised a 390% return in 2004 and savings of several hundred million US dollars (Deffree 2006; von Weizsäcker et al. 1995).

Green consumerism in the electricity market is a possibility that is explicitly mentioned in theoretical models of product differentiation under the heading of price competition (Conrad 2005).[5] Such a favourable characteristic on the demand side would favour the Porter Hypothesis as it leads to a more favourable demand side and thus to a more conducive market structure. However, the evidence of green consumerism working in energy markets is limited in the EU. For example the energy markets in the UK and Germany have been fully deregulated, but demand from household customers for 'green' electricity remains low. Labels guaranteeing the use of specific energy sources or

---

5 Energy markets are characterised mainly by price competition because a high quality of energy supply is generally expected by customers.

standards for energy efficiency are, in this respect, an important device to signal the environmental quality of a differentiated product to environmentally conscious customers.

Establishing the possibility that the basic mechanisms behind the Porter Hypothesis *can* work in energy markets does not imply that the type of innovatory activity proposed in that Hypothesis *actually* takes place. Therefore, in the following, one specific aspect of activities to increase energy efficiency will be analysed with regard to whether the Porter Hypothesis actually holds. This is the case of DSM. DSM has been defined as 'measures taken by an electricity supplier or other party (apart from the electricity consumer) to reduce a consumer's demand for electricity through improvements in the efficiency with which electricity is used' (LE Energy Limited and SRC International ApS 1992) and as the planning and implementation of activities aimed at influencing the use of electricity by customers in ways that result in changes of the utility's load shape that are desired by the utility (Gellings 1996). Tables 6.1 and 6.2 provide an overview of possible DSM activities that satisfy these definitions.

TABLE 6.1 Energy-specific demand-side management (DSM) programmes

Source: adapted from Nadler and Geller 1996

| DSM approach | Example | Appropriate or best application |
| --- | --- | --- |
| Load management | Direct load control, energy management systems; peak control rates | Demand reduction at peak load times; matching of marginal cost to prices |
| Energy audits | Household surveys, commercial and industrial energy audits | Identification of most viable energy-efficiency investments; complement to incentive programmes |
| Appliance labels | Refrigerators, washing machines | Better provision of information; new construction areas or markets where existing practices are inefficient |
| Subsidies and rebates | Energy-efficient appliances, equipment, lighting, etc. | Provision of incentives for energy-saving investments |
| Leasing and loans | Insulation (e.g. in new construction markets and other retrofits) | Customers with limited capital availability; markets where long lead-times are acceptable |
| Performance contracting | Building controls, mechanical contracting, self-generation equipment: installation, operation and maintenance | Overcoming investment barriers due to long payback periods |
| Direct installation | Air-conditioning equipment; well-insulated windows | Low-income customers, when DSM savings are needed quickly |
| Market transformation | Equipment replacement and new construction measures | Areas where DSM savings are needed in the longer term |
| Bidding | Utilities request proposals from outside parties to supply demand-side resources | Transfer of performance and implementation risks from utilities to third parties; inexperienced utilities |

TABLE 6.2 **Non-specific demand-side management (DSM) services**
Source: adapted from Chamberlin and Herman 1996

| DSM service | Example | New marketing method |
|---|---|---|
| Energy solutions | Selection of contractors, equipment installation | Financing, shared savings (performance contracting), contract energy services, packaged with other services, third-party financing |
| Equipment contracting | Owning or maintaining generation plants or energy management systems | Contract energy services, packaged with other services (e.g. power marketing and brokering) |
| Building maintenance | Power quality, reliability and backup services; equipment rental, maintenance and repair | Project management; technical, engineering and design assistance; warranties and building commissioning |
| Cable and communications systems; entertainment services | Internet access, telephone lines | Packaged with other services (e.g. bundling with electricity supply) |
| Other services | Billing and information services, security systems, risk-management products and services; emissions trading services, greenhouse gas management services | Packaged with other services |

The argument has been made that deregulation itself triggers increased DSM levels, resting on the argument that the concept of energy services is based on the notion that customers want to buy the services that energy provides (e.g. heat, light and power) rather than electricity or gas itself. As a result, the lowest-cost solution for light, power or heat is in the customers' interest, in line with the concept of LCP that has been pursued by at least some municipalities on their own (e.g. Stadtwerke Hannover 1998). Assuming this customer preference holds, the provision of demand-side energy-efficiency services would be an important part of product offerings of firms in a less regulated energy market. Chamberlin and Herman (1996) use precisely this argument when they claim that a large market for energy-efficiency services exists in a more competitive environment.

In terms of new business models, although some of the programmes will likely be the same and carried out by the same firms in deregulated energy markets, future DSM opens up opportunities for innovative energy services and new delivery mechanisms provided by specialised energy service providers, since these will succeed only if such service offers more benefits to customers than competing products by traditional utilities.

In addition to that, LCP or contracting by firms or municipal utilities would lead to higher energy efficiency and is essentially a supply-side activity that positively affects market structure. There is some evidence of this approach being successful, but it depends also on other aspects such as changes in the regulation of energy markets. Here, an important role exists for energy agencies, which is another new successful business model: for example, in Hannover, Germany, or Leicester, UK (Results Centre 1993; Stadtwerke Hannover 1998).

Changes in the regulation of energy markets can enable the mechanisms behind the Porter Hypothesis in a more indirect way. However, rather than deregulation of energy markets, this would probably require specific regulations aimed at fostering energy efficiency. Opposed to this, the objective of the liberalisation activities carried out over the past decade on the national energy markets of EU member states were aimed mainly at improving the workings of price competition by decreasing the level of vertical integration present in these markets. By reducing vertical integration, it was intended to limit natural monopolies to the distribution network, whereas competition would be increased in energy production and distribution. While, therefore, the current process of deregulation of energy markets in the EU is unrelated to the objective of increasing energy efficiency, there are nevertheless country-specific examples of regulation that aims for higher energy efficiency. For example, in Germany specific regulation to increase energy efficiency has been enacted (Wenzel 2006).

However, deregulation (i.e. regulation that does not introduce more stringent environmental quality requirements but is largely unrelated) seems also to have enabled new decentralised energy technologies (such as fuel cells, micro turbines, internal combustion engines, combined-cycle gas turbines [CCGT], wind turbines and photovoltaic cells). The 'dash for gas' in the early deregulation of energy markets in the UK is well documented (Watson 1997). In a similar line of argument to that for decentralised energy technologies, it seems also to be possible that deregulation has a positive effect on (energy-specific and non-energy-specific) DSM activities. Assuming inefficiency in highly regulated energy markets, profits should decrease after deregulation.

However, energy suppliers may be able to counter this by incorporating into their business additional services and activities that bring value to their customers and enable the firms to charge a premium price. In this case, after deregulation and incorporation of services that act as a means to compensate for profit erosion from deregulation by moving into higher-value segments energy efficiency could still be improved (e.g. if the services reported in Tables 6.1 and 6.2 are implemented). In this situation, DSM would be an activity that simultaneously increases private and social benefits on a deregulated market, and deregulation could be perceived as the type of regulation that is proposed in the Porter Hypothesis.

Markard and Truffer (2006) provide an example for this based on an empirical analysis of the liberalisation of the European energy-supply industry, which they find has been an external driver to bring about more radical innovation (e.g. in terms of fuel cells, CCGT and electricity from non-fossil sources). In addition to that, other regulation that introduces stringent environmental targets (such as the *Wärmeschutzverordnung* [heat-saving ordinance] in Germany) may additionally motivate the creation of loan schemes that can increase the profits of energy suppliers in the case where DSM activities targeted at meeting stringent targets avoid costly investment in new generation capacity, which is more costly in deregulated markets. In such a case, DSM can also be linked with LCP as, for example, promoted by municipal utilities or specialised energy-service providers. The overall market potential for energy services is likely to be critical to how profits change as a result of increased DSM activities after deregulation.

## 6.7 Discussion and conclusions

This section attempts to summarise the insights gained into the links between SCP and the Porter Hypothesis. The starting point of the analysis was how the Porter Hypothesis may be applicable to fostering SCP, especially with regard to energy efficiency. This addressed the 'energy-efficiency gap' hypothesis and the issue of market barriers to energy efficiency. It also evaluated the different economic models and approaches concerning the behaviour of economic agents in energy markets with special regard to demand-side energy efficiency and addressed one of the most important questions—namely, to what degree firms and customers are only partially energy-inefficient in their behaviour. Answering this question also answers to what extent market failure accounts for inefficiency and thus if it is the dominant source for inefficiency on energy markets. Should market failure not be the main cause, other explanations would have to be explored such as whether the energy-efficiency gap is caused mainly by general anomalies in the behaviour of economic agents: for example, regarding their investment behaviour.

An attempt was then made to link the insights from the debate about energy efficiency to the Porter Hypothesis and to opportunities for firms with regard to new (and more sustainable) business models and additional potential for innovation with regard to energy efficiency. This analysis indicates that deregulation per se is not the type of regulation proposed in the Porter Hypothesis as it does not explicitly tighten standards with regard to environmental quality, particularly with regard to energy efficiency. It was also pointed out that demand-side effects with regard to environmental product differentiation may not be strong and thus may not favourably influence the market structure in the energy-supply industry with regard to SCP. However, for the special case of DSM, a more in-depth analysis revealed that it may well benefit directly from deregulation and bring about exactly the effects proposed in the Porter Hypothesis. Even more so, deregulation may trigger the development and expansion of new business models (e.g. energy agencies, energy service provides or LCP activities by municipal utilities) that are more likely to bring about more sustainable production and consumption.

In addition to this, the analysis also pointed to an additional effect along the lines of the Porter Hypothesis that is, however, more indirect and also not fully in line with Porter's reasoning. This is the positive effect on demand that deregulation had on decentralised energy technologies and especially with regard to the use of CCGT, micro turbines and internal combustion engines (Markard and Truffer 2006). One important issue behind this last effect is that the Porter Hypothesis can be evaluated at different levels: for example, at the level of the individual firm, industrial sector, national economy or economic trading area such as EU or NAFTA (North American Free Trade Agreement). The separation of firms that do and do not benefit firms from mechanisms behind the Porter Hypothesis also opens up possibilities for new business models and for product differentiation, especially when environmentally conscious customers exist (Conrad 2005).[6]

---

6 Nevertheless, it needs to be stated that electricity (but not the underlying energy sources) is a relatively homogenous product and that vertical product differentiation is therefore very difficult, and horizontal product differentiation is at least as difficult, as witnessed, for example, by the challenges faced in Germany to brand electricity with regard to features that highlight the heterogeneity of the underlying energy sources ('green', 'yellow' [referring to the German brand Yello] and other electricity).

It is important to point out that energy efficiency is only one important aspect of energy policy with regard to sustainability. The other is the use of renewable energy sources, which is an equally important issue with regard to the question underlying this chapter, especially as environmental regulation and market deregulation simultaneously influences both energy efficiency as well as the level of use of renewable energy sources for electricity generation in particular and energy provision in general. One would expect that the mechanisms of the Porter Hypothesis work better for renewable energy sources than for energy efficiency, given that regulation in the former area is usually much more aligned to the basic idea behind the Hypothesis, and in fact this seems to be a major reason for positive effects on the adoption of the use of wind turbines and photovoltaic cells. There are various examples of government intervention in terms of stringent regulation that has resulted in an increased level of innovation in renewable energy systems, and liberalisation in the EU seems also to have had an effect here (Markard and Truffer 2006).

An important example is the EEG (*Erneuerbare-Energien-Gesetz*; a renewable energy levy) in Germany, which forces large energy generators to buy decentrally generated electricity (e.g. from water, biomass, wind or solar photovoltaic sources) at prices that make investments in such renewable energy sources viable (basically, by setting a price to be reimbursed to decentralised producers that is higher than that for fossil fuel or nuclear electricity). Effectively, of course, this is a form of subsidy that puts the burden on large energy generators, the benefits of this accruing to small and innovative firms focusing on renewable energy technologies (e.g. Vestas, Enercon and Solarworld). While, in a narrow sense, this result would not be consistent with the Porter Hypothesis, it may still be consistent with it if the sum of the benefits accruing to the smaller firms and the social benefits from an increased use of renewable energies that result from a reduction in environmental externalities is larger than the cost to large electricity generators. Certainly, it is the stringent regulation that was introduced that resulted in the increase in innovatory activities in the smaller firms and in this sense supports the Porter Hypothesis. Here again, the issue of the level at which the Porter Hypothesis is addressed becomes pivotal.

As concerns future research foci, four aspects seem to be particularly relevant. First, although several countries have pursued different policies of re-regulation and liberalisation, there seems to be a certain degree of convergence with regard to the environmental effects of these policies: for example, their effect on energy-efficiency activities. It might be worthwhile to explore further the implications of this in the context of the Porter Hypothesis. In particular, one could focus on the implementation of regulation, as one needs to distinguish between regulation and implementation for evaluating the validity of the Porter Hypothesis. If implementation is inefficient, then it impedes the Porter Hypothesis, even if a tight regulatory regime exists. Likewise, weaker regulation implemented very effectively can bring about the effects proposed by Porter.

Second, the Porter Hypothesis and its link to SCP are embedded in a larger debate about governance systems, which include, for example, integrated product policy regimes. Research has indicated, for example, that a combination of market-based, legal, educational and information instruments are able to reduce housing-related environmental impacts of consumption and that environmentally benign technologies are especially relevant in order to achieve this (OECD 2001). Furthermore, it is argued that a transformation from a government to a governance approach in environmental policy (Scheer and Rubik 2006) will tend to foster this trend. Life-cycle approaches to sustainable consumption are considered to be suitable for a transition to function-based product-service systems, where the functional unit of products or services is no longer assumed to be fixed, which can enable more radical steps to be taken towards SCP (De Leeuw 2006). In the energy-supply industry one of the observed effects of liberalisation has been an increased variety in firms' innovation strategies. Also, in the sense that innovation is more driven by cost considerations and market needs, it has become more radical and market-oriented, overcoming a considerable amount of resistance that existed under the previous monopoly structures in energy supply.

Pressures from regulation as proposed by the Porter Hypothesis can be an important factor in changing governance contexts and in doing so can influence the wider set of policy tools that matter for SCP. If this is the case, then the mechanisms behind the Porter Hypothesis enable more radical innovation, which is considered essential for SCP (Charter 2006). An example of this is the recent EU directive on energy efficiency and energy services (2006/32/EG) that came into force on 17 May 2006 and which requires each EU member state to propose an action plan before June 2007 on how to reduce its annual energy consumption by 9% until 2016 by means of energy services and policy measures (Wuppertal Institute 2006).

Third, case studies on SCP with a focus on energy efficiency in different firms and industries could provide insights into when the Porter Hypothesis is actually applicable, not only in the energy-supply industry but also in other sectors of the economy that are concerned with cost-efficient energy-efficiency improvements. Arguably, in all industries in which energy efficiency can be improved through investment with a positive rate of return, this would represent a situation in which the Porter Hypothesis holds if these investments are triggered by regulation aimed at improved energy efficiency (or, equivalently, carbon efficiency). Examples of such regulations that have been discussed here are the *Wärmeschutzverordnung* (von Weizsäcker *et al.* 1995) and EEG in Germany, but there are also other regulations such as the recent EU directive on energy use in buildings as well as stringent regulation on the standby performance of consumer electronics devices, as planned, for example, by the German government.[7]

Last, a formal analysis focusing on environmentally differentiated products as well as willingness to pay for environmental quality (possibly incorporating reciprocal behaviour of customers and firms) could provide insights into the factors that determine the validity of the Porter Hypothesis under different assumptions. This could also help to derive further insights into the likely behaviour of firms (e.g. in the energy-supply industry) with regard to innovatory strategy aspects or linkages to SCP as possible guiding principles in the governance of firms.

---

7 The International Energy Agency (IEA 2001) estimates that 3–13% of household electricity consumption in OECD countries relates to standby energy use.

# References

Chamberlin, J.H., and P.M. Herman (1996) 'How Much DSM Is Really There? A Market Perspective', *Energy Policy* 24.4: 323-30.

Charter, M. (2006) 'Sustainable Consumption and Production, Business and Innovation: A Discussion Document', in A. Tukker and M.M. Andersen (eds.), *Proceedings of the 'Changes to Sustainable Consumption' Workshop of the Sustainable Consumption Research Exchange (SCORE!) Network*, Copenhagen, Denmark, 20–21 April 2006 (Copenhagen: SCORE! Network): 243-52.

Conrad, K. (2005). 'Price Competition and Product Differentiation when Consumers Care for the Environment', *Environmental and Resource Economics* 31: 1-19.

Deffree, S. (2005) 'The New Black is Green', *Electronic News*, 11 September 2005; www.reed-electronics.com/semiconductor/article/CA6282870?nid=2012 [accessed 17 March 2006].

De Leeuw, B. (2005) 'The World Behind the Product', *Journal of Industrial Ecology* 9.1–2: 7-10.

EEA (European Environment Agency) (2004) *Analysis of Greenhouse Gas Emission Trends and Projections in Europe 2004* (Copenhagen: EEA; reports.eea.eu.int/technical_report_2004_7/en [accessed 25 August 2007]).

—— (2005) *Household Consumption and the Environment* (Copenhagen: EEA)

Gabel, L.H., and B. Sinclair-Desgagné (1999) 'The Firm, Its Routines and the Environment', in T. Tietenberg and H. Folmer (eds.), *The International Yearbook of Environmental and Resource Economics, 1998/1999* (Cheltenham, UK: Edward Elgar): 89-118.

Gellings, C.W. (1996) 'Then and Now: The Perspective of the Man who Coined the Term "DSM" ', *Energy Policy* 24.4: 285-88.

Hemmelskamp, J. (1997) 'Environmental Policy Instruments and their Effects on Innovation', *European Planning Studies* 5.2: 177-94.

Huntington, H.G. (1994). 'Been top down so long it looks like bottom up to me', *Energy Policy* 22.10: 833-39.

——, L. Schipper and A. Sanstad (1994) 'Introduction to the Special Issue: Markets for Energy Efficiency', *Energy Policy* 22.10: 795-98.

IEA (International Energy Agency) (2001) *Things That Go Blip in the Night: Standby Power and How to Limit It* (Paris: IEA).

Jaffe, A.B., R.G. Newell and R.N. Stavins (2002) 'Environmental Policy and Technological Change', *Environmental and Resource Economics* 22: 41-69.

LE Energy Limited and SRC International ApS (1992) *Demand Side Measures: A Report to the Office of Electricity Regulation* (London: Office of Electricity Regulation [OFFER]).

Levine, M.D., E. Hirst, J.G. Koomey, J.E. McMahon and A. Sanstad (1994) *Energy Efficiency, Market Failures, and Government Policy* (Berkeley, CA: Lawrence Berkeley Laboratory).

Markard, J., and B. Truffer (2006) 'Innovation Processes in Large Technical Systems: Market Liberalisation as a Driver for Radical Change?', *Research Policy* 35.5: 609-25.

Moll, H.C., K.J. Noorman, R. Kok, R. Engström, H. Throne-Holst and C. Clark (2005) 'Pursuing More Sustainable Consumption by Analysing Household Metabolism in European Countries and Cities', *Journal of Industrial Ecology* 9.1–2: 259-75.

Montero, J.-P. (2002) 'Permits, Standards, and Technology Innovation', *Journal of Environmental Economics and Management* 33: 23-44.

Moschandreas, M. (1994) *Business Economics* (London: Routledge).

Nadler, S., and H. Geller (1996) 'Utility DSM: What Have We Learned? Where Are We Going?', *Energy Policy* 24.4: 289-302.

OECD (Organisation for Economic Cooperation and Development) (2001) *OECD Environmental Outlook* (Paris: OECD).

—— (2002) *Towards Sustainable Household Consumption* (Paris: OECD).

Patterson, M.G. (1996) 'What Is Energy Efficiency? Concepts, Indicators and Methodological Issues', *Energy Policy* 24.5: 377-90.

Porter, M.E. (1991) 'America's Green Strategy', *Scientific American* 264.4: 96.

—— and C. van der Linde (1995) 'Toward a New Conception of the Environment–Competitiveness Relationship', *Journal of Economic Perspectives* 9.4: 97-118.

Rennings, K., A. Ziegler, K. Ankele, E. Hoffmann and J. Nill (2003) *The Influence of the EU Environmental Management and Auditing Scheme on Environmental Innovations and Competitiveness in Germany: An Analysis on the Basis of Case Studies and a Large-scale Survey* (Mannheim, Germany: ZEW)

Results Centre (1993) 'Leicester, England: Comprehensive Municipal Energy Efficiency Profile 76'; solstice.crest.org/efficiency/irt/76.htm [accessed 3 April 2006].

Roediger-Schluga, T. (2003) 'Some Micro-evidence on the "Porter Hypothesis" from Austrian VOC Emission Standards', *Growth and Change* 34: 359-79.

Romstad, E. (1998) 'Environmental Regulation and Competitiveness', in T. Barker and J. Koehler (eds.), *International Competitiveness and Environmental Policies* (Cheltenham, UK: Edward Elgar): 185-96.

Sanstad, A.H., and R.B. Howarth (1994) ' "Normal" Markets, Market Imperfections and Energy Efficiency', *Energy Policy* 22.10: 811-18.

Scheer, D., and F. Rubik (2006) 'Governance towards Sustainability: Meeting the Unsustainable Production and Consumption Challenge', in D. Scheer and F. Rubik (eds.), *Governance of Integrated Product Policy: In Search of Sustainable Production and Consumption* (Sheffield, UK: Greenleaf Publishing): 10-15.

Simon, H.A. (1945) *Administrative Behavior* (New York: Macmillan).

Singer, P. (2006) 'Significant Cost Savings in Energy Reduction', *Semiconductor International*, 2 January 2006; www.reed-electronics.com/semiconductor/article/CA6302634?pubdate=2%2F1%2F2006&industryid=3032&nid=2012 [accessed 17 March 2006].

Stadtwerke Hannover (1998) *Unsere Energiespar-Ideen rechnen sich! Die LCP-Testphase im Überblick* (Hannover, Germany: Stadtwerke Hannover AG).

Sutherland, R.J. (1996) 'The Economics of Energy Conservation Policy', *Energy Policy* 24.4: 361-70.

Tukker, A., M.J. Cohen, U. de Zoysa, E. Hertwich, P. Hofstetter, A. Inaba, S. Lorek and E. Stø (2006) 'The Oslo Declaration on Sustainable Consumption', *Journal of Industrial Ecology* 10.1–2: 9-14.

United Nations (2002) *Report of the World Summit on Sustainable Development in Johannesburg* (New York: United Nations; www.johannesburgsummit.org [accessed 25 August 2007]).

Von Weizsäcker, E.U., A.M. Lovins and L.H. Lovins (1995) *Faktor Vier: Doppelter Wohlstand, halbierter Naturverbrauch* (Munich: Droemersche Verlagsanstalt).

Watson, J. (1997) *Constructing Success in Electric Power Generation: Combined Cycle Gas Turbines and Fluidised Beds* (PhD thesis; Brighton, UK: Science Policy Research Unit [SPRU], University of Sussex).

Wenzel, D. (2006) 'Der Weg zum 2000-Watt-Bürger: Bundeskanzlerin Merkel will die Energie-Effizienz der deutschen Wirtschaft verdoppeln', *Die Welt*, 15 March 2006: 12.

Wuppertal Institute (2006) 'Neue Richtlinie Endenergieeffizienz und Energiedienstleistungen', *WI-News*, 9 June 2006; www.wupperinst.org/Publikationen/Presse/2006/03_2006.html [accessed 9 June 2006].

# 7
# Marketing in the age of sustainable development

*Frank-Martin Belz*
Technische Universität München, Germany

## 7.1 Marketing: part of the problem or part of the solution?

At the World Summit on Sustainable Development (WSSD) in Johannesburg 2002 all countries committed themselves to promote sustainable consumption and production (SCP) patterns, with developed countries taking the lead (United Nations 2005a). As a follow-up to the World Summit, the United Nations Environment Programme (UNEP) is pursuing the development of a ten-year framework of programmes on SCP, in support of regional and national initiatives (UNEP 2006). The promotion of SCP signifies a shift in environmental policies, which for many years have been focused on the production side. However, despite cleaner production and higher eco-efficiency of end-use appliances, overall environmental impact is still growing. That is why the consumption side has become an important issue in environmental policies as well. Both sides have to be addressed to reduce the overall environmental impact, particularly in industrialised countries. The promotion of SCP is a common task of governments, civil society organisations and the private sector (United Nations 2005a). This chapter deals with the role of business in SCP. It focuses on marketing as a business function that occupies a pivotal position between producers and consumers (Charter 2006). So far, marketing has not been involved or engaged in the SCP debate. The natural environment has hardly been considered by marketing theory and practice. As the natural environment changes as a result of increasing human pressure, the environment will become a key factor for marketing in the future (Belz 2006).

What is the impact of human beings on the natural environment? The overall impact of human activities on the natural environment is the result of three major factors: population, consumption per person and technology. This can be expressed through the IPAT formula (Ehrlich and Holdren 1971):

$$\text{Impact} = \text{population} \times \text{affluence} \times \text{technology} \qquad [7.1]$$

**Population** is the total number of people living on Earth. The world population grew from 2.5 billion in 1950 to 6.5 billion in 2005 and is projected to grow to 9 billion by 2050 (United Nations 2005b).

**Affluence** relates to the amount each person consumes. Based on the consumption per capita, three consumer classes can be distinguished on a global scale: the lower, the middle and the upper consumer classes (Durning 1992). The upper consumer class includes all households whose income per family member is more than $7,500 per year. They may live in climate-controlled, heated buildings, equipped with refrigerators, washing machines, dishwashers and many other electrically powered gadgets. They are able to dine on meat and processed, packaged food and to imbibe soft drinks and other beverages from disposable containers. The dominant modes of transportation in the upper consumer class are automobiles and, increasingly, airplanes. The members of the upper consumer class enjoy a kind of material lifestyle never experienced before in the history of humanity.

The 'democratisation of consumption' started in North America and expanded to Western Europe and Japan in the past half-century. Since the 1990s this consumer culture has been spreading throughout the world. According to the latest estimations, 1.7 billion people belong to the global consumer class, which is about a quarter of the present world population. Although most consumption spending still occurs in more developed regions, the number of consumers is increasing in the less developed regions. Almost half of the global consumer class lives in developing countries (Gardner *et al.* 2004). If the same levels of consumption that several hundred million of the most affluent people enjoy today were replicated across the world tomorrow, the impact on water supply, air quality, climate, biodiversity and human health would be severe. Thus, the ecological question is also a social question of resource equity in a global world with limited resources (Wuppertal Institute 2005). As resource scarcity increases as a result of world population growth and the rise of the global consumer class, equitable distribution of natural capital, goods and services becomes more difficult and makes conflicts over oil, natural gas, water and arable land more likely.

**Technology** determines the amount of resources used to produce each unit of consumption (input side) and the amount of waste or pollution generated by each unit of consumption (output side). A shift from non-renewable resources to renewable resources and an increase in resource productivity and energy efficiency by 'Factor 4' (von Weizsäcker *et al.* 1998) to 'Factor 10' (Schmidt-Bleek 1998) are contributions that can be made to reduce environmental impact. This requires process, product and service innovations as well as system innovations (Elzen *et al.* 2000). Higher resource productivity and energy efficiency in developed countries achievable through new technologies may enable less developed countries to achieve their fair share of resources, fulfil their basic human needs and lead a life of dignity. Therefore, an increase in resource productivity and energy efficiency also has an essential contribution to make to equity, security and peace in the world.

What are the links between marketing and the IPAT formula? Which role does marketing play when it comes to population, affluence and technology? Marketing has relatively little direct effect on population. However, marketing does have an influence on both the level of affluence and the technology that creates it (Peattie 1995). In this respect, marketing plays an ambivalent role: on the one hand, it promotes a consumer society and materialistic lifestyles, which impose problems on the social and natural environments; on the other hand, marketing has the potential to foster sustainable products and services. It can also facilitate new lifestyles that are less materialistic and more sustainable. Thus, marketing is not just part of the problem but also part of the solution. In this chapter it is argued that sustainability marketing can play a key role in the promotion of SCP.

In the following, sustainability marketing is defined (Section 7.2) and a managerial approach to sustainability marketing is developed (Section 7.3), differentiating six steps: socio-ecological problems, consumer behaviour, normative sustainability marketing, strategic sustainability marketing, operational sustainability marketing and transformational sustainability marketing. In addition, six distinctive characteristics of sustainability marketing in comparison with classical approaches of marketing are highlighted (Section 7.4). At the end of the chapter, some conclusions are drawn (Section 7.5).

This chapter focuses on business-to-consumer (B2C) relations and consumer goods marketing as the interface between production and consumption. Investment goods marketing dealing with business-to-business (B2B) relations is beyond the scope of this chapter, although it is considered to be important as well (Mintu-Wimsatt and Bradford 1995).

## 7.2 Definition of sustainability marketing

**Marketing** is generally defined as building lasting and profitable customer relationships (Kotler and Armstrong 2004). Far from 'telling and selling', modern marketing analyses customer needs and wants; develops products that provide superior value; and prices, distributes and promotes them effectively to selected target groups. Two core marketing activities are attracting new customers by promising superior value and keeping current customers by delivering satisfaction. As many consumer goods markets in developed countries are stagnating there has been a shift from attracting new customers and generating sales to retaining current customers and building long-lasting relationships: that is, from transaction marketing to relationship marketing (Christopher *et al.* 1991).

**Sustainable development** is a kind of development that meets the needs of the present without compromising the ability of future generations to meet their own needs (WCED 1987). Meeting the needs of the present means intragenerational equity (i.e. equality between North and South and between the more affluent and the less affluent). Considering the needs of future generations implies intergenerational equity (i.e. equality between one generation and another). Sustainable development requires the most basic human needs of all to be met: that is, the need for clean water, enough to eat, shelter, sanitation, schools and transportation. If basic human needs are met we may enter a new era of economic growth for nations in which the majority is poor. Global sustainable development implies that those who are poor get a fair share of resources and that those who are more affluent adopt lifestyles that need fewer resources and less energy. Sustainable development is a process, whereas sustainability is the possible outcome and ultimate aim. Sustainable development is a process of change that is neither easy nor straightforward. Sustainable development deals with trade-offs, continuously balancing economic, environmental and social goals in a responsible way. In marketing, these kinds of conflicts and trade-offs become more obvious than in any other business function.

**Sustainability marketing** goes beyond conventional marketing thinking. If marketing is about satisfying customer needs and building profitable relationships with customers, sustainability marketing may be defined as building and maintaining sustainable relationships with customers, the social environment and the natural environment. By creating social and environmental value, sustainability marketing tries to deliver and increase customer value. Sustainability marketing aims at creating customer value, social value and environmental value. Similar to the modern marketing concept, sustainability marketing analyses customer needs and wants; develops sustainable solutions that provide superior customer value; and prices, distributes and promotes them effectively to selected target groups. Throughout the whole process, sustainability marketing integrates social and ecological aspects.

Some authors use the term 'sustainable marketing' synonymously to the definition of sustainability marketing given above (Charter *et al.* 2002). However, in everyday language the term 'sustainable' means durable, long-lasting or everlasting. Hence, sustainable marketing is often interpreted as a kind of marketing that builds long-lasting customer relationships effectively—without any particular reference to sustainable development or consideration of sustainability issues. That is why the term 'sustainable marketing' might be misleading and is not used here.

**Green marketing** (Charter 1992; Charter and Polonsky 1999; Ottman 1998; Peattie 1992), **eco-marketing** (Belz 2001; Meffert and Kirchgeorg 1998) and **environmental marketing** (Coddington 1993; Peattie 1995; Polonsky and Mintu-Wimsatt 1995) are concepts closely related to sustainability marketing. Developed during the 1990s, these concepts focus mainly on the natural environment. They deal with the integration of ecological aspects into conventional marketing thinking. Sustainability marketing goes beyond eco-marketing insofar as it also considers the social dimensions—besides the ecological and economic aspects.

## 7.3 Conception of sustainability marketing

The managerial approach of sustainability marketing differentiates six steps, as depicted in Figure 7.1 (Belz 2005, 2006). The first two steps begin with an analysis of the company's situation. In sustainability marketing it is crucial to know not only consumers' needs and wants but also the ecological and social problems of products throughout the product life-cycle, from cradle to grave. The intersection of socio-ecological problems and consumer wants sets the ground for sustainability marketing. It indicates new mar-

## FIGURE 7.1 The concept of sustainability marketing
Source: Belz 2005

ket opportunities for innovative companies. Steps 3–5 describe the implementation of sustainability marketing. Social and ecological criteria are fully integrated into the mission statement, strategies and marketing mix. Hence, sustainability marketing moves from analysis to action. Step 6 is one of the special characteristics of sustainability marketing. It is about the commitment of companies to sustainable development and their active participation in public and political processes in order to change the existing framework in favour of sustainability. In the following, the six steps of sustainability marketing are presented and discussed in more detail.

### 7.3.1 Socio-ecological problems

The point of departure in sustainability marketing is the thorough analysis of social and ecological problems, generally and specifically, regarding products that satisfy customer needs and wants. Here the whole **product life**, 'from cradle to grave' has to be taken into account (i.e. the extraction of raw materials, transportation, production, distribution, use, recycling and disposal of the product). Qualitative and quantitative **life-cycle assessments** (LCAs) are complementary methodologies to analyse the impact of products on the social and natural environments (Fuller 1999). LCA is a **quantitative** method for calculating the impact of a product on the natural environment and human health

(SETAC 1991). A **socio-ecological impact matrix** is a **qualitative** instrument of LCA (Belz and Hugenschmidt 1995). It identifies and categorises the main social and environmental problems of a product throughout the whole life-cycle on the basis of ABC analysis (where A = high impact, B = medium impact and C = low impact).

The relevance and importance of social and ecological issues vary from one product category to the next. Take, for instance, houses, which fulfil human needs for shelter and living and which are one of the most important elements in the environmental load of private consumption (Nijdam *et al.* 2005; Tukker and Jansen 2006). The main product life-cycle phases of houses are planning, resource extraction, transportation, construction, use and final demolition. The relevant environmental and social dimensions include resources, energy, air, water, soil, waste, ecosystems, health and equity. The planning and design of the house has little direct impact on the environmental and social environments. However, indirectly, it does have a substantial impact on these environments. In this phase of the product life-cycle architects and planners decide, in cooperation with builders, how large houses are to be, which materials are to be used, how well insulated the walls and ceilings are to be and what kind of equipment is to be employed to control the climate of the building. The decisions in the first phase influence to a large extent the impact of the following phases, especially in the usage phase, which is of special importance. Houses have a comparatively long lifetime. The energy consumption for controlling the climate of the house is 10–20 times higher than the 'grey energy' (i.e. the energy used to produce the materials and build the house). A large amount of the energy consumption is spent on heating and cooling the house (OECD 2002). The rest of the energy consumption is for heating water, lighting and cooking.

The larger a house the more resources it needs—both for construction and during operation. Since the 1970s the energy efficiency of houses has been improved by use of better insulation for walls, ceilings and windows. However, increasing living space per capita and rising comfort needs partly compensate for energy-efficiency gains (Wilson and Boehland 2005). Another environmental problem is the high rate of use of fresh water in private households, especially in OECD countries, ranging from 100 to 300 litres per capita per day (OECD 2002).

From a health point of view, pollutants in building materials may influence the air quality inside and have toxic or allergic effects on inhabitants. The main problems at the end of the product life-cycle are the vast amounts of waste at demolition, which contain some toxic substances. If waste dumps leak there is the long-term danger that soil and groundwater will become polluted, causing health problems.

When LCA was developed and introduced in companies during the 1990s there were high expectations for its use in marketing, especially communications (Rex 2005). For several reasons these expectations were not fulfilled. One of the reasons is that LCA makes a number of assumptions (e.g. on system boundaries) that influence the results and that are difficult to communicate in simple and plain words to consumers. Another reason is that many marketing decision-makers are simply not familiar with life-cycle thinking and LCA. Marketing scholars are partly responsible for this situation: the natural environment is paid barely any attention in basic and advanced courses of marketing. The concept of the product life-cycle may be prominent in marketing (Kotler and Armstrong 2004), but it is associated with stages of product development, introduction, growth, maturity and decline of products in accordance with sales and profits (Kotler and Armstrong 2004). It is not connected to the biophysical level and the whole product life-cycle of products from cradle to grave. Nevertheless, owing to increasing pressure on the natural environment and the limits of growth (Meadows *et al.* 2004), LCA will eventually become a valuable source of insight into the impacts of products and provide relevant information for marketing decisions. Leading, forward-thinking, companies such as Toyota have already realised that neglecting the impact of products on the social and ecological environments is risky for business and brands and that knowing the impact of products opens up new perspectives and market opportunities.

### 7.3.2 Consumer behaviour

Consumer behaviour includes the purchase, use and post-use of products. The main body of marketing literature is primarily concerned with the first stage, the process and act of purchasing (Kotler and Armstrong 2004). In the context of sustainability, all three stages of consumer behaviour are important. Many significant contributions that consumers can make towards environmental and social quality come in product use, maintenance and disposal or in delaying a purchase or avoiding it altogether (Peattie 1999). **Sustainable consumption** considers ecological and social criteria during all three stages (i.e. the purchase, use and post-use of products). It balances ecological, social and economic criteria responsibly, which is neither an easy nor straightforward task for consumers.

When a customer has a choice between two or more product alternatives, he or she will take the offer that promises the highest **perceived value** according to his or her evaluation of the difference between all the benefits and all the costs of a marketing offer relative to those of competing offers. Customers do not judge product benefits and costs accurately on an objective basis. They act on perceived value, which may differ largely from one customer to another (Kotler and Armstrong 2004).

In general, there are four different types of product **benefits**. The basic benefit is the function of the product. Additional benefits are self-esteem, recognition and edification from 'doing it yourself'. Take, for example, a sports car. Such a car is far more than a mode of transportation (use benefits). It is prestigious and stylish, thus filling the owner with pride (self-esteem) and raising his or her popularity (recognition). Tuning the car is great fun for fanatics, uplifting their spirits (edification). The evaluation is not complete without considering the **costs**, which go beyond the price of a product. Purchase costs, usage costs and post-usage costs have to be taken into account as well. Purchase costs include the cost of searching for a product (search costs), of gathering information on prices and specific features and of comparing the product to alternative marketing offers (information costs) and finally getting the product (transportation costs). Usage costs are often underestimated and not taken into account at the moment of purchase. In the case of long-lasting products such as houses, cars, washing machines and refrigerators, a considerable amount of money is paid for energy during the usage stage. Life-cycle costs are often underestimated by consumers. However, in times of rising resource and energy prices life-cycle costs become increasingly important (Belz 2001).

The individual perception and evaluation of benefits and costs is based on a number of personal and situational factors such as socio-ecological awareness, socio-ecological knowledge, disposable income, peer groups and purchasing situations. Based on individually perceived benefits and costs three different groups can be differentiated (Belz 2001):

- People who are socio-ecologically active ('socio-ecologically actives')
- People who can be socio-ecologically activated
- People who are socio-ecologically passive ('socio-ecologically passives')

The first group has a very high level of socio-ecological consciousness. From their point of view, social and environmental product features go hand in hand with self-esteem and recognition. That is why they are willing to make compromises with other product benefits and/or to accept higher costs. Usually, this group is rather small and represents the lead buyers and users of sustainable products.

The second group has a high level of socio-ecological consciousness. They associate social and environmental product features with some self-esteem and recognition. The members of this group are often willing to pay a higher price for the perceived value added but they are reluctant to make any compromise when it comes to the quality of the product. To a certain extent, this group is open to sustainability innovations. They represent the early adopters.

The third group is not particularly conscious about social and ecological issues. Socio-ecological product features are not perceived as value added. Thus, this group is not willing to make any compromises with respect to performance or price. They represent the average consumer and the late adaptors or laggards in sustainability innovation. Worldwide market research companies keep track of the environmental attitudes and environmental behaviour of consumers.

Since 1990 Roper Starch Worldwide has conducted a survey annually on the environmental attitudes and behaviour of US consumers, based on 2,000 face-to-face interviews which are balanced to the most recent US census. The latest *Green Gauge Report* identifies five consumer segments, each with varying degrees of environmental concern and action (Roper Starch Worldwide 2005):

- True-blue greens: the most environmentally active segment of society (11%)
- Greenback greens: those most willing to pay the highest premium for green products (8%)
- Sprouts: fence-sitters who have been slower to embrace environmentalism (33%)
- Grousers: those uninvolved or disinterested in environmental issues or who feel the issues are too big for them to solve (14%)
- Apathetics: the least engaged group, who believe that environmental indifference is mainstream (34%)

The five consumer segments have been fairly stable in the USA during the past decade: A comparison of results in 1995 with those for 2005 shows that there are hardly any differences in segments and numbers. The share of consumers who are genuinely concerned and engaged with environmental issues has remained pretty much the same (which might be considered as good news or bad news). The first two segments—the true-blue greens and the greenback greens—correspond to the socio-ecologically active and those that can be socio-ecologically activated (approximately 20% of the US population). The other three consumer segments identified by Roper Starch Worldwide seem to represent the socio-ecological passives (about 80% of the US population).

Since 2001 the Natural Marketing Institute (NMI) has conducted primary research on the so-called 'LOHAS consumers'—those who lead 'lifestyles of health and sustainability'. Based on quantitative surveys with more than 8,000 US consumers NMI investigates those for whom environmental, social and healthy lifestyle values play an important role in their buying decisions. The latest NMI survey in 2005 identified four segments (French and Rogers 2005):

- LOHAS: these have a profound sense of environmental and social responsibility and are the most likely of the consumer segments to buy environmentally and socially responsible products (23%)
- Nomadics: these represent the largest segment of consumers, which still show moderate levels of related concerns and select LOHAS behaviour (38%)
- Centrists: these show a lack of LOHAS attitudes and behaviour (27%)
- Indifferents: these are simply not interested in environmental and social issues and are driven by immediate needs, not health or the sustainability of the planet and society (12%)

As the empirical work shows, LOHAS represents a sizable consumer group. Products and services that improve health, safeguard ecosystems, reduce the use of natural resources and develop human potential in a sustainable manner serve the LOHAS consumers. The size of the LOHAS marketplace is estimated to be about $230 billion, including a wide range of products and services such as organic food, dietary supplements, complementary medicine, 'green' buildings, renewable energies, eco-tourism and travel (LOHAS 2008). The LOHAS concept is rather general and vague. It is disputable which consumers and products classify as LOHAS or non-LOHAS. However, the market research conducted by NMI and the communication of LOHAS by Conscious Media via the Internet and national conferences are possible ways of mainstreaming sustainability issues.

Both the *Green Gauge Report* and the LOHAS study provide some interesting insights into marketing decision-makers. They show that around a fifth of the population is socio-ecologically aware and at least partly willing to act on this awareness. This is a sizable consumer group with a lot of purchasing power, representing growing market segments and open market opportunities. Nevertheless, the empirical results should be interpreted with caution by marketers. First, there is the widely known problem of social desirability in surveys. In addition, there is a discrepancy between socio-ecological awareness, intentions and actual behaviour. And, even if consumers do behave according to their social and environmental beliefs in one purchasing decision (e.g. with regard to organic food products), it does not mean they will in another (e.g. with respect to automobiles). On the contrary, socio-ecological consumer behaviour is often inconsistent and contradictory. The following sections show how marketers can narrow the gap between consumer awareness, intentions and behaviour. Based on best practices it presents a systematic approach to market sustainable products and services successfully.

### 7.3.3 Normative sustainability marketing

Visions and values set the normative basis for sustainability marketing on the strategic and operational levels. They are important for the founding of new companies or the reorientation of established companies. Take, for example, Mobility CarSharing, the world's largest car-sharing organisation, founded in Switzerland in 1997. Its vision is to promote combined mobility and to change mobility patterns in a sustainable way (Belz 2000). Or take BP as one of the leading energy companies of the world that is committed to sustainable development. Although its core business is still the exploration, refining and distribution of oil, BP has since the end of the 1990s started to develop and sell renewable energies. The multinational company established a strategic business unit for 'renewable energies' and is already one of the world's leading producers and users of solar cells. The new strategic reorientation based on the vision of sustainable development is also expressed in the new logo of BP, standing for 'Beyond Petroleum'.

In the beginning of the 21st century many companies claim to be committed to sustainable development, assuming economic, ecological and social responsibility. Multinational corporations as well as small and medium-sized companies express their commitment towards sustainability explicitly. One critic sarcastically says:

> It would be a challenge to find a recent annual report of any big international company that justifies the firm's existence merely in terms of profit, rather than 'service to the community'. Such reports often talk proudly of efforts to improve society and safeguard the environment—by restricting emissions of greenhouse gases from the staff kitchen, say, or recycling office stationery—before turning hesitantly to less important matters, such as profits (Crook 2005: 3).

No doubt there is a grain of truth in this critical remark. Nevertheless, corporate sustainability statements, guidelines and principles are vital normative foundations for sustainability marketing if the company and its employees are to live up to it. It is signalling an effort to build up reputation and trust, both internally and externally. On the one hand, sustainability statements and guidelines may help marketing management and employees in strategic and operational decision-making. On the other hand, they send signals to society and market partners along the whole product chain to strive for sustainability.

Well-intended corporate sustainability statements and codes are of little use if they are not integrated in the goal-setting process. Sometimes ecological, social and economic objectives are complementary. The 'win–win–win' rhetoric is based on this assumption and presents some anecdotal evidence for such instances. However, in many cases there are trade-offs between ecological, social and economic objectives that have to be carefully and responsibly solved by decision-makers. Finding the right balance between ecological, social and economic goals is a demanding challenge for marketing and is a continuous process. It also depends on institutions: the more the institution is in favour of sustainability and the more the external effects are internalised the easier it is for decision-makers in marketing to balance the triple bottom line. The socio-ecological objectives in sustainability marketing can be qualitative (e.g. to enhance the use of renewable energies) or quantitative (e.g. to aim for 20% of the total revenue to be achieved with use of renewable energies by 2020). It is problematic if these aims are counterbalanced by evaluation and income systems based on short-term profits.

## 7.3.4 Strategic sustainability marketing

On the strategic level of sustainability marketing there are a number of issues to consider for marketing managers: innovation, segmentation, targeting, positioning and timing. **Sustainability innovations** are a condition *sine qua non* in successful sustainability marketing. Examples of sustainability product and service innovations are 'passive' houses, solar cells, organic food products, fair trade products, hybrid cars and car sharing schemes. If these kinds of sustainability innovations are developed, who are the main target groups? What role do social and environmental aspects play in positioning? What is the unique sustainable selling proposition? When is the right moment in time to enter the market with sustainability innovations?

There are at least three possibilities in the positioning of sustainable products and services (Meffert and Kirchgeorg 1998). First, the socio-ecological dimension plays quite an important role and is communicated as the primary benefit of the product or service, whereas performance and price are secondary. Second, the socio-ecological dimension plays a significant role but is not predominant. It is treated equally to performance and price. Last, the socio-ecological dimension is an integral part of quality and performance. The sustainable positioning depends on a number of influencing factors such as consumer preferences, competitive offers, brand assortment and company size. The first positioning may be suitable for smaller sustainability pioneers following a niche strategy and aiming at the socio-ecological active customer group. Since the sustainable niche is rather limited, it is seldom a viable option for medium-sized or large companies. The second positioning aims at those that can be socio-ecologically activated. If companies manage to combine the socio-ecological dimension with classical buying criteria such as taste, freshness, design, durability and so on to 'motive alliances', this customer group is open to sustainability innovations. In many markets they represent an important, growing, segment. The third positioning is appropriate for companies that aim at the mass market. The social and environmental performance of a product is taken for granted and is not necessarily communicated to the consumer.

In general, sustainable products and services are launched in niches before they eventually reach broader market segments. The mainstreaming of sustainable products and services depends on a variety of external factors (e.g. price systems and existing socio-economic regimes). However, mainstreaming can also be influenced by marketers. In niches, sustainable products and services are optimised according to social and environmental criteria. Thus, the price is usually quite high, the availability is rather low and mass communication is practically non-existent. To market sustainable products and services successfully beyond niches usually means the need to make compromises and optimise the overall quality of those products and services that incorporate social and environmental criteria, to reduce prices, to increase availability and to intensify mass communication.

Many consumer goods markets are characterised by polarisation—that is, the lower-priced segment and the upper-quality segment gain significance and market shares, whereas the middle segment erodes. Customers either demand low-priced products with good value *or* they ask for high-quality products and premium brands that promise a high value added.

What consequences does market polarisation have for the positioning of sustainable products? An obvious opportunity is to position socio-ecological innovations in the quality segment. Take, for example, organic and natural food stores, which put a lot of emphasis on socio-ecological aspects, targeting the socio-ecologically active customer group. Large organic retail chains such as Whole Foods Markets in the USA and Basic in Germany try to appeal to a broader customer group. Their stores are located in urban areas and offer a wide range of organic and natural food products. They appeal mainly to those that can be socio-ecologically activated. The two retail chains grow in stagnant food-markets that are characterised by fierce price and substitution competition.

Another viable positioning strategy for medium-sized and large companies is multi-segment marketing with a clear multi-brand concept. Take, for example, Migros, the largest food retailer in Switzerland, which offers two different retail brands: M-Bio is positioned in the premium segment and fulfils the high quality standards of controlled, certified, organic farming, whereas M-Budget is clearly positioned in the price segment.

For non-governmental organisations (NGOs) it is important to attach importance to the quality as well as to the price segment to upgrade socio-ecological product standards. As a result of public pressure, the Swedish furniture retailer IKEA was made to pay attention to sustainability issues in purchasing and pre-production (IKEA 2004). Similarly, the Swedish textile retailer H&M assumes social and ecological responsibility along the whole textile chain (H&M 2004). Both companies are positioned in the price segment and aim at price-sensitive consumers. These consumers are the socio-ecologically passive who are rarely willing to pay a higher price for socio-ecological benefits. By means of the public pressure the customers of IKEA and H&M get socio-ecological value 'for free'.

From the viewpoint of companies under scrutiny, the main aim is keeping brand image and corporate reputation rather than much socio-ecological differentiation.

Besides targeting and positioning, the question of timing is important: When is the right moment to introduce sustainability innovations to the market? Is it worth entering the market at an early stage and leading the market or is it better to wait and follow at a later stage, depending on the growth and market development? If companies enter the market at an early stage and the technological innovations are not yet perfected or the consumers are not sensitive to the sustainability issues at hand the main challenges consist in continuous product improvements and consumer education (i.e. primary market entry barriers). If companies enter at a later stage and the leading company has already established the reputation of a socio-ecological pioneer and gained significant market shares the main challenge is rather more about fighting the competitor than about informing the consumer (i.e. secondary market entry barriers).

A good example of an early entry and a successful pioneer strategy is provided by Toyota and its hybrid car Prius, which means literally 'to go before'. The Toyota Prius is a hybrid car combining the power of a petrol engine with the efficiency of an electric battery. The electric motor starts the car and operates at low speed. At higher speed the Prius automatically switches to the petrol engine. Under normal driving conditions, the hybrid consumes 4.3 litres of petrol per 100 kilometres and emits 104 grammes $CO_2$ per kilometre—much less than most petrol and diesel cars. The Prius looks stylish and seats five people comfortably. It combines environmental performance with power, convenience and safety. Between 2003 and 2005 the annual sales of Toyota hybrid cars quadrupled from 53,000 to more than 210,000 units, partly attributable to the rise in petrol prices.

## 7.3.5 Operational sustainability marketing

Once a company has decided on its overall sustainability marketing strategies it is ready to begin planning the details of putting it into practice. To implement a sustainability marketing strategy a comprehensive marketing mix has to be developed. The marketing mix first proposed by McCarthy (1960) consists of four Ps: namely, product, place, price and promotion. The classical concept of the four Ps takes the seller's point of view. In the age of customer relationships the four Ps might be better described as the four Cs, consisting of customer solutions, convenience, customer cost and communication (Kotler and Armstrong 2004; Lauterborn 1990). This concept takes the buyer's instead of the seller's viewpoint and is well suited to the idea of sustainability:

- The idea of 'products' tends to put an emphasis on tangible goods, whereas 'customer solutions' include products as well as services (i.e. intangible goods) that satisfy customer needs, substitute for tangible goods and reduce environmental impact
- The idea of 'place' stresses the physical distribution of goods, whereas 'convenience' makes marketers aware that customers want to get products or services as conveniently as possible
- The idea of 'price' takes into account only the amount of money the buyer has to pay to obtain a product or service, whereas 'customer cost' considers the whole life-cycle cost (price, transaction cost, usage cost and disposal cost), which is important in times of rising resource and energy prices
- The idea of 'promotion' means one-way communication from the seller to the buyer, whereas 'communication' operates both ways, which is essential for building trust and credibility

### 7.3.5.1 Customer solutions

In the heart of the sustainability marketing mix are sustainable products and services, which may be defined as products and services that satisfy customer needs and which significantly and continuously improve the social and environmental performance throughout the whole life-cycle in comparison with conventional or competing offers (for a similar definition, see Peattie 1995). This definition has the following characteristics:

- Customer satisfaction. If sustainable products do not satisfy customer needs, they will neither survive nor thrive in the market economy
- Dual focus. Unlike 'green' products, sustainable products have a dual focus on social and/or environmental performance
- Life-cycle orientation. Sustainable products have to take into account the whole life-cycle, from cradle to grave (i.e. from the extraction of raw materials, to transportation, manufacturing, distribution and use through to disposal)
- Significant improvements. Sustainable products have to make a significant contribution to the main environmental and social problems analysed and be identified with use of LCA instruments

- Continuous improvement. Sustainable products are not absolute, but relative, measures, being dependent on the status of knowledge, latest technologies and societal aspirations, which change over time. A product that meets customer needs and that has extraordinary social and environmental performance today may be considered standard tomorrow. Thus, sustainable products have continuously to be improved regarding customer, social and environmental performance

- Competing offers. A product that satisfies customer needs and that proposes environmental and social improvements may still lag behind competing offers. Thus, the offerings of competitors are yardsticks for improvements with regard to customer, social and environmental performance

Often a sharp distinction is made between tangible products and intangible services. However, all products include some kind of service, and all services require some tangible elements (Halme *et al.* 2005; Heiskanen and Jalas 2003). Basically, there are three kinds of sustainable services: product-oriented services, use-oriented services and result-oriented services (Halme *et al.* 2005; Hockerts 1999). Product-oriented services are usually after-sales services such as repair, maintenance, upgrading or take-back of the product. Such offerings might be interpreted as extended producer responsibility (i.e. the company takes responsibility for the product beyond its sale: that is, in the downstream phases of use, re-use, recycling and disposal). Use-oriented services are services that sell the use of a product. The product itself remains in the property of the producer or provider of the use-oriented service. Examples of this type of service are car-sharing schemes and launderettes. Result-oriented services deliver the result of performance. Take, for example, taxi rides, public transport and dry cleaners. In these instances the customer neither owns nor uses the product. He or she receives the 'fruits' of product-service combinations.

### 7.3.5.2 Convenience

Many sustainable products and services are not widely distributed, which is one of the reasons why they remain in niches. They are available only in special shops or via mail, which is highly inconvenient for the majority of potential customers. If sustainable products and services are aimed at larger market segments or mass markets they have to be made widely available by 'multi-channel marketing' (i.e. via direct distribution, online shops, speciality shops, retailers and so on). The mode of transportation is important from an environmental point of view. Generally speaking, ships and trains are more environmentally benign than are planes and lorries. The question of convenience is also crucial for 'retro-distribution' (i.e. the collection of packaging and products at the end of the product's life). To make consumers bring back the used packaging and products the process has to be as convenient and cost-efficient as possible. Often, retailers are good points of return.

### 7.3.5.3 Customer costs

In many advertisements the price is used as a key argument to entice consumers to buy the product. However, in addition to benefits, total customer costs have to be taken into account by companies offering sustainable products and services. This includes the cost of purchase, the cost of use and post-use costs. Some customers, for instance, may be willing to pay a premium price for organic food products but they may not be willing to 'walk the extra mile' and visit the alternative shop at the other end of the city instead of the supermarket around the corner. Most customers are responsive to consumer goods prices but few are aware of the total life-cycle costs, especially costs during the use phase (e.g. the water and energy-consumption costs in the case of washing machines). In times of rising resource and energy prices these types of cost become increasingly important. In addition, it is the task of marketers to make consumers conscious of total life-cycle costs by comparing sustainable products and services with conventional products and services. Take, for example, Mobility CarSharing, mentioned in Section 7.3.3. It makes customers aware of the total costs of ownership of private cars and uses this as an argument in its advertising (Belz 2000).

#### 7.3.5.4 Communication

Social and environmental qualities are often 'credence qualities' that cannot be inspected prior to the purchase or experienced after the purchase (e.g. production by organic farming or purchase through fair trade). Thus the consumer has to trust the information provided by the producer or a third party. That is why credibility and trust are of special importance in the markets for sustainable products and services. Possible ways to inform the consumer and to overcome scepticism are personal declarations by the owner, voluntary self-binding standards, the use of third-party labels and partnerships with governmental organisations as well as with NGOs (Karstens and Belz 2006).

On the one hand, companies have to inform consumers in a simple and credible way about the sustainability issues at stake. On the other hand, companies have to animate consumers to get their attention. This describes the 'communication paradox' of sustainable products and services, that between information and animation, which have to be balanced continuously. A good example is Patagonia, a sportswear and outdoor retailer that has a reputation for very high-quality products and that is also well known for its commitment to the environment. In the 1990s Patagonia introduced post-consumer recycled (PCR) Synchilla fleece, a synthetic fleece made of recycled plastic bottles, and switched to organic cotton (Chouinard and Brown 1997; Meyer 2001). Patagonia advertisements show outdoors sports people such as mountaineers, surfers or snowboarders in wonderful natural surroundings. The pictures are highly emotional and animating. Except for the brand of the company no additional information is given in the advertisements. Detailed information on the superior durability, functionality and performance of PCR Synchilla fleece is given in the catalogue and on the Internet (Patagonia 2006). The environmental performance of PCR Synchilla fleece completes the positioning strategy rather than dominating other criteria (Meyer 2001).

### 7.3.6 Transformational sustainability marketing

Transformational sustainability marketing is about the active participation of companies in public and political processes to change institutions in favour of sustainability (Belz 2001). First, there is a **normative rationale** for such an endeavour. Many owners of family businesses and some managers feel obliged to society and want to be good citizens. In addition, there is also a **strategic rationale**: within the present institutional framework the successful marketing of sustainable products or services is possible but is limited in width and depth. Institutional design fails to set positive incentives for sustainable behaviour, either for producers or consumers. Rather, it allows, and often even exacerbates, unsustainable behaviour. That is why changes in institutions are necessary to expand the intersection between socio-ecological problems and consumer behaviour and to set up the conditions for the successful marketing of sustainable products beyond niches.

Sustainability pioneers and leaders can participate in enlightened self-interest, changing public and political institutions and thus enhancing sustainable development (Bendell and Kearins 2005; Ulrich and Maak 1997). They can help to develop the free-market system towards a socio-ecological market system on a global level. The more that social and political institutions favour sustainable consumption, the easier it is for companies to market sustainable products beyond niches.

The objectives of transformational sustainability marketing are to initiate institutional changes that either set positive incentives for the purchase and use of sustainable products or set negative incentives for the purchase and use of conventional products. Approaches to transformational sustainability marketing are voluntary agreements for minimum socio-ecological standards in industry, the development of sustainability labels in cooperation with NGOs, public support from companies for the ratification and implementation of the Kyoto Protocol and so on.

An example of transformational sustainability marketing is provided by Unilever, which joined forces with the WWF in 1997 to set up the Marine Stewardship Council (MSC), an international non-profit organisation that defines criteria for sustainable fishing, in roundtables and in consultation processes (Karstens and Belz 2006). The MSC label indicates the use of sustainable fishing according to ecological, social and economic criteria. Unilever, as the market leader for frozen fish, with well-known brands such as Iglo, has committed itself to sourcing its fish from sustainable stocks in order to secure future supplies of raw materials. In 2005, 40% of the Iglo frozen fish products met MSC standards and carried the independent MSC label.

Another example of transformational sustainability marketing is provided by BP, which was one of the first oil companies to acknowledge the link between rising $CO_2$ emissions and climate change. Additionally, BP publicly supports the ratification and implementation of the Kyoto Protocol.

It is important to note that transformational sustainability marketing is not simply another form of lobbying to reinforce corporate and business interests. It is about the true commitment of companies to sustainable development and their active participation in public and political processes. It is a discourse with stakeholders to realise institutional frameworks for a market system that is stable, fair and just, serving human beings and a 'good life' (Ulrich and Maak 1997). It remains to be seen and analysed which companies truly support such an endeavour and to what extent such marketing is possible in a global market system with fierce competition.

## 7.4 Special characteristics of sustainability marketing

If we compare sustainability marketing to classical approaches to marketing, in what respects does one differ from the other? What is distinctive about sustainability marketing? There are at least six characteristics of sustainability marketing compared with conventional marketing (Belz 2005, 2006):

- Socio-ecological problems. The analysis and identification of ecological and social problems are points of departure in sustainability marketing. Social and ecological aspects are integrated throughout the whole process of sustainability marketing

- Intersection. Social activists with 'big hearts' put a strong emphasis on the solution of socio-ecological problems but widely neglect consumer wants and demand. They follow a kind of anti-marketing or alternative marketing approach. Mainstream marketing focuses mainly on consumer demand, overlooking the social and ecological dimensions. In contrast, sustainability marketing searches for solutions to socio-ecological problems while meeting consumer demand. It offers sustainable products and services that benefit the individual and the commons

- Normative aspects. Sustainability marketing is based on visions and values. It aims at sustainable and profitable relationships with customers, the natural environment and the social environment. As well as having common marketing goals concerning sales, market share and profits, ecological and social objectives are taken into consideration and balanced in a responsible way

- Information asymmetries. Social and ecological qualities of products are often credence qualities. That is why signalling, credibility and trust are essential in sustainability marketing

- Transformational aspects. Within the existing framework there are few economic incentives to behave in a sustainable way, either for producers or consumers. By engaging in public and political discourses and changing the institutional design in favour of sustainability, companies set the conditions for the successful marketing of sustainable products beyond niches

- Time aspects. Classical marketing is focused on sales and transactions. It is rather short-term in nature and is biased towards the present. Modern marketing represents a paradigm shift from transactions towards relations, which is why it is called 'relationship marketing'. It aims at building lasting relationships with customers in order to produce high customer equity. Sustainability marketing goes much further. It aims at building lasting relationships with customers, with the social environment and with the natural environment. Thus, long-term thinking and 'futurity' are fundamental components of sustainability marketing (Peattie 1999)

## 7.5 Conclusions

Marketing in the age of sustainable development is ambivalent. On the one hand, marketing promotes a consumer society and materialistic lifestyles, which entail serious social and environmental problems. On the other hand, marketing helps to develop and diffuse sustainability innovations, which create customer satisfaction, while considering social and environmental aspects. The main potential of sustainability marketing lies in the promotion of sustainable products and services. This potential has not yet been used, for a number of reasons (e.g. a lack of sustainability innovations, a lack of knowledge on the producer side, a lack of interest on the consumer side, potential and actual conflicts between socio-ecological and consumer benefits and an unfavourable environment with regard to public and political institutions). The mainstreaming of sustainability in the theory and practice of marketing is far from being reality. Here, marketing scholars and practitioners have a crucial role to play, taking on their responsibility to make the resource and energy revolution come true in order to provide a better quality of life for a greater number of people on planet Earth.

## References

Belz, F.-M. (2000) 'Mobility CarSharing: Successful Marketing of Eco-efficient Services', in K.P. Green, P. Groenewagen and P.S. Hofmann (eds.), *Ahead of the Curve* (Dordrecht, Netherlands: Kluwer Academic).

—— (2001) *Integratives Öko-Marketing. Erfolgreiche Vermarktung von ökologischen Produkten und Leistungen (Integrative Eco-Marketing: Successful Marketing of Ecological Products and Services)* (Wiesbaden, Germany: Gabler).

—— (2005) *Sustainability Marketing: Blueprint of a Research Agenda* (Discussion Paper 1: Marketing and Management in the Food Industry; Munich: TUM Business School).

—— (2006) 'Marketing in the 21st Century', *Business Strategy and the Environment* 15: 1-5.

—— and H. Hugenschmidt (1995) 'Ecology and Competitiveness in Swiss Industries', *Business Strategy and the Environment* 4: 229-36.

Bendell, J., and K. Kearins (2005) 'The Political Bottom Line: The Emerging Dimension to Corporate Responsibility for Sustainable Development', *Business Strategy and the Environment* 14: 372-83.

Charter, M. (1992) *Greener Marketing* (Sheffield, UK: Greenleaf Publishing).

—— (2006) 'Sustainable Consumption and Production, Business and Innovation', in *Conference Proceedings: Changes to Sustainable Consumption, Workshop of the Sustainable Consumption Research Exchange (SCORE!) Network*, Copenhagen, 20–21 April 2006, supported by the EU's 6th Framework Programme.

—— and M.J. Polonsky (eds.) (1999) *Greener Marketing: A Global Perspective on Green Marketing Practice* (Sheffield, UK: Greenleaf Publishing).

——, K. Peattie, J. Ottman and M.J. Polonsky (2002) *Marketing and Sustainability* (Cardiff, UK: Centre for Business Relationships, Accountability, Sustainability and Society).

Chouinard, Y., and M.S. Brown (1997) 'Going Organic: Converting Patagonia's Cotton Product Line', *Journal of Industrial Ecology* 1: 117-29.

Christopher, M., A. Payne and D. Ballantyne (1991) *Relationship Marketing. Bringing Quality, Customer Service and Marketing Together* (Oxford, UK: Butterworth Heinemann).

Coddington, W. (1993) *Environmental Marketing: Positive Strategies for Reaching the Green Consumer* (New York: McGraw-Hill).

Crook, C. (2005) 'A Survey of Corporate Social Responsibility', *The Economist*, 22 January 2005: 3-18.
Durning, A. (1992) *How Much is Enough? The Consumer Society and the Future of the Earth* (New York/London: W.W. Norton).
Ehrlich, P.R., and J.P. Holdren (1971) 'Impact of Population Growth', *Science*, March 1971: 1,212-16.
Elzen, B., F. Geels and K. Green (2000) *System Innovation and the Transition to Sustainability* (Cheltenham, UK: Edward Elgar).
French, S., and G. Rogers (2005) 'LOHAS Market Research Review: Marketplace Opportunities Abound', www.lohas.com/journal/trends.html [accessed 25 August 2007].
Fuller, D.A. (1999) *Sustainable Marketing: Managerial–Ecological Issues* (Thousand Oaks, CA/London/New Delhi: Sage Publications).
Gardner, G., E. Asadourian, E. and R. Sarin (2004) 'The State of Consumption Today', in The Worldwatch Institute (ed.), *State of the World 2004. Special Focus: The Consumer Society* (New York/London: W.W. Norton): 3-21.
H&M (2004) *Corporate Social Responsibility Report 2003* (Stockholm: H&M).
Halme, M., G. Hrauda, C. Jasch, J. Kortman, H. Jonuschat, M. Scharp, D. Velte and P. Trindade (2005) *Sustainable Consumer Services: Business Solutions for Household Markets* (London: Earthscan Publications).
Heiskanen, E., and M. Jalas (2003) 'Can Services Lead to Radical Eco-efficiency Improvements? Review of the Debate and Evidence', *Corporate Social Responsibility and Environmental Management* 10: 186-98.
Hockerts, K. (1999) 'Innovation of Eco-efficient Services: Increasing the Efficiency of Products and Services', in M. Charter and M.J. Polonsky (eds.), *Greener Marketing: A Global Perspective on Greening Marketing Practice* (Sheffield, UK: Greenleaf Publishing): 95-108.
IKEA (2004): *IKEA: Social and Environmental Responsibility 2003* (available at www.ikea.com).
Karstens, B., and F.-M. Belz (2006) 'Information Asymmetries, Labels and Trust in the German Food Market: A Critical Analysis Based on the Economics of Information', *International Journal of Advertising* 25: 189-211.
Kotler, P., and G. Armstrong (2004) *Principles of Marketing* (Upper Saddle River, NJ: Pearson, 10th edn).
Lauterborn, R. (1990) 'New Marketing Litany: Four P's Passé; C-words Take Over', *Advertising Age* 1 (October 1990): 19.
LOHAS (2008) 'About LOHAS', www.lohasjournal.com [accessed 21 January 2008].
McCarthy, E.J. (1960) *Basic Marketing: A Managerial Approach* (Homewood, IL: Irwin).
Meadows, D., J. Randers and D. Meadows (2004) *Limits of Growth: The 30-year Update* (White River Junction, VT: Chelsea Green Publishing).
Meffert, H., and M. Kirchgeorg (1998) *Marktorientiertes Umweltmanagement* (*Market-Oriented Environmental Management*) (Stuttgart: Poeschel, 3rd edn).
Meyer, A. (2001) 'What's in it for the customers? Successfully Marketing Green Clothes', *Business Strategy and the Environment* 10: 317-30.
Mintu-Wimsatt, A.T., and D.M. Bradford (1995) 'In Search of Market Segments for Green Products', in M.J. Polonsky and A.T. Mintu-Wimsatt (eds.), *Environmental Marketing: Strategies, Practice, Theory and Research* (New York/London: Haworth): 293-304.
Nijdam, D.S., H.C. Wilting, M.J. Goedkoop and J. Madsen (2005) 'Environmental Load from Dutch Private Consumption: How Much Damage Takes Place Abroad?', *Journal of Industrial Ecology* 9.1–2: 147-68.
OECD (Organisation for Economic Cooperation and Development) (2002) *Towards Sustainable Household Consumption? Trends and Polices in OECD Countries* (Paris: OECD).
Ottman, J.A. (1998) *Green Marketing. Opportunity for Innovation'* (Lincolnwood, IL: NTC Business Books, 2nd edn).
Patagonia (2006) 'Patagonia Outdoor Posters', www.patagonia.com [accessed 25 August 2007].
Peattie, K. (1992) *Green Marketing* (London: Pitman).
—— (1995) *Environmental Marketing Management: Meeting the Green Challenge* (London: Pitman).
—— (1999) 'Rethinking Marketing: Shifting to a Greener Paradigm', in M. Charter and M.J. Polonsky (eds.), *Greener Marketing: A Global Perspective on Greening Marketing Practice* (Sheffield, UK: Greenleaf Publishing): 57-70.

Polonsky, M., and A.T. Mintu-Wimsatt (eds.) (1995) *Environmental Marketing: Strategies, Practice, Theory, and Research* (New York/London: Haworth).

Rex, E. (2005) *Premises for Linking Life-cycle Considerations with Marketing* (Göteborg, Sweden: Chalmers University of Technology).

Roper Starch Worldwide (2005) *Green Gauge Report 2005* (New York: Roper Starch Worldwide).

Schmidt-Bleek, F. (1998) *Das MIPS-Konzept: Weniger Naturverbrauch, mehr Lebensqualität durch Faktor 10* (Munich: Droemer).

SETAC (Society of Environmental Toxicology and Chemistry) (1991) *A Technical Framework for Life-cycle Assessment* (Washington, DC: SETAC).

Tukker, T., and B. Jansen (2006) 'Environmental Impacts of Products: A Detailed Review of Studies', *Journal of Industrial Ecology* 10.3: 159-82.

Ulrich, P., and T. Maak (1997) 'Integrative Business Ethics: A Critical Approach', *CEMS Business Review* 2: 27-36.

UNEP (United Nations Environment Programme) (2006) *10 Year Framework on SCP* (UNEP, Production and Consumption Branch, Sustainable Consumption, www.unpetie.org/pc/sustain/10year/home.htm [accessed July 2006]).

United Nations (2005a) *Plan of Implementation of the World Summit on Sustainable Development* (New York: United Nations).

—— (2005b) *World Population Prospects: The 2004 Revision* (New York: United Nations).

Von Weizsäcker, E., A.M. Lovins and L.H. Lovins (1998) *Factor Four: Doubling Wealth, Halving Resource Use* (London: Earthscan Publications).

WCED (World Commission on Environment and Development) (1987) *Our Common Future* (Oxford, UK: Oxford University Press).

Wilson, A., and J. Boehland (2005) 'Small is Beautiful: US House Size, Resource Use, and the Environment', *Journal of Industrial Ecology* 9.1–2: 277-87.

Wuppertal Institute (Wuppertal Institut für Klima, Umwelt, Energie) (2005) *Fair Future: Begrenzte Ressourcen und Globale Gerechtigkeit (Fair Future: Limited Resources and Global Equity)* (Munich: C.H. Beck).

Part 3
**Design perspective**

# 8
# Review: design for sustainable consumption and production systems*

*Carlo Vezzoli*
Design and Innovation for Sustainability, Italy

*Ezio Manzini*
INDACO-Politecnico di Milano, Italy

## 8.1 Introduction

Transition towards sustainability requires radical changes in the way we produce and consume and, in general, in the way we live. The prospect of sustainability necessarily places the very model of development under discussion. In future decades we must be able to move from a society in which well-being and affluence are measured by the production and consumption of goods to one in which people live better, consuming (much) less. In fact, we need to learn how to live better (the entire population of the planet: the equity principle) and, at the same time, reduce our ecological footprint.

Given the nature and the dimension of this change, it is understood that the transition towards sustainability (and, in particular, towards sustainable ways of living) has to be a wide-reaching social learning process in which a system discontinuity is needed. Therefore a system approach is important in order to seriously tackle the transition towards sustainability; that is, so-called system innovations should take place. From this perspective, the link between the environmental and social dimensions of the problem clearly appears, showing that radical social innovation will be needed in order to move from current unsustainable consumption models to new sustainable ones.

* This chapter is the result of a collaboration between the two authors: Vezzoli wrote Section 8.3–8.6; Manzini wrote Sections 8.1 and 8.2.

In this framework the following chapter reviews how the discipline of design for sustainability has enlarged its scope and field of action over the past two decades: from material and energy low-impact selection to life-cycle design (or ecodesign) of products, eco-efficient (product-service) system design and design for social equity and cohesion.

We describe how this evolution has opened a debate on the role of design itself, a discipline that is already undergoing a redefinition of its (potential) role, as a consequence of other socioeconomic transitions (i.e. service orientation, interconnection, globalisation–localisation [glocalisation]).

The argument will then focus on the emerging hypothesis that a design approach seeking to effectively tackle radical innovation and sustainable consumption should operate (and define its real potential and the way to do so) on a system innovation level. In other words, we require 'design with a strategic approach', as this extension of the design role (field of action) is described in various design schools.

The design research and discussion on this issue is framed within mainstream market dynamics; nevertheless, particular attention is given to promising economic models, referred to as distributed economies (i.e. locally based and network-linked enterprises and initiatives).

Finally, the chapter will discuss present and potential roles for design, and its limitations.

## 8.2 State of the art: understanding the present and potential role of design

When discussing the role of design in the transition towards sustainability, particularly in the design and development of sustainable product and service systems, it is useful to introduce certain general considerations about the nature of design itself and about the evolutionary dynamics that have invested it during its relatively brief history.

The meaning of the word *design* changes according to context (Box 8.1). In an international context, when referring to production and consumption activities, one current 'official' definition is given by the International Council of Societies of Industrial Design (ICSID), which states that 'design is a creative activity whose aim is to establish the multifaceted qualities of objects, processes, services and their systems in whole life-cycles. Therefore, design is the central factor of innovative humanisation of technologies and the crucial factor of cultural and economic exchange' (ICSID 2005).

This definition is official in that the ICSID is the most authoritative international design organisation and is particularly relevant to current debate in that, unlike previous definitions, it includes within the scope of design not only products but also processes and services.[1] In addition, by advancing the idea that design considers products together

1 This definition is in many ways different from the one given by the same body 40 years ago. As Tomàs Maldonado reminds us in his book, *Disegno industriale: un riesame* (Maldonado 1974), the definition of industrial design adopted by ICSID in 1961 (as proposed by Maldonado himself) upheld that 'by industrial design we normally mean the designing of industrially manufactured objects', where the expression 'designing' was to be understood as the activity that leads to 'the coordination, integration and articulation of all the factors that, in one way or another, go to make up the process of shap-

## Box 8.1 Definitions of *design*

Victor Papanek begins his successful book, *Design for the Real World*, by saying 'Every man [sic] is a designer. Everything we do is almost always design, just because design is the basis of every human activity' (Papanek 1970).

Herbert Simon attributes the word *design* with just as wide a meaning when he says, '[design] is concerned with how things ought to be—how they ought to be in order to attain goals and to function' (Simon 1969). As we see, this too refers to an activity that is in its essence a skill proper to all human beings. However, when developing his theory, Simon orientates the meaning of the word principally *towards the technical-scientific dimension of the problem and to the scope of engineering*.

In everyday language, we find definitions such as:

> Design, usually considered in the context of the applied arts, engineering, architecture, and other such creative endeavours, is used as both a noun and a verb. 'Design' as a verb refers to the process of originating and developing a plan for a new object (machine, building, product, etc.). As a noun, 'design' is used both for the final plan or proposal (a drawing, model, or other description), or the result of implementing that plan or proposal (the object produced).*

In this chapter, we refer to an activity developed within the ambits of the 'applied arts, engineering, architecture, and other such creative endeavours'. So, though not explicitly stated, the idea of design is linked not solely to a technical dimension but also to communicative, social and aesthetic aspects of the issues to be addressed.

---

\* Wikipedia: en.wikipedia.org/wiki/Design [accessed 24 January 2005].

with 'their systems in whole life-cycles', it makes significant reference to issues raised by the environmental question.

However, although advanced, the definition proposed by the ICSID still reflects design history and its traditional, privileged relationship with the manufacturing industry of the past century and, more generally, with a heavily product-oriented production culture and idea of well-being. In fact, design was born into this context and developed as a highly product-oriented activity—an activity that has been very successful in solving some concrete everyday problems but which, at the same time, has been equally successful in enhancing consumption.

The role of design as consumption promoter continued when the economy changed and moved towards the present one where communication appears to be more important than the products themselves. In this new context, from the few 'big stars' to the many 'normal' professionals, designers have been able to develop powerful communi-

ing a product' (Maldonado 1974). As we can see, this is already a wide, systemic interpretation (one that includes not only the product itself but also the processes required to bring it into being; but it is still totally product-oriented. Today, in our view, it would be more appropriate to move away from this product-oriented definition to a more solution-oriented one (e.g. see the concepts of product-service systems in Charter and Tischner 2001; Manzini and Vezzoli 2001; Van Halen *et al.* 2005; see also the literature on transformation design, introduced by the Design Council with the RED Paper 02 [RED 2006; see also Burns *et al.* 2006]).

cation skills and have become the main actors in promoting new ideas of well-being and ever-increasing desire—ideas of well-being and desire that, at the end of the day, unfortunately appear to be more and more material-intensive and energy-intensive.

In view of the present growing environmental and social problems, the result is that design has to be considered as 'part of the problem' more than as 'part of the solution'. To escape from this destiny, design has to reorient its skills and capabilities, in both problem-solving and communication, towards sustainability. But, more than anything, it has to redefine the object of its action and move from the traditional product-oriented approach to a solution-oriented one—a major change that will certainly not be easy but which, for several reasons that we will outline in the next paragraph, has a good chance of taking place.

## 8.2.1 Towards 'solution-oriented' design

Today we know that society and industry have changed (and are still changing rapidly). We can see a network and knowledge economy emerging. Our production culture and our ideas of well-being itself are changing (Pine and Gilmore 1999; Rifkin 2000). Design, too, must abandon the product-oriented nature that has characterised it up to now.

This necessity becomes still more urgent if, given the considerations outlined here, we add the question of sustainability, with the pressing need to redefine our ideas of well-being (Manzini 2005) and of economy (Mont 2002; Stahel 1997) from a sustainability perspective. Now we know that this requires us to shift our focus of attention from products alone to product and service (product-service) systems and to solutions.

When designing, we must learn to initiate proceedings not with the products but with the problems for which these products and services may perhaps be the solution—that is, we must develop a 'solution-oriented' design that breaks its ties with the past and fits into the framework of an emerging knowledge society and, we hope, into a new culture and practice of sustainability.

This statement is more than just wishful thinking: this new idea of design is already taking shape and is beginning to emerge in the most advanced discussion, as exemplified in the following extract from a RED document, from the UK Design Council Research Centre:

> A new design discipline is emerging. It builds on traditional design skills to address social and economic issues. It uses the design process as a means to enable a wide range of disciplines and stakeholders to collaborate. It develops solutions that are practical and desirable. It is an approach that places the individual at the heart of new solutions, and builds the capacity to innovate into organisations and institutions (RED 2006).

This view, too, like that proposed by the ICSID, comes from an authoritative organism. The Design Council is one of the oldest organisations in the design world, by tradition cautious to attribute new meaning to the concept of design. This is precisely why the extract quoted is so provocatively stimulating: it gives a very open view of what we must understand by design and of what design can do. Above all, as far as our present interest is concerned, it shows a vision that is not dominated by a product-oriented mentality.

The Design Council tells us that the new design discipline 'builds on traditional design skills to address social and economic issues'; it does not refer to products, services or other artefacts. It speaks of problems to resolve: problems that may require the conception and development of 'solutions that are practical and desirable': new solutions, the purpose of which is to build 'the capacity to innovate into organisations and institutions'. In other words, according to the Design Council, the ultimate aim of this new design discipline is to foster processes of diffuse co-designing in organisations and institutions.

Moving on from these considerations we can conclude this section by proposing the following updated working definition of design: design is a reflective activity aiming to solve problems by developing solutions, where by 'solution' we mean any artefact able to resolve a problem, offered in a 'practical and desirable' way, in other words, in such a way that it can be produced (within a given technical-economic system) and appreciated (in a more socially acceptable framework).

## 8.3 State of the art: insights into the contribution of design in the transition to sustainable consumption and production

This section reviews how the discipline of design for sustainability has enlarged its scope and field of action over time, as observed by various authors (Charter and Tischner 2001; Karlsson and Luttrop 2006; Rocchi 2005; Ryan 2004; Vezzoli 2006).

Historically, since the environment question was raised during the second half of the 20th century, the approach of humanity has moved from damage-remedy policies (an 'end-of-pipe' approach) to actions increasingly aimed at prevention. In other words, we have moved from action and research focused exclusively on de-pollution systems, to research and innovation efforts that aim to reduce the cause of pollution at source (or, more generally, to reduce the environmental impact).[2] In this framework it is useful to trace briefly some fundamental levels in the ongoing interpretation of sustainability by the design world in order to gain a better understanding of the role of design research and practice in relation to sustainable consumption.

A first level on which numerous theorists and academics have been working is the selection of resources, advocating the use of materials and energy sources that have a low environmental impact. Fundamental requirements have been and still are that resources should be non-toxic, recyclable, biodegradable and renewable.

Since the second half of the 1990s attention has moved partially to the product level, to the design of products with a low environmental impact, usually referred to as product life-cycle design, ecodesign or design for environment (Brezet and van Hemel 1997; Heskinen 2002; ISO 2002; Keoleian and Menerey 1993; Manzini and Vezzoli 1998; van Nes and Cramer 2006; Ryan 2003; Sun 2003; Tischner *et al.* 2000; van Hemel 2001). In

---

2 The watchwords of the United Nations Environmental Programme (UNEP) and other institutions became 'cleaner production', defined as 'the continual redesigning of industrial processes and products to prevent pollution and the generation of waste, and risk for mankind and the environment' (UNEP 2000); to this has been more recently added the concept of 'sustainable consumption'.

those years, the environmental effects attributable to a product and how to assess them became clear.³ In particular two main concepts were introduced.

First, the concept of life-cycle thinking: from product design to the design of product life-cycle stages—that is, all the activities needed to produce the materials and then the product, plus those to distribute, use and, finally, dispose of it are considered as a single unit.

Second, the concept of functional thinking was recontextualised from an environmental point of view—that is, to design and evaluate the environmental sustainability of a product, beginning from its function rather than from the physical product itself.

Over the past few years, starting with a more stringent interpretation of sustainability—which tells us (Section 8.1) we must make radical changes to our production and consumption models—attention has moved partially to design for eco-efficient system innovation and therefore to a wider dimension than that of the single product (Bijma *et al.* 2001; Brezet 2003; Charter and Tischner 2001; Cooper and Sian 2000; Goedkoop *et al.* 1999; Hockerts 1998; Lindhqvist 2000; Manzini and Vezzoli 2001; Mont 2002; Scholl 2006; Stahel 1997; UNEP 2002; Zaring 2004). Within the wide debate on the definition of system innovation (see Chapters 18 and 22), design researchers have usually referred to the so-called product-service system (PSS). Among the several converging definitions, the one given by the United Nations Environment Programme (UNEP 2002: 4) says that a system innovation (referred to as a PSS) is 'the result of an innovative strategy that shifts the centre of business from the design and sale of (physical) products alone, to the offer of product and service systems that are together able to satisfy a particular demand'.

In this context, it has been argued (Vezzoli 2003a) even that the design conceptualisation process needs to shift from functional thinking to satisfactional thinking in order to emphasise and to be more coherent with the enlargement of the design scope from a single product to a wider system fulfilling a given demand of needs and wants (i.e. satisfaction). (On the use of 'satisfaction' as key terminology, see the following paragraphs and Marks *et al.* 2006; Meadows *et al.* 2006.)

Still more recently, design research has opened discussion on the possible role of design for social equity and cohesion (Carniatto and Chiara 2006; Carniatto *et al.* 2006; Crul 2003; Crul and Diehl 2006; EMUDE 2006; Guadagnucci and Gavelli 2004; Leong 2006; Maase and Dorst 2006; Mance 2003; Manzini and Jégou 2003; Margolin 2002; Penin 2006; Razeto 2004; Rocchi 2005; Tischner and Verkuijl 2006; Vezzoli 2003a; Weidema 2006⁵) and hence a potential role for a design directly addressing various aspects of social equity and cohesion, aiming at a 'just society with respect for fundamental rights and cultural diversity that creates equal opportunities and combats discrimination in all its forms' (EU 2006).

In fact, this four-dimensional interpretation of design for sustainability:

---

3 What the term 'environmental requirements of industrial products' means becomes clear following studies and new methods of assessing the environmental impact of input and output between the technosphere, geosphere and biosphere. Among others, the most accepted approach is life-cycle assessment (LCA).

1. Selection of resources with a low environmental impact
2. Design of products with a low environmental impact
3. Design for an eco-efficient system
4. Design for social equity and cohesion

does not necessarily represent a chronological evolution, neither does it define precise boundaries between one dimension and another. Nevertheless, it may be useful for a schematic understanding of the contribution of design to sustainability.

We will now see and discuss further how some of these dimensions seems to be more consistent with the radical changes required by sustainability.

## 8.4 Model of change: a system design approach to sustainability

As mentioned above, by starting with a stringent interpretation of sustainability that requires a systemic discontinuity in production and consumption patterns (as we are doing here), attention has moved partly to a possible role for design for eco-efficient system innovation and for improvements to social equity and cohesion: wider dimensions than that of considering a single product or the materials and energy it consumes. This is the effective model of change the design world should work for, seeking to identify the corresponding (design) skills. We will focus on these two dimensions in the following paragraphs.

### 8.4.1 Design for eco-efficient system innovation

Some authors have observed that the criteria for product life-cycle design meets obstacles in traditional supply models of product sale, especially when one is adopting a more stringent interpretation of environmental sustainability that requires a systemic discontinuity in production and consumption patterns (Cooper and Sian 2000; Goedkoop *et al.* 1999; Lindhqvist 2000; Manzini and Vezzoli 1998; Stahel 2001). For most design researchers, a more significant ambit in which to act to promote radical changes for sustainable consumption seemed to be the widening possibilities for innovation beyond the product, particularly with regard to innovation of the system as an integrated mix of products and services that together lead to the 'satisfaction' of a given demand for well-being (Bijma *et al.* 2001; Brezet 2001; Charter and Tischner 2001; Goedkoop *et al.* 1999; Manzini and Vezzoli 2001). Vezzoli (2003a) has argued the need to shift from a 'functional' to a 'satisfactional' approach (per unit of satisfaction). The term 'satisfaction' is proposed to emphasise the enlargement of the design scope from a single product to the system of products and services (and related stakeholders) that together fulfil a given demand in terms of needs and wants—in fact, for a given demand for satisfaction. The use of this terminology is in agreement with that of other authors. Meadows *et al.* (2006), in a 30-year update of their well-known book, *Limits to Growth*, commissioned by the Club of Rome, use 'satisfaction' in a formula to evaluate the limits of growth, modelling the consequences of a rapidly growing world population and finite supplies of resource.[4] Marks *et al.* (2006) argue that among the various indicators measuring personal well-being in the framework of a transition towards sustainability, satisfaction seems to be preferable.

These are innovations leading potentially 'to a system minimisation of resources, as a consequence of innovative stakeholder interactions and related converging economic interests' (UNEP 2002).[5] Commonly referred to in this context as PSSs, system innovations are shifting the centre of business from the design and sale of (physical) products alone to the offer of product and service systems that are together able to satisfy a particular demand. In other words, it is to some extent a shared opinion to talk about eco-efficient system innovation as that deriving from a new convergence of interest between the different stakeholders: innovation not only at a product (or semi-finished) level, but, above all, as new forms of partnership and interaction between different stakeholders, belonging to a particular value chain or value constellation (Normann and Ramirez 1995).

In reality, this interpretation of system innovation forms part of the foundations and criteria already expressed in life-cycle design (UNEP 2002). However, when this approach was adopted, it emerged even more clearly (as the basic assumption) that it was the reconfiguration of the system that constituted the starting point towards achieving certain results. The environmental value must in any case be assessed on the overall effects of the life-cycle of the products and services that make up the system on offer.

It has been observed that not all system innovations (PSSs) are eco-efficient (Cooper and Sian 2000; Mont 2004; Scholl 2006; Tucker and Tischner 2006; UNEP 2002; Zaring 2004). This means that when we design new systems (that have the potential to be radically sustainable) it is of key importance to adopt appropriate methods and tools that steer us towards a sustainable solution.

The introduction of system innovation (or PSS) into design (for sustainability) has led researchers to work on defining new skills of a more strategic nature (see Section 8.2 and the references quoted therein). The consequence of this understanding has been the identification of key issues. First, design must learn to develop environmentally sustainable products and services together. Second, an issue somewhat new to today's design culture and practice, design must learn to promote and facilitate new configurations (of partnership and interaction) between different 'stakeholders' to find innovative solutions able to lead to a convergence of economic and environmental interests. Last, there is a need for an ability to operate or facilitate a participatory design process among entrepreneurs, users, non-governmental organisations (NGOs), institutions and so on.

---

4 Resource and energy use per year = (number of people) × (satisfaction per person per year) × (resource and energy use per satisfaction)
5 To clarify this concept we can take the example used in a UNEP publication (UNEP 2002): given the 'satisfaction' in having clean clothes, we need not only a washing machine but also detergent, water and electricity (and the services that supply them) as well as maintenance, repair and disposal services. So, when we talk about system innovation, we mean an innovation that involves all the different socioeconomic stakeholders in this satisfaction system: the washing machine and detergent producers, the water and electricity suppliers and those responsible for maintenance and disposal.

To visualise the mode of approach it may be useful to draw a parallel (Vezzoli and Manzini 2006) with the design questions that more typically concern a 'traditional' designer, who in designing a product defines the technical, performance and aesthetic characteristics of its components and its connections in order to describe the configuration of the product components that are characterised by materials (with specific performance functions) and by their connection systems. In this way a systems designer for sustainability must imagine and promote innovative types of 'connections' (through partnership and interaction) between appropriate 'components' (socioeconomic stakeholders) in a system responding to a particular social demand for satisfaction. In other words, the components of a satisfaction system are characterised by socioeconomic stakeholders (with their own skills and abilities) and by the interaction occurring between them (partnerships, or, more generally, interaction). Therefore, in order to design a system configuration one must understand who are the best socioeconomic stakeholders to involve (the best 'components') and what are the best interrelationships to enter into (what 'connections' to use).

Calling to mind what was said in Section 8.2, these skills are part of so-called strategic design. Such considerations are leading towards a convergence of system design for eco-efficiency with both strategic design and life-cycle design. Therefore, it has been argued (Brezet 2001; Manzini and Vezzoli 2001) that design for environmental sustainability must use and integrate the methods and tools of strategic design (and vice versa). As far as design practice is concerned, the first design methods and tools have recently been developed (see Section 8.5).

### 8.4.2 Design for social equity and cohesion

As mentioned above, a debate has started on the potential responsibility for design in terms of improvements to social equity and cohesion. This is a role for design that addresses the sustainability of social equity and cohesion directly rather than indirectly as a potential result of a radical reduction in the resources available in industrialised contexts (as usually studied within the approach to system design for eco-efficiency).

Research on this topic involves taking into account (as in the assumptions of the concept of sustainable development) the so-called equity principle (UN 1992), whereby every person, given a fair distribution of resources, has a right to the same environmental space—that is, to the same availability of global natural resources or, better, to the same level of satisfaction that can be had from these in different ways. When the issue of sustainable consumption crosses that of socio-ethical sustainability, the spectrum of implications, of responsibilities, extends to several different issues such as the principles and rules of democracy, human rights and freedom, the achievement of peace and security, the reduction of poverty and injustice, improved access to information, training and employment, and respect for cultural diversity, regional identity and natural biodiversity (UN 2002). The EU Sustainable Development Strategy (EU 2006) defines a similar concept: here, social equity and cohesion is the promotion of 'a democratic, socially inclusive, cohesive, healthy, safe and just society with respect for fundamental rights and cultural diversity that creates equal opportunities and combats discrimination in all its forms'.

Some authors have argued that the socio-ethical and environmental dimensions are closely linked (Crul and Diehl 2006; Mance 2003; Rikfin 2002; Sachs and Tilman 2007;

Sachs *et al.* 2002; Vezzoli and Manzini 2006), when looking at the promising economic model that goes under the name of distributed economies. In short, on the one hand, a decentralised infrastructure 'fed' by renewable sources would reduce environmental impact and, on the other, could facilitate a democratisation of resources and energy, enabling individuals, communities and nations to reclaim their independence while accepting the responsibility that derives from their reciprocal interdependence (self-sufficiency and interdependence).

Before describing the debate on a potential role for design that directly addresses social equity and cohesion, it is worth noting that this is not just a matter of eradicating poverty but, more widely, of facilitating an improvement in the quality of life, which is not a matter of concern only for developing and emerging countries (Marks *et al.* 2006).

Furthermore, a socio-ethical dimension has started to be recognised as an issue to deal with in the global market, even by companies in industrialised economies, because the stakeholders are part of their supply chain. In this context it is worth mentioning corporate social responsibility (CSR), social accountability (Social Accountability 8000 [SA8000])[6] and sustainability reporting guidelines (GRI 2006).

Within the design debate, it has to be remembered that the theoretical contributions on sustainable consumption are not necessarily all recent. By the end of the 1960s, for various reasons, the theory and culture of design anticipated a critique of consumption patterns, or at least some of the leading figures in the culture of design acted as spokespeople for issues relating to the responsibility of designers for consumption patterns (although in different ways and not directly and exclusively associated with environmental impact). We can recall the criticism of consumer society made in denouncements by some exponents of 'radical design' (Birelli *et al.* 1972; Dalisi 1972), on the one hand, and the reaction of Maldonado(1970), on the other, who appealed to a new 'design hope'. The question of designer (social) responsibility was again brought up at the beginning of the 1970s (though never resolved in its implications for design practice). Papanek and Maldonado express similar positions, with regard to the role of consumption. Papanek (1970) writes: 'design can and must become a means for young people to take part in the transformation of society'.

Recently, the design area has opened (or reopened) the discussion on the possible role of design with respect to general questions associated with the various forms of social injustice. We can observe new, although sporadic, interest on the part of design research to move into this territory, to trace its boundaries and to understand its possible implications (Carniatto *et al.* 2006; Leong 2006; Maase and Dorst 2006; Manzini and Jégou 2003; Margolin 2002; Penin 2006; Rocchi 2005; Tischner and Verkuijl 2006; Vezzoli 2003a; Weidema 2005). This is an extremely vast and complex issue and its implications for design have so far been little analysed (and are difficult to face without falling into easy, and hardly constructive, moralism). However, it would seem necessary to address this question when dealing with sustainable consumption.

Within this context, the 'real' effort started in 2000 by UNEP is symptomatic. UNEP set up a group of international researchers (from industrialised, emerging and developing countries)[7] to disseminate worldwide the concept of system innovation and to start

---

6 See www.sa-intl.org.

7 The work involved a group of researchers (including the authors of this chapter) from several countries in the more or less industrialised world; it was set up in 2000 and ended with a publication in 2002 presenting the main achievements (UNEP 2002).

exploring the issue, which can be summed up in the following question: is system innovation (PSS) also applicable in emerging and developing contexts? The question was put simply because what had been so far studied, said and acquired on the development of PSSs concerned only the environmental and economic aspects and mature industrialised contexts; it did not refer to the socio-ethical dimension or to emerging and developing countries or contexts. This question has been the forerunner of another: (if the answer to the first is affirmative) can a system approach favour the social equity and cohesion qualification of these contexts as well as their eco-efficiency, and, if so, with what particular characteristics?

The response of the above-mentioned international group of experts to these questions was the creation of the following hypothesis: 'PSS (system innovation) may act as a business opportunity to facilitate the process of social-economical development of emerging context—by jumping over or by-passing the stage characterised by individual consumption/ownership of mass produced goods—towards the more advanced service-economy "satisfaction-based" and low resources intensive' (UNEP 2002: 3).[8] This hypothesis has also been examined in a series of case studies, collected by the group engaged by UNEP (e.g. see Box 8.2).

### Box 8.2 Greenstar e-commerce and community centres

Of the case studies examined by the United Nations Environment Programme we report here that of Greenstar solar e-commerce and community centres, with bases, so far, in India (Parvatapur), Jamaica and Ghana. These centres are modular, scalable, highly portable 'stations' delivered to villages in the developing world as e-commerce centres. They are solar-powered and connected to the Internet through a satellite or digital modem. Residents of remote rural communities can sell their wares worldwide and become shareholders in Greenstar.

A spin-off from this UNEP project was the setting-up in 2003 of an international network of design higher education institutions (HEIs), the Learning Network on Sustainability (for the period December 2007 to December 2010, funded by the Asia Link Programme, European Commission),[9] covering industrialised and emerging countries.

---

8 The following opportunities for environmental and socio-ethical sustainability, in development and system innovation, have been highlighted (UNEP 2002). First, if PSSs are eco-efficient at the system level it means that they may represent opportunities, at least at a macro level, for a context with fewer economic possibilities to respond more easily to unsatisfied social demands. Second, PSSs are focused more on the context of use, because they do not sell products only but open relationships with the end-user. For this reason, an increased offer in these contexts should trigger greater involvement of (more competent) local, rather than global stakeholders, fostering and facilitating a reinforcement of the local economy. Third, since PSSs are more labour- and relationship-intensive they can also lead to an increase in local employment and a consequent dissemination of skills. Last, since the development of PSSs is based on the building of system relationships and partnerships they are consistent with a democratic re-globalisation process.

9 The Learning Network on Sustainability (LENS) involves the following (the first seven are partners in the Asia Link Programme): Shritshi School of Art, Design and Technology, India; Department of Industrial Design, Faculty of Architecture, King Mongkut's Institute of Technology, Thailand; University of Art and Design, Helsinki, Finland; Faculty of Design, Politecnico di Milano University, Italy;

The network scope has been, since its establishment, to explore the above hypothesis further within the increasingly interconnected and multicultural context of globalisation, paying particular attention to the promising economic model of distributed economies mentioned above. Examples of this are distributed energy generation and distributed computing (peer-to-peer); each of these has the potential to reduce environmental impact while facilitating the democratisation of access to resources. In terms of sustainable consumption and system innovation, the research–education hypothesis adopted by the network is that of a 'system design for sustainability approach with a role in the promotion/development of local-based and network-structured initiatives'. Compared with the 2002 UNEP hypothesis, the LENS (Learning Network on Sustainability) revised hypothesis is as follows: 'a system Innovation (PSS approach) may act as a business opportunity to facilitate the process of social-economical development in an emerging context—by jumping over or by-passing the stage characterised by individual consumption/ownership of mass produced goods—towards a more "satisfaction-based" and low resource intensity advanced service-economy, [from here the added part] characterised by the development of local-based and network-structured enterprises and initiatives, for a sustainable re-globalisation process aiming at a democratisation of access to resources, goods and services' (www.lens.polimi.it [accessed 25 August 2007]).

Another relevant HEI network founded by the European Commission on design for sustainability is the 'Emerging User Demands for Sustainable Solutions' (EMUDE) network,[10] which has formulated a similar hypothesis. Its aim is to explore the potential of social innovation as a driver for technological and production innovation, in view of sustainability. More than 140 cases throughout Europe have been collected and made available. The resulting picture shows the regeneration of the local social fabric based on an enlarged 'family', common local resources, active neighbourhood life, additional circles of relationships and enthusiastically participating people. Within this promising context, the understanding of the project on a design role level is that 'strategic design could contribute to the dissemination of these solutions making them more accessible to a larger audience but, at the same time, trying to keep their initial relational qualities' (EMUDE 2006).

Within system innovation theories we can find coherence with the well-defined and mature research group working on so-called 'transition management for sustainable consumption and production', for whom a process of 'visioning' and experimentation is essential, where niche experiments with new practices and systems could form stepping stones to potential future sociotechnical innovations (see Chapters 18 and 22).

---

Academy of Art and Design, Tsinghua University, China; Indian Institute of Technology, New Delhi, India; School of Architecture and Urbanism, Universidade de São Paulo, Brazil; School of Design, Hong Kong Polytechnic University, China; Department of Industrial Product Design, Istanbul Technical University, Turkey; Industrial Design Department, Technical University of Delft, The Netherlands; Industrial Design Department, Federal University of Parana in Curitiba, Brazil; Faculty of the Built Environment, Industrial Design Programme, University of New South Wales, Australia; and Design Academy, Eindhoven, The Netherlands.

10 The 'Emerging User Demands for Sustainable Solutions' (EMUDE) network is a European design higher education network funded by the European Commission that involves the following universities: Academy of Fine Arts, Krakow, Poland; ENSCI, Les Ateliers, Paris, France; Estonian Academy of Fine Arts, Tallinn, Estonia; Politecnico di Milano, Milan, Italy; School of Design, Glasgow School of Art, Glasgow, Scotland; School of Design, University of Applied Science, Cologne, Germany; Technische Universiteit Eindhoven, The Netherlands; University of Art and Design, Helsinki, Finland.

A common element of the two networks is that the possible role for design (a possible design operativeness) has to be researched on a system or strategic level. This ability can be described as the ability to: promote or facilitate new locally based network-structured sustainable enterprise (Vezzoli and Manzini 2006): that is, to find scenarios for building up partnerships or interaction between different stakeholders in order to aim at sustainable value production, and to facilitate participatory design among different stakeholders, defining relationships and offering alternative systems (products, services and communication). The required skills, abilities and tools may be derived from those recently developed for eco-efficient system design and will necessarily have to be reinterpreted and adapted to the specific requirements of social equity and cohesion.

On the role of design for sustainability to include a socio-ethical dimension, especially regarding that which concerns developing economies, other authors (Crul and Diehl 2006; under the UNEP umbrella of the Division Of Technology, Industry and Economics) have argued that, in developing economies, because of limited awareness of the concept, more immediate technical support is needed to introduce the design for sustainability concept (even on a product innovation level). However, successful implementation of design for sustainability requires working in partnership.

With regard to social impact other authors (Weidema 2005) are investigating the option of extending product LCA beyond the environmental impact to the social (and economic) impact, which is, in principle, more closely linked to the product innovation level. In fact, these studies also lead to system innovation, since they focus on the definition of equitable roles for stakeholders (social indicators).

Finally, system design for environmentally, socially and ethically sustainable innovations represents a ground on which new design roles (as well as methods and tools) have already been defined, as well as being a debated and complex ground on which it is possible to find compelling disciplinary answers to the hypotheses of new potential design roles.

The next section will analyse the related limitations and potential of the role of design.

## 8.5 Limitations and potential of the role of design

First, we describe the limitations in terms of what the designers (and their educators) are really doing in relation to the more or less consolidated potential positive contribution they could make regarding a reorientation towards sustainable consumption and production. It is important to note that designers who are unaware of sustainable criteria and methods (or who may be regarded as irresponsible) may in fact contribute to increased negative impacts by designing products and/or services with a high environmental impact.

In fact, designers have a role to communicate more or less directly through the product, the services or the system. In other words, they may make a contribution in influencing sustainable (or unsustainable) consumption patterns. When speaking about sustainable consumption it is of particular significance to underline that designers may help launch products and services on the market that are culturally or aesthetically obsolescent (or, more generally, of high environmental impact)—that is, products that are designed to have a short life-span and so increase resource flows in production and consumption to fulfil a given demand for 'satisfaction' over time. This negative contribution to sustainable consumption may also be amplified by other communication and brand mechanism designs, with the result of enhancing unsustainable user behaviour.

Having said this, let us describe what the design world is actually doing in relation to making a (more or less consolidated) potential positive contribution towards reorienting towards sustainable consumption and production. For a succinct understanding of the limitations and potential of design for sustainability (as discussed in the previous two sections), in research, education and practice, we can chart, on the one hand, the level of disciplinary consolidation (derived from design research achievements) and, on the other, the level of dissemination of this (in design practice and education).

In Figure 8.1, we find the new research frontiers depicted in the bottom left-hand corner (0% consolidation and dissemination); and in the top right-hand corner (100% consolidation and dissemination)—the point towards which we should steer the various dimensions of the discipline (i.e. towards a high degree of consolidation and widespread dissemination in [education and] practice).

FIGURE 8.1 **The state of the art in terms of consolidation and dissemination limits and potential of different design for sustainability dimensions**

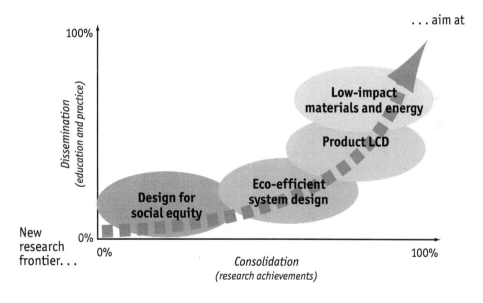

LCD = life-cycle design

Let us refer to the four dimensions we used in Sections 8.3 and 8.4 to schematise the evolution of sustainability into the design discipline: selection of resources with low environmental impact, product life-cycle design or ecodesign, design for eco-efficient systems (PSSs) and design for social equity and cohesion.

In this chart, we can position the selection of low-impact material and energy supplies and the life-cycle design or ecodesign of the product at a good level of consolidation, a discreet level of penetration in education but at a low level in practice (especially when considering the life-cycle design or ecodesign of the product). In fact, at present, as far as the life-cycle design of product is concerned, there is a clear theoretical framework and clear (environmental) requirements as well as available criteria, methods and tools (Crul and Diehl 2006; Diehl and Brezet 2003; Simon et al. 1998; Tischner et al. 2000; Brezet and van Hemel 1997; Vezzoli and Sciama 2006). In terms of dissemination, appropriate methods and tools are starting to be diffused in design HEIs but are not yet well diffused (in terms of integrated methods and tools) in design practice (Tukker and Eder 2000; Costa and Vezzoli 2005).

Let us now focus on eco-efficient system innovation design and design for social equity and cohesion, since these are the most significant ambits the design world should approach and work in, seeking to identify the correlated (design) skills. As far as eco-efficient system design is concerned, the level of consolidation is inferior (compared with low-impact product life-cycle design or ecodesign) and the design practice is, logically, far more sporadic. The first design methods and tools have recently been developed as outcomes of EU-funded projects as part of the 5th Framework Programme, such as:

- PROSECCO (product and service co-design [PSCD] process)
- HiCS (Highly Customerised Solutions; Manzini et al. 2004)
- MEPSS (Method for PSS Development; van Halen et al. 2005)

Examples include design tools for the development of sustainability design-orienting scenarios (SDOS) and service notation tools for the strategic convergence of different stakeholders towards sustainable solutions (Jégou 2006) as well as qualitative brainstorming tools focused on the devising of partnerships and interaction between eco-efficient valid stakeholders (Vezzoli and Tischner 2005).[11]

Nevertheless, further research in design is needed to obtain: (a) a better understanding of system design for eco-efficiency potential and limits (where the designer competences can really come into their own) and (b) related methods and tools for design (and system assessment). However, we are now starting to see the first courses being inserted in design curricula of some European HEIs.[12]

With regard to design for social equity and cohesion, little has been elaborated, either at an operative level or in terms of defining the real design potential and boundaries of action (this is a new design research frontier; see Section 8.4 for a hypothesis on potential roles for design).

In fact, very few are the tools and methods developed to orient the system design process towards socio-ethically sustainable solutions.[13] One such tool, however, is found

---

11 Criteria that achieve and determine (or otherwise) system eco-efficiency have been drawn up inside a design tool called the Sustainability Design-Orienting (SDO) toolkit, developed as part of the MEPSS research. Free use and access is available at: www.mepss-sdo.polimi.it [accessed 25 August 2007].
12 Among others, the Technical University at Delft in The Netherlands is running a course titled 'PSS Design', and the Politecnico di Milano is running a course titled 'System Design for Sustainability'.
13 Among such tools we could mention the section on socio-ethical sustainability in the Sustainability Design-Orienting (SDO) toolkit (www.mepss-sdo.polimi.it [accessed 25 January 2008]), being part of the MEPSS research.

in the results of the EU-backed MEPSS research, which attempted to bring the complexity of the social equity and cohesion issue in line with a possible design activity (Tischner and Verkuijl 2006). A design-for-sustainability practical approach for developing economies has been developed as a joint effort between the UNEP and Technical University of Delft (Crul and Diehl 2006), proposing a step-by-step approach, considering environmental, socio-ethical and economic aspects.

As a consequence of the dearth of such tools and methods there are obviously very few teaching proposals. In synthesis, this is indeed a very complex and as yet unconsolidated area of knowledge where the real possibility for design action has to be defined and the conditions for this to happen have to be built up.

Along this complex path, as the HEI design networks EMUDE and LENS (see Section 8.4.2) are witnessing, there is a growing opinion that design HEIs could have a key role to play: in this task they should rethink themselves as places for advanced research and education as well as diffusers of sustainable ideas.

Some authors (Penin 2006; Vezzoli 2006) have moved even further forward by proposing that HEIs could be promoters of 'incubators', whose aim should be to create enterprises that involve several local stakeholders and that promote long-term local development, with local, regenerative, low-resource-intensity solutions. Penin has adapted some strategic design tools for this end and has tested them in a project for the Brazilian network of Popular Cooperative Technological Incubators (ITCP). These are part of the university corpus and have the aim of creating new jobs for groups of people with low incomes, from a perspective of local development and in a context of solidarity economy, through the dissemination of technical and scientific know-how produced within the university.

## 8.6 Conclusions

When talking about design for sustainable consumption and production it is common to assume that radical changes have to happen and that these may be better pursued through system innovation. We have seen how part of the pathway has been trodden in design research, especially for what concerns eco-efficient system design, though much less for social equity and cohesion.

We anticipate fundamental years for the understanding and consequent promotion of new and credible design roles, methods and tools to contribute to the development of sustainable consumption and production system innovations—innovations that are not only technical but also social (and ethical), as the transition towards sustainable development requires. In fact, the potential role of design for sustainable consumption and production has contributed to putting (traditional) design patterns themselves under discussion and has set a still very open and challenging debate on how the design discipline could tackle this issue. Indubitably, this is a challenge that the 'design world' must come to terms with in a more and more cross-disciplinary arena, and indeed this is one of the leading principles of the SCORE! project.

# References

Appadurai, A. (1996) *Modernity at Large: Cultural Dimensions of Globalisation* (India: Oxford University Press).

Baumann, Z. (2000) *Liquid Modernity* (Cambridge, UK: Polity Press; Oxford, UK: Basil Blackwell).

Beck, U. (1999) *Che cos'è la globalizzazione: Rischi e prospettive della società planetaria* (Rome: Carocci).

Bijma, A., M. Stuts and S. Silvester (2001) 'Developing Eco-efficient Product-Service Combinations', in *Proceedings of the 6th International Conference 'Sustainable Services and Systems: Transition towards Sustainability?'*, Amsterdam, The Netherlands, October 2001.

Birelli, G., C. Caldini, F. Fiumi and P. Galli (1972) '9999', in *Casabella* 361.

Brezet, H. (2001) *The Design of Eco-efficient Services: Methods, Tools and Review of the Case Study Based 'Designing Eco-efficient Services' Project* (report; The Hague: VROM).

—— (2003) *Product Development with the Environment as Innovation Strategy: The Promise Approach* (Delft, Netherlands: Delft University of Technology).

—— and C. van Hemel (1997) *Ecodesign: A Promising Approach to Sustainable Production and Consumption* (Paris: UNEP).

Burns, C., H. Cottam, C. Vanstone and J. Winhall (2006) *Transformation Design, Creative Commons* (London: Design Council).

Carniatto, V., and E. Chiara (2006) 'Design for a Fair Economy', in *Proceedings of P&D Conference*, Cutritiba, Brazil, 2006.

——, F.V. Carneiro and D.M.P. Fernandes (2006) 'Design for Sustainability: A Model for Design Intervention in a Brazilian Reality of Local Sustainable Development', in *Proceedings of the International Design Conference*, Dubrovnik, Croatia, 2006.

Castells, M. (1996) *The Rise of the Network Society* (Oxford, UK: Basil Blackwell).

Charkiewicz, E. (2002) 'Changing Consumption Patterns in Central and Eastern Europe', in United Nations Environment Programme (UNEP), *Industry and Environment Review* (Paris: UNEP).

Charter, M., and U. Tischner (eds.) (2001) *Sustainable Solutions: Developing Product and Services for the Future* (Sheffield, UK: Greenleaf Publishing).

—— et al. (2004) *Eco-design and Environmental Management in the Electronics Sector in China, Hong Kong and Taiwan* (Farnham, UK: The Centre for Sustainable Design/Pera Innovation).

Cooper T. (2005) 'Slower Consumption: Reflections on Product Life Spans and the "Throwaway Society" ', *Journal of Industrial Ecology* 9.1–2.

—— and E. Sian (2000) *Products to Services* (Friends of the Earth/Centre for Sustainable Consumption, Sheffield Hallam University, UK).

Costa, F., and C. Vezzoli (2005) 'Formazione e domanda di professionalità ambientali nel settore del disegno industriale: Relazione sullo stato dell'arte dell'insegnamento della disciplina dei Requisiti ambientali dei prodotti industriali e della presenza di competenze progettuali-ambientali nelle aziende e negli studi di design', in *Proceedings Design didattica e questione ambientale*, DIS-INDACO Politecnico di Milano, Milan, 2005.

Cottam, H., and C. Leadbeater (2004a) *Health: Co-creating Services* (London: Design Council, RED Unit).

—— and C. Leadbeater (2004b) *Open Welfare: Designs on the Public Good* (London: Design Council).

Cova, B. (1995) *Au-delà du marché: quand le lien importe plus que le bien* (Paris: L'Harmattan).

Crul, M. (2003) *Ecodesign in Central America* (Delft, Netherlands: Delft University of Technology).

—— and J.C. Diehl (2006) *Design for Sustainability: A Practical Approach for Developing Economies* (Paris: United Nation Environmental Programme/Technical University Delft).

Dalisi, R. (1972) 'La tecnica povera in rivolta', *Casabella* 365.

Deaglio, M. (2004) *Postglobal* (Rome: Laterza).

Diehl, J.C., and J.C. Brezet (2003) 'Ecodesign Methodology, Tools and Knowledge Transfer', in E. Corte-Real, C.A.M. Duarte and C. Rodrigues (eds.), *Senses and Sensibility in Technology: Linking Tradition to Innovation through Design* (Lisbon: IADE).

—— and J.C. Brezet (2004) 'Design for Sustainability: An Approach for International Development, Transfer and Local Implementation', in *EMSU 2004: Proceedings of Environmental Management for Sustainable Universities (EMSU)*, Monterrey, Mexico, 2004.

Douglas, M., and B. Isherwood (1979) *The World of Goods: Towards an Anthropology of Consumption* (New York: Basic Books).
EMUDE (Emerging User Demands for Sustainable Solutions) (2006) *Final Report, 6th Framework Programme (Priority 3-NMP), European Community*, www.sustainable-everyday.net.
EU (European Union) (2003a) *Study on External Environmental Effects Related to the Life-cycle of Products and Services: Final Report* (BIO Intelligence Service and O2 France; Brussels: European Commission Directorate General Environment, Directorate A: Sustainable Development and Policy Support).
—— (2003b) 'Directive 2002/95/CE for the Prohibition and Limitation of Use of Lead, Mercury, Cadmium, Chromium (VI) as well as Some Flame Retardants within Electric and Electronic Equipment', *Official Journal of the European Union*.
—— (2005) 'Directive 2005/32/CE: Establishing a Framework for the Setting of Ecodesign Requirements for Energy-using Products and Amending Council Directive 92/42/EEC and Directives 96/57/EC and 2000/55/EC of the European Parliament and of the Council', *Official Journal of the European Union*.
—— (2006) *Renewed Sustainable Development Strategy* (Publication 10117/06; Brussels: Council of the European Union).
Florida, R. (2002) *The Rise of the Creative Class, and How it is Transforming Work, Leisure, Community and Everyday Life* (New York: Basic Books).
Geels, F. (2002) *Understanding the Dynamics of Technological Transitions: A Co-evolutionary and Sociotechnical Analysis* (PhD thesis; Enschede, Netherlands: University of Twente).
—— and R. Kemp (2000) *Transitions from a Socio Technological Perspective* (Report for VROM: Enschede, Netherlands: University of Twente; Maastricht, Netherlands: MERIT).
Geyer-Alley, E. (2002) 'Sustainable Consumption: An Insurmountable Challenge', in *Industry and Environment Review* (Paris: United Nations Environment Programme [UNEP]).
Giddens, A. (2000) *Runaway World: How Globalisation is Reshaping Our Lives* (London: Profile Books).
Goedkoop, M., C. van Halen, H. te Riele and P. Rommes (1999) *Product Services Systems: Ecological and Economic Basics* (Report 1999/36; The Hague: VROM).
GRI (Global Reporting Initiative) (2006) *G3 Sustainability Reporting Guidelines* (Amsterdam: GRI).
Guadagnucci, L., and F. Gavelli (2004) *La crisi di crescita: Le prospettive del commercio equo e solidale* (Milan: Feltrinelli).
Heskinen, E. (2002) 'The Institutional Logic of Life-cycle Thinking', *Journal of Cleaner Production* 10.5.
Hockerts, K. (1998) 'Eco-efficient Service Innovation: Increasing Business-Ecological Efficiency of Products and Services', in M. Charter (ed.), *Greener Marketing: A Global Perspective on Greener Marketing Practice* (Sheffield, UK: Greenleaf Publishing): 95-108.
ICSID (International Council of Societies of Industrial Design) (2005) 'Definition of Design', www.icsid.org.
ISO (International Organisation for Standardisation) (2002) *Environmental Management: Integrating Environmental Aspects into Product Design and Development* (ISO/TR 14062: 2002[E]; Geneva: ISO).
Jégou, F. (2006) 'Service Notation Tools to Support Strategic Conversation for Sustainability', in *Proceedings: Perspectives on Radical Changes to Sustainable Consumption and Production (SCP), Sustainable Consumption Research Exchange (SCORE!) Network*, Copenhagen, April 2006.
Joore, P. (2006) 'Guide Me: Translating A Broad Societal Need into a Concrete Product Service Solution', in *Proceedings: Perspectives on Radical Changes to Sustainable Consumption and Production (SCP), Sustainable Consumption Research Exchange (SCORE!) Network*, Copenhagen, April 2006.
Karlsson, R., and C. Luttrop (2006) 'EcoDesign: What Is Happening? An Overview of the Subject Area of EcoDesign', *Journal of Cleaner Production* 14.15-16.
Keoleian, G.A., and D. Menerey (1993) *Life-cycle Design Guidance Manual: Environmental Requirements and the Product System* (Washington, DC: US Environmental Protection Agency).
Klein, N. (2002) *Fences and Windows: Dispatches from the Front Lines of the Globalization Debate* (New York: Picador).
Landry, C. (2000) *The Creative City: A Toolkit for Urban Innovators* (London: Earthscan Publications).
Leong, B. (2006) 'Is a Radical Systemic Shift toward Sustainability Possible in China?', in *Proceedings: Perspectives on Radical Changes to Sustainable Consumption and Production (SCP), Sustainable Consumption Research Exchange (SCORE!) Network*, Copenhagen, April 2006.

Lewis, H., J. Gertsakis, T. Grant, N. Morelli and A. Sweatman (2001) *Design + Environment: A Global Guide to Designing Greener Goods* (Sheffield, UK: Greenleaf Publishing).

Lindhqvist, T. (2000) *Extended Producer Responsibility in Cleaner Production* (PhD thesis; Lund, Sweden: International Institute for Industrial Environmental Economics [IIIEE], Lund University).

Maase, S., and K. Dorst (2006) 'Co-creation: A Way to Reach Sustainable Social Innovation?', in *Proceedings, Perspectives on Radical Changes to Sustainable Consumption and Production (SCP), Sustainable Consumption Research Exchange (SCORE!) Network*, Copenhagen, April 2006.

Maldonado, T. (1970) *La speranza progettuale: Ambiente e società* (Turin: Einaudi).

—— (1974) *Disegno industriale: Un riesame* (Milan: Peltrinelli).

Mance, E. (2003) *La rivoluzione delle reti: L'economia solidale per un'altra globalizzazione* (Bologna, Italy: EMI).

Manzini, E. (2005) 'Enabling Solutions for Creative Communities: Social Innovation and Design for Sustainability', *Design Matters* 10.

—— and F. Jégou (2003) *Sustainable Everyday: Scenarios of Urban Life* (Milan: Edizioni Ambiente).

—— and C. Vezzoli (1998) *Lo sviluppo di prodotti sostenibili* (Rimini, Italy: Maggioli editore).

—— and C. Vezzoli (2001) 'Strategic Design for Sustainability', in *TSPD Proceedings*, Amsterdam, The Netherlands, September 2001.

——, L. Collina and S. Evans (eds.) (2004) *Solution-Oriented Partnership* (Cranfield, UK : Cranfield University).

Marchand, A., and S. Walzer (2006) 'Designing Alternatives', in *Proceedings: Perspectives on Radical Changes to Sustainable Consumption and Production (SCP), Sustainable Consumption Research Exchange (SCORE!) Network*, Copenhagen, April 2006.

Margolin, V. (2002) The Politics of the Artificial (Chicago: University of Chicago Press).

Marks, N., S. Abdallah, A. Simms and S. Thompson (2006) *The (Un)Happy Planet. An Index of Human Well-being and Environmental Impact* (London: New Economics Foundation/Friends of the Earth).

Meadows, D., D. Meadows and J. Randers (2006) *Limits to Growth: The 30-year Update* (White River Junction, VT: Chelsea Green).

Mont, O. (2002) *Functional Thinking: The Role of Functional Sales and Product Service Systems for a Functional Based Society* (Research Report for the Swedish Environmental Protection Agency; Lund, Sweden: International Institute for Industrial Environmental Economics [IIIEE], Lund University).

—— (2004) *Product-Service Systems: Panacea or Myth?* (PhD thesis; Lund, Sweden: International Institute for Industrial Environmental Economics [IIIEE], University of Lund).

Normann, R., and R. Ramirez (1995) *Le strategie interattive d'impresa: Dalla catena alla costellazione del valore* (Milan: Etas Libri).

Papanek, V. (1970) *Design for the Real World* (Milan: Arnaldo Mondatori Editore).

Penin, L. (2006) *Strategic Design for Social Sustainability in Emerging Contexts* (PhD thesis; Milan: Politecnico di Milano).

—— and C. Vezzoli (2004) 'Campus: "Lab" and "Window" for Sustainable Design Research and Education: The DECOS Educational Network Experience', in *Proceedings of EMSU 2004: Environmental Management for Sustainable Education*, Monterrey, Mexico, June 2004.

Pine, J.B., and J.H. Gilmore (1999) *The Experience Economy: Work is Theatre and Every Business a Stage* (Boston, MA: Harvard Business School Press).

Razeto, L. (2004) *Le imprese alternativ: Principi e organizzazione delle economie solidali* (Bologna, Italy: EMI).

Ray, P.H., and S.R. Anderson (2000) *The Cultural Creatives: How 50 Million People are Changing the World* (New York: Three Rivers Press).

RED (2006) *Transformation Design* (RED Paper 02; London: Design Council).

Rifkin, J. (2000) *The Age of Access: How to Shift from Ownership to Access in Transforming Capitalism* (London: Penguin Books).

—— (2002) *The Hydrogen Economy: The Creation of the Worldwide Energy Web and the Redistribution of Power on Earth* (New York: Penguin).

Rocchi, S. (2005) *Enhancing Sustainable Innovation by Design: An Approach to the Co-creation of Economic, Social and Environmental Value* (PhD thesis; Rotterdam: Erasmus University).

Rocha, C., R. Frazao, M. Zackrisson and K. Christiansen (2006) 'The Use of Communication Tools and Policy Instruments to Facilitate Changes towards Sustainability', in *Proceedings: Perspectives on Radical Changes to Sustainable Consumption and Production (SCP), Sustainable Consumption Research Exchange (SCORE!) Network*, Copenhagen, April 2006.

Ryan, C. (2003) 'Learning from a Decade (or so) of Eco-Design Experience: Part 1', *Journal of Industrial Ecology* 7. 2.

—— (2004) *Digital Eco-sense: Sustainability and ICT: A New Terrain for Innovation* (Melbourne: Lab 3000).

—— (2006) 'Eco-innovative Cities', in *Proceedings: Perspectives on Radical Changes to Sustainable Consumption and Production (SCP), Sustainable Consumption Research Exchange (SCORE!) Network*, Copenhagen, April 2006.

Sachs, W., and S. Tilman (eds.) (2007) *Fair Future: Limited Resources and Global Justice* (London: Zed Books).

—— et al. (2002) *The Jo'burg-Memo: Fairness in a Fragile World. Memorandum for the World Summit on Sustainable Development* (Berlin: Heinrich Böll Foundation; www.boell.de).

Sangiorgi, D. (2005) 'Interaction Story Board', in C. van Halen, C. Vezzoli and R. Wimmer (eds.), *Methodology for Product Service System: How to Develop Clean, Clever and Competitive Strategies in Companies* (Assen, Netherlands: Van Gorcum).

Scholl, G. (2006) 'Product Service Systems', in *Proceedings: Perspectives on Radical Changes to Sustainable Consumption and Production (SCP), Sustainable Consumption Research Exchange (SCORE!) Network*, Copenhagen, April 2006.

Sen, A. (1999) *Development as Freedom* (New York: Anchor Books).

Simon, H.A. (1969) *The Sciences of the Artificial* (Cambridge, MA: MIT Press).

Simon, M., et al. (1998) *Ecodesign Navigator: A Key Resource in the Drive Towards Environmentally Efficient Product Design* (Cranfield, UK: Manchester Metropolitan University/Cranfield University/EPSRC).

Soumitri, V., and C. Vezzoli (2002) 'Product Service System Design: Sustainable Opportunities for All. A Design Research Working Hypothesis: Clothing Care System for Kumaon Hostel at IIT Delhi', in *Proceedings: Ecodesign International Conference*, New Delhi, 2002.

Stahel, W. (1997) 'The Functional Economy: Cultural Change and Organisational Change', in D.J. Richards (ed.), *The Industrial Green Game* (Washington, DC: National Academic Press).

—— (2001) 'Sustainability and Services', in M. Charter and U. Tischner (eds.), *Sustainable Solutions: Developing Products and Services for the Future* (Sheffield, UK: Greenleaf Publishing): 151-64.

Stiglitz, J. (2002) *Globalization and Its Discontents* (New York: Norton).

Sun, J. (2003) 'Design for Environment: Methodologies, Tools, and Implementation', *Journal of Integrated Design and Process Science* 7.1.

Tischner, U., et al. (2000) *Was ist EcoDesign?* (Basel, Switzerland: Birkhäuser Verlag).

—— and M. Verkuijl (2006) 'Design for (Social) Sustainability and Radical Change', in *Proceedings: Perspectives on Radical Changes to Sustainable Consumption and Production (SCP), Sustainable Consumption Research Exchange (SCORE!) Network*, Copenhagen, April 2006.

Tukker, A., and P. Eder (2000) *Eco-design: European State of the Art* (Brussels: ECSC/EEC/EAEC).

—— and U. Tischner (eds.) (2006) *New Business for Old Europe: Product-Service Development, Competitiveness and Sustainability* (Sheffield, UK: Greenleaf Publishing).

UN (United Nations) (2002) *World Summit on Sustainable Development: Draft Political Declaration, Submitted by the President of the Summit*, Johannesburg, 2002.

—— (1992) *Report of the United Nations Conference on Environment and Development, Rio de Janeiro, 3–14 June 1992, Annex I, Rio Declaration on Environment and Development, UN General Assembly*, Rio de Janeiro, 1992.

UNDSD (United Nations Division for Sustainable Development) (2001) *Indicators of Sustainable Development: Guidelines and Methodologies* (New York: UNDSD).

UNEP (United Nations Environment Programme) (2000) *Achieving Sustainable Consumption Patterns: The Role of the Industry* (Paris: UNEP IE/IAC).

—— (2002) *Product-Service Systems and Sustainability: Opportunities for Sustainable Solutions* (Paris: UNEP, Division of Technology Industry and Economics, Production and Consumption Branch).

UNESCO (2005) *Our Creative Diversity* (Paris: UNESCO Publishing).
UNFPA (United Nations Population Fund) (2006) 'State of World Population, 2006', Thoraya Ahmed Obaid, Executive Director, UNFPA.
Van Halen, C., C. Vezzoli and R. Wimmer (eds.) (2005) *Methodology for Product Service System: How to Develop Clean, Clever and Competitive Strategies in Companies* (Assen, Netherlands: Van Gorcum).
Van Hemel, C.G. (2001) 'Design for Environment in Practice: Three Dutch Industrial Approaches Compared', in *4th Ntva Industrial Ecology Seminar and Workshop: Industrial Ecology, Methodology and Practical Challenges in Industry*, Trondheim, Norway, June 2001.
Van Nes, N., and J. Cramer (2006) 'Product Life-time Optimisation: A Challenging Strategy towards More Sustainable Consumption Patterns', *Journal of Cleaner Production* 14.15–16.
Vezzoli, C. (2003a) *Systemic Design for Sustainability* (Cumulus working paper; Helsinki: University of Art and Design [UIAH]).
—— (2003b) 'A New Generation of Designer: Perspective for Education and Training in the Field of Sustainable Design. Experiences and Projects at the Politecnico di Milano University', *Journal of Cleaner Production* 11.1.
—— (2006) 'Design for Sustainability: The New Research Frontiers', *Da Vinci Journal* 3.2.
—— and E. Manzini (2006) 'Design for Sustainable Consumption', in *Proceedings: Perspectives on Radical Changes to Sustainable Consumption and Production (SCP), Sustainable Consumption Research Exchange (SCORE!) Network*, Copenhagen, April 2006.
—— and L. Penin (eds.) (2005) *Designing Sustainable Product-Service Systems for All* (Milan: Libreria Clup; clup@galactica.it).
——and D. Sciama (2006) 'Life-cycle Design: From General Methods to Product Type Specific Guidelines and Checklists: A Method Adopted to Develop a Set of Guidelines/Checklist Handbook for the Eco-efficient Design of NECTA Vending Machines', *Journal of Cleaner Production* 14.15–16.
—— and G.V. Soumitri (2002) 'Sustainable Product-Service System Design for All: A Design Research Working Hypothesis', in *Proceedings of the International Conference on Ecodesign (IEEP)*, New Delhi, India, November 2002.
—— and U. Tischner (eds.) (2005) 'Sustainability Design-orienting Toolkit (SDO-MEPSS)', www.mepss-sdo.polimi.it [accessed 25 August 2007].
Weidema, B.P. (2005) *The Integration of Economic and Social Aspects in Life-cycle Impact Assessment* (Paper 2-0; Copenhagen: LCA Consultants).
Zaring, O. (2004) *Creating Eco-efficient Producer Services* (Göteborg, Sweden: Göteborg Research Institute).
Zupi, M. (ed.) *La globalizzazione vista dal sud del mondo* (Rome: Laterza).

# 9
# Design for (social) sustainability and radical change

*Ursula Tischner*
Econcept, Agency for Sustainable Design, Cologne, Germany

## 9.1 Introduction: is design part of the problem or part of the solution?

The role of design in a society that faces more and more socioeconomic and environmental problems is changing. Designers cannot do what they have been used to doing over the previous 100 years: there is no sense in designing superfluous products for saturated markets. Instead, design must fill the gap between production and consumption in a way that leads to real problem-solving and radical changes towards more sustainability of production and consumption systems. Designers must become facilitators between consumers and producers or create new systems of (co-)production and (co-)design that fulfil needs and solve real problems with the maximum benefits for consumers, producers and the natural environment.

Today, without any sustainability considerations, many companies have started conquering the mass markets of so-called developing countries such as India and China because of the billions of potential customers in the future. Despite the fact that so-called developing or emerging countries cannot simply adopt the industrialised countries' lifestyles without catastrophic consequences for the natural and social environment, this kind of 'globalisation' or 'new colonialism' is happening nonetheless and governments can hardly control such developments.

Thus we urgently need intelligent and sustainable solutions that fulfil the needs of people, especially those of the world's poor, that create equal opportunities for socio-cultural and economic development, particularly in less industrialised countries, and that enable people to live with an acceptable quality of life without compromising the carrying capacity of our planet.

## 9.2 What to do?

### 9.2.1 Introduction

Indeed, instead of continuously trying to improve and restyle products for saturated markets and wealthy societies, more and more designers are starting to tackle real problems and are searching for feasible ways to contribute to real solutions. There are design projects for 'the bottom of the pyramid' (BOP) that aim to create greater quality of life and economic opportunities, especially in so-called developing countries. There are educational programmes on humanitarian and sustainable design.[1] There are also many creative initiatives directed at groups in society that suffer under human-made or natural disasters,[2] and there are more and more designer networks that aim at creating eco-efficient and sustainable solutions and experience exchange.[3] However, this quest for design to become part of the solution instead of creating (global) environmental and social problems is far from being a mass movement, and many activities are still relatively idealistic and may be thought to lack a little in professionalism. And some tools and methodologies are still lacking, in particular:

- Evaluation tools and methods to include the sociocultural dimension of sustainability in design activities and that go beyond ecodesign and eco-efficiency
- Tools and methods to leapfrog to drastically improved more sustainable innovation and that go beyond incremental improvements
- Tools and methods for sustainability-oriented market research
- Tools and methods to involve customers, consumers and other stakeholders more directly in design activities (i.e. to encourage participatory design processes)
- Tools and methods to create innovations throughout the complete value chain (i.e. upstream towards the suppliers and downstream to end-of-life actors)

The sustainable design research community works to close these gaps and fill in the blanks. The following sections introduce activities and results that were generated

- In a German research project on sustainable system innovations, 'The Sustainable Office',[4] by econcept,[5] the Agency for Sustainable Design and the Institute for Ecological Economy Research (IÖW)

1 See www.designacademy.nl [accessed 25 August 2007], on Master Man and Humanity.
2 See www.design4disaster.org [accessed 25 August 2007].
3 See www.o2.org and www.ecodesign.at [accessed 25 August 2007].
4 The Sustainable Office research project ran from 2002 to 2005. A final report of the project was published in July 2005 (Konrad et al. 2005). The project was funded by the German Federal Ministry of Education and Research (BMBF).
5 See www.econcept.org [accessed 25 August 2007].

- In SusProNet and MEPSS (Method for Product-Service System Development), two European research and network projects on product-service systems (PSS; the most important findings of SusProNet are summarised in Tukker and Tischner 2006)[6]

Experiences from these types of research project suggest that for designers to work in a professional way and to contribute to the sustainability of production and consumption systems it is necessary to get an overview about the social, environmental and economic sustainability of current systems first. For this, tools are still relatively new or lacking. Thus, multidisciplinary and interdisciplinary teams or groups of stakeholders must come up with creative visions and solutions regarding how these systems can be changed in favour of greater sustainability. Participatory planning processes and co-design processes need to be organised, starting with consumer or customer needs and aiming at maximum social, environmental as well as economic sustainability. In the end the results can even be completely new business models and radical system innovations that would never occur if only one company were to try to develop new solutions.

In 'The Sustainable Office' project these kinds of methods and tools were applied, and results show that the role of designers in these processes was much more as facilitators of creativity and of group dynamic democratic processes and much less the egocentric design guru that knows how the world should look.

### 9.2.2 From incremental to radical and system innovations

In scientific and sociopolitical discussions on sustainability the term 'system innovation' is frequently used (see Bierter 2002; Maßelter and Tischner 2000; Sauer 1999). With its origins in innovation research, system innovation is considered to be a decisive factor in the sustainable transition of modern societies. Yet much is unknown about what system innovations are and how they are different from other innovation types. This factor was the starting point for the conceptual and analytical part of the research project on The Sustainable Office. The aim was to develop a criteria-based classification scheme for the determination of innovation type and sustainability potential.

Innovation types generally are distinguished as incremental or radical innovation (e.g. Clark and Staunton 1989; Freeman and Perez 1988). **Incremental innovations** are relatively small changes of processes and products. They take place more or less continuously in all industry sectors, depending on a combination of demand, sociocultural factors and technological possibilities. Often they are not the result of directed research and development (R&D) but rather the result of

- Learning by doing, particularly from findings and suggestions coming from engineers, based on experience and normal problem-solving activities, or
- Learning by using, based on initiatives and suggestions of users

**Radical innovations** occur discontinuously and are normally the result of targeted R&D of companies, (related) universities or other research institutes. They are unequally divided over industry sectors and over time. Examples are the development of nylon, nuclear energy and the birth control pill (Freeman and Perez 1988).

6 See www.suspronet.org and www.mepss.nl [accessed 25 August 2007].

The concept of *system innovation* allows further differentiation between incremental and radical innovations. In this concept, changes occur in technological systems that affect several industries and can lead to completely new industry sectors (Freeman and Perez 1988). They are based on a combination of radical and incremental innovations, connected to organisational and managerial innovations, which affect more than one or a few companies and lead to new sociotechnical configurations. Examples are the development of synthetic materials (Freeman and Perez 1988), electrification and centralised energy supplies (Kemp 2000).

A further differentiation is made for innovations that are based on behavioural changes and on changes in use patterns: so-called *behavioural innovations* (Hoogsma 2000).

Some examples of the four innovation types—incremental, radical, behavioural and system—are given in Box 9.1.

### 9.2.3 Innovation types and sustainability potential

Now, it is essential to identify the sustainability potential of the different innovation types. Traditionally, two different positions can be found. One assumes that incremental innovations bring significant environmental improvements (Clausen *et al.* 1997; Clayton *et al.* 1999). The other position suggests that small improvements connected to incremental innovations are not enough and that, for substantial improvements in the state of the environment, radical technical innovations are needed (Kemp 2000; Weaver *et al.* 2000).

FIGURE 9.1 Differentiation of innovation types and their sustainability potential

Source: Konrad *et al.* 2005

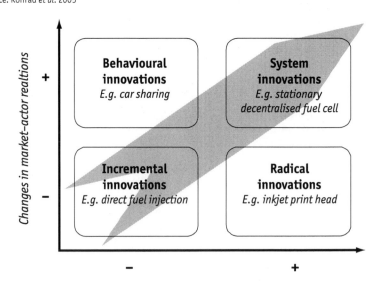

**Box 9.1 Examples of the four innovation types**

- Incremental innovations
    - Network living
    - UMTS (Universal Mobile Telecommunication Network) and WLAN (Wireless Local Area Network)
    - Airbag acceleration sensor
    - High-temperature materials for gas turbines
    - Lightweight steel constructions
- Radical innovations
    - Inkjet print heads
    - Micro-engines
    - Keyhole surgery
    - Exhaust-gas catalytic converters
    - Adaptronic materials
- Behavioural innovations
    - Bike-sharing schemes
    - Employee mobility systems
    - Emissions trading
    - Energy contracting
    - E-commerce (online bookselling)
    - Electronic paper
- System innovations
    - Electronic road pricing (the German truck toll system)
    - Personal rapid transit systems
    - Magnetic levitation (maglev) trains
    - Stationary decentralised fuel cells for home energy supply
    - Solar power generation (photovoltaic cells)
    - Stirling motors
    - Energy-efficient homes
    - Multimodal human–computer interaction
    - Online trade (or barter) of audio files
    - Micro-electro-mechanics
    - Smart labels
    - Soil-resisting glass coatings based on nano materials
    - Polytronics

The fact, that in many industry sectors the environmental benefits caused by incremental technological improvements is more than compensated for by the industry's growth makes focusing on more radical improvements indispensable. However, technological innovation alone does not seem to be enough because of the potential rebound effects (e.g. use of a highly efficient eco-car as the second or third car in a household). Therefore, user-oriented changes, such as a switch to non-ownership consumption types (e.g. car sharing)—also discussed under the heading PSS—are needed to move towards overall improvements in production and consumption systems (see Hirschl *et al.* 2001; Tukker and Tischner 2006).

Against this background the following assumption was formulated in The Sustainable Office project: positive sustainability effects on a larger scale are generated by those innovations that both change the market–actor relations and create changes in knowledge, technology and organisation (i.e. that connect organisational transformations, social innovations and the intelligent use of new technologies). These innovations are hypothetically called **sustainable system innovations** and are illustrated in Figure 9.1.

### 9.2.4 Criteria for system innovation and sustainability

#### 9.2.4.1 How to identify system innovations

To create a methodology to recognise and distinguish system innovations from other innovation types a set of criteria was developed in The Sustainable Office project that is based on the two general dimensions of 'knowledge, technology, organisation' and 'market–actor relationships'. For the two dimensions sub-indicators were formulated to check for each innovation under analysis the size of changes in each sub-category and to conclude accordingly if the innovation is a system innovation or not:

- Dimension 1 (knowledge, technology, organisation): has the innovation
    - Led to a (fundamental) shift in knowledge?
    - Been characterised by a multidisciplinary approach?
    - Used a recombination of technical elements?
    - Used a combination of products and services
    - Led to a change in infrastructure, organisational processes and structures?

- Dimension 2 (market–actor relationships): has the innovation
    - Altered the market structure?
    - Involved a combination of actors and networks?
    - Involved the participation of actors from civil society?
    - Led to a change in the context of use?

Those innovations identified as system innovations answered 'yes' to at least four sub-criteria questions in the first dimension and to at least three sub-criteria in the second dimension.

### 9.2.4.2 How to identify the sustainability potential of system innovations

The method for the sustainability assessment in the Sustainable Office project was based on the concept of three dimensions to sustainability: the environmental, economic and sociocultural dimensions (see UNDSD 2001):

- Environment: the elements of this dimension may be
  - Environmentally friendly use of resources
  - Eco-efficiency
  - Protection of ecosystems
- Economy: this dimension may see
  - Sustaining or growth of financial assets
  - Protection or improvements in competitiveness
  - Improvements in market mechanisms
- Sociocultural: this dimension may be characterised by
  - The existence of equal opportunities
  - Improvements in employment rates and social security
  - The creation of opportunities for (personal) development
  - Improvements in quality of life and health

The sustainability potential of an innovation can be estimated by using a similar procedure to that used for determining the innovation type: for the three dimensions, qualitative scores are given, using criteria such as those listed above. An innovation can have positive, negative or no influence on any of the three dimensions. When the scoring for a dimension is not possible because the innovation results in both positive and negative influences in the same dimension, and no decision can be made on which influence outweighs the other, the particular dimension is marked as ambivalent.

Positive scores on all three dimensions result in a strong sustainability potential. Positive scores on two dimensions, with one dimension being neutral or ambivalent, results in a medium sustainability potential. Two neutral or ambivalent scores and only one positive score for the dimensions results in a weak sustainability potential. All other combinations imply that there are no improvements in sustainability and that the innovation is ambivalent with regard to reaching the goal of sustainable development.

A similar qualitative sustainability screening tool was developed in the SusProNet project, as described in Table 9.1.

Both the case studies and the demonstration projects in The Sustainable Office project have been assessed regarding their type of innovation and sustainability potential using the methods described above. Some results are presented below.

## TABLE 9.1 Product-service system screening tool as developed in SusProNet

Source: Tukker and Tischner 2006

| Aspect | Score[a] |
|---|---|
| **a) Economic or profit aspects** | |
| How profitable or valuable is the solution for the providers (these can be a consortium of companies), including cost of production, cost of capital and market value of the solution for the provider(s)? Is it cheaper to produce than the competing product? | |
| How profitable or valuable is the solution for customers or consumers? Are there concrete, tangible savings in time, material use, etc. for the customer? Does it provide 'priceless', intangible added value such as esteem, experiences, etc., for which the customer is willing to pay highly (in comparison to a traditional product system)? | |
| How difficult to implement and risky is the solution for the providers? Can a promised result be measured and delivered with a high probability, or has the client a high and uncontrollable influence on the costs? When is the return on investment expected? | |
| How much does the solution contribute to the ability to sustain value creation in the future? Does it give the consortium that puts the PSS on the market now and in the future a crucial and dominant position in the value chain? | |
| TOTAL | |
| **b) Environmental or planetary aspects** | |
| How good is the solution in terms of material efficiency (including inputs and outputs or waste)? | |
| How good is the solution in terms of energy efficiency (energy input and recovery of energy without transportation)? | |
| How good is the solution in terms of toxicity (including input–output of hazardous substances and emissions without transport)? | |
| How good is the solution in terms of transport efficiency (transportation of goods and people, including transport distances, transportation means, volume and packaging)? | |
| TOTAL | |
| **c) Social and people aspects** | |
| Does the PSS contribute to quality of work in the production chain (environment, health, safety; enriching the life of workers by providing learning opportunities, etc.)? | |
| Does the PSS contribute to the 'enrichment' of life of users (by providing learning opportunities, enabling and promoting action rather than passiveness, etc.)? | |
| Does the PSS contribute to intragenerational and intergenerational justice (equal wealth and power distribution between societal groups, North–South, and not postponing problems to the next generation, etc.)? | |
| How much does the solution contribute to the respect of cultural values and cultural diversity (e.g. customised solutions, contributing to the social well-being of communities, regions, etc. [cultural values])? | |
| TOTAL | |
| *Results: main aspect*[b] | |
| a) Economic/profit | |
| b) Environmental/planet | |
| c) Social/people | |

a 1 = 'better', 0 = 'equal', –1 = 'worse'

b Score is between +4 and –4

## 9.3 How to do it: 'The Sustainable Office', a pilot project

### 9.3.1 Introduction

In the German research project, The Sustainable Office, an interdisciplinary team of managers, engineers, designers, architects and other stakeholders within the office area were invited by econcept to develop sustainable (system) innovations. The challenge was to invent new solutions for the office of the future that extend beyond pure products and include social, organisational and institutional changes and, as much as possible, positive sustainability effects.

In the course of the project a trend analysis about the needs of office workers, today and in the future, took place and a definition of The Sustainable Office was jointly formulated. Then, in different workshops, using different methods and tools, innovation groups created ideas that were evaluated, and, finally, six promising innovations were chosen and elaborated. The whole process was facilitated by the design experts of econcept.

The starting point of the innovation project was a clearly visible trend: the shift from stationary, rather material, reproducing, individual office work towards more virtual, non-stationary, creative office work. Other important trends are demographic changes in the office workforce, dematerialisation, the use of virtual working and information and communications technology (ICT) and so on.

Taking these trends and developments into account for designing an office that is prepared for the future, it is necessary to plan with a process orientation and to focus on the changing needs of office workers. This means, for example, the introduction and/or design of the following elements:

- Information: access to networks and knowledge databases everywhere
- Communication: meeting points, which allow both formal and informal communication
- Interaction: areas equipped with versatile technology that support teamwork
- Organisation: optimal coordination of all areas
- A place to retreat: a customised place that can be used for solitary work
- Storage: infrastructure for (virtual) storage
- Vitalisation: catering, exercise, entertainment, sport, play, fun and all vital interests

### 9.3.2 Sustainability and the office

In the office sector discussion of sustainability began in relation to health, productivity and eco-efficiency issues. However, this does not cover all the dimensions of a sustainable office environment. Therefore the definition of a sustainable office (see Box 9.2) and the set of sustainability criteria were further specified for all the three dimensions of sustainability (environment, economy and sociocultural issues) in The Sustainable

Office project. These criteria were used later on to inspire and evaluate the innovative ideas and solutions (see Box 9.3).

**Box 9.2 Definition of The Sustainable Office in the research project**

A **sustainable office** is an office environment that:
- Cost-effectively provides optimal conditions for productive work (profit)
- Is healthy and allows the user to fulfil his or her needs (physical and psychological), offering a satisfying combination of business and private life
- Is eco-efficient in terms of input (space, materials, energy) and output (waste, emissions) over the whole life-cycle and reduces the need for transportation (people or goods)

**Box 9.3 Criteria of The Sustainable Office in the research project**

### Economic criteria

- Cost per workplace (as low as possible): for example
    - Reduce the cost for the office building and infrastructure
    - Increase the use time, longevity, flexibility and efficiency of office equipment and building
    - Value stability
    - Increase the value of the building and infrastructure as well as the return on investment, amortisation and so on
- Work productivity (as high as possible): for example
    - Increase work stability
    - Lower rates of illness
    - Ensure and improve employees' qualifications
    - Ensure infrastructure (e.g. ICT, information and knowledge-management systems) is productive
- Long-term planning, risk minimisation and risk management: for example
    - Seek multi-functionality of space and infrastructure, enabling diverse, flexible uses
    - Avoid misinvestment and miscalculation
    - Pursue proactive consideration of legislative and political conditions
    - Look at longer-term horizons for planning

### Environmental criteria

- Material efficiency: to include, over the whole life-cycle,
    - The building

- Equipment
- Consumables
- etc.
- Energy-efficiency: to include (but excluding transport and mobility)
  - Heating
  - Light
  - ICT
  - etc.
- Space efficiency: to include
  - The building
  - The interior
  - The workplace
  - etc.
- Hazardous substances and emissions: to include consideration of (but excluding transport, mobility)
  - Indoor climate
  - Noise pollution
  - Electro-smog
  - etc.
- Re-use and recycling possibilities, waste and disposal, including consumables, the building and equipment, etc.
- Transport and mobility

## Socio-ethical criteria

- Healthy physical work environment:
  - Ensure equipment and infrastructure make work easy and comfortable
  - Look at ergonomics
  - Ensure safety in the workplace
- Healthy psychological work environment:
  - Ensure workload is adequate to level of qualification
  - Provide adequate rewards
  - Remove unhealthy stress and bullying
  - Ensure staff feel motivated and a sense of satisfaction in their work
  - Provide opportunities for self-determination
- Fair wages and quality of jobs:
  - Ensure accessibility and permeability for different qualifications
- Socially responsible corporate culture:
  - Ensure equity
  - Make the workplace barrier-free for employees with disabilities or special needs and respect different ages and cultures

- Encourage participation
- Foster an innovative and creative climate
● Social infrastructure and services: for example
    - Provide education and training opportunities
    - Show respect for family background of employees (childcare opportunities)
    - Allow a healthy combination of job and private life
    - Offer services at the workplace (food, shopping, laundry, etc.)
    - Provide excellent mobility and transportation opportunities

### 9.3.3 Process: methodology and tools

The work in the demonstration projects was done within the framework of several workshops and innovation teams. Over the course of the project many (creativity) tools were used; some were developed for the project, some had been successfully applied before (e.g. moderated discussions, Metaplan-method,[7] progressive abstraction, scenario building, role-playing games, brainstorming and brain-writing). Tools developed or adapted specially for the project were, for example, the sustainability portfolio diagram, the sustainability evaluation checklist (see Box 9.3), use of a system map or blueprint and sustainability scenario-building. Table 9.2 shows the different phases of the demonstration projects alongside a timescale and a listing of the tools used. All workshops were conducted by econcept and all results were made available to the participants and were published on the project website.[8]

### 9.3.4 Some results: sustainable innovations for the office

#### 9.3.4.1 Revital facade

Under the assumption that conservation of existing office buildings is economically and environmentally more efficient than the construction of new buildings, this innovation offers a method to prevent the demolition of old buildings and to revitalise existing office buildings. The concept is based on facade modules that enable technical and visual upgrading and renovation of existing office buildings in a highly cost-efficient way. The starting points for the development of this idea were office buildings dating from the 1950s to the 1980s that are in need of renovation, have a frame construction and had become unappealing to investors or tenants because of shortcomings related to the normal ageing process and by the changing *Zeitgeist*.

The modules can also be used in new buildings. The modular facade can include heating, ventilation, decentralised air conditioning, electrical and ICT cables, lighting and shade elements and solar modules, according to the specific needs in the building. Unlike a classical renovation, the installation of the modules is possible without much interruption of working processes within the building.

---

7 For a brief introduction of the Metaplan method, see www.12manage.com/methods_schnelle_metaplan.html [accessed 25 August 2007].
8 See www.nachhaltigesbuero.de [accessed 25 August 2007].

TABLE 9.2 Overview of the development phases in the innovation projects, tools used and timing

| Development phases | Tools | Timing (months) |
|---|---|---|
| **Kick-off meeting** | | |
| **Phase 1: in-depth analysis** of the office sector, needs, stakeholders and sustainability issues; also, formation of the innovation teams | Trend analysis; stakeholder matrix; consumer research | 1–4 |
| **Start-up workshop** In this workshop the goals, sustainability criteria, planning and responsibilities, etc. were discussed | Keynote presentations; open discussions; adoption of the demonstration project blueprints; use of Metaplan (brain-writing, brainstorming, clustering and selection of ideas) | |
| **Phase 2: idea generation** | Brain-writing; defining obstacles and drivers; clustering of ideas; moderated discussion; progressive abstraction ('What's it really about?'); scenario and role playing; brainstorming; rough sustainability assessment of the ideas using a portfolio diagram; creation of a standard for defining the innovation ideas | 6–11 |
| **Halftime workshop** Exchange of experiences and reflection on each other's ideas | Description and presentation of the system innovation ideas with a prepared format; assessment of the ideas using a sustainability checklist; assessment of the marketability and feasibility combined with sustainability with use of a portfolio diagram; moderated discussions | |
| **Phase 3: detailing the concepts** In this phase five teams detailed the selected system innovation ideas. Each team was managed by one or two participants and was supported by one of the research partners | Workshops in smaller groups, making 2D and 3D impressions, simulations or animations; interviews and surveys with potential partners and users, expert workshop with external experts; standard Powerpoint and reporting formats; orienting life-cycle assessments | 13–22 |
| **Phase 4: implementation tests** After the detailing of the five innovation concepts, all concepts were tested and prepared for a potential implementation | Marketing plan; business plans; simulation of the demonstration projects in scenarios; presentation at a trade fair; assessment of market opportunities: obstacles, drivers, sustainability factors; building of an extended stakeholder network; final reports | 23–27 |
| **Final workshop** The results of the demonstration projects were presented in February 2005 in an open conference | Description and presentation of the system innovation ideas with a prepared format | |

## 172 SYSTEM INNOVATION FOR SUSTAINABILITY 1

FIGURE 9.2 **Revital facade; this upgrading example (right) shows one possible makeover.**

Source: Schmidt Reuter GmbH Cologne

**Sustainability evaluation**

The main benefits of the Revital facade compared with alternatives such as demolition and new development, complete renovation or similar one-off facade renovations are the savings on total investments and the lower investment risks. Short reconstruction time and high cost-efficiency as well as extended use-time and the increased attractiveness of the buildings lead to increased profitability.

Environmental benefits are the prevention of demolition, an environmentally conscious renovation and decreased energy use.

### 9.3.4.2 Rent-o-box

Traditional stationary work is steadily being replaced by flexible work. Office work is becoming increasingly more virtualised and the infrastructure more decentralised. For modern 'working nomads' a complete and energy-autarkic mobile rent-o-box was designed to fulfil the needs for time-flexible office space: users can rent it for shorter or longer periods; it offers space for two complete desks and for meetings of up to six persons (within the room). The Rent-o box is a complete, self-sufficient, mobile office unit that can be placed almost anywhere. The box offers the most modern dematerialised technological infrastructure for concentrated individual and team work or for creative relaxation. It can be booked online or via telephone through an operator service (similar to car sharing). This operator also organises maintenance (cleaning, repair) and the supply of goods (operating resources, beverages, snacks). Preferably, the Rent-o-box should be positioned at traffic hubs or frequently visited locations or at locations where no infrastructure is available (e.g. in developing countries). It may be placed outdoors or indoors, including the top of buildings or at construction sites.

FIGURE 9.3 **Rent-o-box: (a) entrance side, (b) window side and (c) interior**
Source: econcept, Agency for Sustainable Design

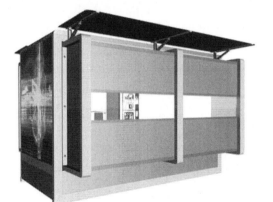

**Sustainability evaluation**
The prefabricated mobile Rent-o-box solution substitutes and complements the availability of office buildings and enables efficient and flexible use of space. The box is very resource-efficient (utilising solar panels, rainwater and fuel cells) and saves its users the purchase of their own office equipment by offering complete and modern office infrastructure. Thus it satisfies the needs of modern flexible office workers in a sustainable way, as shown by a life-cycle assessment.

First calculations show that the concept is profitable when users pay between €10 and €20 per person per hour. Important income is generated by video screen advertisements on the outside of the box. The operator expenses are minimised because of the extensively automated concept. The maintenance expenses are minimal as a result of the partly self-cleaning materials and sensible use of automation.

### 9.3.4.3 Interaction analysis

To virtualise processes in small and medium-sized enterprises (SMEs) in a more human way, the interaction analysis offers a consulting concept that focuses on human interaction and the creativity and well-being produced by communication. In the interaction analysis all encounters occurring in a central office are quantitatively and qualitatively assessed. The result shows which contacts are necessary to achieve high creativity and motivation, and which can be virtualised and decentralised. Thus a new interaction culture can be created that is very efficient and that is convenient for the employees at the same time. The aims are to offer mainly traditional SMEs sensible access to ICT and to strengthen their attractiveness for highly qualified employees.

**Sustainability evaluation**
This concept can lead to an improvement along all three sustainability dimensions. It focuses on improving communication and on the well-being of workers (and also includes flexible working places and schedules). The results of aligning office architecture and technology with new demands may be a reduced need for space (and therefore also a decrease in rent), equipment and travelling hours. Optimised processes and motivated employees can improve work productivity. In addition, by creating flexible work options a better combination of job and private life is possible.

### 9.3.4.4 Sustainable Design of Business Management (sdbm)

The Sustainable Design of Business Management (sdbm) is a business model for organising services and equipment mainly for SMEs. Companies rent not only office space but also services such as IT equipment and administration, a central copying station, childcare, professional training and catering. The idea is to provide services for a bundle of small companies that normally could not afford to offer such services to their employees.

The sdbm offers fully furnished and technically equipped office environments for flexible use. All fields of activities outside the operative business of a company are supported. The concept includes measures and services to combine business and private life (e.g. childcare, shopping, education and personal development).

**Sustainability evaluation**
The social sustainability and economy of office work can be improved by the sdbm. The bundling of services also allows optimisation of the following aspects (either in terms of costs or quality): cost per workplace, cost of energy, cost for maintenance, work productivity, company image and networking. Environmental aspects that are likely to be improved are material efficiency (less workspace needed), the life-span of equipment and energy use.

### 9.3.5 Conclusions of the project 'The Sustainable Office'
All of the developed innovations within The Sustainable Office project were evaluated with use of a qualitative sustainability assessment and showed positive sustainability effects. For two innovations (the Revital facade and the Rent-o-box) an orienting life-cycle assessment was carried out. Overall, both innovations showed moderate to high sustainability improvements.

Thus the innovation projects can be regarded as a success in terms of creating office-related, potentially marketable and more sustainable solutions. However, according to the set of criteria developed for determining the type of innovation, only one innovation is a true system innovation: the sdbm.

Possible explanations for this result are as follows:

- Although the demonstration projects included multidisciplinary development teams, one individual participant with his or her individual competences and interests guided each project. Against this background, the projects barely managed to cover every dimension of a system innovation

- During the demonstration projects it became clear that searching for sustainable solutions for the office area made more sense than striving for system innovations. The goal of reaching sustainability was challenging enough and the concept easier to understand than the concept of system innovation. Overall, it turned out that the defined concept of system innovation was too inflexible for practical use

Thus, the research project was able to prove that, with the chosen approach (methods and tools) and a multidisciplinary stakeholder team, it is possible to develop sustainable innovations for a specific field of interest. However, the hypothesis that system innovations automatically lead to more sustainable solutions could not be verified.

From the accompanying analysis and evaluation of case studies in the project, as carried out by the Institute for Ecological Economy Research (IÖW), it became clear that the positive sustainability effects of system innovations generally depend on many conditions and that sustainability has to be an inherent element in the development of system innovations right from the beginning to ensure a high sustainability potential.

## 9.4 Overall conclusions

### 9.4.1 No golden path but high potential

The concept of system innovations is a promising approach for the development of sustainability-oriented solutions, but system innovations are not automatically sustainable. They should be designed with use of tools that integrate sustainability criteria and should be realised with an implementation process that controls the environmental, economic and social impacts.

The qualitative criteria-based evaluation schemes for the determination of the innovation type and its sustainability potential proved to be useful and can be used by future application-oriented sustainability research in order to enhance the implementation of sustainable innovations.

Another decisive factor regarding sustainability is how the system innovation will be used: for example, will the users of glass with soil-resistant nano-coatings actually change their cleaning behaviour so that they clean less? Here, the delivery of a product-service system that combines the new technological solution with a window-cleaning service that is focused on cleaning efficiency might solve the problem. This example shows how important it is for processes aiming at developing sustainability include customers, consumers, users and other societal actors in the innovation process.

### 9.4.2 Supportive measures

The development, implementation and diffusion of sustainable system innovations are associated with many complex challenges, such as knowledge gaps and highly unpredictable processes. Individual actors are usually not capable of overcoming all these issues. Supporting measures are needed both on a conceptual and on a practical level (e.g. at the design research, politics and economy and society levels). Politics should stimulate measures to be taken, and the economy and society should be open-minded and educated on the concept and development process.

Designers and design researchers have an important role to play in:

- Getting involved with the subject in their daily work, experimenting with it in their projects and using the available and introduced tools for sustainable system innovations, as discussed above

- Expanding the collection of system innovation case studies, also focusing on specific characteristics (e.g. on which models of interdisciplinary cooperation or which types of networks are helpful for system innovations)

- Supporting and studying the diffusion of system innovations and the obstacles and drivers involved

- Initiating studies to analyse the control mechanisms that can influence the formation and development of system innovations, on a governance, market and hierarchical level

- Supporting the development and application of further methods and tools for sustainable system innovations (e.g. partnership management tools that are focused on system innovations, or trend research tools looking at broader production-consumption systems)

- Suggesting this approach to their clients and colleagues and educating others about it

- And, maybe most importantly, moving the position of the designer up from the consultant at the end of the innovation process who designs a nice shape for a solution generated by others (e.g. engineers) to the, somewhat impartial, manager of the system innovation process who involves several partners and facilitates and directs it

Finally, educational programmes are missing and are needed urgently to provide knowledge to a range of different disciplines (such as marketing, product development, design and management) about the subject of sustainable system innovations and product-service systems.

# References

Bierter, W. (2002) 'System-Design, radikale Produkt- und Prozessinnovationen', in G. Altner, B. Mettler-von Meibom, U.E. Simonis and E.U. von Weizsäcker (eds.), *Jahrbuch Ökologie 2002* (Munich: Verlag C.H. Beck): 171-87.
Clark, P., and N. Staunton (1989) *Innovation in Technology and Organisation* (London/New York: Routledge).
Clausen, J., U. Petschow and J. Behnsen (1997) *Umwelterklärungen als Innovationsbarometer. Eine explorative Fallstudie in der Lebensmittelindustrie* (Publication 114/97; Berlin: Institut für ökologische Wirtschaftforschung).
Clayton, A., G. Spinari and R. Williams (1999) *Policies for Cleaner Technology: A New Agenda for Government and Industry* (London: Earthscan Publications).
Freeman, C., and C. Perez (1988) 'Structural Crisis of Adjustment, Business Cycles and Investment Behaviour', in G. Dosi, C. Freeman, R. Nelson, G. Silverberg and L. Soete (eds.), *Technical Change and Economic Theory* (London/New York: Pinter Publishers): 38-66.
Hirschl, B., W. Konrad, G. Scholl and S. Zundel (2001) *Nachhaltige Produktnutzung: Sozial-ökonomische Bedingungen und ökologische Vorteile alternativer Konsumformen* (Berlin: edition sigma).
Hoogsma, R. (2000) *Exploiting Technical Niches* (unpublished dissertation; Enschede, Netherlands: University of Twente).
Kemp, R. (2000) 'Incremental Steps and their Limits: Integrated Product Policy and Innovation', *Ökologisches Wirtschaften* 6: 24-25.
Konrad, W. (1997) *Politik als Technologieentwicklung, Europäische Liberalisierungs- und Integrationsstrategien im Telekommunikationssektor* (Frankfurt am Main/New York: Campus Verlag).
——, U. Tischner, M. Hora, D. Scheer and M. Verkuijl (2005) *Endbericht, das nachhaltige Büro, Praxis und Analyse systemischer Innovationsprozesse* (Heidelberg/Cologne: IÖW/econcept).
Maßelter, S., and U. Tischner (2000) *Sustainable Systems Innovation: Final Report* (Cologne: econcept, for Bundesministerium für Bildung Forschung und Wissenschaft [BMBF]).
Sauer, D. (1999) 'Perspektiven sozialwissenschaftlicher Innovationsforschung: Eine Einleitung', in D. Sauer and C. Lang (eds.), *Paradoxien der Innovation: Perspektiven sozialwissenschaftlicher Innovationsforschung* (Frankfurt am Main/New York: Campus Verlag): 9-22.
Tukker, A., and U. Tischner (eds.) (2006) *New Business for Old Europe: Product-Service Development, Competitiveness and Sustainability* (Sheffield, UK: Greenleaf Publishing).
UNDSD (United Nations Division for Sustainable Development) (2001) *Indicators of Sustainable Development: Guidelines and Methodologies* (New York: UNDSD).
Weaver, P., L. Jansen, G. Van Grootveld, E. Spiegel and P. van, Vergragt (2000) *Sustainable Technology Development* (Sheffield, UK: Greenleaf Publishing).

# 10
# Social innovation and design of promising solutions towards sustainability
## Emerging demand for sustainable solutions (EMUDE)

*François Jégou*
Strategic Design Scenarios, Belgium

## 10.1 Introduction

Our Western consumption society extended to the world scale will lead to an ecological and social disaster in the short to medium term. If we look at our daily lives, everywhere we see unsustainable practices consuming energy and materials and producing pollution. But, if we were to observe more closely, we could also see emerging new practices alternative to the mainstream; for example:

- Production activities based on local resources and skills
- Healthy, natural forms of nutrition
- Self-managed services for the care of children and the elderly
- New forms of exchange
- Alternative mobility systems to replace the monoculture of individual cars
- Socialising initiatives to bring cities to life
- Networks linking consumers directly with producers

# 10. SOCIAL INNOVATION AND DESIGN OF PROMISING SOLUTIONS TOWARDS SUSTAINABILITY

These solutions are the result of action by groups of particularly inventive, enterprising people who have been able to identify objectives and find suitable tools and organisational forms for achieving them. More than niche social innovations, these cases reveal potential emerging models for a daily life more in line with sustainable development.

This chapter is based on the results of the European-funded research project EMUDE (Emerging User Demands for Sustainable Solutions)[1] and its development through the activities of the Sustainable Everyday Project.[2] Its aim is to present a strategic design activity to support the development and dissemination of these promising solutions inspired by the observation of social innovation initiatives.

First, Section 10.2 presents an overview of the EMUDE project process. Second, Section 10.3 describes six patterns of 'ways of doing' that emerge from the analysis of the collected cases. Third, Section 10.4 presents through a series of examples the possible contributions of a strategic design activity to facilitate access to the cases' solutions, keeping their original relational qualities. Last, some conclusions will be presented in Section 10.5.

## 10.2 The EMUDE research process

EMUDE is a programme of activities funded by the European Commission. The aim is to explore the potential of social innovation as a driver for technological and production innovation in view of sustainability.

The research process was based on the hypothesis that designers could play the role of 'antennas' detecting, through diffused social innovation, promising signals towards sustainability, as shown in Step 1 in Figure 10.1.

Students of the following European design schools were involved in looking for and documenting cases of social innovation:

- Academy of Fine Arts, Krakow, Poland
- ENSCI Les Ateliers, Paris, France
- Estonian Academy of Fine Arts, Tallinn, Estonia
- Politecnico di Milano, Milan, Italy
- School of Design, The Glasgow School of Art, Glasgow, Scotland
- School of Design, University of Applied Science, Cologne, Germany
- Technische Universiteit Eindhoven, Eindhoven, The Netherlands
- University of Art and Design, Helsinki, Finland

---

1 EMUDE (Emerging Users Demands for Sustainable Solutions) is a Specific Support Action (SSA) funded by the European Commission (NMP-2002-505645).
2 The Sustainable Everyday Project is an initiative of the Politecnico of Milano and SDS_Solutioning. It consists of a programme of dissemination events and exhibitions and a web-based 'catalogue of cases' and 'scenario laboratory' on new forms of organising daily life towards sustainability: www.sustainable-everyday.net [accessed 25 August 2007].

FIGURE 10.1 The EMUDE research process working as a signal amplifier identifying promising signals (Step 1), reinforcing them (Step 2) and re-emitting them into the system in the most suitable ways and forms (Step 3)

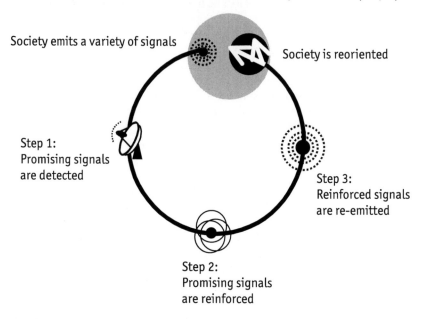

Students from each school were provided with an investigation toolkit to support them in conducting interviews, taking pictures, analysing the collected material and posting it online in a standard format.

More than 140 cases throughout Europe were collected and made available to the EMUDE research consortium, which consisted of the following universities and research institutes:

- Politecnico di Milano, INDACO Department Coordinator
- National Institute for Consumer Research (SIFO), Norway
- Netherlands Organisation for Applied Scientific Research (TNO)
- Strategic Design Scenarios (SDS)
- Doors of Perception
- Philips Design
- Joint Research Centre, Institute for Prospective Technological Studies (JRC–IPTS)
- Central European University (CEU), Budapest Foundation
- Consumers International (CI)
- United Nations Environment Programme (UNEP DTIE)

Analysis of the collected material was made by various consortium members, with different areas of expertise, in order to filter down and reinforce the most promising cases (Step 2 of Fig. 10.1). In particular, sociological approaches allow a better understanding of the motivations and identify profiles of the subjects or communities promoting and participating in the identified cases. A qualitative assessment was conducted in order to estimate the potential of each case both to reduce environmental impact and to regenerate social fabric.

A total of 73 cases were shortlisted and clustered into different typologies of everyday living solutions seen as being promising in terms of sustainability. Six such 'ways of doing', from the user point of view, in terms of his or her day-to-day life, emerge from the sample of cases available:

- Family-like Services, provided through equipment available in the household and common family skills
- Community Housing, based on habitat infrastructure to facilitate the sharing of domestic services and resources
- Extended Houses, where some of the household functions are fulfilled through collective infrastructure in the neighbourhood
- Elective Communities, where circles of people organise to provide each other with mutual help
- Service Clubs, based on open workshop places and the involvement of passionate amateurs in supporting and encouraging newcomers
- Direct Access Networks, where people organise to get products and services, cutting out intermediaries

These 'ways of doing' will be presented in detail in Section 10.3.

Possible improvements were suggested in order to facilitate the access to each type of solution, to increase the number of potential users and to support their dissemination. This twofold process was based on research of potential supporting technologies, on the one hand, and on the development of the initial case into a replicable product-service system (PSS) addressing larger audiences, on the other. The results of this last part will be presented in the last section.

Finally, the resulting promising daily life patterns and their related improved solutions were presented in a highly visualised format in order to contribute to a social conversation towards more sustainable everyday life (Step 3 in Fig. 10.1).

The complete visual scenario is shown by a matrix (Fig. 10.2) where, vertically, for each of the six ways of doing examples of improved solutions have been developed and, horizontally, eight different user profiles shows how these solutions may be integrated into different lifestyles.

Finally, for each of the 24 solution samples, one of the users explains his or her motivation to use it and the benefits he or she draws from it (Fig. 10.3).

## 182 SYSTEM INNOVATION FOR SUSTAINABILITY 1

FIGURE 10.2 EMUDE matrix: six new 'ways of doing' and eight potential adopters

|  | Nathalie | Jarkko | Erek |
|---|---|---|---|
| **Family-like Service**<br>Foot Bus<br>Home Laundry<br>Micro-nurseries<br>Family Take-away | | | |
| **Extended Houses**<br>Multi-user laundry<br>Collective Rooms<br>Co-housing<br>Car Sharing | | | |
| **Community Housing**<br>Party Place<br>Open Handyshop<br>Washing Restaurant<br>Kid House | | | |
| **Elective Communities**<br>Neighbourhood Library<br>Kids Clothing Chain<br>Active Shopping List<br>Living Corn Book | | | |
| **Service Club**<br>Shopping Club<br>Wood Atelier<br>Green Gardening<br>Second Hand Atelier | | | |
| **Direct Network Access**<br>Regional Market<br>Country Meal<br>Product Time Sharing<br>eStop | | | |

## 10. SOCIAL INNOVATION AND DESIGN OF PROMISING SOLUTIONS TOWARDS SUSTAINABILITY  *Jégou*  183

One of the purpose of the EMUDE research was to show the possible result of a strategic design activity applied to social innovation. An attempt has been made to both cover different categories of collaborative services and show their adoption among different user profiles. As shown in this matrix, six different 'ways of doing' displayed vertically combined with eight different fictional personae horizontally. Where they intersect, 24 examples of possible collaborative services are distributed across the matrix. Each persona is involved in three services demonstrating how different ways of fulfilling everyday activities in a new and more sustainable way can combine with the motivations, expectations and constraints of different potential adopters.

FIGURE 10.3 Examples of possible contributions of strategic design activities to facilitate access to the case-study solutions while keeping their original relational qualities: [a] eStop and [b] Kids Clothing Chain (overleaf)

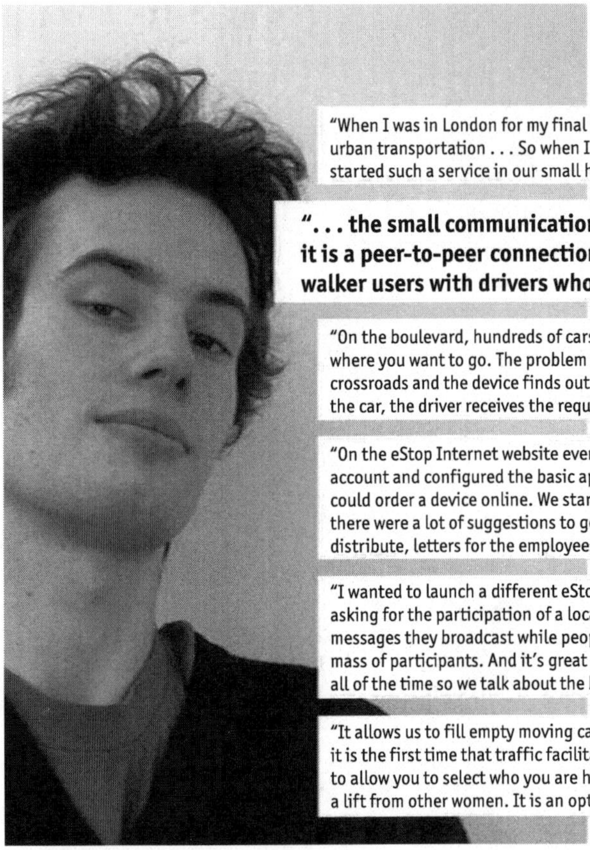

## eStop

"When I was in London for my final internship, we were members of the eStop to move beyond urban transportation . . . So when I came back to Finland with a group of friends we also started such a service in our small home town . . ."

". . . the small communication device allows us to do urban hitchhiking: it is a peer-to-peer connection system; it means that it just connects walker users with drivers who are going in the same direction . . ."

"On the boulevard, hundreds of cars are passing: there must be one that's going more or less where you want to go. The problem is knowing which one! With eStop you just stand at the crossroads and the device finds out if there are any vehicles that could give you a lift. Then, in the car, the driver receives the request on his own device and invites you to jump in."

"On the eStop Internet website everything is there to build a new network. I opened an account and configured the basic application to adapt it to our town. Then, whoever wanted could order a device online. We started with the friends of our friends and on the website there were a lot of suggestions to get the service started: sponsorship raising, leaflets to distribute, letters for the employees of the local companies, stickers to put on the cars . . ."

"I wanted to launch a different eStop more focused on young people . . . So we had the idea of asking for the participation of a local radio station that broadcasts local bands. Through the messages they broadcast while people are in their cars, we immediately reached the critical mass of participants. And it's great when we eStop someone: he listens to this station almost all of the time so we talk about the bands they have been playing . . ."

"It allows us to fill empty moving cars. The more cars are participating the more effective it is: it is the first time that traffic facilitates sustainability! And there are even enough participants to allow you to select who you are hitchhiking with. Lots of women, for example, prefer to get a lift from other women. It is an option that promotes confidence from the outset . . ."

a

FIGURE 10.3 (continued)

## Kids Clothing Chain

"When I arrived in Belgium, trying to clothe my two kids with just one salary . . . I didn't know anybody but some mothers of the school were using the Kids Clothing Chain."

"... **the system re-uses the principle of the clothes box exchanged between brothers and sisters. Everybody can belong to it in the area. The children inherit from the other children as if they belonged to a big family ...**"

"To participate, the simplest thing is to bring your own clothes as a subscription. We choose a box with a selection of same-age clothes. At the end of the season, when the clothes are used up we replace them. Each successive user is identified on the box and has to take care of the content if he wants to use it again as exchange money in the system."

"On the front of the box people often stick a picture of their child, usually with the clothes he likes most . . . it helps to choose . . . We can also add comments: 'it lacked warm pullovers for winter' . . . or 'my daughter really liked the small print sweater'. That gives the system its identity: the clothes are not anonymous. Others have loved them and one day we'll no doubt see them again worn by another kid in the area."

"We may not like all the clothes that are in a box . . . But even if we don't wear them all, we really buy much less. Economically it is beyond comparison . . . We no longer have all those too-small clothes being accumulated. They are re-used immediately until they become worn out."

b

## 10.3 New ways of doing: emerging everyday life models—promising in terms of sustainability

From the most promising cases in terms of sustainability, different 'ways of doing' in daily life emerged.

### 10.3.1 Family-like Services

This cluster is based on solutions developed from traditional household activities such as looking after children, preparing meals, washing clothes or playing host to a relative. The most significant examples are:[3]

- Nurseries based at home, organised by a young mother looking after two or three babies belonging to others as well as her own child

- Senior couples, with a spare room in their apartment after their children have left, who accommodate a student

- People who regularly organise meals at their home, as a restaurant, and receive guests

The principle is to make use of the existing infrastructure in the household (available space, appliances used with only a low intensity, etc.) and of the common skills gained through family life (taking care of others, cooking, shopping, etc.). The family circle extends to include more subjects (mostly singles living in the neighbourhood), such as an elderly person with reduced autonomy, a student on his or her own, or couples absorbed in their careers who need to find professional services but who at the same time appreciate the relational quality of being in some way 'adopted' by a family.

### 10.3.2 Community Housing

This second cluster focuses on forms of organisation made possible by households that are physically near (i.e. in the same village or in the same building) in order to facilitate the sharing of resources and to provide mutual help. Typical examples are:[4]

- In a rural context, eco-villages that organise in cooperatives to promote ecological building or to produce organic food or renewable energy

- In an urban context, co-housing schemes, where the habitat is designed to facilitate the sharing of gardens, children's playrooms, dinning rooms, recreational places, bedrooms for occasional guests, equipped areas for washing of clothes and cars for collective use

---

3 In the EMUDE research, these three examples correspond, respectively, to the following cases: Nurseries at Home (Milan, Italy), Lodging a Student at Home (Milan, Italy) and Livingroom Restaurant (Eindhoven, The Netherlands).
4 In the EMUDE research, these two types of examples correspond to the cases of Findhorn Eco-village (UK); and De Kersentuin (Sustainable Housing and Living) and the urban scheme, Aquarius (Eindhoven, The Netherlands).

Spaces halfway between domestic and rented, or shared goods with a status between personally owned and leased, complement the private household. They are managed by the community under a formal agreement and they make family daily lives easier, more functional and more efficient.

### 10.3.3 Extended Houses

This third cluster covers the existence of services within walking distance where part of the function of the household can be fulfilled. For example:[5]

- A municipality may make available an area of public land in the city for use as vegetable plots for people who do not have gardens (e.g. allotments)
- The elderly community may organise a collective space in the neighbourhood to take meals together
- A private initiative may offer open recreational space for children to meet and play together

The common denominator in these examples it that they go beyond the physical boundaries of the household. Peripheral places in the neighbourhood with a status between public shops frequented mostly by local clients and private places collectively managed are invested in for family life. The area is then perceived as an extension of the household, resulting in a more friendly and lively atmosphere in the streets.

### 10.3.4 Elective Communities

This fourth cluster is based on small and medium-sized circles of people interested in exchanging services or goods on a local basis. Typical examples of theses circles are:[6]

- Local exchange trading systems organising mutual help on the basis of the time spent in helping
- People in the same neighbourhood involved in the cleaning and maintenance of the area and in the improvement of living conditions
- Larger networks exchanging second-hand books and organising discussion groups online between readers

Elective communities combine the opportunity of physical proximity that facilitates personal relationships, spontaneous exchanges and trust building and sharing an affinity motivated by a common material interest, a convergence of views or set of objectives. Beyond families and friends, they constitute additional layers of social fabric, providing not only tangible services but also a strong feeling of belonging, identity and support.

---

5 In the EMUDE research, these three examples correspond, respectively, to an orchard in North Park (Milan, Italy), Meerhoven Senior Club (Eindhoven, The Netherlands) and Cafézoid (Paris, France).
6 In the EMUDE research, these three examples correspond, respectively, to the following cases of Ayrshire LETS (Local Economic Trading System) (UK), Neighbourhood Shares (Eindhoven, The Netherlands) and *Buchticket* (Cologne, Germany).

### 10.3.5 Service Clubs

This fifth cluster covers forms of open clubs where members provide mutual support in the fulfilment of a particular activity. Widespread examples of these clubs are:[7]

- Bicycle repair workshops, offering tools, infrastructure and spare parts to their members
- Places dedicated to the renovation and customisation of furniture, using the competences and equipment in one woodwork shop
- Places renovating spare building parts and making them, along with advice on how to use them, available to people building their own house

Similar organisations exist that focus on, for example, gardening, cooking, do-it-yourself and sewing. These clubs are 'open' in the sense that they work on the basis of free involvement: senior members provide help to new participants who, after they feel more confident in the activity, will themselves offer support to other newcomers. The need for permanent staff is reduced to the minimum required to ensure maintenance of the structure.

### 10.3.6 Direct Access Networks

This last cluster includes all kinds of direct relationships between producers and consumers both for the purchase and for the exchange of goods. Typical examples are:[8]

- Purchase groups, where neighbourhood communities organise to buy their food in bulk directly from local farms, getting both quality products and good deals
- Reciprocal to the above, local biofood producers take the initiative to deliver orders directly to the final clients in the city
- Between customers, second-hand exchange systems are organised through radio programmes or web portals

Unlike the previous clusters, these networks may cover large areas, such as an entire town for an exchange pool, or the nearby countryside for regional food delivery. But, even though they may be more widely spread, the relationships are always characterised by direct contact, with no or few intermediaries, and by person-to-person agreements based on mutual trust.

---

7 In the EMUDE research, these three examples correspond, respectively, to +bc (Milan, Italy), Mööblikom (Tallinn, Estonia) and materjalid.net (Tallinn, Estonia).
8 In the EMUDE research, these three examples correspond, respectively, to Gruppo d'Acquisto Solidale (Milan, Italy), Food Link Van Group (Skye, Scotland, UK) and the 'Exchange Corner' of Radio Krakov (Krakow, Poland).

## 10.4 Strategic design towards a synthesis of quality and access

### 10.4.1 Increasing user accessibility

In presenting the collection of cases from which the above six patterns of 'ways of doing' emerged the EMUDE research does not pretend to offer more than a glimpse of the richness of social innovation and temporary typologies used to structure a vision of a more sustainable everyday life. Nevertheless, these cases do outline an emerging social landscape characterised by a high level of relational quality between people: beyond the benefits of each solution, the resulting image shows the regeneration of the local social fabric based on an enlarged family, common local resources, an active neighbourhood life, additional circles of relationships and passionate participating people.

The last part of this chapter attempts to show how strategic design may contribute to the dissemination of the observed solutions, making them more accessible to a larger audience while, at the same time, trying to keep the relational qualities.

From the previous six 'ways of doing', a sample of solutions has been developed, inspired by the clustered cases (Fig. 10.2). Five main design strategies were used to facilitate access to the solutions:

- Enhance local visibility
- Fluidify management
- Reduce cognitive cost
- Offer different levels of involvement
- Support collective use

In the following I will attempt to frame these strategies and illustrate them with examples taken from the solutions developed in the EMUDE research.

#### 10.4.1.1 Enhance local visibility

The very process of looking for cases in the EMUDE research revealed the existence of many initiatives hidden in the social fabric. At the same time, it demonstrates their lack of visibility to those not directly participating in them but who might like to get involved. Thus, as a first step, facilitation of access to these solutions means their existence must be relayed through use of simple communication tools.

As an example of such a tool, '100 Meters Around' is an Internet platform connecting demand and supply of informal services within a walking distance around the household or the workplace. This service would allow the promoter of a solution to advertise his/her offer and, reciprocally, a potential user to review what opportunities are available within reach.

Along the same lines, more traditional communication systems, still focused at a local level, could also increase visibility of local initiatives. 'Regional Market' is a TV shopping programme broadcast by a local network. It is dedicated to goods and services produced in the locality that are not advertised via regional or national media.

### 10.4.1.2 Fluidify management

The cases outlined in Section 10.3 are self-managed and run via the personal effort put in by their organisers; they therefore require participants to dedicate time and organisational skills. These initiatives are often similar to micro-enterprises offering services in terms of the tasks to be performed or the number of participants needed to coordinate those tasks. But, as they have emerged from a bottom-up social dynamic, only a few of them are able to activate suitable management tools and dedicated technological support. Facilitation of access to such solutions will thus require forms of participation supported by proper infrastructure in order to reduce the time needed to be dedicated to the service.

For instance, eStop (Fig. 10.3[a]) is a peer-to-peer communication device facilitating hitchhiking in urban areas, organised through a club. The system matches people on foot with drivers passing by. It is based on a small digital map of the town concerned, divided into zones, which enables users to indicate where they drive and where they want a lift. The device plays the role of a broker for a user on foot waiting at a traffic light, sending his or her message to only those drivers that are likely to give him or her a lift because they are going in the same direction.

### 10.4.1.3 Reduce anticipation

More than being time-consuming in their implementation and management in daily life, promising cases require forethought from their users. Taking part in a car-sharing system involves booking it in advance. Acquiring second-hand books involves taking part in an exchange network and waiting for an available copy. Buying organic products directly at the local farm, through fair trade routes, as a purchasing group, means ordering and planning purchases in advance. In a mainstream culture of immediate consumption of personalised products and instantaneous services, the requirement of planning in advance and anticipating needs is a heavy constraint for users. To improve access to such solutions there is a need to reduce the cognitive cost on users.

For example, the loan of domestic tools is a natural thing to do, but at a neighbourhood scale it means the need to track successive users, check that tools are given back in good order, bargain for maintenance or substitution in case of a problem and so on. Product time sharing is a personal leasing service proposed when one buys new goods such as tools or specific domestic appliances that are not used all the time: simply by applying for the service the new owner becomes the leaser of the products. He or she gains access to a professional infrastructure for leasing goods, including a means to track a fleet of products, a reservation and booking system, the management of financial rewards, etc.

### 10.4.1.4 Offer different levels of involvement

As already mentioned, the success of many of the observed cases is partially a result of the constant participation of the people taking part in the solution. This deep involvement produces high relational qualities and a strongly socialised atmosphere. However, such involvement is very demanding: it is a particular material and affective choice that has to be made. To facilitate access to the solutions means opening up to less personal involvement and leaving it as a matter of choice how much participants become involved in supplying the solution. At the same time, the success and continued existence of the solutions become less dependent on participants' continued involvement.

The way Service Clubs work is significant in terms of plurality of access. Unlike the case studies outlined in Section 10.3, which are based on amateurs providing mutual help, structures such as the Second Hand Fashion Atelier offer various means of participation: people can book occasionally to use the sewing machines or they can ask a particular service of the professional sewer who organises the workshop. The core organisation of the solution remains on the basis of passionate amateurs fostering mutual help and socialisation but the structure behaves also as a commercial enterprise offering a range of services and charging for those services according to the willingness of users to participate more or less actively.

### 10.4.1.5 Support collective use

Sharing and collective use are a part of many of the observed case studies. This common use of resources requires a lot of attention in managing places and products collectively, in organising time sharing, ensuring maintenance of products where there are multiple users, and providing conciliation in case of conflict between people sharing the same device. Providing access to solutions based on sharing means providing support to compensate for the burden of self-managed collective resources.

For instance, the Neighbourhood Library is based on collective neighbourhood access to the books people have on their living-room shelves, with the solution benefits from the organisation and management system of a professional library. Participants are registered members and they choose which of their books to place on the database. An online catalogue is available, and books are tracked through a computerised booking system, with return dates and optional rewards. The sharing of books becomes, then, an easy and fruitful operation.

## 10.4.2 Preserving relational qualities

The five design strategies that have been presented in Sections 10.4.1.1–10.4.1.5 are not exhaustive of the possibilities for enlarging access to the solutions suggested by the promising cases observed. Further development is needed to complete them but they already give an idea of the improvement in terms of functionality that a strategic design approach can bring to social innovations.

The final part of this chapter will be dedicated to the other side of the strategic design approach performed within the framework of the EMUDE scenario. It will present a sample of five design guidelines used to maintain in the new solutions the relational quality initially present in the promising case.

In other words, providing access to these solutions for a larger audience, facilitating participation, reducing time and organisational constraints and providing professional support to streamline the processes involved may lead to the generation of more commercially oriented services. The risk is that we divorce the solutions from their substantial humanistic and relational content, which is the essence of their perceived quality.

The following five patterns should support the strategic design process to avoid this problem as far as possible:

- Promote availability
- Keep the relational scale
- Enhance the semi-public status
- Provide support for participation
- Build trust-based relationships

Sections 10.4.2.1–10.4.2.5 will attempt to frame these patterns and to illustrate them through examples taken from solutions developed in the EMUDE research.

### 10.4.2.1 Promote availability

In the observed cases, there is a large range of relationships between subjects and resources, from individual ownership to use of collective goods. Second-hand clothes and books as well as unused equipment are owned by someone, but, at the same time, this person does not need them anymore and may agree, within a socialisation process, to offer them for free or in exchange for a form of reward.

These resources are available. This state of availability (compared with what is intensively or even regularly used day to day) is promoted and put forward to enhance social quality in the solutions.

For instance, the Kids Clothing Chain (see Fig. 10.3[b]) solution reproduces the pattern of passing a box of used clothes between children. The network supports families in the neighbourhood in finding a 'provider' with older children and a 'client' with younger children in order to exchange clothes, with mutual respect and trust and observing simple rules for the maintenance and renewal of clothes. The solution promotes a state of availability, establishing exchange as a default for children's clothes.

### 10.4.2.2 Keep the relational scale

The observed cases work at a local scale. They emerge from bottom-up initiatives. They disseminate more than they grow and generally keep to a reasonable scale around a local place and a small number of people. An exchange shop, a recreational space for children, a repair workshop— all these work within a walking distance. A purchase group or exchange system are easier to organise and maintain between a few people, so the solutions developed tend to keep this relational scale based on direct face-to-face relationships.

The Micro-Nursery for instance, is based on the reasonable number of children that a mother can take care of. In the solution, the National Childhood Organisation in the country where the solution would be implemented would support the initiative to make it as professional and reliable as possible, providing a service kit with support, training, assistance and so on. The quality of relationships in the solution and its feasibility is mainly attributable to the fact that the nursery does not exceed the size of a large family.

### 10.4.2.3 Enhance the semi-public status

Contrary to the organisation of mainstream society, showing a strong limit between public and private, the observed cases exhibit an in-between status. An orchard in a public park, a purchase group, a café for older citizens—these are all examples of semi-public (or semi-private) places, frequented by regular users but open to others, collectively managed but generating a feeling of intimacy.

Such a semi-public status for resources enhances the relational quality between the people who use or organise those resources. They are neither together nor isolated; they are voluntarily in a relationship.

All the solutions developed within the community housing and the extended home clusters 'play' with this intermediary status. The existence of collective rooms in a condominium is a typical example. They extend each apartment with a TV room, a shared office, a children's area or additional bedrooms. These places are private, collective and shared. The perceived feeling is of 'fuzzy boundaries' of personal space overlapping with the space of the others, providing occasions for socialisation.

### 10.4.2.4 Support the expression of people skills

Interpersonal skills play a central role in most of the case studies. In contrast to service industries—where the sophisticated design of infrastructure allows the employment of less- or low-qualified people—an increase in the qualifications or ability of the people involved and the transfer of know-how between those people is part of the final aim of many initiatives. Combining both, the solutions developed tend to support the material part of service provision in order to free people of low-interest tasks and to leave them more space to express their skills and transfer their competence.

Shopping For You, for instance, is not a shopping service centred on delivery, as in most online supermarkets. A mother of a family, a good cook, offers her services to compose seasonal menus, to select good products and to combine them with a wide knowledge of recipes. She offers a solution for four or five households based on the provision of a selection of ingredients and related cooking advice. She uses an electronic organiser to receive and manage multiple demands, give orders, keep track of expenses and so on. She then can dedicate herself to expressing her skills shopping in the street market and exchanging with her clients.

### 10.4.2.5 Build trust-based relationships

In most of the observed cases, the construction of mutual trust is based on the relationship between individuals. It is mainly because participants meet physically and collaborate that such initiatives are viable. Professional services on a large scale are based on the opposite: trust is based on the institution and its longevity; relationships with the service staff are codified and anonymous.

The relational quality of the developed solutions is generally based on a mix of the two approaches. The Micro-Nursery is not only a face-to-face agreement between people but also the result of training and the overseeing of each nursing family by a qualified third party from an institution such as the National Childhood Organisation. Similarly, one can follow a diet through the Country Meal Subscription; the food is provided by an organic farm but the diet is checked by a professional doctor. In these two solutions, trust is based on a complementary relationship with a skilled person and a professional figure.

## 10.5 Conclusions

In conclusion, this tentative characterisation of the strategic design approach taken in a sample of solutions developed from observed case studies reveals a certain convergence. All the strategies to facilitate access as well as the related patterns to enhance the relational quality between people are based on the same intermediary vision combining social dynamics and technical support infrastructure. Participation of the user is fluidified through technical support. The cognitive cost of organising collective solutions is made easier through support and management infrastructures. Involvement of participants is real but gradual. Groups respect the human and local scale. Resources are owned but available. Places are somewhere between public and private. Trust is both relational and institutional. Solutions are semi-finished and services are enabling.

## Bibliography

Manzini, E., and F. Jégou (2003) *Sustainable Everyday, Scenarios of Urban Life* (Milan: Edizione Ambiente).
Manzini, F., and E. Manzini (2008) *Collaborative Services, Social Innovation and Design for Sustainability* (with contributions from P. Bala, C. Cagnin, C. Cipolla, J. Green, T. van der Horst, B. de Leeuw, H. Luiten, I. Marras, A. Meroni, S. Rocchi, P. Strandbakken, E. Stø, J. Thakara, S. Un, E. Vadovics, P. Warnke and A. Zacarias; Milan: Edizioni Poli.design, forthcoming).
Meroni, A. (ed.) (2007) *Creative Communities: People Inventing Sustainable Ways of Living* (with contributions from P. Bala, P. Ciuccarelli, L. Collina, B. de Leeuw, F. Jégou, H. Luiten, E. Manzini, I. Marras, A. Meroni, E. Stø, P. Strandbakken and E. Vadovics; Milan: Edizioni Poli.design, March 2007).

# 11
# Eco-Innovative Cities Australia
## A pilot project for the ecodesign of services in eight local councils

*Chris Ryan*
University of Melbourne, Australia

## 11.1 Introduction

Policies and programmes for sustainable development have undergone significant shifts in focus in the more than three decades since the first global conference on the environment in Stockholm 1972 (Ryan 2004). While described in different ways from different theoretical perspectives within the literature, the broad sweep of those shifts is clear: the focus has moved from production and the firm, to products and consumption and then to structural issues that shape systems of production and consumption.

During the past decade or so, policies and programmes have sought to influence the design of key elements of the production–consumption system: production processes (cleaner production), products (ecodesign or life-cycle design), the infrastructure of distribution systems and end-of-life disposal, as well as the organisational logic of the system as a whole. The need for dramatic ('Factor $X$') reductions in the resource intensity of the economy have become paramount, to achieve an environmentally sustainable existence in the face of population and economic growth in industrial economies and in the rapidly developing regions of Asia. Research and policy development is increasingly emphasising the need for *systemic* change though eco-innovation.

Thus there is growing interest in coordinated policy and governance processes for sustainable **systems innovation** and for initiatives that aim to shape, stimulate and manage a transition to more sustainable 'configurations' of the systems through which we satisfy our needs for housing, food, mobility, communications, entertainment and so on

(Smith et al. 2005; Tukker 2005). Whether discussed in terms of changing 'patterns of consumption and production', or 'sociotechnical regimes',[1] eco-innovation has to address the interrelationships of technological and social practice.

At the same time, research and practice in ecodesign has moved beyond 'product re-engineering' to changing 'product systems', with a greater focus on influencing patterns of consumption. The need for ever greater decoupling of economic value from resource consumption points to the need to 'dematerialise consumption' (to reduce material flows within the economy). This has led, for example, to a reconceptualisation of 'the product' as a 'product-service system' (PSS). From this perspective, the market value of products is seen to derive from their function as (individualised, on-demand) 'service-providing' machines. This reconceptualisation has opened up a new domain for eco-design, with greater systems implications: new service–product mixes and new service–product relationships that are able to satisfy consumer needs in new ways that have (potentially) greatly reduced resource demands (Manzini and Vezzoli 2002; Mont 2004; Ryan 2004b; Tukker and Tischner 2006). Many PSS strategies propose the transformation of established relationships of economic exchange, with, for example, consumers gaining access to products that remain the property of producers.

In the design of sustainable systems and in new policy approaches and governance, developments in information and communications technology (ICT) are playing a significant role. Accurate monitoring (and effective communication) of environmental performance is widely recognised as a fundamental driver of political change and policy development over the past few decades (Ryan 2004b). Effective monitoring, feedback and communication systems, for example, are recognised as important at the product level (to communicate measures of performance to the market [through labels, etc.] or to the user, to guide use patterns), in the management of the product chain (for life-cycle impact assessment), for system logistics (efficient transport and reduced travel paths) and for assessment of total system performance (e.g. on greenhouse gas emissions). Possibly the most significant contribution of ICT will prove to be our enhanced ability to explore, model and visualise alternative system configurations. Various authors have pointed to the important role of visions of technological expectations in framing sociotechnical problems as well as for motivating action to address them (e.g. Smith et al. 2005). Our growing capacity to gather, model and visualise data (even for complex systems) has undoubtedly increased our confidence in approaching change from a systems perspective and increased our interest in exploring alternative spaces of possibilities.

---

1 Broadly, a sociotechnical regime can be considered as a set of rules and structures embedded in the system of production processes, products, infrastructure, institutions, skills, knowledge, cultural values, aspirations, ways of handling artefacts, organising people and viewing problems, through which societal needs are satisfied (after Rip and Kemp 1998; Geels 2002; Smith et al. 2005).

## 11.2 Framing the approach to the Eco-Innovative Cities programme

### 11.2.1 Introduction

Eco-Innovative Cities (EiC), a pilot programme for eco-innovation within local councils in Melbourne, Australia, is a product of the changes in policy thinking referred to above. It builds on the development of knowledge, methods and practice in ecodesign (or 'design for environment') over the past decade or so. The development of the EiC programme has been influenced by a desire to explore four issues that arise from literature and practice:

- Balancing the social and the technical in developing eco-innovation
- Understanding the potential for *local* action for eco-innovation
- The appropriateness of the service sector as an innovation target
- The potential to use new information technology systems to support networking and the diffusion of new eco-innovation initiatives

These issues are briefly elaborated below.

### 11.2.2 Balancing the social and the technical as sites for change

Mainstream approaches to eco-innovation tend to privilege technical and material change over changes in lifestyles, culture, social behaviour and organisation, even when the latter domain is acknowledged as important. Innovation programmes frequently assume that the primary agent for transformation will be technology, that significant system change will result only from a significant scientific or technological 'breakthrough'. Social and organisational changes are never considered irrelevant, but within a technology focus paradigm they are often relegated to the category of issues that will affect the 'adoption' of new (technical) innovation.

However, this simple technology-led perspective is at odds with the knowledge and understanding from the past decade or so of work in ecodesign, from which a broad consensus seems to have emerged about a progressive widening of the focus for design (Ryan 2003, 2004a; Tischner *et al.* 2000). This thinking links the broadening of the design focus—beyond the material and the technical aspects of the product to the cultural, behavioural and organisational systems that establish the function and value of the product—to the increasing scale of environmental improvement. In ecodesign research and practice the focus of intervention has shifted from production and products to 'changing patterns of consumption' and thus to the need to 'reshape' individual and social behaviour, cultural values, lifestyles, aspirations and so on.

The EiC programme was developed with an explicit aim to innovate in the social as well as the technological spheres. The programme defines eco-innovation as a combination of:

- Changes in the organisation of infrastructure and systems of production and consumption

- The adoption of new patterns of consumption or lifestyles or changes in consumer behaviour
- The application and use of new technology, products or materials or the innovative recombination of existing technologies

### 11.2.3 The 'local' as a niche for innovation

Environmental policy and sustainable development programmes have to consider the issue of the appropriate scale at which processes and action should be targeted and deployed. There are economic and political reasons why structural change has to deal with things on a large scale—national, regional, global. However, even large-scale, technology-focused programmes have to acknowledge the critical role of community values and attitudes in determining their success or failure. Attitudes and values are not homogeneous, and programmes developed with a commitment to community engagement and consultation have had to be sensitive to the views of different communities of interest, whether defined by socioeconomic factors, cultural and religious beliefs, and so on, or spatially, by local networks and conditions and a shared local identity.

An interesting question to be explored is whether the common environmental adage is meaningfully reversible—that to think local and act global is as relevant as to think global and act local. Are there sets of localised conditions that provide a 'niche' for new innovative solutions, because sociotechnical regimes have had to respond to different pressures and opportunities from those that apply at more macro scales? If such niches do produce new local innovations, could these be relevant at the macro scale; could they be 'scaled up' to produce larger structural change?

Models of innovation typically begin with a process for the generation of a wide range of new concepts, followed by some other process of filtering that acts to select suitable concepts for further development. Within the firm this is usually presented as a managed process (e.g. Verloop and Wissema 2004). Recent theories of sociotechnical systems change have used a quasi-evolutionary model to describe innovative change, starting with a diversity of system variations that are then subject to adaptive pressures, or 'selection' by 'market forces, existing physical infrastructure and institutional and societal factors' (Tukker 2005: 66). This evolutionary model is used to explain the process by which radical system changes occur and to develop policies and programmes able to manage transitions to new sociotechnical regimes that are seen as more environmentally, socially and economically desirable (Kemp *et al.* 1998).

Variations in sociotechnical systems can arise in niches where different social, cultural, technical or market conditions either exist or can be established. The EiC programme aims to test the idea that local councils—and local communities defined by their democratic structures of government—can act as such a niche for new eco-innovation with wider social relevance. The programme is explicitly focused towards local action, the development of local solutions and the networking of a series of local projects, working with eight local governments in Melbourne, Australia. This reflects a commitment in the sustainability policies of the State government of Victoria. It also reflects a growing interest in 'the local' as a site for eco-innovation, an interest that has grown with the production of some high-profile and influential (even 'iconic') products initiated by local governments (see Section 11.4.2.3).

## 11.2.4 Eco-innovation and the service sector

Even though PSS theory and experimentation has introduced the idea of 'services' as part of the ecodesign agenda, little attention has been given to the ecodesign or eco-innovation of 'services' per se, in spite of the dominant role of the service sector within most industrialised economies. In part this probably reflects the historical origins of design for cleaner production and cleaner products, with its roots in engineering and chemicals or manufacturing professions; in part it is explained by the artificial and contradictory nature of the classification of 'the services sector' in industrial economies, which often includes domains that are heavily based on products, technology or materials, such as the 'transport' sector. Whatever the underlying causes, there is little reported work on the application of ecodesign and eco-innovation methodologies and processes to 'real' (labour-based) services. Yet the potential for significant results from the application of a variety of 'preventative environmental strategies' in the service sector has long been recognised (e.g. Graedel 1998; Kisch 2000).

The EiC programme takes the ecodesign or eco-innovation of services as the primary focus of a practical project. The approach involves the adaptation of methodologies developed for PSS and places considerable emphasis on envisaging alternative systems and mapping new possibility spaces. Because many of the services delivered by local councils focus on social and cultural activities and essential lifestyle-support systems (such as heath, recreation, childcare and care for the aged) the project has set out to explore solutions that balance social and individual behaviour and lifestyle elements with changes to material infrastructure (such as roads and drainage) and technologies.

## 11.2.5 New collaborative (ICT-based) systems

One of the results of the development of networked communications systems—the Internet—is the software to support new communities of interest that need not be localised in a spatial sense. Software systems have also enriched the scope for the projection of ideas in a way that encourages creative engagement of those with common interests. The EiC programme seeks to test whether a variety of online collaborative systems can support a design or innovation programme and realistically assist in the diffusion of new ideas and innovative concepts. The development of the Wiki—as a collaborative system for the development of a common, distributed, knowledge base—has provided open-source software systems for sharing knowledge. Testing the value of such systems in this context was an additional goal for the programme.

# 11.3 The Eco-Innovative Cities programme: overview

## 11.3.1 Introduction

The programme, a one-year 'pilot', is viewed as a practical exploration of the potential to innovate service provision at the local government level. The aim is to challenge the current delivery of council operations and decision-making processes, to shift from a focus on waste and recycling to one on designing more sustainable systems. The eight participating councils have already made significant commitments to reducing the emission of greenhouse gases and many have established green purchasing programmes. They have not, until now, explored the redesign of their service provision as a way of reducing resource consumption.

In the project to date, the participating councils have identified some core services that are important both in terms of their social, political and economic role as well as in terms of their contribution to resource consumption. Working on a two-monthly cycle, using a service redesign process developed for this programme, two service areas ('libraries' and 'the local street') have so far being tacked by each council. The results of those activities is communicated between all the participants for shared learning. At the end of the trial programme a resource 'kit' will be produced and a larger (possibly multi-state) programme will be launched. The programme commenced in November 2005 as a one-year pilot scheme.

### 11.3.2 Participants

EiC is funded by a statuary authority, Sustainability Victoria, and the research and practice is part of a close collaboration between the Australian Centre for Science Innovation and Society (ACSIS), at the University of Melbourne, and the International Council for Local Environmental Initiatives (ICLEI), Australia and New Zealand.

#### 11.3.2.1 Australian Centre for Science Innovation and Society

ACSIS is a new national centre created in 2005, based in Victoria, at the University of Melbourne. ACSIS was established to bring together researchers, business, government and the community to provide a network space in which emerging technology, science and innovation can be 'shaped' to achieve optimum social, economic and environmental outcomes.

- As a research hub, ACSIS assesses the potential value of investing in different emerging technologies, based on a process of mapping their potential social, environmental and economic outcomes
- As an innovation hub, ACSIS will work closely with its stakeholders to explore innovative new technological solutions with triple-bottom-line outcomes
- As a knowledge-sharing hub, ACSIS actively explores methodologies and processes for informing the public and framing public debates so that appropriate technological choices are made and implemented

#### 11.3.2.2 International Council for Local Environmental Initiatives: Local Governments for Sustainability, Australia and New Zealand

ICLEI Local Governments for Sustainability is an international association of local governments and national and regional local government organisations that have made a commitment to sustainable development. More than 470 cities, towns, counties and their associations worldwide form ICLEI's growing membership. ICLEI works with these and hundreds of other local governments through international performance-based, results-oriented campaigns and programmes. ICLEI in Australia and New Zealand (ICLEI–A/NZ) is hosted by the City of Melbourne and was established in September 1999.

### 11.3.2.3 The councils (Melbourne, urban and peri-urban)

Local councils are the third tier of government in Australia (along with the State and Federal government). There are 722 local government bodies accountable to a diverse range of metropolitan, regional, rural and indigenous communities. The population and geographic size of councils across the country differ greatly (the largest has 900,000 residents; the average is 26,400). There are about 6,600 elected councillors with an average of just under 10 councillors per council. Local councils spend around Aus$17 billion each year, providing an increasingly broad range of infrastructure and economic and community services to residents. The roles and responsibilities of local government in Australia differ somewhat from state to state, but their functions cover broadly:

- Infrastructure and property services, including local roads, bridges, footpaths, drainage, waste collection and management
- Provision of recreation facilities, such as parks, sports fields and stadiums, golf courses, swimming pools, sport centres, halls, camping grounds and caravan parks
- Health services such as water and food inspection, immunisation services, toilet facilities, noise control and meat inspection and animal control
- Community services, such as childcare, care and accommodation for the aged, community care and welfare services
- Building services, including inspections, licensing, certification and enforcement, planning and development approval
- Cultural facilities and services, such as libraries, art galleries and museums
- Household waste collection and water and sewerage services

## 11.4 The context

### 11.4.1 The State of Victoria: high consumption footprint

Although being the smallest in terms of land area, in population terms Victoria is the second largest of the Australian mainland states, with a population of 4.8 million. It is the most densely populated, generating 25% of the country's gross domestic product (GDP), with an ecological footprint of approximately 8 'global hectares', around 5% higher than that of Australia as a whole.[2]

---

2 Victoria's footprint is three times higher than its calculated biocapacity (5.4 ha; EPA Victoria 2005).

Recent studies have highlighted the critical resource consumption drivers for Victoria: population and economic growth (particularly a reliance on exports of primary produce and mineral resources); lifestyle (material consumption per head; particularly housing size and house occupancy rates and the scale of car ownership) and fossil-fuel-based energy provision (particularly related to transport and the generation of electricity from brown coal). The impact of global warming is expected to exacerbate these pressures.[3] Consumption pressures for the state of Victoria include:

- End-use energy consumption. Electricity use has an average growth rate of around 2% per annum and a relatively new summer peak demand profile increasing at closer to 3% (NRE 2002); the latter is expected to grow as a result of greenhouse effects. Transport accounts for a significant portion of energy consumption

- Water consumption. Rainfall in Victoria has decreased significantly over the past 50 years, with predictions of a future 5% decrease by 2020 (CSIRO 2004). Current annual water consumption in Victoria is about 5,800 gallons, with about 77% used for agricultural irrigation. Both irrigation and urban and industrial consumption increased by 50% between 1984 and 1997 (CSIRO 2004)

- Waste and material flows. Australia's per capita level of material flow is very high by world standards, with domestic waste per capita second only to the USA among OECD countries (at 620 kg per year). The generation of solid waste in Victoria increased by 60% between 1993 and 2002

- Transport, households and urban development. The increasing resource intensity of Victorian lifestyles can be seen in the consumption of household appliances (with the average household now having 30 appliances, compared with 6 in 1954). In 2004 Australians wasted almost Aus$5 billion of unconsumed food (Hamilton et al. 2005). Victorian transport relies heavily on roads and private vehicle use, with the number of vehicle registrations increasing by 20% in the past ten years and with rates of private vehicle ownership similar to those in the USA

## 11.4.2 Concern about environmental issues

### 11.4.2.1 Public attitudes

Surveys suggest that the Victorian public views health and education as more important than the environment and that levels of environmental concern appear to have plateaued. Broadly, this follows a trend seen in Australia as a whole.[4] Some of this undoubtedly reflects the sceptical stance of the Federal Australian government towards

---

3 In each decade since 1950, Victoria's average maximum temperature has increased by 0.11°C, and the minimum by 0.07°C. In 2002, Victoria's total greenhouse gas emissions were 117 million tonnes, with per capita emissions of 24 tonnes (the Australian average being 28 tonnes per person), among the highest in the OECD countries.
4 In 2004 57% of adults reported concern about the environment, down from 75% in 1992.

key environmental issues such as greenhouse warming.[5] When asked to rank what they saw as the top priority of various policy issues by government, only 8% named the environment (60% saying that the government considered the environment as least important; DSE 2004). However, public attitude surveys need to be treated with some caution in light of the enthusiastic take-up of kerb-side recycling schemes[6] and of residential water-conservation programmes.

### 11.4.2.2 Industry attitudes and action

Surveys of industry and business in Australia also show a generally low priority given to environment as a driver for investment or new strategic development. A 2002 review of a range of such surveys concluded that industry and business have consistently given a low priority to the environment for current and short-term strategic planning (Greene 2002).

The KPMG *International Survey of Corporate Responsibility Reporting 2005* found that reporting rates in Australia are lower than in most of the countries surveyed (quoted in DEH 2005).

### 11.4.2.3 Local government attitudes and action

The picture of environmental concern and action at the level of local government (LG) in Australia appears as something of a contrast to that of industry and the general public. Past research has identified some unique features of LG in terms of approaches to programme delivery. Council staff can be broadly split into three groups: 'insiders', who view LG workers as 'creatures and servants of the local'; 'outsiders', who see LG workers as 'creatures and servants of the State'; and 'facilitators', who understand both of these perspectives and actively pursue a role as a bridge (Wild River 2005). One result of these differences in perspective is that action at the LG level is often more dynamic, giving expression to different interests and priorities than those that operate at the higher levels of government.

Where potential environmental issues and actions at State and Federal level are often 'evaluated' in strongly economic terms (typically in terms of 'payback periods') there is a sense that, at the LG level, assessment of environmental objectives and outcomes are more mediated, with social outcomes, 'lifestyle values' and longer-term views given higher priorities. There are popular and successful local sustainability programmes initiated and supported by local councils.[7] Australia's most recent and certainly most innovative 'green' building is the new council offices for the city of Melbourne.[8] Other local councils are also investing in innovative 'green' buildings for offices or locally provided facilities such as libraries. Local councils have supported the infrastructure of success-

---

5 Australia has refused to ratify the Kyoto Protocol and it is only recently that the government has been forced to moderate its early public stance that global warming was scientifically uncertain and that curbing greenhouse gases was not economically in the country's best interests.
6 Victorians have increased their household recycling rate from 14% in 1993 to 34% in 2004.
7 Such as the widespread 'sustainability street' programme; see www.voxbandicoot.com.au/SustainabilitySt [accessed 25 August 2007].
8 The building, known as CH2, opened in August 2006: see www.melbourne.vic.gov.au/info.cfm?top=171&pg=1933 [accessed 25 August 2007].

ful kerb-side recycling and other resource-recovery systems. The Cities for Climate Protection (CCP) programme of ICLEI has over 210 local governments from around Australia participating—a coverage of over 80% of the Australian population. CCP empowers local governments to cut greenhouse gas emissions.[9]

The importance of LG is summed up by Wild River (2005: 48): 'every sustainability issue is a local sustainability issue . . . LG support for sustainable development initiatives can be a key factor in their success. LGs can also powerfully constrain sustainability outcomes if they oppose or fail to actively support them'.

## 11.5 The project: aims, process and methodology

### 11.5.1 Aims

EiC is a pilot project. Its aim is to optimise an 'eco-service design' methodology for later use in the context of local government. EiC aims to get councils to:

- Understand the idea of 'sustainable consumption' and to recognise that the concept has relevance to the provision of services
- Change their views on the process of continuous improvement in service provision to encompass issues of resource consumption
- Recognise that issues of resource consumption will require moving beyond continuous, incremental improvements in the current delivery and operation of council services to more radical ecodesign and 'eco-innovation'

The term 'eco-innovation' refers to dramatic reductions in the eco-footprint of services.[10]

### 11.5.2 Process

Selected CCP councils were offered an opportunity to sign up to the project, to commit to the work and to appoint an officer as the EiC coordinator. The coordinators that have been selected are generally those whose roles in their councils are focused on environmental (or sustainability) activities. Coordinators were given some initial training in the methodology (at a half-day workshop) and then supported by monthly 'issue papers' produced by the research team. They are responsible for 'running' the methodology within their councils, coordinating the input of council staff and reporting monthly on progress, learning and results, via teleconferences and through online 'Wiki' forums (see below). The initial workshop was used to define the selection of service areas that would

---

9 In 2005 CCP LGs reported emission reductions of 1.55 million tonnes of carbon dioxide ($CO_2$) equivalent, exceeding the previous year's figures by more than 22%. Total emission reductions over the life of the programme now exceed 5 million tonnes of $CO_2$ equivalent.
10 The term 'eco-footprint' which is used as a measure of the environmental impact of consumption (for example, see www.epa.vic.gov.au/ecologicalfootprint/calculators/default.asp) is used here in a looser sense to imply the total 'size' of the resource demands of the services in question.

be the case studies for this first year, settling on 'libraries' and 'streets', with 'libraries' as the starting project.

As the programme proceeded it was apparent that the action expected from the EiC coordinators was too demanding and difficult to achieve and it has therefore been altered to include workshop support from the ACSIS and ICLEI staff.

### 11.5.3 The design methodology

#### 11.5.3.1 Introduction

The EiC methodology builds on past experience, on the part of the author, in product ecodesign, and recent work in eco-service design and PSSs.[11] The process is built around workshops and uses the 'constructive tension' between two ways of viewing the subject matter: an 'incremental improvement' approach and a (radical) 'redesign' approach. These are described as follows:

- Incremental improvement accepts the structure of the existing service and aims to make it more eco-efficient by changing some aspects to reduce resource consumption. This may involve changing products (such as vehicles), extending the life of other products (such as books), redesigning buildings (e.g. to reduce energy consumption), or adjusting the current service 'system' (the logic or connection between different parts of the service, such as the way that books are purchased)

- Service redesign starts from a critical reflection on the needs the service is intended to fulfil and looks for different ways in which those needs could be met, other than through the existing service. The service redesign approach is intended to be the more 'blue sky' and more 'innovative' approach, a chance to 'rethink' the whole idea. In theory, at least, although not always in practice, for an existing service the process of service redesign can deliver greater overall improvement in resource consumption than can the process of incremental improvement

A step-by-step process aims to place these two approaches in constructive tension so as to increase the chance that the outcomes will be innovative—delivering solutions that are both novel and achievable. A workshop on incremental improvement is followed later by another workshop on system redesign, after which the analysis and solutions generated from the incremental improvement approach are revisited. The two 'possibility spaces' represented by the two approaches are kept as separate as possible; in practice, the separation occurs by allowing time between the associated workshops and by encouraging the coordinators to set up events with different 'characters' and purposes. The methodology emphasises the positive value of 'conceptual tensions'.

---

11 The methodology is based on work to appear in the forthcoming UNEP ecodesign manual, *D4S: Design for Sustainability: A Global Guide* (Crul and Ryan 2008) in particular from the chapters on 'Design Oriented Scenarios' (Manzini *et al.* 2008), 'Product Service System Design' (Ryan *et al.* 2008) and 'Quick Start Guide' (Ryan 2008).

### 11.5.3.2 The incremental approach

This involves the following steps:

1. Select a team; prepare a background report (a service dossier)
2. Run a 'workshop':
   - Draw a life-cycle service-system map
   - Draw up a stakeholder matrix
   - Identify resource inputs and (waste) output
   - 'Vote' on priority areas for action
   - Brainstorm initial ideas for improvement
   - Deliver a report and a proposal for actions to be investigated
3. Quantification of inputs and outputs (first-level approximation of data)
4. First-level evaluation of improvement options
5. Search for other improvement options and evaluate them

The **service dossier** is a short 'background analysis' of the service, produced as a way of gathering information and drawing people into the workshop process.

The **life-cycle service-system map** (LCSSM) map is a simple visual representation of the life-cycle 'flow' of the service. It is assumed that the workshop process will work better a visual 'map' is created (even if it is not graphically elegant) to identify all the components of the system—the connections, movements and relationships—necessary for the service to operate.

The **stakeholder matrix** (SM) is collectively developed as a simple table of all the stakeholders involved in the service, identifying the characteristics of the relationship of each stakeholder to the service.

**Priorities for action** are determined through 'voting', with a simple ranking given to each item.

**Ideas for improvement** proceed through brainstorming sessions, using 'sticky notes' for each idea, with these quickly posted around the walls and grouped into related categories.

A simple four-cell **business feasibility eco-matrix** is used as check for considering the business case for various improvement options. The cells are:

- Cell 1: significant environmental gains, technically and economically difficult
- Cell 2: limited environmental gains, technically and economically difficult
- Cell 3: limited environmental gains, technically and economically feasible
- Cell 4: significant environmental gains, technically and economically feasible

### 11.5.3.3 The service redesign approach

By its nature the service redesign approach is less amenable to a structured set of steps. There are, however, some activities and tools that can facilitate the process and add considerably to the outcomes. The methodology includes the following:

1. Develop a 'working definition'(or multiple definitions) of the purpose of the service, taking input from a range of people. Such definitions should:
    - Be from the community perspective
    - Identify the 'need' that is being fulfilled, trying to 'get beyond' the obvious, aiming for a statement that encapsulates the richness of the service and one that looks at it from a 'different perspective'
    - Accommodate some divergence of views
2. Brainstorm new service ideas (again, using 'sticky notes')
3. Elaborate a small range of alternative services (developing an LCSSM and SM for each)

Participants are encouraged to try to produce an advertising poster for each of the elaborated services—a simple visualisation of the services for users.[12] Finally, the potential value of each new service possibility is evaluated using the four-cell matrix. Then participants 'revisit' the outcomes of the incremental approach.

### 11.5.3.4 Collaboration: workshops, teleconferences and Wiki

The research process is focused strongly on collaboration—both within the participating councils and between the EiC team leaders. Team leaders undertake internal workshops and then report back to the other team leaders via teleconferences and also on a more ongoing basis via a Wiki online discussion and collaboration forum. ICLEI has found, from its experience, a format and process for teleconferences that allow them to be a very effective mechanism for collective learning. Rather than bringing these dispersed participants together for feedback and discussion, initial face-to-face workshops have been followed with one-hour teleconferences each month. These meetings do not focus on the outcomes of the service-design workshops (which are communicated through the Wiki online collaborative space) but rather on the experience of the process and the methodology. Teleconferences become a way of generating discussion around any practical difficulties and any new practical solutions or adaptations of the methodology.

Most recent projects of this kind have made some effective use of online (web or email) discussion spaces. With the advent of open-source Wiki software and document managers a new and much more collaborative system was available to the project.[13] The issue papers, which had first been produced in email and print form, now appear only on the Wiki site. The progressive discussion in the teleconferences is also entered directly into the Wiki.[14]

---

12 The step forms an important part of the design-oriented scenario technique of Manzini et al. (2008) from the UNEP publication D4S: Design for Sustainability: A Global Guide (Crul and Ryan 2008).
13 See 'Docu-wiki', at wiki.splitbrain.org/wiki:dokuwiki, and 'DocManager', at wiki.docmgr.org/index.php/DocMGR [accessed 25 August 2007].
14 A welcome page is publicly accessible: mel.iclei.org/eic-wiki/welcome [accessed 25 August 2007].

## 11.6 Review and outcomes

### 11.6.1 Introduction

At this time (July 2006) the participants have completed their initial test of the methodology through the exploration of the service of 'libraries' and a collection of services that come together in 'the street'. Some obvious and some unusual concepts and aims for libraries are as follows:

- The library buildings should be ecodesigned, as exemplars and showcases
- The utilisation rate of libraries should be increased: for example, by combining local council libraries with those of local schools
- Libraries should be used as a resource centre for information about sustainable lifestyles and sustainable consumption, inducing more informed behaviour
- The scope of what is provided for sharing through libraries should be widened, from books to other important consumer articles such as tools
- Libraries could be taken into public–private partnerships: for example, as 'distributed' collections in cafés, reducing the distance travelled for users

### 11.6.2 Practical limitations: constraints on the methodology

The project work focused around 'the street' really highlighted what initially appeared as a barrier to working through the methodology in the library exercise. In complex services (or collections of services) it is difficult to obtain the necessary cooperation and time from people involved across what are often separate domains of responsibility. While participants reported that the methodologies are thought-provoking and stimulated new ideas in service delivery, the difficulties of working within the day-to-day operations of councils meant that the methodology has had to be further simplified. This reflects the short time available and the degree of systems thinking and silos.

#### 11.6.2.1 Time availability

All council officers are busy with multiple priorities and projects, so the project has had to prove its value. Workshops have had to be limited to two hours in duration (to be able to attract participation from enough staff from relevant areas). It was not possible to rely on any significant research activity outside the workshop time. The most frustrating limitation proved to be in moving from the (more creative) exploration of current conditions and resource use to the systematic evaluation of priorities for action. The strong need was to arrive (quickly) at (creative) solutions and identify actions that were sufficiently relevant to current or impending items on the council business plan. The result was a lack of time to systematically prioritise resource areas and a tendency to consider all areas as equal targets for action; the recycled content of materials in street furniture, for example, became as important as fossil fuel consumption from vehicles, and so on. There were also only limited possibilities for 'radical rethinking' and little chance for the intended 'creative tension' between the incremental and the radical.

### 11.6.2.2 Systems thinking and silos

Council staff were not used to standing back from an area and considering system connections. Perhaps the biggest success of these workshops (and the one that has drawn most positive evaluation) was that the process brought staff together across different, otherwise 'siloed', domains of interest and responsibility. There were comments at the workshops such as: 'we've never met before to discuss these things'; 'we've never been properly enabled to look at the systems of connection between our different areas of programme delivery'. The workshops were successful in the sense that they demonstrated system connections and the value of system-wide thinking and solutions. For example, no one had apparently considered the relationship between rainwater collection from buildings discharged to the street and the size and maintenance of drainage and final discharge of water flows. Staff were most likely to treat the rainwater from roofs as a given, considering the potential of household, commercial or retail retention of water though on-site tanks, as 'out of their sphere of influence' (an echo of the' outsider–insider' split referred to in Section 11.4.2.3).

## 11.6.3 The emergent idea of 'resource-conscious design'

For many people involved in this programme (attending the workshops) this was essentially an introduction to design or 'design thinking'. However, there was almost always a small number of people involved from an area identified in some way as 'engineering', and it was clear that they often felt quite comfortable with the idea of a design approach to services. At one of the 'street' workshops an extensive discussion on the design process was led by a council engineer. Out of this came the term 'resource-conscious design' (RCD) as a description of what the programme or methodology was aiming to deliver. RCD as a term was raised in subsequent workshops (with other councils) and seemed to generate positive attention and support. It is an interesting term, placing the focus on the idea of 'being conscious of resource consumption' in all design (and business) decisions. It was described by one participant as 'putting on a different pair of glasses in strategic planning'. One council is subsequently pursuing the idea of a 'resource conscious design policy'.

## 11.6.4 The problem of outsourcing and avoiding the 'liability trap'

Most councils have long ago adopted a policy of outsourcing some of their work. In some cases whole departments of council have been 'spun off' to form new service companies (e.g. a number of councils have outsourced the design work on roads and related infrastructure). It seems (from the viewpoint of those engineers still in councils) that the outside design consultants will always overspecify so as to avoid any potential for liability arising from future engineering or structural problems. The programme is collecting examples of what this means in practice and how it affects innovation and experimentation.

# References

Crul, M., and C. Ryan (eds.) (2008) *D4S: Design for Sustainability: A Global Guide* (Paris: United Nations Environment Programme, forthcoming).

CSIRO (Commonwealth Scientific and Industrial Research Organisation) (2004) *Environmental Sustainability Issues Analysis for Victoria* (Melbourne: Department of Sustainability and Environment).

DEH (Department of Environment and Heritage) (2005) *The State of Sustainability Reporting in Australia 2005* (Canberra: Commonwealth of Australia DEH).

DSE (Department of Sustainability and Environment) (2004) *Environment and Sustainability Community Research: Final Report* (Melbourne: TQA Research).

EPA (Environment Protection Authority) Victoria (2005) *The Ecological Footprint of Victoria: Assessing Victoria's Demand on Nature* (Global Footprint Network; Sydney: EPA Victoria/University of Sydney, 14 December 2005).

Geels, F.W. (2002) 'Technological Transitions as Evolutionary Reconfiguration Processes', *Research Policy* 31.8–9 (December 2002): 1,257-74.

Graedel, T. (1998) 'Life-cycle Assessment in the Service Industries', *Journal of Industrial Ecology* 1.4: 57-70.

Greene, D. (2002) *Industry Attitudes to the Environment* (Canberra: Department of Environment and Heritage, Government of Australia).

Hamilton, C., R. Denniss and D. Baker (2005) *Wasteful Consumption in Australia* (DP-77; The Australia Institute; www.tai.org.au).

Kemp, R., J. Schot and R. Hoogma (1998) 'Regime Shifts to Sustainability through Processes of Niche Formation: The Approach of Strategic Niche Management', *Technology Analysis and Strategic Management* 10.2 (June 1998): 175-95.

Kisch, P. (2000) *Preventative Environmental Strategies in the Services Sector* (PhD thesis; Lund, Sweden: International Institute for Industrial Environmental Economics [IIIEE], Lund University).

Manzini, E., and C. Vezzoli (2002) *Product-Service Systems and Sustainability: Opportunities for Sustainable Solutions* (Paris: United Nations Environment Programme).

——, F. Jégou and A. Meroni (2008) 'Design Oriented Scenarios', in M. Crul and C. Ryan (eds.), *D4S: Design for Sustainability: A Global Guide* (Paris: United Nations Environment Programme, forthcoming).

Mont, O. (2004) *Product-Service Systems: Panacea or Myth?* (PhD thesis; Lund, Sweden: International Institute for Industrial Environmental Economics [IIIEE], University of Lund).

NRE (Department of Natural Resources and Environment) (2002) *Energy for Victoria* (Melbourne: NRE).

Rip, A., and R. Kemp (1998) 'Technological Change', in S. Rayner and E.L. Malone (eds.), *Human Choices and Climate Change. Vol. 2. Resources and Technology* (Columbus, OH: Battelle).

Ryan, C. (2003) 'Learning from a Decade (or So) of Eco-design Experience: Part 1', *Journal of Industrial Ecology* 7.2: 9-12.

—— (2004a) 'Learning from a Decade (or So) of Eco-design Experience: Part 2', *Journal of Industrial Ecology* 8.4: 6-9.

—— (2004b) *Digital Eco-sense: Sustainability and ICT: A New Terrain for Innovation* (Melbourne: Lab 3000).

—— (2008) 'Quick Start Guide', in M. Crul and C. Ryan (eds.), *D4S: Design for Sustainability: A Global Guide* (Paris: United Nations Environment Programme, forthcoming).

——, U. Tischner and C. Vezzoli (2008) 'Product Service System Design', in M. Crul and C. Ryan (eds.), *D4S: Design for Sustainability: A Global Guide* (Paris: United Nations Environment Programme, forthcoming).

Smith, A., A. Sterling and F. Berkhout (2005) 'The Governance of Sustainable Socio-technical Transitions', *Research Policy* 34: 1,491.

Tischner, U., E. Schmincke, F. Rubik and M. Prösler (2000) *How to Do Ecodesign?* (Frankfurt: Verlag Form Praxis).

Tukker, A. (2005) 'Leapfrogging into the Future: Developing for Sustainability', *International Journal of Innovation and Sustainable Development* 1.1–2: 65-84.

—— and U. Tischner (eds.) (2006. *New Business for Old Europe: Product-Service Development, Competitiveness and Sustainability* (Sheffield, UK: Greenleaf Publishing).

Verloop, J., and J.G. Wissema (2004) *Insight in Innovation: Managing Innovation by Understanding the Laws of Innovation* (Boston, MA: Elsevier).

Wild River, S. (2005) 'Enhancing the Sustainability Efforts of Local Governments', *International Journal of Innovation and Sustainable Development* 1.1–2: 46-64.

# 12
# Is a radical systemic shift toward sustainability possible in China?

**Benny C.H. Leong**
Hong Kong Polytechnic University, China

## 12.1 Two trivial incidents

Last summer I travelled from Hangzhou to Guangzhou. Arriving at Hangzhou airport 90 minutes early to enjoy a proper lunch before boarding the plane, I went to a self-service canteen. It was a fine place to eat: the price was reasonable, it was neat and comfortable, the choice of food was wide and the menu was well presented. However, there was a specific condition attached to eating: the wasting of food was prohibited. To reinforce this, warning signs advised customers to 'take only what you need or pay for the leftovers' (Fig. 12.1). Despite the visibility of these signs, many people walked out of the restaurant leaving plenty of unconsumed food on their tables (Fig. 12.2), but there was no condemnation of this whatsoever.

Half an hour later, I boarded the plane and settled into my seat next to the window. The plane was quite full, and most of the passengers were Chinese. A well-dressed Chinese man who I took to be a native of Hangzhou arrived to take the seat beside me. I took no notice of him, pondering as I was the incident that I had just witnessed, until he switched on the personal spotlight above him to read. This action would not have been surprising had it been night and the lights of the cabin turned off, but they were not, and it was a sunny day. There was more than enough light. A moment later I realised that more than 80% of the personal spotlights were switched on (Fig. 12.3).

These incidents may seem insignificant, but they have wider implications for sustainability:

FIGURE 12.1 The warning signs on the display table

FIGURE 12.2 Unconsumed food left on one of the tables

FIGURE 12.3 On the plane from Hangzhou: more than 80% of the spotlights were switched on inside the cabin

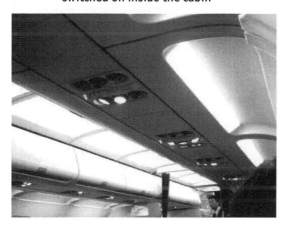

- These people seem avaricious, and appear to share the trait of irresponsible exploitation of resources

- These people are native Chinese, a group that represents one-fifth of the world's population. An emerging wave of unrestrained consumption is therefore foreseeable

## 12.2 The dilemma for China

China is famous for its recent rapid growth in consumerism. Retail sales of consumer goods topped US$157.4 billion in the first two months of 2006 alone and are forecast to grow at a rate of 13.2% per annum.[1] China is not only the 'factory of the world'; it is also becoming the largest market for commodities and one of the most ravenous consumers of energy.[2] China is already the world's second-largest producer of greenhouse gases, which contribute to global warming and the imbalance of the world's climate. China's carbon dioxide ($CO_2$) emissions will have easily overtaken those of the current biggest greenhouse gas emitter, the USA, by 2020 (*Sydney Morning Herald* 2005), and will dwarf any cuts in emissions that the rest of the world might make. At the same time, rapid industrialisation has engendered a highly polluted environment that has cost China dearly.[3] Hence, it is desirable for China to slow its industrialisation and rate of consumption.

However, the rapid economic development and concomitant rise in energy consumption in China cannot be easily decelerated. That would not only spawn massive unemployment and a decline in gross domestic product (GDP) but also create inflation and result in a loss of capital. Any hiccup in China's economic progress would also be likely to affect the world economy.[4]

This is the dilemma that is currently facing China's ruling regime. As the pace of environmental deterioration in China accelerates, innovative solutions that are capable of simultaneously conserving the country's natural resources and energising the economy are urgently being sought. The developed West has moved toward the replacement of

---

1 Retail sales of consumer goods were US$509 billion in 2003, and US$569 billion in 2004 (NBSC 2006; *People's Daily* 2004).
2 According to the latest publication of the IEA (International Energy Agency), China consumed around 7 mb/d (million barrels per day) in 2006. Yet the world's energy needs will be over 50% higher in 2030 than today (demand will reach 116 mb/d, with China consuming 14%, or 16.24 mb/d of the total). China and India together account for 45% of the increase in global demand in this scenario (IEA 2007).
3 Roughly 70% of China's rivers and lakes are polluted (Yardley 2005). China's emissions of sulphur dioxide ($SO_2$) were the highest in the world in 2004, and about 30% of the country is affected by acid rain. In 1997, the World Bank reported that 178,000 premature deaths are caused each year in China by poor air quality. With 346,000 hospital admissions and 6.8 million outpatient visits due to respiratory ailments, the cost of air pollution has been put at US$43 billion a year (*Sydney Morning Herald* 2005).
4 For example, the average price of shoes and clothing in the USA has fallen by 35% in real terms over the past ten years as a result of the importation of cheaper goods from China. This has pushed wages down and lowered inflation and interest rates. What, then, would happen if the export of manufactured goods and importation of raw materials in China decreased (*Economist* 2005c)?

its outdated and unsustainable economy with function-based (service-led) sustainable consumption and production (SCP) practices, which are increasingly being seen as a possible and promising solution for China's development dilemma.

In pursuit of creative, sustainable solutions for China, two series of design workshops and a joint research programme were initiated from 2000 to 2004 to promote the concept of solution-based design.[5] These endeavours aimed to generate an initial understanding of the difficulties and possibilities of introducing a radical systemic shift toward sustainability through a switch from product-based to function-based production and consumption.

## 12.3 Learning from the workshops and collaborative research

The first series of design workshops in 2000 yielded three major findings.

- The idea of shifting from 'product' to 'solution' design to minimise environmental cost was, in theory, well understood by Chinese students (the emerging consumers of China)

- Many of the participants were doubtful about the actual usefulness and relevance of the 'visionary ideal' of solution-driven design for sustainability in China

- The localisation of the design for sustainability (DfS) concept was one of the major concerns

To experiment with a hypothetical design process[6] that connects the practice of sustainable product-service systems (SPSS) and industrial design, and the issue of DfS localisation, the Learning Network on Sustainability (LENS) programme (see below and footnote 5) and a second series of design workshops were initiated in 2002 and 2003, respectively. One of the objectives was to find the causes of doubt about the introduction of sustainable services in China. In the LENS programme, a network of various universities from different countries was formed to experiment with transcultural design for sustainability (Vezzoli 2006). This taught us that, other than discussion of the environmental, economic and social aspects of sustainability, the debate over 'cultural' and 'political' differentiation as barriers to concept promotion is essential in China.

For the second series of workshops, an introductory discussion session was held. A sample sustainable service solution called 'Sunwash Laundry' (Fig. 12.4) was intro-

---

5 The joint research programme named LENS (Learning Network on Sustainability) was initiated by Carlo Vezzoli in 2002. The two series of design workshops were held in conjunction with Ezio Manzini and Elaine Anne, Director of Kaizor Innovation Ltd, in Beijing, Guangzhou, Hunan, Wuxi and Hong Kong in 2000 and 2004 to promote the concept of design for sustainability and system–product design in China.
6 This hypothetical design process is called 'system–product design' (SpD). It is devised specifically to identify and develop products or systems that provide sustainable solutions based on a product-service system approach (Leong 2002).

## FIGURE 12.4 'Sunwash Laundry': a sample sustainable service solution

1. On the rooftop of the residential area is a laundry service centre . . .
2. High-speed elevators in the building provide means of transport for residents to and from the rooftop . . .
3. The laundry centre is managed and operated by well-trained staff . . .
4. People can socialise with neighbours while taking a break at the cozy teahouse . . .
5. Or sometimes one might prefer the door-to-door laundry delivery service.

duced, and a specific question about the solution was projected onto a screen or wall: 'will this sustainable service solution work in this city?' The participants were then asked to write down on sticky notes all of the factors that might influence the introduction of the service to the community.

The sticky notes revealed the following concerns about the introduction of solution-based sustainable services.

- There is a foreseeable conflict between new ideals and old customs
- The possession of physical products is still much preferred
- Sharing is not favoured, and shared facilities or devices are not valued
- There is a lack of trust and communication among people in general
- Service quality and staff are very much distrusted
- Personal hygiene is a great concern
- Privacy and information security is very much an issue

However, there was insufficient information from this initial feedback to draw any sensible conclusions. To substantiate and verify the findings, additional investigation was carried out.

## 12.4 The context of the limits and barriers to sustainable consumption and production

Four specific contextual aspects of the limits or potential barriers to SCP—political, economic, cultural and social—were studied and analysed.

## 12.4.1 Political appropriation

A matrix can illustrate four possible political scenarios—open–authoritarian, closed–authoritarian, closed–democratic and open–democratic—in which political and economic regimes interact with the society and environment of a country (Fig. 12.5).

FIGURE 12.5 **Matrix of four political–economic scenarios**

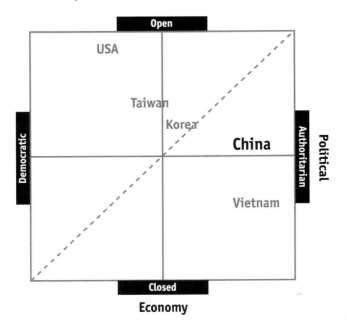

Zarsky and Tay (2000) have argued that 'whether an economy is open or closed, a politically controlling state is likely to engender high levels of environmental degradation'. Although China shifted to a relatively open economy in the late 1970s, its controlled polity and absence of constitutional rules on economic policy leave room for appropriation and exploitation by local officials at the province, city and township levels.

### 12.4.1.1 Policy abuse

The promotion policy by local officials is based on successful economic leadership in the region under control. Hence, local officials impose measures to boost local revenue at all costs, including the discouragement of 'green' initiatives and the prevention of moves toward social and cultural sustainability. For example, in Shenzhen car pooling is treated as tax evasion and is banned.[7] In Guangzhou in 2002, the government insisted on constructing an inner ring road in the city centre to ease traffic congestion and support the

---

7 A few office workers in Shenzhen organised themselves to share the use of a private car and the cost of its maintenance. However, instead of providing incentives and encouragement, the government decided that the car owner's act of charging the other users constituted tax evasion (Tong 2006).

developing automobile manufacturing industry, which it thought would increase the local GDP. Today, with the number of private cars growing exponentially in Guangzhou, traffic congestion is occurring again in the city centre. The ring road project, which came at the tremendous cost of the demolition of culturally rich old districts, as well as public discontentment as a result of forced resettlement, brought only a temporary solution to the problem.

### 12.4.1.2 Misdirected initiatives

In October 2005, the central government initiated a policy to narrow the widening gap between rural and urban China by promoting the construction of a 'new socialist countryside'. The aim of this initiative was to defuse unrest and turn the country's 800 million rural dwellers into an economically strong force. For local officials of Pinggu on the outskirts of Beijing, this meant bulldozing the houses of peasants and replacing them with weekend villas for city dwellers. This has merely served to intensify the disagreement between rural residents and local officials over the long unsettled issue of land ownership.[8]

These are the consequences of the central government's drive for continual economic growth without control over policy application. The growing mistrust and tension between the general public (especially in rural areas) and the ruling regime will certainly retard further economic reform and cripple any future progress toward sustainability.

## 12.4.2 Economic reality

There are three fundamental economic issues that any systemic reform initiative for a sustainable China must confront: the material-based economy, rising consumerism and the wealth disparity.

### 12.4.2.1 Material-based economy

The unflagging 8.7% average annual growth in GDP in China has been the result of product-based production and consumption over the past 25 years. China is likely to remain heavily dependent on its manufacturing industry to sustain this level of growth.[9]

### 12.4.2.2 Rising consumerism

With the steady rise in income, Chinese consumers, in contrast to a decade ago, now thirst for 'taste-driven' goods. DVD players, mobile phones, digital devices and comput-

---

8 Many peasants state that local bureaucrats have effectively become landlords by seizing land under the auspices of local government and exploiting it themselves (*Economist* 2006).
9 China's manufacturing sector has been developing rapidly in the last decade; it is now known as the 'workshop of the world', producing two-thirds of all photocopiers, microwave ovens, DVD players and shoes, over half of all digital camera and two-fifths of personal computers (*Economist* 2004).

ers are now desirable gadgets that Chinese consumers long to possess.[10] Materialistic possession as a sign of 'wealth' is a very strong idea in modern China.

### 12.4.2.3 Wealth disparity

There is a sharp difference in income between urban and rural dwellers in China. Urban Chinese earn between three and six times more than rural Chinese.[11] This great disparity has created a 'digital divide' that prevents the development and adoption of a service economy based on information and communications technology (ICT).[12]

## 12.4.3 Cultural constraints

China was a predominantly agricultural country for a long time, and its culture originates from a so-called 'peasant' community, with the following main attributes:[13]

- National order and social stability were upheld and enforced by the *jia* (family). For example, under the Ming Dynasty (1368–1644 CE) the family was identified as the official unit of the country. In fact, the rule of China until modern times was managed by two basic entities: the millions of kinship-based rural communities and the central administrative body of literate officers under the ruler

- The family was also treated as an economic unit called the *hu* (household), which practised self-sufficient and self-reliant peasant-based production. Parental management and ruling, which was founded on the notion of family hierarchy, prevailed as the main form of organisation

- As Chinese society was formed by peasant communities based on kinship, a basic belief system developed that held that the resources of the world are in fixed supply and that resource competition among family entities was inevitable. This promoted a suspicious attitude toward strangers, and social lives revolved around tightening relationships with familiars rather than the formation of bonds with anyone else (Bond 1991). The Western concept of 'society' or 'community' did not exist (Lin 2002)

---

10 In the 1990s, the six most sought-after large basic household products were televisions, washing machines, refrigerators, air conditioners, cameras and videocassette recorders. By the turn of this century they were taste-driven goods and electronic gadgets, such as cars, digital cameras, mobile phones, computers and DVD players (McEwen *et al.* 2006).
11 The richest 10% own 45% of the country's wealth, whereas the poorest 10% own only 1.4%. The average disposable income of Shenzhen residents in 2003 was 23,900 yuan, whereas that of Ningxia residents was 4,912 yuan (*China Daily* 2005a).
12 According to a recent survey by the China Internet Network Information Center (CNNIC), by June 2007 the *netizen* (Chinese residents who are aged six or above and have accessed the Internet in the past six months) population in rural China had reached only 5.1% compared to 21.6% in urban areas (CNNIC 2007). The application of ICT in a product-service system is believed to allow the substitution of some of the more environmentally burdening processes and activities, such as transportation and publishing, and is thus beneficial for the development of a 'lighter' and 'greener' economy.
13 The word *peasant* is used in this context rather than *farmer*, which refers to the more commercial and industrial practice of agriculture (Lau and Yang 1998).

- Peasant communities also cultivated Confucianism; this promotes, through a kind of pragmatism, secular living the best as one can as well as self-realisation and actualisation through one's affiliations (Woo 1995). With the centuries of social instability and poverty that have been experienced in China, that pragmatism is still embraced and endorsed by most people[14]

These attributes have created a system with a weak (lacking objective rules and institutions) and pragmatic (short-sighted and advantage-seeking) form of management that has long haunted Chinese society and its economy.[15]

### 12.4.4 Social displacement

According to the latest official projections, China's urban population will more than double, to 1.1 billion, by the middle of this century.[16] Every year, 10–12 million rural dwellers migrate to the cities (*People Daily* 2005a). To this newly formed urban population, the city is an alien place, and 'society' is a strange concept, because the close bonds of kinship are what make up their inherited concept of community.[17] In the past, most Chinese recognised a five-stage process of self-actualisation: the rectification of one's own heart, self-cultivation, family regulation, the right governing of the country and the pacification of the world (Fig. 12.6).

However, the unfamiliar environment in which these migrants find themselves serve to break their link with this process of self-realisation and demands the formation of a new world-view. Disconnected from familiar social bonds, mistrustful and lacking in discipline in their daily lives, these new city dwellers may well develop self-interested and antisocial behaviour (Fig. 12.7).

---

14 Pragmatism in China even extends to the issue of courtship. A social study revealed that 'the listing of desired traits in a spouse emphasises very practical considerations, such as the prospective spouse's income, ability to provide housing, etc. Romantic considerations like intense feelings and a sense of spiritual partnership are relatively less important' (Bond 1991).
15 Lieberthal notes that 'Chinese managers continue to fail in the critical tasks of systems integration and optimisation . . . it will take decades to overcome the ingrained systemic weaknesses that prevent them from designing efficient enterprises' (Lieberthal and Lieberthal 2004: 8-9). Similar problems have persisted for centuries, and they hinder China's social reformation and development (Liang 1987; Lin 2002).
16 The Chinese Academy of Science (CAS) predicts that 75% of the population will be urban by the middle of the century, and views the movement to cities as the only way to address many of China's rural problems, including the loss of arable land and rising unemployment. The CAS projects that China's urban population will more than double by 2050, to 1.1 billion (Li 2005).
17 The renowned linguist, philologist and writer Lin Yu Tang once said that 'public consciousness, community service and civil awareness are all new terms for Chinese . . . They only care about their family, but not society' (Lin 2002: 159).

FIGURE 12.6 Five-stage Confucian process of self-actualisation, or realisation

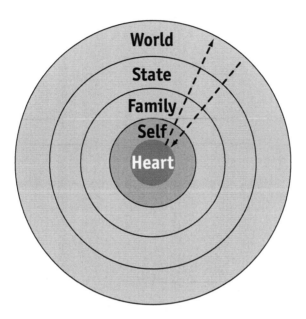

FIGURE 12.7 Social displacement in an alienated society

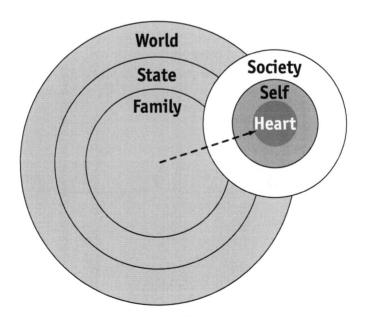

### 12.4.5 Summary

Thus, in addition to the workshop findings, the contextual (macro) and personal (micro) factors that could discourage the adoption of SCP practices in China can be summarised as follows:

- A controlled polity with a lack of constitutional rules on state policy
- Exploitation of economic policy by local bureaucrats at the cost of environmental degradation
- Rising consumerism and materialistic consumption
- A digital divide and a low rate of Internet access, crippling the development of ICT-enhanced services
- Rapid urbanisation and a rapid rise in the number of urban migrants, creating social displacement
- A weak conception of 'civil society' among most urban migrants
- A prevalence of pragmatism, which engenders self-interested and advantage-seeking behaviour that may be antisocial
- A general lack of mutual trust and a distrust of the service system
- A preference for the possession of items rather than the sharing of facilities

These factors seem to be the complete opposite of the criteria that are necessary for the development of sustainable service-based business, such as a relatively democratic polity, an active and strong civil society, an open and developed market economy with an enabled business environment (partnership-oriented, service-driven, stakeholder-responsive and customer-focused, with a high level of environmental consciousness), trust in society and the community and a developed ICT infrastructure (Angel and Rock 2000; Manzini *et al.* 2004; Mont 2000). These criteria have not even been fully met by most developed countries in the West, but in China most are non-existent. This leads to the question of whether a radical shift toward SCP practices will be possible in China in the foreseeable future.

## 12.5 The signs of change

A few signs of change can already be discerned, in political will, economic environment and consumer attitudes.

### 12.5.1 Political will

In the 11th Five-year Plan, which was promulgated by the central government in 2005, one of the focuses was to create a more balanced mode of development (*People Daily* 2005b; Miles 2005). The government promised an increase in environmental spending to US$169 billion, or 1.6% of GDP, and a 'green GDP'.[18] In addition, various cities have been undertaking 'greening' initiatives. Beijing, Shanghai and Guangzhou, for example, have launched plans to improve their environments.[19]

### 12.5.2 Economic environment

Under the World Trade Organisation (WTO) agreement, by 2006 foreign financial service firms will be permitted to provide an array of banking services in China. This should help to push access to capital away from state-owned enterprises and toward the private sector. A freer market economy will also deter the government from channelling individual savings to finance the expansion of state-owned enterprises (*Harvard Business Review* 2004).

### 12.5.3 Consumer attitudes

The increase in wealth in China is cultivating a 'greener' and more critical middle class and urban population.[20] For example, last winter hundreds of people living on the outskirts of Beijing protested against plans for a new factory because they feared that it would pollute the neighbourhood. In addition, a recent survey confirmed that 74% of urban consumers claimed that they would reject environmentally unfriendly products and services (Hollyway Consulting 2004).

## 12.6 In search of an appropriate strategy

Although the signs of change outlined in Section 12.5 are positive signs of a shift toward SCP practices in China, appropriate transition strategies have yet to be identified. In the following, two initial strategies are proposed to help businesses and non-governmental organisations (NGOs) overcome the possible barriers to the promotion of SCP in China.

---

18 Green GDP is the balance after imputed environmental cost and environmental resource protection expenditure are deducted from actual GDP. This highlights the interaction between the environment and the economy and was initiated by the State Environmental Protection Administration (SEPA) in 2003 and confirmed by Wen Jiabao, Premier of the State Council, in 2004. Test runs of green GDP have been promised for several regions, which should include Shanxi province. One study used green GDP measurements to show that Shanxi's economy has hardly grown at all in the past two decades (Bezlova 2005; Buckley 2004).
19 In January 2005, the policy document 'General City Planning for Beijing Municipality (2004–2020)' was approved by the State Council. Beijing was the first municipality to make being a 'liveable city' one of its objectives (Yuan 2005). In early 2005, Shanghai's administration set aside Chongming Island for ecological and recreational purposes, such as eco-tourism and organic agriculture. Also in 2005, the City Urban Greening Committee of Guangzhou started its 'garden in the air' programme, through which 60,000 m$^2$ of roofs will be covered with plants (*China Daily* 2005b).
20 The average household income in China rose by 30% between 1997 and 2004 (McEwen *et al.* 2006).

### 12.6.1 Overcoming the economic divide by making technology more accessible

The application of ICT is significant for successful sustainable solution-based businesses because it boosts the utilisation of such businesses. As China is still a developing country, with a massive deprived rural population and a large number of urban migrants, its rates of access to the Internet and information are low (12.3% in 2007, compared with 71.4% in the USA [IWS 2007]). However, it is hard to imagine the government increasing information access by funding computing centres or networks, because this would require great expenditure on infrastructure and could engender corruption among bureaucrats.[21] It would appear that a radical economic transformation toward a service-based economy in China is unattainable. However, rather than reaching for the impossible goal of system alteration, there is an appropriate piece of communication technology that could be utilised to promote bottom-up development—the mobile phone. There are several advantages in the application of the mobile phone to support SPSSs in China:

- It is handy and comparatively affordable (at present, the cheapest handsets are about US$30; Standage 2005)
- It is already accessible, with a penetration rate of around 26% in 2005, which is projected to reach 33% in 2007 (Market Research.com 2006)
- It requires no fixed network and no related infrastructure
- Its potential for service utilisation has been proven: cashless transactions have been made in Zambia and other African countries through mobile phones (*Economist* 2005a)
- It can be used by illiterate and literate people alike and is thus preferable to the alternative of a US$100 laptop[22]
- Telephone operatives would be required, which would create jobs for the millions of unemployed in China
- It would enable long-term growth, as an extra 10 phones per 100 people in developing countries increases GDP growth by an estimated 0.6% (*Economist* 2005a)

The mobile phone has great potential to facilitate China's economic transformation and advancement, and alternative business models and service systems should be developed to improve its utilisation.

---

21 The United Nations proposed a similar initiative—the Digital Solidarity Fund—in March 2005 to 'enable excluded people and countries to enter the new era of the information society'. However, its value has been queried by economists and social critics (*Economist* 2005a).
22 A hand-cranked laptop worth roughly US$100 was exhibited at the World Summit on the Information Society (WSIS) in Tunis during 2005. The product is expected to be in the hands of schoolchildren in poorer countries by late 2006. The One Laptop Per Child project is sponsored by Google, AMD, News Corporation, Red Hat, Brightstar, and Nortel, with the laptops built and designed by the MIT Media Lab and the Design Continuum (Wikipedia 2005).

## 12.6.2 Alter social behaviour through cultural bias

Any radical shift toward a sustainable China will require a massive social movement. The transition to an efficient, sustainable society is less a technological and more a cultural issue: the challenge is to counteract the prevailing climate of avarice and irresponsible social behaviour. The findings from the workshop series and other studies indicate that this behaviour is a product of culture, but culture could also be used to nurture the opposite behaviour of sustainable consumption. For instance, because of cultural norms and relative economic deprivation, Chinese people are inclined to care about:

- Wealth, or being 'money-minded', which originates from the cultural norm of pragmatism
- Health, or being concerned about the body, which derives from the cultural notion of the five cardinal relations, or *wu lun*[23]

The concern about wealth (economy) and health ('micro' ecology) could be developed into a strategy to promote sustainable solution-based initiatives in China. The the-

FIGURE 12.8 **The relevance matrix**

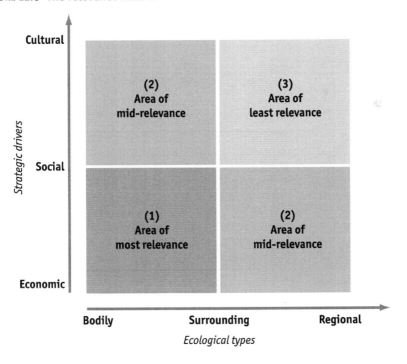

23 According to Confucius, one's existence is defined by five cardinal relations: those between emperor and officer, father and son, husband and wife, siblings, and friends. One's physical being must also be taken care of as a tribute to one's parents (Bond 1991; Sun 1992).

FIGURE 12.9 A retail stall of the Greendotdot Company

ory of relevant ecology[24] and the relevance matrix (Fig. 12.8), which were developed to help devise, analyse and verify various sustainable solutions, could be used to aid the development of such a strategy. The matrix comprises a vertical axis of three strategic drivers—economic, social and cultural—and a horizontal axis of three ecological types—bodily (the most intimate 'ecology'), surroundings (immediate environment) and regional.

Three areas are defined by the two axes in the relevance matrix:

- Area of most relevance
- Area of mid-relevance
- Area of least relevance

It is interesting to reveal how 'cultural bias' works in the area of most relevance of the matrix. Business solutions that fall within this area are expected to gain greater public recognition and acceptance and hence have a greater chance of success. Several business operations that focus on bodily health and economic gain have been quietly established and well received by consumers in Hong Kong. These include voluntary organisations such as the group purchase scheme, Organic Vegetable Centre, and business organisations such as Hong Kong Organic Country and Greendotdot Company (Fig. 12.9).

The aforementioned strategy could also be employed to promote sustainable service businesses that may not be directly related to bodily and economic issues, if marketers can visualise and promote the obvious personal gains and losses to the target consumers (economic gains or worsening health) in both quantitative and qualitative terms.

## 12.7 Conclusions

Can we now answer the question of whether a radical systemic shift toward sustainability is possible in China? One could be sceptical and focus on the irresponsible consumption that I experienced in Hangzhou, but successful sustainable businesses give a glimmer of hope that a systemic shift toward a 'greener' sustainable economy is not entirely impossible.

To aid the success of such a radical system shift in China, design should take on an operative role to assist the implementation of sustainable innovations. It should not be limited or isolated in research, education or practice, because design for sustainability must deal with possible cultural and social change. Designers should work collectively with other disciplines to find better system solutions. Hence, 'co-creation' will be a new

---

24 The word *ecology* here incorporates both human subjectivity and natural objectivity and is related to the realms of humanity, surroundings and the regional environment, which merge into a global ecology. The word *relevant* refers to the relevance of a particular matter to a given set of people. For example, remote environmental problems may be perceived to be intimate and relevant only if obvious personal gains and losses (such as economic gains or worsening health) that result from these problems can be visualised by an individual (Leong and Manzini 2006).

keyword for design alongside its characteristic roles of conceptualisation, visualisation and communication.

The next 15 years will be decisive for China, and perhaps for the world, as the 'environmental tipping points' of the globe become more threatened.[25] As a leading energy consumer and greenhouse gas emitter, China will certainly be one of the key players in shaping the future ecology of the world.

However, one should be aware that any possible adoption of systemic SCP in China should be less about forging a consensus on a *universal* theory for promotion than the development of a particular design strategy and its adoption in various contexts. This involves indigenous knowledge about *cultural* and *political* differentiation, which is very controversial in China.

## References

Angel, D.P., and M.T. Rock (eds.) (2000) *Asia's Clean Revolution: Industry, Growth and the Environment* (Sheffield, UK: Greenleaf Publishing).

Bezlova, A. (2005) 'China goes for "green GDP" ', Inter Press Service News Agency, www.ipsnews.net/africa/interna.asp?idnews=30849 [accessed 25 August 2007].

Bond, M.H. (1991) *Beyond the Chinese Face: Insights from Psychology* (New York: Oxford University Press).

Buckley, L. (2004) The Globalist "Green GDP": The Real Chinese Revolution?', *The Globalist*, 23 September 2004, www.theglobalist.com/DBWeb/StoryId.aspx?StoryId=4156 [accessed 25 August 2007].

*China Daily* (2005a) 'Income Gap in China Widens in First Quarter', *China Daily*, June 2005; www.chinadaily.com.cn/english/doc/2005-06/19/content_452636.htm [accessed 25 August 2007].

—— (2005b) 'Guangzhou to Build Garden-in-air', *China Daily*, October 2005; www.chinadaily.com.cn/english/doc/2005-10/17/content_485647.htm [accessed 25 August 2007].

CNNIC (China Internet Network Information Center) (2007) 'Survey Report on Internet Development in Rural China 2007', August 2007; www.cnnic.cn/html/Dir/2007/09/13/4795.htm [accessed 27 January 2008]

*Economist* (2004) 'The Dragon and the Eagle: A Survey of the World Economy', *The Economist*, 2 October 2004: 3-24.

—— (2005a) 'Behind the Digital Divide', *The Economist Technology Quarterly*, 12 March 2005: 16-17.

—— (2005b) 'From T-shirts to T-bonds'. *Special Report: China and the World Economy: 'How China Runs the World Economy'*, 30 July–5 August 2005: 65-67.

—— (2006) 'Fat of the Land: But only for a select few', *A Survey of China*, 25–31 March 2006: 5-9.

*Guardian Unlimited* (2005) 'Environmental Tipping Point', special report, *Guardian Unlimited*, August 2005; www.guardian.co.uk/climatechange/story/0,12374,1546824,00.html [accessed 25 August 2007].

Hollyway Consulting (2004) *The 60 New Thinking Methods for SME* (Beijing: China Textile Publishing).

IEA (International Energy Agency) (2007), press releases, 7 November 2007; www.iea.org/textbase/press/pressdetail.asp?PRESS_REL_ID=239 [accessed 27 January 2008].

IWS (Internet World Stats: Usage and Population Statistics) (2007) 'World Internet Users: November 2007'; www.internetworldstats.com/stats.htm [accessed 27 January 2008].

Lau, C.-C. and C.K. Yang (1998) *The Chinese Society: From No Change to Great Changes* (Hong Kong: Chinese University Press, in Chinese).

---

25 Environmental tipping points are delicate thresholds, such as in global warming, where a slight rise in the Earth's temperature can cause a dramatic change in the environment that itself triggers a far greater increase in global temperatures, which is probably irreversible (*Guardian Unlimited* 2005).

Leong, B.D. (2002) How Will the Concept of "Design for Sustainability" Revive Industrial Design Practice in China and the Rest of World?', *1st China–USA Joint Conference on Design Education*, Beijing, China, 14–17 May 2002 (China Machine Press): 122-27.

—— and E. Manzini (eds.) (2006) *Design Vision on the Sustainable Way of Living in China* (Guangzhou, China: Lingnan Art Publishing).

Li, Z.J. (2005) 'Rapid Urbanization Catching Experts' Attention', World Watch Institute, 13 November 2005; www.worldwatch.org/node/64 [accessed 27 January 2008].

Liang, Z.M. (1987) *The Essence of Chinese Culture* (Hong Kong: Joint Publishing, in Chinese).

Lieberthal, K., and G. Lieberthal (2004) 'The Great Transition', *Harvard Business Review* on 'Doing Business in China' (Boston, MA. Harvard Business School Press): 1-30.

Lin, Y.T. (2002), *My Country and My People* (Hong Kong: Joint Publishing, in Chinese, republished from the 1935 edn).

Manzini, E., L. Collina and S. Evans (2004) *Solution Oriented Partnership: How to Design Industrialised Sustainable Solutions* (European Commission GROWTH Programme; Cranfield University, UK).

Market Research.Com (2006) 'China Mobile Market 2006', www.marketresearch.com/product/display.asp?ProductID=1212255&xs=r [accessed 25 August 2007].

McEwen, W., X.F. Fang, C.P. Zhang and R. Burkholder (2006) 'Inside the Mind of the Chinese Consumer', *Harvard Business Review*, March 2006: 68-76.

Miles, J. (2005) 'Hu, Wen, How? China's leaders will worry more about protecting power at home than projecting it abroad', *The Economist: The World in 2006* (special 20th edition): 66-67.

Mont, O. (2000) *Product-Service Systems (Final Report), February 2000* (Lund, Sweden: International Institute of Industrial Environmental Economics, Lund University).

NBSC (National Bureau of Statistics of China) (2006) 'The total retail sale of consumer goods increased 12.5% in the first two months', March 2006; www.stats.gov.cn/english/newsandcomingevents/t20060320_402311456.htm [accessed 25 August 2007].

*People Daily* (2004) 'Total Retail Sales of Consumer Goods', *People Daily Online*, January 2004; english.peopledaily.com.cn/data/figures/2004/trs1.html [accessed 25 August 2007].

—— (2005a) 'China to spend 15 trillion yuan for urbanization in 50 years', *People Daily Online*, March 2005; english.people.com.cn/200503/04/eng20050304_175500.html [accessed 27 January 2008].

—— (2005b) 'China's development mode to change in "11th Five-Year Plan" period', *People Daily Online*, October 2005; english.peopledaily.com.cn/200510/09/eng20051009_213463.html [accessed 27 January 2008].

Standage, T. (2005) 'Connecting the Next Billion', *The Economist: The World in 2006* (special 20th edition): 117.

Sun, L.K. (1992) *The Deep Structure of Chinese Culture* (Hong Kong: Zheyan Publishing, in Chinese).

*Sydney Morning Herald* (2005) 'China clears the way for Kyoto trade', *Sydney Morning Herald*, 15 February 2005; www.smh.com.au/news/Environment/China-clears-the-way-for-Kyoto-trade/2005/02/14/1108229928909.html?oneclick=true [accessed 25 August 2007].

Tong, H.W. (2006) 'Preface', in B.D. Leong and E. Manzini (eds.), *Design Vision on the Sustainable Way of Living in China* (Guangzhou, China: Lingnan Art Publishing): 4-6.

Vezzoli, C. (2006) 'Educating Designers to Transcultural Creative Thought for Sustainability', paper presented at the *Engineering and Product Design Education Conference*, Salzburg University of Applied Sciences, Salzburg, Austria, 7–8 September 2006.

Wikipedia (2005) 'Laptop', en.wikipedia.org/wiki/$100_laptop [accessed 25 August 2007].

Woo, H.K.W. (1995) *The West in Distress: Resurrecting Confucius's Teachings for a New Cultural Vision and Synthesis* (Hong Kong: Chinese University Press, in Chinese).

Yardley, J. (2005) 'China's next big boom could be the foul air', *The New York Times: Week in Review*, October 2005; www.nytimes.com/2005/10/30/weekinreview/30yardley.html [accessed 25 August 2007].

Yuan, R. (2005) 'Livable City Standards and Development in China', *China & World Economy* (Beijing: Chinese Academy of Social Sciences, IWEP) 13.5 (September/October 2005): 104-13.

Zarsky, L., and S.C. Tay (2000) 'Civil Society and the Future of Environmental Governance in Asia', in D.P. Angel and M.T. Rock (eds.), *Asia's Clean Revolution: Industry, Growth and the Environment* (Sheffield, UK: Greenleaf Publishing): 128-54.

# Part 4
# Consumer perspective

# 13
# Review: a multi-dimensional approach to the study of consumption in modern societies and the potential for radical sustainable changes

*Eivind Stø, Harald Throne-Holst, Pål Strandbakken and Gunnar Vittersø*
SIFO, Norway

## 13.1 Introduction

This chapter addresses three central discourses in, or aspects of, the general field of consumption and the environment. It concerns (1) the relationship between needs and wants, (2) the tension between rational models of consumer behaviour and models based on routines and habits and (3) the question of political consumption. The aim is to develop a multi-dimensional understanding of the role of consumers and consumption in modern societies, relevant to the discussion on sustainable consumption. Further, this understanding will be used to develop a model for analysing and catalysing changes in a sustainable direction.

Our main contribution comes from the perspective of the sociology of consumption, but we want to consider insights from economical and/or psychological traditions or paradigms in order to arrive at an integrated approach. We believe that the concept of the rational actor is necessary, but not sufficient to analyse environmentally relevant consumption practises. This might alienate economists as well as sociologists. The three discourses, however, all point to differences of approach between and within, for example, sociology and economics. Thus, interdisciplinary struggles over definitions and approaches are to be expected.

We have seen a number of shifts in the focus and in the rhetoric of the environmental discourse. A key shift occurred in 1987 with the introduction of the notion of 'sustainable development', which aimed at reconciling environmental concerns with the concept of economic growth (WCED 1987). Another shift was from nature to society, with a parallel shift of emphasis from the natural sciences to the social sciences (Stø et al. 2005; Strandbakken 1995). We have also seen a shift in environmental concern from pollution and biodiversity to eco-efficiency and energy use, often employing the concepts of Factor 4 or Factor 10 (von Weizsäcker et al. 1997). Simultaneously, the centre of attention might have shifted from production processes and manufacturing, to consumption and lifestyles. The United Nations 10 Year Framework of Programmes on Sustainable Consumption and Production (UN 2002), the revised EU Sustainable Development Strategy (EU 2006) and manifestos by scientists and non-governmental organisations (NGOs) (ANPED/Eco-Forum/EEB 2004; Tukker et al. 2006) all identified consumption as a central challenge to achieving sustainable development. The transition is not complete, but the consumption side of environmental issues is no longer as neglected as it used to be.

We have also seen a shift from consumption as an economic and material category, to consumers as economic and political actors. One reason for this is that the eco-efficiency of products might have been countered by increased consumption (Throne-Holst et al. 2007). The result is that the environmental impacts of consumption in many areas have increased, in spite of the dissemination of energy-saving innovations. The economic literature labels this the rebound effect (Greening et al. 2000; Musters, 1995), but a similar phenomenon is also observed by other disciplines, but with other names and explanations (Uiterkamp 2000).

Technical and political progress has been achieved in the environmental field. This does not mean, however, that environmental problems are solved or that industrial pollution is under control, at least not from a global perspective. There is still a need for technical innovations in production processes as well as a rise in demand for increased environmental quality for consumer products, as ecosystems around the globe remain under threat. But the rebound effect illustrates that the main challenge in the future will probably be on the consumption side, and we will succeed in countering it only if individuals are mobilised in their roles as both consumers and voters in a bottom-up approach.

## 13.2 State of the art: understanding consumer behaviour

### 13.2.1 Introduction

We build this chapter around three theoretical or pragmatic themes and discourses, all relevant to the political discussion on sustainable consumption:

- The relationship between needs and wants: use value, exchange value and the meaning of consumption (compare also the contribution of Scholl to this book, in Chapter 14)

- The tension between rational models of consumer behaviour and models based on routines and habits (compare also the contribution of Schultz and Stieß to this book, in Chapter 16)

- Consumers as actors in markets and politics: political and ethical consumption

### 13.2.2 The relationship between needs and wants: use value, exchange value and the meaning of consumption

The first issue we want to highlight in the contemporary consumption discourse is the rather neglected question of the relationship between needs and wants, reintroduced in the book *How Much Is Enough?* (Durning 1992).

This need–want relationship was nearly 'forgotten' because research showed that need is a problematic concept for consumption (Campbell 1998; Douglas *et al.* 1998), while other traditions have challenged the value-for-money models. Baudrillard emphasised the symbolic values of consumption: 'The fundamental conceptual hypothesis for a sociological analysis of consumption is not use value, the relation to needs, but symbolic exchange value' (Baudrillard 1998: 30). This phenomenon was recognised by Veblen (1899) more than 100 years ago, and more recently in the writings of Bourdieu (1984) and in the postmodernistic 'tradition' (Featherstone 1991). Here, consumption is closely linked to the identity of modern individuals, which goes far beyond needs and the use value of products (Douglas and Isherwood 1979). Thus, many researchers are uneasy with the concepts of needs and use values. They search for other points of departure because of the reductionist character and the moralistic bias of the needs approach. Theorising from needs also carries a potentially unpleasant political message. It tends to blame consumers alone for unsustainable development—it is they who 'over-consume' beyond what they really need.

It is, however, problematic to replace the concept of needs completely with wants and desires. After all, there are no limits to personal wants (e.g. see the book *Greed* by the US anthropologist Robertson [2001]). Without need-based concepts it is hardly possible to answer the question of 'how much is enough?' (Jackson 2004; Jackson and Michaelis 2003; Jackson *et al.* 2004). For sustainability this is important because there are physical limits to human activity, recognised by the vast majority within the scientific community (Meadows *et al.* 1972, 2004). This is the main argument behind the rethinking of basic human needs. This is also the point of departure in the Brundtland Report on sustainable development, defined as a development that meets the needs of the present generation without compromising the ability of future generations to meet their own needs (WCED 1987). We hence reconsider the need–wants dichotomy below, in part drawing on the recent work of Jackson *et al.* (2004).

The classical psychological literature on needs was highly influenced by Maslow (1943) in the 1950s and 1960s. Maslow distinguished between material needs (at the bottom of his hierarchy), social needs and self-realisation needs (at the top of his pyramid). His controversial argument is that material needs have to be, at least partially, satisfied before needs higher in the hierarchy may become important sources of motivation. Within sociology, these arguments have been developed further by Ingelhart (1977, 1990) in his discussion about 'postmaterialism'.

Maslow has been criticised on three points. First, anthropologists have shown that in many societies there is a compromise even between basic survival needs and moral and spiritual needs. Second, closely linked to this argument, the hierarchy appears to deny access to the satisfaction of higher needs in less developed countries, and legitimises dictatorship (Galtung 1990)—according to this argument, the people in these countries will have to wait for the satisfaction of their basic needs before any further needs can be met. Last, the hierarchical approach overemphasises the individual nature of needs satisfaction and understates the importance of society and culture. This argument is highly relevant for consumption levels and patterns of consumption in modern societies.

This criticism led to a redefinition of human needs. A very influential scheme was developed by Max-Neef (1992), discerning nine sets of 'axiological' needs (subsistence, protection, affection, understanding, participation, identity, idleness, creation and freedom) against four 'existential' categories (being, doing, having and interaction) in a 36-cell model.

Max-Neef also suggested distinguishing between needs and satisfiers. A need is dualistic: it is a deprivation in the sense of something being lacking; it is a potential to the extent that it mobilises the subject. Satisfiers, by contrast, represent different forms of being, having, doing and interaction which contribute to the actualisation of needs. Needs are finite, few and classifiable; satisfiers are culturally determined and numerous.

Against this background, to what degree are the essential critiques of needs theory still relevant? We side here with Jackson, who is critical of the neoclassical approach, substituting needs with wants, preferences and desires. But there are still arguments from anthropology, sociology and cultural studies that the modern needs approach has to take into account:

> Material commodities are important to us, not just for what they do, but for what they signify: about us, about our lives, our loves, our desires, about our successes and failings, about our hope and our dreams. Material goods are not just artefacts. Nor do they offer purely functional benefits. They derive their importance, in part at least, from their symbolic role in mediating and communicating personal, social, and cultural meaning not only to others but also to ourselves (Jackson *et al.* 2004: 21-22).

It could also be argued that most of the criticism of need theory is more relevant to discussion about satisfiers than about basic needs. It is naïve to ask consumers to give up their wants and desires without offering them new dreams to dream.

### 13.2.3 The tension between rational models of consumer behaviour and models based on routines and habits

Psychologists, as well as other social scientists, have engaged in the study of consumer behaviour in relation to sustainability. While the social psychologists emphasise the role of information in changing individual attitudes and consequently behaviour, the focus within environmental sociology has usually been either on individualistic or on more structural theoretical models. In this section we will look at the attitude–behaviour model from psychology before considering various contributions from the field of environmental sociology.

The most influential theory of attitudes and behaviour is the theory of reasoned action (Ajzen and Fishbein 1980). According to this, behaviour should be predicted from actors' attitudes and intentions. But the relationship is mediated by people's perceived behavioural control. This means that, if a person believes he or she is unable to adapt the behaviour in question, he or she is unlikely to try to do so. Behavioural intentions could be predicted from attitudes, subjective norms and perceived behavioural control.

The point of departure is that most actions of social relevance are under volitional control. The central element of the model is that behavioural intentions are the immediate determinant of behaviour. Intentions are based on two dimensions: (1) attitudes toward the behaviour and (2) subjective norms. By subjective norms Ajzen and Fishbein refer to the perception of the social pressure put on the individual to perform or not to perform the behaviour. The attitudes are more specifically determined by a set of behavioural beliefs, concerning the possible consequences of the behaviour and by the evaluation of these beliefs.

The critique against them is developed along two dimensions. First, it is argued that individual consumers do not behave in the rational way the model presupposes. Second, the context of social behaviour is missing in the model: consumers are not necessarily only individuals. They belong to households or communities with values and norms and they act within a political and economic context created by businesses and political authorities. Their model has also received criticism from other social psychologists. Eiser (1994: 22) challenges what he calls the 'causal dependence': 'a view of behaviour as the product of a rational will'. Eiser differentiates between discrete decisions and habitual behaviour. He claims that Ajzen and Fishbein's model is well suited to predict discrete decisions 'isolated in time and space', but that 'empirical difficulties arise when trying to apply the model to the prediction of habitual behaviour' (1994: 23). The global attitudes in the model depend on specific beliefs, but Ajzen and Fishbein do not give a satisfactory explanation as to which of these beliefs are relevant and why.

This critique seems relevant, but we are aware of the potential strength of Azjen and Fishbein's model, as long as it is used for the product-related choices of the marketplace. This perhaps points to another weakness in their approach: for economic psychology in general, consumers are to a large degree synonymous with customers. But this is not always the case: consumers have interests and attitudes beyond the purchase phase.

Some critics claim that by focusing on individual behaviour the responsibility for sustainable development is taken away from politics and placed on the individual consumer. Other contributions question the effect of specific policy measures, such as consumer information, in changing consumer behaviour (Shove 2003; Vitterso 2003). Bauman (1999) takes these critical arguments even further by stating that the influence of ordinary citizens as well as national governments is constantly diminishing.

We also have to take into account that consumers are not atomistic actors in the market; they belong to communities with specific values and local cultures. Miller (1998) argues that consumption is not primarily an individual activity but is framed by specific cultural and social contexts within and outside the household. From this perspective, norms, habits and routines are decisive factors explaining consumption practices. Following the same lines, Gronow and Warde (2001) have opened a debate relevant to our research questions. They claim that, during the 1990s, the focus of consumer research went from spectacular to ordinary consumption. Consumption is mainly about the everyday life of ordinary consumers, and this should be better reflected in contemporary research. We must not forget that consumption in modern societies is, to a very large degree, the consumption of ordinary products, with few opportunities for excitement. Much environmentally significant consumption is bound up with routines and habits.

Gronow and Warde are inspired by anthropology, where the routines of everyday life have always played an important part. Their other main contribution is to link the modern studies of consumption to classical sociology, especially to the work of Max Weber. They also undertake a very interesting discussion on a distinction made by Bourdieu, where they question both the rigidity and the hierarchy of tastes.

### 13.2.4 Consumers as actors in markets and politics: political and ethical consumption

Finally, we want to consider the relationship between market and politics. Here, consumers are regarded as actors not only in the market of consumer goods and services but also in various political arenas (Stø et al. 2005). Political and ethical consumption and boycotts are not a new phenomenon. However, in late modernity, the complicated relationship between our role as consumers and citizens has been revived by the shift in political paradigm from government towards governance (Olsen 1992).

Both Giddens (1991) and Beck (1992) analyse processes of individualisation in the modern society. Contrary to traditional societies, where individual choice was limited and individuals were subordinated to the collective order of social norms, the modern person is able to make his or her own decisions with a greater degree of independence from norms and social structures than before. One of these processes is political consumerism, or individual collective action as Micheletti (2003) calls this phenomenon. Consumption turns into politics when consumers choose market arenas to influence decisions made by governments and companies and mobilise other consumers to take part in this activity. This concept is closely linked to ethical consumption, where consumers make some of the same decisions without necessarily involving other consumers (Terragni et al. 2006).

Micheletti (2003: 15) sums up five basic reasons that theoretically justify conceiving consumption as politics. First, this is an arena that can be used by people to express themselves when other arenas have failed. Second, political consumption is used to set the political agenda of other actors. Third, it is possible to influence the value decisions made in private corporations. It is really a difficult task through other channels. Fourth, consumption offers individuals market-based tools such as boycotts and 'buycotts'. At last, consumption is becoming more political because the political landscape has changed completely. Trust in the traditional political system and channels is decreasing, as is voting turnout in most countries. People are still concerned about the future, however, but perhaps are more oriented towards sub-politics and single-issue cases; and, here, the market opens up new opportunities.

Many classify political consumption as a typical left-wing phenomenon, mainly dealing with child labour and questions of fair trade, but this is not necessarily the case: the concept is used for any activity when the market is used as a political arena. Some would claim that one of the most effective consumer boycotts was the Nazi campaign against Jewish-owned shops in Germany in the 1930s.

We identify two main arguments against the promotion of political consumption: political consumption is said to destroy the market, to be a market perversion. Within the free-trade paradigm there is no place for other considerations than traditional value-for-money evaluations. If you bring in social or political aspects, you reduce the efficiency of the market, nationally and globally. And it is argued that political consumption destroys politics. Market activity threatens political democracy by introducing an individual economic power into collective political arenas (Eriksen and Weigård 1993; Jacobsen and Dulsrud 2007). In the political system we all have one vote. In the market our 'voting' influence increases with how much money we spend, so it is by nature undemocratic (Persky 1993). Thus, both arguments conclude that the political and economic systems should be kept separate and not mixed.

Another relevant perspective is the 'dream society', developed by the director of the Copenhagen Institute for Future Studies, Jensen (1999). It is a rather controversial theory, where the title of the phenomenon—the dream society—may be problematic to many readers and researchers. Presenting a broad canvas, Jensen maps the history of humankind beyond the information society. The heroes of society are no longer hunters and farmers, not even factory workers, but information workers. In the dream society the paramount image is history, and the hero is the person who is able to tell a good story!

The central message from Jensen is that consumers link stories to various goods and services. Consumers prefer the good stories; they want to look at themselves in the mirror without shame. This is the slowly prevailing trend, that moves constantly like a glacier. If businesses around the world are not prepared for this, they will lose, according to Jensen. They will lose maybe even without ever knowing why, because they are producing and marketing their goods in the same traditional way. Consumers will, to an increasing extent, be concerned about the ethical, political and environmental aspects of production and distribution of consumer goods. Are the producers relying on child labour? Do they take care of the environment, locally and globally? What about their human rights record?

From our perspective, the active and politically conscious consumer represents an enlargement of individual and collective political influence (Kjærnes and Stø 1996; Micheletti *et al.* 2004). However, for this to work, consumers must be helped to make informed political choices in the market of goods and services. In order to fill this information gap, a number of labels have been developed at the national, regional and global levels. The history of voluntary environmental labels covers almost 30 years, starting with the Blue Angel in Germany. Success varies from one labelling scheme to another, and there are also considerable differences between product groups. According to recent research, not surprisingly consumers trust labels with third-party certification (Type I labels) more than labels developed by businesses (Type II labels) (Rubik and Frankl 2005). Other important labels cover fair trade, with Max Havelaar the most important, internationally. In forestry, we have two global certification labels, that of the Forest Stewardship Council (FSC) and the Programme for the Endorsement of Forest Certification schemes (PEFC). The European Union has developed mandatory energy labels to guide consumers' choice in the market and to move producers and importers in the desired direction (Blue Angel,[1] Max Havelaar,[2] FSC,[3] PEFC,[4] European Energy Label[5]).

The question of labels relates to all the issues we have touched on. First, they presuppose knowledge and some degree of rationality among consumers. Second, they fit well into the ordinary consumption perspective, because they help consumers with everyday decisions. Last, the symbols also express the meaning of consumption, and are relevant to ethical and political consumption.

## 13.3 State of the art: insights into how to change consumer behaviour to sustainable consumption and production

### 13.3.1 Introduction

Behind these three theoretical–empirical consumption discourses, we might identify a number of one-dimensional drivers for change—drivers that are relevant to potential changes towards sustainable consumption. In this section we discuss these one-dimensional drivers of change separately. These 'drivers' are probably best regarded as a kind of Weberian 'ideal type', synthesising a single aspect of a phenomenon that never appears empirically in its 'pure' form. In the next section we will integrate them into a more elaborate model of change. We have identified the following single-dimensional drivers for change from current to sustainable consumption:

- Change in **values**
- Changes in **attitudes**
- Change in **knowledge**
- Changes in the **symbolic** aspects of consumption and of individual identity
- Changes in **habits** and **routines**
- The existence or non-existence of **windows of opportunity**

There are apparently two levels in the discussion within each approach to change. The first is the claim that changes in this particular dependent variable will lead to changes in sustainable consumption, as the crucial independent variable. The second level is how and when these changes in the dependent variable can take place. What political, economic and juridical measures are needed, and indeed possible, to put forward in late modernity?

1 www.blauer-engel.de/englisch/navigation/body_blauer_engel.htm [accessed 23 January 2008].
2 www.maxhavelaar.ch/en [accessed 23 January 2008].
3 www.fsc.org [accessed 23 January 2008].
4 www.pefc.org/internet/html [accessed 23 January 2008].
5 eur-lex.europa.eu/smartapi/cgi/sga_doc?smartapi!celexapi!prod!CELEXnumdoc&lg=EN&numdoc =31992L0075&model=guicheti [accessed 23 January 2008]

### 13.3.2 Changes in consumer values

Change in consumer values is strongly related to the needs–wants discussion in Section 13.2.2.2. Once we use needs as a point of departure, we are drawn into a moral discussion where over-consumption is condemned as unsustainable (Jackson *et al.* 2004: 24). In some cases this leads to attempts to 'downshift' consumption by individuals. But it also leads to calls to regulate the consumption of others, which have provoked reaction from the needs-sceptics. Who has the legitimacy to decide the levels of consumption of others? (Douglas and Isherwood 1979).

The need-sceptics may have a strong argument if it were to concern only the regulation of consumption at an individual level. But individual conspicuous consumption, or over-consumption, at the household level is not the key problem. It is also a matter of unsustainable production and consumption on an aggregate level. It becomes a moral question and a political question when we accept that there are limits to growth, which leads to questions of how to share wealth. This dilemma is usually not addressed by the needs-sceptics.

A future sustainable society will have to be developed with regard to tensions between needs and wants. The challenge is set by Jackson *et al.* (2004: 15), who note that '[. . .] it is particularly vital to be able to identify which bits of consumption contribute to human needs satisfaction, and which simply operate as pseudo-satisfiers and destroyers'. Our argument here is not to replace use value and exchange value with symbol value and identity in research and policy. Rather, we argue that a combination of these perspectives is necessary, as consumption under modern or postmodern conditions is complex.

But how should we change the consumer values in the desired direction? And how do we replace wants with a concern for needs? This is a serious challenge. The popular answer to this question is often that we have to start in schools and even nursery schools. Other alternatives are to mobilise positive role models within pop culture, sports and the media to change values. This is difficult, however, not least because we are opposing major economic and social drivers and trends in modern societies.

### 13.3.3 Changes in attitudes

The possibilities for changing consumer behaviour have been a central issue for policy-makers and researchers since the early 1990s, when sustainable consumption was put on the political agenda. In the political as well as in the scientific discourse we find tensions between different views on behavioural change. The main conflict is observed in the emphasis on the role of the individual consumer in both consumer responsibilities and possibilities for changing behaviour. Below we will discuss the opportunities for changing attitude in more detail. The aim is not to contribute in general to the complex discussions about the relationship between values and attitudes. We stress only that attitudes are usually the more specific part and values the more general and constant part of the cognitive and psychological structure of human beings.

Many economic–psychological studies are conducted to examine why some people have pro-environmental behaviour whereas others do not, and how consumers might be persuaded to behave more pro-environmentally. These studies tend to concentrate on people's attitudes towards the environment. Most studies, however, find at best only weak correspondence between reported attitudes and actual behaviour. Kaiser *et al.* (1999) propose that the weak correlation between attitudes and behaviour are a result of:

- A lack of a unified concept of attitude among researchers
- Differences in the way attitude and behaviour are measured
- Some researchers failing to recognise constraints on people's behaviour

As mentioned previously, the most important model in this attitude–behaviour paradigm or tradition is the theory of reasoned action (Ajzen and Fishbein 1980). Ajzen (1991) later developed the model to include behaviour that is not under complete volitional control in his theory of perceived behavioural control, as a new, third, independent determinant of intention and behaviour. Other authors have developed the initial Ajzen and Fishbein model further by introducing moderators of the attitude–behaviour relationship, where one important element is the involvement with the attitude object and another is the strength of the attitudes. Central variables include attitude accessibility, experience with the attitude object, attitude confidence and the amount of information available in memory.

All these elements have only contributed to the improvement of Azjen and Fishbein's original model That is why we chose to regard it as a paradigm and not merely as another theory. Any number of intermediate variables might be added in order to increase the explanatory and predictive value without changing the fundamental logic of the scheme.

The improvements were necessary because the model has failed to predict consumer behaviour satisfactorily. This is still the main critique against the Azjen and Fishbein model: even the improved model is too simple to predict the complicated behaviour of modern consumers (Spaargaren 1997; Stø and Strandbakken 2005a).

### 13.3.4 Changes in knowledge about products and processes

Another key variable is consumer knowledge and information, often regarded as the answer to the question about how to change the attitudes of ordinary consumers and citizens. While education is regarded as the means to change values, consumer information is designed to fill the information gap. Well-informed consumers are supposed to act differently from consumers without knowledge about products, markets and processes.

Based on this approach NGOs and public authorities have carried out information campaigns directed towards general or specific consumer groups in order to change their behaviour; and businesses have contributed with green marketing and environmental claims. A tool with some success is the various eco-labelling schemes (Rubik and Frankl 2005). Knowledge about eco-labels has increased significantly: for instance, in Germany and in the Nordic countries during the 1990s (Stø and Strandbakken 2005b; Strandbakken 1995). Correspondingly, market shares of paper products and detergents carrying the White Swan label have increased substantially and are reasonably high in Nordic countries. This is also the case for paint and paper products with the Blue Angel label in Germany. The main elements in these successes are that the labels are based on life-cycle assessment (LCA) approaches and are managed by independent eco-labelling bodies.

Eco-labelled alternatives are often more expensive than conventional products. Their market potential is often measured within a cognitive approach identified as consumers' willingness-to-pay. The willingness to spend money on certain products is used as a measurement for consumers' values and attitudes. This willingness is dependent on a large number of well-known factors, such a price, quality, availability, social context and the social and economic background of the consumers.

Godfrey (2002) gives an overview of the willingness-to-pay literature for environmental products. Most studies show that consumers are willing to pay the premium price for environmental products, but not always. The knowledge of consumer acceptance for paying a higher price for environmentally friendly products appears to be inconsistent and inconclusive. However, a number of studies have concluded that consumers are willing to pay 5% more for ordinary 'green' product alternatives, and even more for specialised products (Godfrey 2002: 54-55).

### 13.3.5 Changes in the symbolic aspects of consumption

The three drivers discussed so far originate from more or less rational-choice approaches, focusing on use value, exchange value and the relationship between price and quality. We have, however, tried to show that modern consumption cannot be understood with such approaches alone.

Goods are produced and consumed not (only) for use value or functional (needs) reasons in modern affluent societies but also, equally, for emotional or cultural reasons. In the future, products will have to appeal to our 'hearts' in addition to our 'heads'. We have used the dream society argument of Jensen (1999) to illustrate this approach. His main idea is that industries and business that fail to take account of this change in the long run will lose in the competition, as it is no longer a market for physical products; it is as a market for dreams.

His key example is relevant to ethical discussions about consumption and the environment: the production of eggs. Eggs are produced in many ways and qualities. Consumers supposedly do not want their eggs to be produced in narrow cages; they want the birds to live more natural lives, to have access to earth and sky, as in the old days. To a large degree they are also willing to pay the premium price of 15–20% extra for eggs produced in this way. The reason for this, according to Jensen, is that consumers prefer the story of the free-range hens to the story of the conventional industrially produced eggs. If you are served an omelette based on eggs from free-range hens, together with organic bread, you will probably be told the story of the food. If, on the other hand, you are served conventional products, there is no story to tell.

We have taken the argument from Jensen a bit further. In the 'dream society' consumers and citizens, to some extent, use the market as a political arena. Consumers use boycotts or 'buycotts' to influence governments' and companies' market practice, human rights records, trade union rights, child labour, fair trade and the environmental impact of their production and distribution. Thus, one central tenet of political and ethical consumption is that actions by consumers are not necessarily linked to the objective properties of the product. Parts of their actions are based on evaluations beyond the traditional value-for-money paradigm: on social, political, ethical and environmental aspects in the production process and its surroundings. In addition, the market is used as a political arena because this could be the most effective way to reach the companies and political authorities in question. Sometimes this is the sole channel for potential political influence, because the traditional channels are closed.

### 13.3.6 Changes in habits and routines

So, consumer behaviour should not be understood only as rational choices in the market. To include ethical and political variables is necessary, but not sufficient. In addition we have to include routine and habit, and we should understand the difference between selection and choice. In shopping praxis, consumers make selections of goods and services every time they shop, but it is not always a matter of conscious choice (Warde 1997).

Gronow and Warde (2001) have emphasised the importance of ordinary consumption. They use the history of consumer research to place their contribution within the main trends of the social sciences. The sociology of consumption has developed in phases. Critical theory dominated the 1960s and 1970s, inspired by Marxism, Neomarxism and the Frankfurt School. Consumption should fill basic needs, and consumers were regarded as victims of a manipulative advertising industry. The main concept was compensatory consumption. The postmodernist contributions of the 1980s and 1990s changed this picture completely. Consumption was no longer mainly a question of basic needs but rather a matter of more spectacular consumption. Consumers were not victims but active searchers for new identities and symbols in the market of consumer goods and services. The main concept now was conspicuous consumption.

The new platform of Gronow and Warde will build not only on a negative critique of postmodernism and critical theory; central elements of these 'schools' will survive in a new paradigm of consumers and consumption in a new century. They are, however, opposed to the elitism of critical theory and the one-sided focus on spectacular consumption in postmodernist contributions to consumer research. Whereas the rational and reflexive perspectives on consumer behaviour emphasise conscious reasoning and individual action, the message from the daily-life perspective is that action and practices result from routines and habits more than from individual active choice.

Halkier, building on the work of Giddens and Beck, modified this and stated that in daily life you will observe a mixture of rational, intended behaviour and more routine practices, so that 'a sharp distinction between reflexive and routinised consumption practices is impossible to sustain in empirical analysis' (Halkier 2001: 27).

This implies that it is decisive to create routines that stimulate sustainability, but this is not easy in everyday life where there are various obligations and time constraints.

### 13.3.7 The existence of windows of opportunities

The potential for change can also be discussed in relation to windows of opportunity. Svane (2002) used this concept to discuss environmental questions in his work on sustainable housing in Nordic countries. His basic idea is that in everyday life it is difficult for consumers to change habits, even if they are well informed and highly motivated. However, when people make various fundamental changes in their life they are susceptible to changes in other aspects as well. Potential situations of opportunity (or 'windows' of opportunity) are when people change dwelling place, change workplace or occupation, get married or divorced, have children, etc. In the need–opportunity–ability (NOA) model of Vlek *et al.* (1997), concepts such as ability and opportunity are applied to describe some of the same phenomena on the individual level. In a recent Norwegian study (Throne-Holst *et al.* 2008) actual situations of opportunities were more important for energy saving than were attitudes among consumers.

However, windows of opportunity can also be created by market actors and by political authorities at the local and national level. In the EU project ToolSust we assessed the potential and possibilities for sustainable changes among consumers in five European cities (Stø 2004) Special focus was placed on the framework created by businesses and local authorities. Within the energy sector consumers had some degree of freedom. With regard the issue of heating their homes, they could reduce the indoor temperature, insulate windows and doors, install heat exchangers or use energy-efficient electrical equipment. However, the main decisions concerning energy supply were taken at the municipal level. As far as recycling was concerned, households had to conform to systems created by local authorities; similarly, the retail system in the cities was the main explanation for what means of transport were used for shopping. The possibility of choosing organic food was limited in most cities studied in the project. All in all, windows of opportunity were rather narrow.

This means that if the positive values, attitudes, knowledge and symbolic meanings that are developing among consumers should be transformed into sustainable behaviour, the windows of opportunity have to be expanded. This will be discussed more in detail when we turn to the limitation of the consumption perspective. First we will try to integrate the single-dimensional approaches discussed above.

## 13.4 Synthesis: models of change

So far we have presented a series of one-dimensional models for consumer change, all with strengths and weaknesses. But what is the relationship between them, and is it possible to integrate them into one, not too complicated, model?

Our point of departure for integrating the models is the understanding that consumption is a process, as illustrated in Figure 13.1. From the consumer's point of view this process includes at least four phases: (1) planning, (2) buying, (3) using and (4) disposal (Throne-Holst *et al.* 2007). However, the consumption process is also closely linked to the production and distribution of goods and services. This may seem naïve and obvious to the reader, but the literature on consumption shows that this is not necessarily so.

When we discuss the environmental impact of consumption it is imperative to include the entire consumption process. There might be a theoretical agreement on this statement, but, empirically, various disciplines have concentrated on separate parts of this process.

In the marketing literature the focus has been on the planning and buying phases. The strength of this approach is the close link between shopping behaviour and production

**FIGURE 13.1 A model of the production–consumption process**

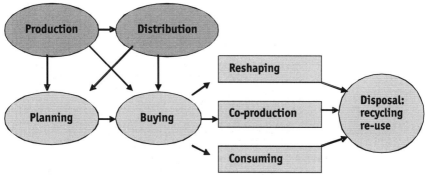

and distribution. It is important to know what various groups of consumers buy and why they prefer some products and brands over other alternatives. However, the value chain too often ends with the purchase; the use phase is largely neglected. It has not been considered interesting to find out what households and consumers do with their purchased products, to look at how they use them.

Relevant to the sustainability discussion, the normative goals in this tradition have been to help consumers choose environmentally friendly products in the market rather than the more polluting or energy-consuming products. Labels, specific environmental information and 'green' marketing are regarded as the main tools.

It is worth noting that a similar emphasis on the purchasing activity of individual consumers is found in modern studies of political and ethical consumption (Micheletti 2003). The centre of attention has been on the development of eco-labels, fair-trade labels and other relevant instruments for bringing information to consumers and the subsequent introduction of the labelled products into the consumer market. We will return to this later, but it is striking how even this approach has neglected the use phase. However, some attention has been paid to the disposal phase.

To some extent this focus on shopping behaviour is also found in modern consumer organisations. Their magazines emphasis 'best buys' and 'value for money', based on the traditional consumer right to choose. However, the consumer movement has a long tradition of advising consumers in the use phase. One example could be the introduction of freezers in modern households in the 1950s and 1960s, where consumer organisations played an active role in educating consumers (Strandbakken 2006). Today, this kind of information is less popular; it is often regarded as moralistic. It also contributes to shifts in the responsibility from producers and retailers towards households and consumers, and that is not welcomed among consumer organisations.

Within the sociological and anthropological tradition, more emphasis has been put on the use phase; that is the strength of this position. Consumption is to a large degree a matter of using products in different ways. Even though most of the goods in modern societies are bought in the market, we reshape, modify and adjust them to our own values, needs and wants. There is also a substantial degree of production taking place in modern households, as illustrated in Figure 13.1. This is related to the preparation of food, the maintenance and production of clothes, the renovation of houses and other household-related work. In economic terms such activity is very important for households, even though it is not measured in the gross national product (GNP).

Translated to sustainability, the main focus should perhaps be on the sustainable use of products. For many products and activities the use phase is more environmentally important than the buying phase, such as for washing machines and the washing process. It is probably more important *how* you do your laundry than what washing machine you have bought and what detergent you use (Rubik and Frankl 2005). The decisive variables are temperature, the filling of the washing machine and frequency of use. Other more culturally related values could be tolerance for dirt and what it means to be clean (Klepp 2003, 2005). However, the weakness of this tradition is that the importance of the market phase has, to some extent, been underestimated, at least in practice. Cultural studies tend to decouple consumption from production (Fine 2002: 58).

If we return to Section 13.3, it is easy to see that there is a relationship between values and attitudes, and also between attitudes and knowledge. This is the rational part of the model. In addition, symbolic aspects of consumption and routines and habits are closely linked to basic values in society. This constitutes the other element in the model. For both liaisons the fruitful starting point in the model seems to be consumer wants. Both links are necessary to avoid moralism in debates over sustainable consumption. But we also have to link the discussion to basic needs, only then is it possible to deal with the crucial question: how much is enough?

On the other side, we have argued that we have to create windows of opportunity for sustainable consumption behaviour. Some of these situations of opportunity are based on changes at the individual or household level; political authorities, NGOs, industries and retailers create others. The model is illustrated in Figure 13.2.

Our main message is that we need a multi-dimensional understanding of consumer behaviour in general (see Section 13.2). This understanding is also relevant to attempts at changing consumer behaviour towards sustainability, discussed in Section 13.3, even though some of the single-dimensional explanations have contributed substantially to understanding drivers for change. Our second message is that we have to combine models dealing with rational choices with models more concerned with habits, routines and the symbolic aspects of consumption. Our final message is the relevance of windows of opportunity. These windows can be created not only by consumers themselves but also by political authorities and businesses.

FIGURE 13.2 A multi-dimensional model for change in consumer behaviour

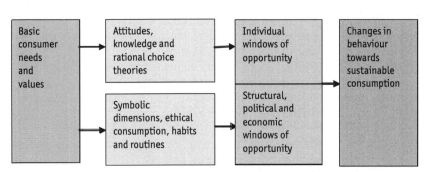

## 13.5 Limitations to consumer-oriented approaches

We have argued for the importance of consumption and the role of consumers in the process towards sustainable development. However, even within the multi-dimensional model there are limitations to such an approach. As we see it, there are at least three possible shortcomings, and we have already touched on them in the discussion above:

- First, in many models consumers are regarded as atomistic individuals decoupled from the social and cultural context
- Second, there is a strong tendency to decouple consumption from production and distribution of goods and services. This tendency varies from one model to another
- Last, in many of the models, political decisions at the local, national and global level do not play a significant part

The case for windows of opportunity meets some of the constraints highlighted under the last two points. We will have a closer look at them by referring to a model we have used to study the role of consumers in environmental successes (Stø et al. 2005). The model has been developed with regard to two central academic and political discussions:

- The relationship between market and state in modern welfare societies
- The relationship between micro and bottom-up activities, on the one hand, and macro and top-down initiatives on the other

TABLE 13.1 Individual influence through micro and macro strategies in markets and politics

|  | Micro | Macro |
|---|---|---|
| **Politics** | • Voters:<br>– Voting behaviour<br>– Membership of political parties, organisations<br>• Articulation:<br>– Opinion polls<br>– Articles in newspapers<br>– Word of mouth<br>• Influence over other voters<br>• Political consumers | • Political parties<br>• Consumer organisations<br>• Environmental organisations<br>• Influence:<br>– Government<br>– Parliaments<br>– Taxes<br>– Legal regulation<br>– Norms and education |
| **Markets** | • Consumers:<br>– Shopping behaviour<br>– Complaints process<br>– Individual boycotts<br>– Word of mouth<br>• Influence other consumers | • Consumer organisation:<br>– Organised boycotts<br>– 'Name and shame' or praise<br>• Influence:<br>– Industry and trade<br>– Product development<br>– Competition |

Based on these two dimensions we developed a model that distinguishes between individual and collective actions in markets and politics (Table 13.1). On the one hand, consumers might choose between individual and collective strategies; on the other hand, they can act strictly as a consumer in the market or move the questions beyond markets, into politics (Stø et al. 2005). This is an 'ideal-type' model that does not do justice to hybrid solutions—beyond market and state—such as the 'the third sector' and new governance models (Pestoff 1999). We will now briefly discuss each cell in the table.

### 13.5.1 Micro strategies and the market

Consumers make more or less conscious choices in the market of goods and services. Consumers might choose certain products instead of others, based on customs, routines and their notions of quality, price and (sometimes) political dimensions, such as the environment. Influence increases with affluence.

### 13.5.2 Macro strategies and the market

Consumers influence the market through consumer organisations and by organising campaigns to influence the decisions of other consumers in a positive or negative direction. Lately, the Internet has been an effective tool in such activity. In the literature, consumer boycotts are most common, but 'buycotts' are also identified among consumer groups.

The goal of these individual and collective activities by consumers is to influence the strategic decisions and the market supply of the other stakeholders in the market—mainly industrial companies and retailers.

### 13.5.3 Micro strategies and politics

The political counterpart is the secret-ballot political election. Political institutions at the local, regional and national level are elected by secret voting where one person has one vote. Through elections citizens influence the policy at the local and national level; voters choose one party or candidate in preference to other alternatives.

### 13.5.4 Macro strategies and politics

One of the main tasks of consumer organisations is to have a collective influence on the decisions made by parliaments and governments in relation to taxes, legislation and education. These political decisions constitute the framework for individual consumer behaviour in the market for goods and services.

### 13.5.5 Conclusion

Our main conclusion is that activity at the micro level is important, but the model has to be expanded to include macro perspectives within both the political and the market arenas.

## 13.6 Conclusions

In the concluding part of this chapter we will expand the argument on the dualism of consumption. Consumption and consumers have a crucial part to play in development towards a sustainable future. Without mobilising individuals in their roles as consumers and voters in a bottom-up approach we will not succeed in reaching the goal of sustainability. Furthermore, the political tools should not be limited to the shopping phase; more attention has to be put to the use phase. In this combination of bottom-up and top-down initiatives it is important to avoid moralism, and our strategy must take account of the aforementioned tension between rational choices and routines and between needs and wants.

The responsibility for the achievement of sustainable development should not be put on the shoulders of consumers alone. We need to develop a stakeholder approach where the behaviour of industries, retailers, NGOs and political authorities at the local, national, regional and global levels are decisive. Only when committed actors work together will we be able to achieve sustainability. Thus, within such a framework created by political authorities and businesses, consumers have a specific part to play. This is also the point of departure in the SCORE! project, combining perspectives and disciplines in a multi-dimensional approach.

Environmental goals will not be reached if the solutions to the problems remain at the individual level alone. In order to solve problems related to the environment, it is necessary to influence the strategic decisions taken by the main actors in the market and in politics. Thus, we need to go from the micro to the macro level, and, simultaneously, to move from the market into politics (Thøgersen 2005). Consumer policy empowers individuals to change lifestyle in at least two different ways—not only by reducing personal constraints and limitations but also by loosening some of the external constraints that make change difficult.

Ethical consumption is on the rise in many countries. However, it is striking that most of the discussion about political, ethical and social consumption focuses very narrowly on the shopping phase. It is also striking that the main theme is boycotts or 'buycotts' of certain brands, labelled products or products from specified countries. The level of consumption—that is, the question of over-consumption—is not regarded as a moral, ethical or political question in the literature.

A more radical ethical approach might be to buy fewer products and thus reduce the level of consumption (Vittersø et al. 1998). We have, however, an interesting example of the opposite view, in President Bush's message to the US people after 11 September 2001. Individuals were frightened, and they reduced their normal participation in everyday activities, including shopping. This was a threat to the US economy and the message from the president and his wife was clear: do not stop consuming! Consumption was regarded as a patriotic activity and contributed to the fight against terrorism.

# References

Ajzen, I. (1991) 'The Theory of Planned Behaviour', *Organisational Behaviour and Human Decision Processes* 50: 179-211.
—— and M. Fishbein (1980) *Understanding Attitudes and Predicting Social Behaviour* (Englewood Cliffs, NJ: Prentice-Hall).
ANPED (Northern Alliance for Sustainability)/Eco-Forum/EEB (European Environmental Bureau) (2004) 'Ostend NGO Statement towards Sustainable Consumption and Production Patterns', *EU Stakeholder Meeting*, Ostend, 24–26 November 2004; www.eeb.org/activities/sustainable_development/NGO-Ostend-Statement-24-11-2004.pdf [accessed 25 March 2007].
Baudrillard, J. (1998) *The Consumer Society: Myths and Structures* (London/Thousand Oaks, CA/New Delhi: Sage Publications).
Bauman, Z. (1999) *In Search of Politics* (Cambridge, UK: Polity Press).
Beck, U. (1992) *Risk Society: Towards a New Modernity* (London: Sage Publications).
Bourdieu, P. (1984) *Distinction: A Social Critique of the Judgement of Taste* (London: Routledge).
Campbell, C. (1998) 'Consumption and the Rhetorics of Need and Want', *Journal of Design History* 11.3: 235-46.
Douglas, M., and B. Isherwood (1979) *The World of Goods: Towards an Anthropology of Consumption* (London: Penguin Books).
——, D. Gasper, S. Ney and M. Thompson (1998) 'Human Needs and Wants', in S. Rayner and E. Malone (eds.), *Human Choice and Climate Change: Volume 1* (Washington, DC: Battelle Press).
Durning, A.T. (1992) *How Much Is Enough? The Consumer Society and the Future of the Earth* (The WorldWatch Environmental Alert Series; New York: W.W. Norton).
Eiser, J.R. (1994) *Attitudes, Chaos, and the Connectionist Mind* (Oxford, UK: Basil Blackwell).
Eriksen, E.O., and J. Weigård (1993) 'Fra statsborger til kunde: Kan relasjonen mellom innbyggerne og det offentlige reformuleres på grunnlag av nye rolle' (in Norwegian), *Norsk Statsvitenskapelig Tidsskrift* 9.2: 111-31.
EU (European Union) (2006) *Renewed Sustainable Development Strategy* (Publication 10117/06; Brussels: Council of the European Union; ec.europa.eu/environment/eussd [accessed 25 August 2007]).
Featherstone, M. (1991) *Consumer, Culture and Postmodernism* (London: Sage Publications).
Fine, B. (2002) *The World of Consumption* (London: Routledge).
Galtung, J. (1990) 'International Development in Human Perspective', in J. Burton (ed.), *Conflict: Human Needs Theory* (New York: St Martin's Press).
Giddens, A. (1991) *Modernity and Self-identity: Self and Society in the Late Modern Age* (Stanford, CA: Stanford University Press).
Godfrey, S. (2002) *An Analysis of the Trade-offs and Price Sensitivity of European Consumers to Environmentally Friendly Food and Beverage Packing Using Conjoint Methodology* (Lausanne, Switzerland: University of Lausanne).
Greening, L.A., D.L. Greene and C. Difiglio (2000) 'Energy Efficiency and Consumption: The Rebound Effect, A Survey', *Energy Policy* 28.6–7: 389-401.
Gronow, J., and A. Warde (eds.) (2001) *Ordinary Consumption* (London: Routledge).
Halkier, B. (2001) 'Routinisation or Reflexivity? Consumers and Normative Claims for Environmental Consideration', in J. Gronow and A. Warde (eds.), *Ordinary Consumption* (London: Routledge): 25-44.
Ingelhart, R. (1977) *The Silent Revolution* (Princeton, NJ: Princeton University Press).
—— (1990) *Culture Shift in Advanced Industrial Societies* (Princeton, NJ: Princeton University Press).
Jackson, T. (2004) *Motivating Sustainable Consumption* (Guildford, UK: University of Surrey, Centre for Environmental Strategy).
—— and L. Michaelis (2003) *Policies for Sustainable Consumption* (a report to the Sustainable Development Commission; portal.surrey.ac.uk/pls/portal/docs/PAGE/ENG/RESEARCH/CES/CESRESEARCH/ECOLOGICAL-ECONOMICS/PROJECTS/FBN/POLICIES.PDF [accessed 22 January 2008]).
——, W. Jager and S. Stagl (2004) *Beyond Insatiability: Needs Theory, Consumption and Sustainability* (Guildford, UK: University of Surrey, Centre for Environmental Strategy).

Jacobsen, E., and A. Dulsrud (2007) 'Will Consumers Save the World? The Framing of Political Consumerism', *Journal of Agricultural and Environmental Ethics* 20.5: 469-82.
Jensen, R. (1999) *The Dream Society* (New York: McGraw-Hill).
Kaiser, F., S. Wolfing and U. Fuhrer (1999) 'Environmental Attitude and Ecological Behaviour', *Journal of Environmental Psychology* 19: 1-9.
Kjærnes, U., and E. Stø (1996) 'The Political Legitimacy of Consumer Interests', paper presented at the *ESA Working Group, Sociology of Consumption*, Tallinn, Estonia, 29-31 August 1996.
Klepp, I.G. (2003) 'Clothes and Cleanliness: Why We Still Spend as Much Time on Laundry', *Ethnologia Scandinavia* 33: 61-74.
—— (2005) 'Demonstrations of Feminine Purity', in G. Hagemann and H. Roll-Hansen (eds.), *Twentieth-century Housewives: Meanings and Implications of Unpaid Work* (Oslo: Unipub Oslo Academic Press).
Maslow, A.H. (1943) 'A Theory of Human Motivation', *Psychological Review* 50: 370-96.
Max-Neef, M.A. (1992) 'Development and Human Needs', in P. Ekins and M.A. Max-Neef (eds.), *Real-Life Economics: Understanding Wealth Creation* (London/New York: Routledge).
Meadows, D.H., D.L. Meadows, J. Randers and W.W. Behrens III (1972) *The Limits to Growth* (New York: Universe Books).
——, J. Randers and D.L. Meadows (2004) *Limits to Growth: The 30-year Update* (White River Junction, VT: Chelsea Green Publishing)
Micheletti, M. (2003) *Political Virtue and Shopping* (New York: Palgrave Macmillan).
——, A. Follesdal and D. Stolle (eds.) (2004) *Politics, Products and Markets* (New Brunswick, NJ/London: Transaction Publishers).
Miller, D. (1998) *A Theory of Shopping* (Ithaca, NY: Cornell University Press).
Musters, A.P.A. (1995) *The Energy–Economy–Environment Interaction and the Rebound Effect* (Internal Report ECN-I-94-053; Petten: Netherlands Energy Research Foundation [ECN]).
Olsen, J.P. (2002) 'Reforming European Institutions of Governance', *Journal of Common Market Studies* 40: 581.
Persky, J. (1993) 'Retrospectives: Consumer Sovereignty', *Journal of Economic Perspectives* 7.1: 183-91.
Pestoff, V.A. (1999) *Beyond the Market and State* (Brookfield, VT: Ashgate).
Robertson, A.F. (2001) *Greed: Gut Feelings, Growth and History* (Cambridge, UK: Polity Press).
Rubik, F., and P. Frankl (2005) *The Future of Eco-labelling: Making Environmental Product Information Systems Effective* (Sheffield, UK: Greenleaf Publishing).
Shove, E. (2003) *Comfort, Cleanliness and Convenience* (Oxford, UK: Berg).
Spaargaren, G. (1997) *The Ecological Modernisation of Production and Consumption: Essays in Environmental Sociology* (PhD thesis; Landbouw Universiteit Wageningen).
Stø, E. (ed.) (2004) *The Involvement of Consumers to Develop and Implement Tools for Sustainable Households in the City of Tomorrow: Final Report to the European Commission* (Oslo: National Institute for Consumer Research [SIFO]).
—— and P. Strandbakken (2005a) 'Background: Theoretical Contributions, Eco-labels and Environmental Policy', in F. Rubik and P. Frankl (eds.), *The Future of Eco-labelling: Making Environmental Product Information Systems Effective* (Sheffield, UK: Greenleaf Publishing): 16-45.
—— and P. Strandbakken (2005b) 'Eco-labels and Consumers', in F. Rubik and P. Frankl (eds.), *The Future of Eco-labelling: Making Environmental Product Information Systems Effective* (Sheffield, UK: Greenleaf Publishing): 92-119.
——, H. Throne-Holst and G. Vitterso (2005) 'The Role of Consumers in Environmental Successes', in K.G. Grunert and J. Thøgersen (eds.), *Consumers, Policy and the Environment: A Tribute to Folke Ölander* (New York: Springer): 325-56.
Strandbakken, P. (2006) *Produktlevetid og miljø (Life-span of Products and the Environment)* (PhD thesis; University of Tromsø, Norway and the National Institute of Consumer Research [SIFO]).
—— (1995) *Bærekraftig forbruk (Sustainable Consumption)* (Report 1-1995; Lysaker, Denmark: National Institute for Consumer Research [SIFO], in Norwegian, with a summary in English).
Svane, Ö. (2002) *Nordic Households and Sustainable Housing: Mapping Situations of Opportunity* (Nordic Report in the TemaNord-series, TemaNord:523; Copenhagen: Nordic Council of Ministers).

Terragni, L., E. Jacobsen, G. Vitterspø and H. Torjusen (2006) *Etisk-politisk forbruk en oversikt* (Publication 1-2006; Oslo: National Institute of Consumer Research [SIFO], in Norwegian).

Thøgersen, J. (2005) 'How May Consumer Policy Empower Consumers for Sustainable Lifestyles?', *Journal of Consumer Policy* 28: 143-78.

Throne-Holst, H., E. Stø and P. Strandbakken (2007) 'The Role of Consumption and Consumers in Zero Emission Strategies', *Journal of Cleaner Production* 15.13–14: 1,328-36.

——, P. Strandbakken and E. Stø (2008) 'Identification of Households' Barriers to Energy Saving Solutions', *Management of Environmental Quality* 19.1: 54-66.

Tukker, A., M.J. Cohen, U. de Zoysa, E. Hertwich, P. Hofstetter, A. Inaba, S. Lorek and E. Stø (2006) 'The Oslo Declaration on Sustainable Consumption', *Journal of Industrial Ecology* 10.1–2: 9-14.

Uiterkamp, A.J.M.S. (2000) 'Energy Consumption: Efficiency and Conservation', in B. Heap and J. Kent (eds.), *Towards Sustainable Consumption: A European Perspective* (London: The Royal Society): 111-15.

UN (United Nations) (2002) *Plan of Implementation of the World Summit on Sustainable Development* (New York: UN).

Veblen, T. (1899) *The Theory of the Leisure Class* (Great Mind Series; London: Prometheus Books, 1998 edn).

Vitterspø, G. (2003) *Environmental Information and Consumption Practices: A Case Study of Households in Fredrikstad* (Professional Report No. 4-2003; Oslo: National Institute for Consumer Research).

——, P. Strandbakken and E. Stø (1998) *Grønt husholdningsbudsjett: Veiledning til et mindre miljøbelastende forbruk* (Report 7-1998; Lysaker, Denmark: National Institute of Consumer Research [SIFO]).

Vlek, C., W. Jager and L. Steg (1997) 'Modellen en strategieën voor gedragsverandering ter beheersing van collectieve risico's', *Nederlands Tijdschrift voor de Psychologie* 52: 174-91.

Von Weizsäcker, E., A.B. Lovins and L.H. Lovins (1997) *Factor Four: Doubling Wealth, Halving Resource Use* (Report to the Club of Rome; London: Earthscan Publications).

Warde, A. (1997) *Consumption, Food and Taste* (London: Sage Publications).

WCED (World Commission on Environment and Development) (1987) *Our Common Future* (Brundtland Report; Oxford: Oxford University Press).

# 14
# Product-service systems
Taking a functional and a
symbolic perspective on usership

*Gerd Scholl*
Institut für ökologische Wirtschaftsforschung (IÖW), Germany

## 14.1 Introduction

Product-service systems (PSSs) have been a major concern of research into sustainable consumption for more than ten years (e.g. Bartolomeo *et al.* 2003; Goedkoop *et al.* 1999; Mont 2004; Stahel 1994). Usually, PSSs are divided into services providing added value to the product life-cycle, such as maintenance and upgrading, services providing enabling platforms for customers, such as renting or leasing, and services providing final results to the customers, such as mobility services or delivery of warmth (UNEP 2002). In this chapter, we will focus on PSSs that replace ownership of products. Obenberger and Brown (1976: 82) have denoted this category of consumption as 'usership'. They define it as 'all types of consumption in which the consumer does not possess legal title to the product'.

Although usership has been addressed by marketing research on a few occasions (e.g. Berry and Maricle 1973; Durgee and O'Connor 1995; Johnson *et al.* 1998; Obenberger and Brown 1976), there is still a lack of analysis of the consumer's response to the substitution of possessions by services. Only a few studies have addressed this issue in the context of sustainability (e.g. Behrendt *et al.* 2003; Hirschl *et al.* 2003; Meijkamp 2000; Schrader 1999).

In the following sections we will take a closer look at the conditions under which consumers might be inclined to accept loss of ownership in consumption. We start with a discussion from a functional perspective that is informed by neo-institutional economics. This analysis yields important insights into the nature of the transaction process between producers and consumers. It abstracts, however, from the social and symbolic qualities of consumption, which are examined in the second part of the chapter and derived from current consumer research and socio-psychological thinking. After briefly introducing the idea of consumption as a social process, we pay attention to the different facets of symbolic meaning related to the ownership of consumer goods. This analysis provides a basic framework against which the symbolic meaning of usership is approached. In the concluding part of the chapter we will elaborate on the relationship between the two perspectives within the discussion on radical changes towards sustainable consumption.[1]

## 14.2 Product-service systems from a functional perspective

The functional perspective on PSSs is based on neo-institutional economics. This school of micro-economics is a further development of neoclassical economics under more realistic suppositions (e.g. Coase 1984). It is assumed that (1) people strive to maximise their individual utility when deciding on different consumption alternatives, (2) market information is not complete and information processing capacities are limited ('bounded rationality') and (3) individuals tend to display opportunistic behaviour, in particular in the case of asymmetric information.

Neo-institutional economics includes property rights theory, information economics, transaction cost theory and principal–agent theory.

### 14.2.1 Property rights theory

The theory of property rights assumes that the behaviour of market actors is strongly influenced by the way in which property rights are allocated among market participants (e.g. Alchian 1965). Ownership is at the core of property rights theory. It represents a set of different single rights (Silver 1989):

- The **right to use** designates that individuals are free to decide on the way, the frequency and the time and place of use of their private property
- The right to use is practically guaranteed by the **right to exclude third parties** from the use of property without approval from the owner
- Furthermore, owners have the **right to change** their property the way they like. This includes unintentional wear and tear and also the conscious alteration of the material properties of goods (e.g. painting a chair a new colour)

---

1 The author is deeply grateful to his colleague Wilfried Konrad at IÖW for several discussions he had with him on the issues presented.

- In addition, owners hold the **right to sell and dispose of** their private property. The latter, however, may not be regarded as a 'right' in its true sense, since disposing of discarded goods is very often associated with some time and/or financial efforts. In contrast, the possibility of selling your own property has recently gained some practical relevance within consumption activities as a result of to the emergence of Internet platforms such as eBay

- Last, ownership encompasses the **right to appropriate any income** generated by the selling or renting of property

In the case of consumer goods, property rights may, at least partly, turn into duties (Schrader 2001). The owner has to store his or her private property correctly in order to preserve its usability and protect third parties against unwanted contact with it. This challenge becomes greater the smaller the available space—think, for instance, about the problem of finding a parking place in densely populated urban areas. Another duty results from the fact that the consumer has to maintain the owned goods to ensure their proper functioning and find suitable (normally costly) ways of disposing of the good in case his or her wish to use the good has terminated. And he or she has to bear all the fixed and variable costs for purchase, storage, use, maintenance and disposal of the property.

This brief discussion reveals that, on the one hand, ownership provides a number of rights and direct benefits to the consumer. On the other hand, owning things generates duties and risks that may reduce the utility of ownership. Hence, changing the allocation of property rights between supplier and customer—that is, purchasing the right to use only and leaving all other property rights with the service provider—may be beneficial to the consumer if ownership duties outweigh ownership rights (e.g. Bagschik 1999; Hockerts 1995).

### 14.2.2 Information economics

The analysis of information economics addresses the different qualities of market goods (e.g. Nelson 1970; Stigler 1961). **Search qualities** refer to those properties of goods that can be assessed prior to the market transaction (e.g. the processing speed of a computer or the opening hours of a tool rental facility). **Experience qualities** derive from those characteristics that can be evaluated during use of the good only (e.g. the taste of a pizza or the driving quality of a shared car). Those qualities that can be assessed neither *a priori* nor during use are called **credence qualities** (e.g. the environmental impacts of a launderette).

In general, services—as opposed to physical products—are characterised by a large amount of experience and credence qualities (Zeithaml 1981). Owing to the ensuing assessment problems, the customer normally bears a bigger purchasing risk in the case of services than in the case of material goods. For example, in contrast to using one's own pair of skis one can never be sure in advance that rented ski equipment will have similar operating characteristics and, hence, give similar pleasure to the user.

But market actors can take measures to reduce the structural information asymmetries in service markets. On the one hand, suppliers can engage in **signalling** strategies (e.g. Spence 1973); that is, they can start communicating credible quality information to their potential clients. Advertising, warranties and the use of independent certification schemes are examples of signalling strategies. Suppliers will take signalling measures as long as the costs for signalling do not exceed the cost of offering services of an inferior quality: for example, the cost of product take-back and loss of reputation (Göbel 2002). Car-sharing companies in Germany, for instance, employ a certain kind of signalling strategy. They use the national eco-label (the Blue Angel), which is well known among German consumers and has been available to car-sharing services since 1999. On the other hand, consumers can do a market **screening**: for example, through personal communication with current clients, product trials, price comparisons and by reading product tests. However, in case of consumer-oriented PSSs, market screening is not an easy task, since there are only a few clients one could ask to exchange experiences, since product trials are yet not very common, and product tests of PSSs are not widespread.

### 14.2.3 Transaction cost theory

The exchange of market goods is at the core of the transaction cost approach (e.g. Alchian and Demsetz 1972; Coase 1937). The costs arising during coordination of these exchange processes are called transaction costs. They include the costs for initiating, agreeing on and controlling the contractual arrangements between market participants, such as the effort required to find suitable contractors, the time needed for negotiations or the measures taken to check compliance with agreed quality objectives, quantities, prices, etc. The theory assumes that transaction costs diminish with increasing number of transactions (Williamson 1990).

Very often, transactions are more complex in the case of PSSs as compared with mere purchasing arrangements (e.g. Schrader 2001). Finding a suitable supplier may be more costly because of the novelty of the offer and the assessment problems mentioned in Section 14.2.2. Contractual negotiations may also be more costly if, for instance, securities have to be deposited. In addition, control costs may accrue, for example, for booking the item to rent, for scrutinising its proper functionality and for transporting it to the place of use and returning it to the rental company. These costs of transaction will decrease, however, the more often the service is used.

Hence, transaction costs can be an important impediment to consumer acceptance of PSSs. Their analysis, however, discloses possible ways to make usership more attractive. Decentralising product pools, introducing delivery services, optimising the user interface (e.g. through online booking and automatic settlement of accounts) are means of reducing the costs of transaction and, thereby, of improving the benefits of usership as compared with ownership.

### 14.2.4 Principal–agent theory

In contrast to transaction cost theory, which addresses the nature of the exchange between producer and consumer, principal–agent theory is about the behaviour of market actors in situations of asymmetric information (e.g. Arrow 1985). From this perspective, the principal faces a lack of information and the agent is inclined to display opportunistic behaviour. Opportunistic behaviour may result from hidden characteristics in the offer (i.e. properties that are known solely to the agent [e.g. quality]) and from hidden actions of the agent (i.e. activities that are necessary to produce the offer and that cannot be assessed by the principal in advance). These two types of uncertainty are especially relevant to the market acceptance of PSSs.

In the first case, of **hidden characteristics**, the supplier has an incentive to offer a service of lower quality if this provides a larger profit margin. This requires that the customer cannot recognise this weakness in the offer prior to purchase. One can assume, however, that suppliers of leases or rentals do not always exploit this theoretical advantage, since customer transactions occur more often than in the case of one-off product purchase; thus, at the end of every use period the customer may choose the exit option (Franck *et al.* 1999: 34).

The second case, of **hidden actions**, can be illustrated with the example of the sharing of consumer goods. The user takes the role of the agent and the supplier the role of the principal. Assuming that the user cannot be sanctioned for inadequate behaviour, he or she has an incentive to reduce his or her efforts to use the good properly and may tend overuse it. The supplier may respond in three different ways to this 'moral hazard' (Franck *et al.* 1999). She or he may try to immunise the rented object through a particularly robust product design (*ex ante*), she or he may seek to monitor the intensity of use during product utilisation (*in situ*) or she or he may attempt to detect misuse afterwards (*ex post*). However, the first option is costly and the second option very often is not technically feasible. Only *ex post* detection appears to be a suitable means to prevent the risk of moral hazard—provided that product faults from manufacturing and maintenance can be unequivocally separated from wear and tear arising from improper use.

### 14.2.5 Conclusions from the functional analysis

The discussion of PSSs from a neo-institutional perspective has revealed that renting and leasing might be beneficial alternatives to product ownership, particularly if ownership duties weigh heavily. It can also be concluded that market actors need to take up signalling and screening measures to reduce the risks resulting from the experience and credence qualities of services. Finally, we have seen that PSSs need to be designed so as to minimise the cost of the transaction and the risks of improper use.

The viewpoint of neo-institutional economics, however, has a number of limitations. Individual preferences are treated as given and constant. Hence, their formation and alteration is not part of the behavioural model. Additionally, decision-making is modelled as a cognitive process, so that affective rationales for consumption cannot be grasped with this approach. Finally, neo-institutional analysis neglects the fact that consumption is a social process and that consumer goods bear symbolic meaning. We will address this latter aspect in the following sections.

## 14.3 Product-service systems from a symbolic perspective

### 14.3.1 The symbolic meaning of consumption

The functional or utilitarian perspective assumes that a consumer will processes information effectively to maximise the utility that can be drawn from consuming a certain good or service. This perspective ignores, however, that people consume goods and services for reasons other than their functional properties. Symbolic qualities also play a vital role when people buy and consume goods.

Thorstein Veblen was among the first to show that people consume in order to preserve or achieve a superior social status (Veblen 1899). But products can be more than just status symbols. They can convey age, gender, ethnicity and other symbolic information (Belk 1995). They can be regarded as 'symbols for sale' (Levy 1959) and contribute to the construction of our modern culturally constituted societies, 'because they are a vital and tangible record of cultural meaning that is otherwise intangible' (McCracken 1986: 73). It does not matter, here, whether we, as Bourdieu (1984) argues, regard consumption as the manifestation in which social class structure is reproduced ('we consume what we are') or whether we look at it, as claimed by Baudrillard (1998), as mere consumption of signs and symbolic value through which social stratification emerges ('we are what we consume').[2] Rather, the decisive issue is to acknowledge that commodities possess symbolic meaning and that this specific quality of consumption items plays an essential role in creating and maintaining personal and social identities.

**Identity** is the answer to the question, 'Who am I?' (Blasi 1988). It is the image the individual has of his- or herself. It represents its personal and social characteristics. It is the mirror that reflects the self to the individual and the medium that signifies the self towards others (e.g. Baumeister 1986). Identity is quite close to the notion of **self concept**, which is the 'beliefs a person holds about his or her own attributes, and how he or she evaluates these qualities' (Solomon 2002: 150). It is made up of multiple role identities, such as gender roles, occupational roles, family roles and so on (e.g. Laverie *et al.* 2002).

People articulate their identities, to themselves and to others, not solely, but also, through their material possessions (e.g. Beaglehole 1932; Belk 1988; Csikszentmihalyi and Rochberg-Halton 1981). In their symbolic roles possessions are the expressions of who somebody is—that is, they reflect personal characteristics, such as personal values and personal history, and symbolise categorical evidence, such as group affiliations and social position (Dittmar 1992). Hence, ownership is strongly tied to a specific symbolic

---

2 The heated debate over the degree of freedom available to people in their consumption activities, as it is put forward in sociologies of consumption and (material) culture studies, cannot be fully reproduced here. While (post-)structuralist authors argue that consumption is (still) socially structured (e.g. Bourdieu 1984; Holt 1998), postmodernist writings have emphasised individual autonomy in appropriating lifestyles as a means of constructing and expressing desired identities (e.g. Featherstone 1991). More recently, the postmodernist or voluntaristic position has also been criticised with the argument that consumption is largely structured by everyday life and is thus characterised by a 'balance between creativity and constraint' (Mackay 1997: 10), guided by routines and social conventions and framed by sociotechnical systems (e.g. Campbell 1996; Gronow and Warde 2001; Shove 2003).

meaning contained in and conveyed through the material artefacts possessed (empirical evidence from a representative consumer survey is given, e.g., in Hirschl *et al.* 2003).

But what will happen to this symbolic meaning, when we turn from ownership to usership (Fig. 14.1)? In order to shed light on this question we first draw attention to the symbolic roles of possessions and then address the symbolic meanings conveyed in usership.

FIGURE 14.1 **From ownership to usership**

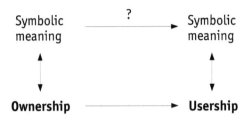

## 14.3.2 The symbolic meaning of ownership

According to Richins (1994), possessions bear a private and public meaning. The first dimension mirrors the subjective meanings an object holds for a particular individual and the second dimension stands for the collectively agreed-on elements of meaning associated with an object. The two dimensions are 'distinct but related entities' (Richins 1994: 517). Hence, in the following discussion we separate them mainly for analytical reasons. In practice, the symbolic interaction between, on the one hand, the object and the individual (intrapersonal dimension) and, on the other hand, the object, the individual and larger social entities (interpersonal meaning) will always be intertwined: a BMW sports car might be a sign of superior social status and at the same time provide a feeling of privacy and autonomy to its owner.

### 14.3.2.1 Intrapersonal meaning

There is plenty of evidence that possessions may serve as **symbols of control**. Furby (1978), for instance, has shown in a number of studies that ownership is an expression of a person's ability to control the environment. Belk (1988) argues that people regard material objects as part of their 'extended self' to the extent to which they are able to exercise control over them. And Hunt *et al.* (1990) have revealed that individuals who believe life is a product of external forces rather than a result of personal endeavours more often tend to have materialistic attitudes that, according to Belk (1985), often go hand in hand with a tendency to maintain ownership of one's possessions.

From another perspective, possessions can be considered as the **symbolic containers of our memories**. As they accompany us through our lives they become 'a symbolic, yet concrete, collage of our personal history' (Dittmar 1992: 89). Possessions can be memories in a direct sense, such as a souvenir, a photograph album or an heirloom, and they can indirectly serve as triggers for memorable experiences, such as the first car one drove after receiving one's driving licence, or the desk one sat at when having written one's master's thesis.

Besides being the material record of our personal history, possessions can be a means of creating continuity, especially in periods of change. Studying child development, Winnicott (1971) worked out the notion of **transition objects** for artefacts, such as cuddly toys, which provide security and stability in an environment that, from the infant's point of view, alters rapidly. This stabilising role is relevant in adulthood as well (e.g. Kamptner 1989). In times of transition, for example, when moving to a new place, when finalising one's professional life or when starting a family, the continuous use of owned objects may render the adjustment to new circumstances easier (e.g. Friese 1998).

Moreover, possessions can be a means of **symbolic self-completion** (Wicklund and Gollwitzer 1982); that is, they can be used to compensate for the experience of an incomplete self-definition. It could be shown, for example, that unsuccessful students of business administration disproportionately often adorned themselves with the symbols of success, such as briefcases, business suits and flamboyant wristwatches. Through reciprocal processes of 'self-symbolisation' and 'registering' (Wicklund and Gollwitzer 1982), individuals use symbols of their desired identities hoping that these 'completed' identities will be affirmed by others.

This mechanism of registering alludes to the reflexive nature of the identity-shaping process: self-completion through the use of possessions requires feedback from relevant others. This triadic interaction between the individual, the object and social groups will be discussed in the next section.

### 14.3.2.2 Interpersonal meaning

Since self-perception of an individual is, among others, dependent on the achieved locus within the social hierarchy, *status symbols* are an essential part of the individual's identity. They work like 'trophies' providing the owner with information about his or her accomplishments (Csikszentmihalyi and Rochberg-Halton 1981). Material things can turn to signifiers of status when they are rare, expensive, visible and sought-after (Csikszentmihalyi and Rochberg-Halton 1981).

As mentioned in Section 14.3.1, Veblen (1899) addressed this issue fairly early on. He analysed the consumption behaviour of the US *nouveaux riches* at the turn of the 20th century. He found that this group displayed its superior social position through consumption of expensive goods with little or no utility—a phenomenon he referred to as 'conspicuous consumption'. Veblen's observations suggest that 'all consumers attempt . . . to "keep up with the Joneses", mimicking or emulating the consumption habits of their social superiors' (Clark *et al.* 2003: 16).

Emulation, however, implies that the signifying effect of status symbols is temporary in nature—reinforced by the greater abundance of, and access to, material goods (e.g. Blumberg 1974). Moreover, the assignment of status value to products, in particular in modern consumer societies, has become a dynamic process (e.g. Holt 1998), so that status validation increasingly turns to an 'exercise in "applied semiotics" . . . [in which] the status-conscious actor must engage in an ongoing quest to identify those specific consumption events that will successfully yield social honour' (Solomon 1999: 75).

The locatory function of possessions is not confined, however, to the vertical dimension of status. They may also serve, at a horizontal level, to indicate membership of certain social groups (e.g. Solomon 1983). Entering a group means taking a specific role. The various roles an individual adopts while acting in different social entities contribute to the manifestation of his or her self concept. With regard to the meaning of material possessions in this process of self-definition Laverie *et al.* (2002) found that a specific role identity is more important for shaping the individual's identity when more identity-relevant possessions are available to enact the identity well.

Group novices will more often tend to rely on the typical product symbols of a role and attach more importance to owning these symbols than in-group members (Solomon 1983, 1999) owing to the fact that material symbols can be appropriated much more easily than other signifiers of group membership, such as proper manners and adequate skills. This mechanism is especially relevant in periods of role transition and identity development, as, for instance, in adolescence.

### 14.3.2.3 Conclusions

This brief overview has revealed that through their symbolic meaning possessions can convey different aspects of the owner's identity to that person and to others. The symbolic meaning is generated both through interaction between the individual and the object and through interaction between the individual, the object and larger social entities. In a symbolic sense, possessions are a means simultaneously to ensure personal continuity and coherence, support individual autonomy, give a sense of uniqueness and provide social affiliation.

This branch of research does not include, though, the idea of consumption without ownership. As a consequence, the terms 'ownership' and 'possessions' are used largely synonymously without any specification of the possibly temporary nature of possessing a thing—as in a renting or leasing arrangement.

## 14.3.3 The symbolic meaning of usership

We have argued above that ownership evokes symbolic meaning by contributing to the creation and maintenance of personal and social identities. But what is left, symbolically, when possessions have turned to a PSS, when your own car has been replaced by access to a car-sharing fleet, when your own washing machine has been substituted by a launderette and/or a textile-cleaning service or when you have traded in your own pair of skis for the on-demand purchase of rented skis?

One can suggest that the provision of a symbolic sense of control is far more difficult with ownership-replacing services than with possessions. Furthermore, it is hard to imagine that renting and leasing services or services providing final results may work as a container of personal memories in the same way that permanently used goods may do—the 'container' being largely intangible. This argument also applies to the symbolic quality of possessions as transition objects or means of symbolic self-completion. But it may be less applicable to the interpersonal dimension of symbolic meaning—that is, one can principally envisage that usership might be employed to signify social position and/or group affiliation. Imagine a consumer expressing her or his 'green' attitude by membership in a car-sharing organisation or a customer of ski rentals communicating his or her skills to peers by frequently using and perfectly mastering new, cutting-edge equipment.

Hence, exploring possible ways to enhance the symbolic meaning of usership means (1) to regain the intrapersonal symbolic qualities that have 'disappeared' along with the loss of ownership and (2) to strengthen the interpersonal symbolic qualities that might also have been diluted in this transition. With regard to the first concern, we will focus on the aspect of symbolic control, because the other issues, such as symbolic transition or self-completion, appear to be far less attainable by means of an essentially intangible offer.

To build a sound basis for the discussion of the two issues, we refer to a conceptualisation of a service as a three-dimensional process (e.g. Meyer 1991):

- Service as a resource. This perspective focuses on the internal factors necessary to produce the service. They encompass tangible elements, such as facilities, infrastructure and machines, and intangible assets, such as the skills of employees
- Service as a process. During the process of service delivery, internal factors are combined with the external factor (i.e. the customer). The nature of the service encounter can be distinguished:
    - With respect to the type of external factor brought in (i.e. the customer him- or herself, as in healthcare services or in a taxi ride): person-oriented services
    - With respect to the commodity of the customer (as in repair services or in tax consultancy): goods-oriented services
- Service as a result. This dimension stresses the result of the service delivery process. One can differentiate between a direct result (e.g. a fixed car) and an indirect result (e.g. recovery of the car owner's mobility)

In the following, we use this approach to identify possible service-related factors shaping the symbolic meaning of usership. In Table 14.1, the three service dimensions—resources, process and results—are juxtaposed with two layers of symbolic meaning: the symbolic sense of control provocable in and through usership at an intrapersonal level and the categorical symbolic meaning that helps to locate the individual socially at an interpersonal level. The different factors included in the cells of the matrix are discussed in the following sections.

TABLE 14.1 Service-related factors shaping the symbolic meaning of usership

|  | Service resources | Service process | Service results |
| --- | --- | --- | --- |
| **Symbolic control (intra-personal)** | Access Scope | Interaction Physical surroundings | Mastery of objects |
| **Categorical symbolism (inter-personal)** | Service brand | Service brand | Service brand |

### 14.3.3.1 Symbolic control in and through usership

A perceived lack of control influences the behaviour of the customer in the service encounter (Bateson 1992). It may be one important source of dissatisfaction with the service performance (e.g. Bateson 1992; Gupta and Vajic 2000).

When moving from ownership to usership people give up full control over the commercial good. To compensate for this confinement, service resources should be designed so as to restore options of control for the customer. This can be achieved by maximising ease of access to the service (e.g. by offering flexible opening hours, online booking and delivery services). Moreover, a sense of control can be rebuilt by extending the scope of the service offer. This may encompass various measures, such as increasing the amount of available rental goods to ensure that demand can be met properly, enlarging the variety of rental goods to enhance fit with customer needs (as car-sharing services do with a fleet covering small cars and limousines as well as cabriolets and vans) or offering rental goods of superior quality that provide higher performance and so on.

When it comes to the process of service delivery the quality of the interaction between the employee and the customer is crucial for re-establishing symbolic control. Notwithstanding the huge variety of possible interaction modes in usership—imagine a ski rental company, a launderette or a nappy service—a competent and friendly appearance in the contact personnel represents a generic response to the perceived loss of control. Moreover, Smith and Houston (1983) found that customers transpose repeated service encounters into stereotyped mental scripts. The scripts guide customers' experience of the service and can be regarded as a strategy of cognitive control. Hence, matching the processes in the service encounter with the mental scripts of the customers may further enhance the perceived control of the process. One can conclude that standardisation of the service process over time is another means to regain a sense of control in usership.

A similar effect might stem from the physical surroundings of the service. If customers perceive greater personal control in the 'servicescape' (Bitner 1992), the pleasure of the service experience is increased, and this in turn makes service entry more attractive to customers. Environmental dimensions, such as clear signage and good ventilation, in combination with proper spatial layout and good functionality of the servicescape, may all contribute to perceptions of personal control (Bitner 1992).

Another factor possibly evoking a sense of symbolic control in usership is mastery of the object. In particular, in services providing enabling platforms it is vital that the goods delivered for temporary use are self-explanatory and easy to operate. In the case of complex items, and also when customers are unfamiliar with the rental good, clear product design and plain instructions, provided either by the personnel or by simple manuals, can avoid a perceived loss of control and help customers master the object more quickly.

### 14.3.3.2 Categorical symbolism in and through usership

When we looked at the interpersonal symbolic meaning of possessions we argued that individuals express their social identities, among others things, through acquisition and use of material possessions. By so doing they seek to locate themselves—and are located by others—in social–material terms. Implicitly, this mechanism presumes that symbolic meaning is constructed socially. Meaning evolves from interaction of the individual with other individuals and with social institutions, such as the media. As a consequence, the symbolic meaning of possessions varies historically and socioculturally—as we have noted above with the increasingly volatile meaning of status symbols.

This discussion has already suggested that the structure of modern societies is not adequately represented by the idea of vertical stratification based on socioeconomic criteria, such as education, age, income, etc. Instead, modern societies are built from several different lifestyles by which people communicate their identities to themselves and others and articulate which groups they do (and do not) belong to and what social position they occupy or strive for (e.g. Chaney 1996).

Shaping categorical symbolism in and through usership implies finding ways to better connect a PSS with the lifestyles of the envisaged target groups. This connection can be facilitated by a dedicated service brand. In general, a brand works as a symbolic resource carrying or portraying a set of social meanings associated with the consumption of the branded good (e.g. Elliott and Wattanasuwan 1998). From the customer's point of view, a brand provides orientation, generates trust and conveys prestige. From the supplier's point of view, a service brand helps to profile the offer and communicate desired product images (Bruhn 2001).

Creating a brand for usership services involves similar measures as in general service brand management. Focusing on the service resources, branding can be obtained by the brand quality of the service-related products (e.g. von der Oelsnitz 1997), by choosing a brand name that is distinguishable, relevant, memorable and flexible (Berry and Parasuraman 1992), by applying the trademark to all physical evidence of the service according to a coherent corporate identity (Meffert and Bruhn 2003) and so on. The brand identity may further be visualised during the service encounter by uniform clothing of the service personnel. As regards the service results, the establishment of a brand can be supported by providing branded gifts and souvenirs or by offering service-related products for sale, such as travelling bags sold in train shops (e.g. Stauss 2001). Thereby, customers are enabled to maintain symbolic contact with the service provider and to communicate service consumption towards relevant others.

## 14.4 Overall conclusions

The above discussion has shown that consumer acceptance of PSSs is a complex issue. The switch from ownership to usership can provide a considerable increase in resource productivity; it does, however, require far-reaching change in current consumption patterns.

The analysis of PSSs from a functional perspective provides evidence of the potential benefits of a shift to usership. By consuming services instead of products consumers may be released from the duties linked with ownership, such as the need to provide storage space or the necessity to properly maintain the owned goods. Furthermore, a number of recommendations as to the proper design and implementation of PSSs can be derived from the analysis within neo-institutional economics. In order to meet customer acceptance PSSs should be designed so as to minimise the costs of transaction (e.g. by decentralising product pools, by introducing delivery services and by optimising the user interface). Suppliers should find ways to credibly convey the quality of the service offer (e.g. by employing independent labelling schemes or by providing warranties). Finally, improper use of shared goods should be prevented by installing suitable monitoring and detection systems.

The viewpoint of neo-institutional economics does neglect, though, that consumption of products and services is a social process. Correspondingly, the symbolic perspective reveals that material goods are part of the social fabric of our lives and that ownership essentially contributes to identity formation and social cohesion in modern consumer societies. Therefore, starting from the different layers of symbolic meaning that may emanate from owning consumer goods this chapter has presented an exploration into the symbolic properties connected with or connectable to PSSs. Possible ways to recover the symbolic meaning that is 'lost' when giving up ownership of material goods have been sketched. Giving clients a sense of control in the service interaction and creating a service brand that offers a meaningful service experience to the client are important approaches to enhance the symbolic meaning of usership.

Both perspectives are important to consider if one wants to develop feasible forms of usership that substantially contribute to more sustainable consumption patterns. But acknowledging the social quality of consumption is especially vital for achieving radical changes towards sustainability. The example of PSSs reveals that present consumer behaviour, which is based largely on traditional structures of ownership, is loaded with symbolic meaning. And this symbolic meaning also has to change or be reformulated if consumption is to alter radically.

# References

Alchian, A.A. (1965) 'Some Economics of Property Rights', *Il Politico* 30: 816-92.
—— and H. Demsetz (1972) 'Production, Information Costs and Economic Organisation', *American Economic Review* 62: 777-95.
Arrow, K.J. (1985) 'The Economics of Agency', in J.W. Pratt and R.J. Zeckhauser (eds.), *Principals and Agents: The Structure of Business* (Boston, MA: Harvard Business School Press): 37-51.
Bagschik, T. (1999) *Gebrauchsüberlassung komplexer Konsumgüter* (Wiesbaden, Germany: Deutscher Universitätsverlag).
Bartolomeo, M., D. dal Maso, P. de Jong, P. Eder, P. Groenewegen, P. Hopkinson, P. James, L. Nijhuis, M. Örninge and G. Scholl (2003) 'Eco-efficient Producer Services: What Are They, How Do They Benefit Customers and the Environment and How Likely Are They to Develop and be Extensively Utilised?', *Journal of Cleaner Production* 11: 829-37.
Bateson, J.E.G. (1992) 'Perceived Control and the Service Encounter', in J.E.G. Bateson (ed.), *Managing Services Marketing* (Forth Worth, TX: Dryden Press, 2nd edn): 123-32.
Baudrillard, J. (1998) *The Consumer Society: Myths and Structures* (London: Sage Publications, first published 1970).
Baumeister, R.F. (1986) *Identity: Cultural Change and the Struggle for Self* (New York/Oxford, UK: Oxford University Press).
Beaglehole, E. (1932) *Property: A Study in Social Psychology* (New York: Macmillan).
Behrendt, S., C. Jasch, J. Kortman, G. Hrauda, R. Pfitzner and D. Velte (2003) *Eco-service Development: Reinventing Supply and Demand in the European Union* (Sheffield, UK: Greenleaf Publishing).
Belk, R.W. (1985) 'Materialism: Trait Aspects of Living in the Material World', *Journal of Consumer Research* 12.3: 265-80.
—— (1988) 'Possessions and the Extended Self', *Journal of Consumer Research* 15: 139-68.

—— (1995) 'Studies in the New Consumer Behaviour', in D. Miller (ed.), *Acknowledging Consumption: A Review of New Studies* (London/New York: Routledge): 58-95.
Berry, L., and K.E. Maricle (1973) 'Consumption without Ownership: Marketing Opportunity for Today and Tomorrow', *MSU Business Topics*, Spring 1973: 33-41.
—— and A. Parasuraman (1992) *Service-Marketing* (Frankfurt/New York: Campus).
Bitner, M.J. (1992) 'Servicescapes: The Impact of Physical Surroundings on Customers and Employees', *Journal of Marketing* 56: 57-71.
Blasi, A. (1988) 'Identity and the Development of the Self', in D.K. Lapsley and F.C. Power (eds.), *Self, Ego, and Identity: Integrative Approaches* (New York: Springer): 226-42.
Blumberg, P. (1974) 'The Decline and Fall of the Status Symbol: Some Thoughts on Status in a Post-industrial Society', *Social Problems* 21.4: 490-98.
Bourdieu, P. (1984) *Distinction: A Social Critique of the Judgement of Taste* (London: Routledge & Kegan Paul).
Bruhn, M. (2001) 'Die zunehmende Bedeutung von Dienstleistungsmarken', in R. Köhler, W. Majer and H. Wiezorek (eds.), *Erfolgsfaktor Marke. Neue Strategien des Markenmanagements* (Munich: Vahlen Verlag): 213-25.
Campbell, C. (1996) 'The Meaning of Objects and the Meaning of Action: A Critical Note on the Sociology of Consumption and Theories of Clothing', *Journal of Material Culture* 1.1: 95-107.
Chaney, D. (1996) *Lifestyles* (London: Routledge).
Clark, D.B., M.A. Doel and K.M.L. Housiaux (eds.), (2003) *The Consumption Reader* (London/New York: Routledge).
Coase, R.-H. (1937) 'The Nature of the Firm', *Economica: New Series* 4: 386-405.
—— (1984) 'The New Institutional Economics', *Journal of Institutional and Theoretical Economics* 140: 229-31.
Csikszentmihalyi, M. (2001) 'Why We Need Things', in D. Miller (ed.), *Consumption: Critical Concepts* (London: Routledge): 485-93.
—— and E. Rochberg-Halton (1981) *The Meaning of Things: Domestic Symbols and the Self* (Cambridge, UK: Cambridge University Press).
Dittmar, H. (1992) *The Social Psychology of Material Possessions: To Have Is to Be* (Hemel Hempstead, UK: Harvester Wheatsheaf).
Durgee, J.F., and G.C. O'Connor (1995) 'An Exploration into Renting as Consumption Behaviour', *Psychology and Marketing* 12.2: 89-104.
Elliott, R., and K. Wattanasuwan (1998) 'Brands as Resources for the Symbolic Construction of Identity', *International Journal of Advertising* 172: 131-45.
Featherstone, M. (1991) *Consumer Culture and Postmodernism* (London: Sage Publications).
Franck, E., T. Bagschik, C. Opitz and T. Pudack (1999) *Strategien der Kreislaufwirtschaft und mikroökonomisches Kalkül* (Stuttgart, Germany: Schäffer-Poeschel Verlag).
Friese, S. (1998) 'Zum Zusammenhang von Selbst, Identität und Konsum', in M. Neuner and L. Reisch (eds.), *Konsumperspektiven: Verhaltensaspekte und Infrastruktur* (Berlin: Duncker & Humblot): 35-53.
Furby, L. (1978) 'Possessions in Humans: An Exploratory Study of its Meaning and Motivation', *Journal of Social Behaviour and Personality* 6: 49-65.
Göbel, E. (2002) *Neue Institutionenökonomik: Konzeption und betriebswirtschaftliche Anwendungen* (Stuttgart, Germany: Lucius & Lucius).
Goedkoop, M.J., C.J.G. Van Halen, H.R.M. Te Riele and P.J.M. Rommens (1999) *Product Service Systems: Ecological and Economic Basics* (The Hague: Ministry of Housing, Spatial Planning and the Environment [VROM]).
Gronow, J., and A. Warde (eds.) (2001) *Ordinary Consumption* (London: Routledge).
Gupta, S., and M. Vajic (2000) 'The Contextual and Dialectical Nature of Experiences', in J.A. Fitzsimmons and M.J. Fitzsimmons (eds.), *New Service Development: Creating Memorable Experiences* (Thousand Oaks, CA: Sage Publications): 33-51.
Hirschl, B., W. Konrad and G. Scholl (2003) 'New Concepts in Product Use for Sustainable Consumption', *Journal of Cleaner Production* 11: 873-81.

Hockerts, K. (1995) *Konzeptualisierung ökologischer Dienstleistungen* (DP-29; St Gallen, Switzerland: Institut für Wirtschaft und Ökologie).
Holt, D.B. (1998) 'Does Cultural Capital Structure American Consumption?', *Journal of Consumer Research* 25: 1-25.
Hunt, J.M., J.B. Kernan, A. Chatterjee and R.A. Florsheim (1990) 'Locus of Control as a Personality Correlate of Materialism: An Empirical Note', *Psychological Reports* 67: 1,101-102.
Johnson, M.D., A. Herrmann and F. Huber (1998) 'Growth through Product-sharing Services', *Journal of Service Research* 1: 167-77.
Kamptner, L.N., (1989) 'Personal Possessions and their Meanings in Old Age', in S. Spacapan and S. Oskamp (eds.), *The Social Psychology of Aging* (Newbury Park, CA: Sage Publications): 165-96.
Laverie, D.A., R.E. Kleine III and S. Schultz Kleine (2002) 'Re-examination of Kleine, Kleine and Kernan's Social Identity Model of Mundane Consumption: The Mediating Role of the Appraisal Process', *Journal of Consumer Research* 28: 659-69.
Levy, S.J. (1959) 'Symbols for Sale', *Harvard Business Review* 37: 117-24.
McCracken, G. (1986) 'Culture and Consumption: A Theoretical Account of the Structure and Movement of the Cultural Meaning of Consumer Goods', *Journal of Consumer Research* 13: 71-84.
Mackay, H. (1997) 'Introduction', in H. Mackay (ed.), *Consumption and Everyday Life* (London: Sage Publications): 1-12.
Meffert, H., and M. Bruhn (2003) *Dienstleistungsmarketing: Grundlagen–Konzepte–Methoden* (Wiesbaden, Germany: Gabler).
Meijkamp, R. (2000) *Changing Consumer Behaviour through Eco-efficient Services: An Empirical Study on Car Sharing in The Netherlands* (Delft, Netherlands: Delft University of Technology, Design for Sustainability Programme).
Meyer, A. (1991) 'Dienstleistungs-Marketing', *Die Betriebswirtschaft* 51.2: 195-209.
Mont, O. (2004) *Product Service Systems: Panacea or Myth?* (PhD thesis; Lund, Sweden: University of Lund).
Nelson, P. (1970) 'Information and Consumer Behaviour', *Journal of Political Economy* 78: 311-29.
Obenberger, R.W., and S.W. Brown (1976) 'A Marketing Alternative: Consumer Leasing and Renting', *Business Horizons*, October 1976: 82-86.
Prentice, D.A. (1987) 'Psychological Correspondence of Possessions, Attitudes, and Values', *Journal of Personality and Social Psychology* 53: 993-1,003.
Richins, M.L. (1994) 'Valuing Things: The Public and Private Meaning of Possessions', *Journal of Consumer Research* 21: 504-21.
Schrader, U. (1999) 'Consumer Acceptance of Eco-efficient Services. A German Perspective', *Greener Management International* 25: 105-21.
—— (2001) *Konsumentenakzeptanz Eigentumsersetzender Dienstleistungen, Konzeption und empirische Analyse* (Frankfurt am Main: Peter Lang).
Shove, E. (2003) *Comfort, Cleanliness, and Convenience: The Social Organisation of Normality* (Oxford, UK/New York: Berg).
Silver, M. (1989) *Foundations of Economic Justice* (Oxford, UK/New York: Basil Blackwell).
Smith, R., and M. Houston (1983) 'Script-Based Evaluation of Satisfaction with Services', in L. Berry, L. Shostak and G. Upah (eds.), *Marketing of Services* (Chicago: American Marketing Association).
Solomon, M.R. (1983) 'The Role of Products as Social Stimuli', *Journal of Consumer Research* 10: 319-29.
—— (1999) 'The Value of Status and the Status of Value', in M.B. Holbrook (ed.), *Consumer Value: A Framework for Analysis and Research* (London/New York: Routledge): 63-84.
—— (2002) *Consumer Behaviour. Buying, Having and Being* (Upper Saddle River, NJ: Person Education).
Spence, A.M. (1973) *Market Signalling: Information Transfer in Hiring and Related Processes* (Cambridge, MA: Harvard University Press).
Stahel, W. (1994) 'The Utilisation-Focused Service Economy: Resource Efficiency and Product-life Extension', in B. Allenby and D. Richards (eds.), *The Greening of Industrial Ecosystems* (Washington, DC: National Academy Press): 178-90.
Stauss, B. (2001) 'Markierungspolitik bei Dienstleistungen: "Die Dienstleistungsmarke"', in M. Bruhn and H. Meffert (eds.), *Handbuch Dienstleistungsmanagement: Von der strategischen Konzeption zur praktischen Umsetzung* (Wiesbaden, Germany: Gabler, 2nd edn): 549-70.

Stigler (1961) 'The Economics of Information', *Journal of Political Economy* 69: 213-25.
UNEP (United Nations Environment Programme) (2002) *Product-Service Systems and Sustainability: Opportunities for Sustainable Solutions* (Paris: UNEP).
Veblen, T. (1899) *The Theory of the Leisure Class* (Great Minds Series; London: Prometheus Books, 1998).
Von der Oelsnitz, D.D. (1997) 'Dienstleistungsmarken: Konzepte und Möglichkeiten einer markengestützten Serviceprofilierung', *GfK Jahrbuch der Absatz und Verbrauchsforschung* 1: 66-89.
Wicklund, R.A., and P.M. Gollwitzer (1982) *Symbolic Self-completion* (Hillsdale, NJ: Erlbaum).
Williamson, O.E. (1990) *Die ökonomische Institution des Kapitalismus: Unternehmen, Märkte, Kooperationen* (Tübingen, Germany: Mohr).
Winnicott, D.W. (1971) *Vom Spiel zur Kreativität* (Stuttgart, Germany: Klett-Cotta).
Zeithaml, V.A. (1981) 'How Consumer Evaluation Processes Differ between Goods and Service', in J.H. Donnelly and W.R. George (eds.), *Marketing of Services* (Chicago: American Marketing Association): 186-90.

# 15
# Social capital, lifestyles and consumption patterns

*Dario Padovan*
University of Torino, Italy

## 15.1 Introduction

The continuous social, economic and environmental transformations that are a feature of European cities force us to reflect on the meaning of the quality of life and on the challenges that will be faced in the future. The aim of this chapter is to analyse the strategies that could be adopted to improve the quality of life of city and town dwellers in terms of ecological sustainability.

This chapter illustrates a study carried out in the city of Padova, in which the existing relationships between the quality of the urban environment, the nature of social capital and the consumption patterns in two different areas of the city were analysed. The data gathered by means of focus groups that were set up in the two areas could help in developing a model of sustainable urban living.

The study highlighted citizens' broad knowledge, on the basis of which the quality of the natural capital present in the urban environment deeply influences their own quality of life. Furthermore, the study has been able to identify the nature of social relations, which are a crucial element not only in 'living better' but also in establishing strategies to deal with environmental risks.

The concept of social capital becomes a means of explaining a considerable part of the social actions and transactions of individuals' daily lives. This study evaluates the potential of social capital as a vector to promote mechanisms aimed to drive the consumption patterns and lifestyles of householders toward sustainability.

## 15.2 Sustainable urban life and social capital

For people, social relations and bonds, which have been built, developed and also broken over time, play a crucial role. The quality of these relationships depends, in part, on economic conditions, the social stigma that may be attributed to the neighbourhood and the spatial distribution of personal and collective spaces. Usually, we think that in such housing spaces 'bonding' social capital prevails. But, as we shall see below, to produce improvements in quality of life and social cohesion people often need a 'bridging' social capital. Our research on two such neighbourhoods shows that people are endowed with a mix of 'bonding' and 'bridging' connections. It means that social changes, also in the field of consumption, are possible.

### 15.2.1 What is social capital?

Crucial in implementing the research has been the concept of social capital. In recent years, increasing interest has been shown in the notion of 'social capital'. The term captures the idea that social bonds, social norms, trust and other social features play an important part in sustainable livelihoods.

Despite its topicality, the concept has a long history in sociology (Portes and Sensenbrenner 1993). Ferdinand Toennies reflected on sentiments and motives that draw people to each other, keep them together and induce them to join action. Émile Durkheim discussed the importance of 'value introjection', the idea that values, moral imperatives, and commitments precede contractual relations, driving individuals to behave in ways accepted by the collectivity. Georg Simmel wrote about 'reciprocity transactions', the norms and obligations that emerge through personalised networks of exchanges: for instance, favours between neighbours or information between brokers. Max Weber captured the idea of 'enforceable trust', the idea that formal institutions and informal groups use different mechanisms for ensuring observance with shared rules of conduct—using legal–rational mechanisms (bureaucracies) or substantive–social mechanisms (kinship or community).

Pierre Bourdieu (1979), James Coleman (1988, 1990) and Robert Putnam (1993, 2000) have given a clear theoretical framework for the social capital concept. Bourdieu suggested that social capital is a network of relationships, which is the product, intentional or unintentional, of social investment strategies aimed at the building and reproduction of durable and useful social relationships able to offer material and symbolic benefits. These relationships enlarge the individual or collective actors' action capabilities and, if extended enough, the social system's action capabilities too. For Coleman, social capital is an important resource for individuals, as it greatly affects their ability to act and their perceived quality of life. It is not lodged either in the actors themselves or in physical implements of production but in the structure of relations between actors. In short, social capital is a public good shared by a number of individuals. Robert Putnam stressed the role of civic participation in implementing democracy and social cohesion. Putnam defines social capital as 'features of social organisation, such as trust, norms and networks, that can improve the efficiency of society by facilitating coordinated actions' (Putnam 1993: 169).

## 15.2.2 Forms of social capital and networks

The idea of social capital is strongly associated with that of 'networks of relations'. People connect through a series of networks, and these networks tend to determine attitudes, beliefs, identities and values as well as access to resources, opportunities and power (Field 2003). Obviously, different types of networks are at work. Strong bonding ties give particular groups a sense of identity and common purpose but, without 'bridging' ties that transcend various social divides, bonding ties can become a basis for the pursuit of narrow interests and can actively exclude outsiders. A restricted radius of trust within a tightly knit group, such as family members or closed circles of friends, can promote forms of social interaction that are inward-seeking and less oriented to trust and cooperation at the wider community level (Portes and Landolt 1996). Cross-cutting ties between groups open up different opportunities to all members, and in so doing they also build social cohesion. Mark Granovetter (1973) emphasised the 'strength of weak ties', highlighting the importance of those ties that run beyond the immediate circle of small family or neighbourhood dwellers, giving actors richer resources to achieve a better quality of life. For certain individuals or groups, these kinds of networks can create a competitive advantage in pursuing their ends. Ronald Burt has drawn attention to the fact that actors who bridge between sub-groups have access to unique resources and information that makes them powerful brokers in a system (Burt 1992). Three basic forms of social capital have been identified (Woolcock 1999):

- Bonding social capital: relations between family members, close friends and members of ethnic groups
- Bridging social capital: relations with distant friends, associates and colleagues
- Linking social capital: relations between different social strata in a hierarchy, where power, social status and wealth are accessed by different groups

Woolcock (2001) relates 'linking social capital' to the capacity of individuals and communities to control resources, ideas and information from formal institutions beyond the immediate community radius. In this respect, social capital is no different from other forms of capital: it may be used to serve different ends, not all necessarily desirable for the community at large.

## 15.2.3 Sources of social capital

Social capital is characterised by different kinds of social conditions. Following in part the scheme of Pretty and Ward (2001) I have identified the following:

- Participation in networks and groups
- Reciprocity and exchange
- Trust and confidence
- Social norms
- The commons

### 15.2.3.1 Participation in networks and groups

Key to all uses of the concept of 'social capital' is the notion of more or less interlocking networks of relationships between individuals and groups. People engage with others through a variety of formal and informal associations that are both voluntary and equal. Individuals acting on their own cannot generate social capital. It depends on a propensity for sociability, a capacity to form new associations and networks. The nature of connections, networks and groups and the quality of people's involvement in those is a vital aspect of social capital.

### 15.2.3.2 Reciprocity and exchange

Social capital does not imply the immediate and formally accounted exchange of the legal or business contract but a combination of short-term altruism and long-term self-interest. The individual provides a service to others, or acts for the benefit of others at a personal cost, but in the general expectation that this kindness will be returned at some undefined time in the future. In a community where reciprocity is strong, people look after each other's interests. Reciprocity and exchanges increase trust. It contributes to the development of long-term obligations between people, which can be an important aspect of achieving positive environmental outcomes.

### 15.2.3.3 Trust and confidence

Trust entails a willingness to take risks in a social context based on a sense of confidence that others will respond as expected and will act in mutually supportive ways, or at least that others do not intend harm. Trust is a lubricant, oiling the wheels of a variety of social and economic mechanisms. It reduces the transaction costs between people and so liberates resources. Instead of having to invest in monitoring others, individuals are able to trust them to act as expected. It can also create a social obligation—trusting someone engenders reciprocal trust. The presence of the norm of trusting reduces the uncertainties present in social life, ensuring social equilibrium for individuals when they are facing everyday risks. In this sense, trust becomes an indicator of social sustainability.

### 15.2.3.4 Social norms

Common rules and social norms are the mutually agreed norms of behaviour that place group interests above those of individuals. They give individuals the confidence to invest in collective or group activities, knowing that others will do so too. Individuals can take responsibility and can ensure their rights are not infringed. Social norms are sometimes called 'the rules of the game', or 'the internal morality of a social system', 'the cement of society'. They reflect the degree to which individuals agree to mediate or to control their own behaviour. Social norms are usually unwritten but commonly understood formulae for determining what patterns of behaviour are expected in a given social context and for defining what forms of behaviour are valued or socially approved.

### 15.2.3.5 The commons

The combined effect of trust, networks, norms and reciprocity creates a good community, with shared ownership over resources known as 'the commons'. 'The commons' refers to the creation of a pooled community resource, owned by no one but used by all (Goldman 1998; Ostrom 1990). The short-term self-interest of each, if unchecked, would render the common resources overused and in the long term would be destroyed. Only where there is a strong ethos of trust, mutuality and effective informal sanctions against 'free-riders' can the commons be maintained indefinitely and to the mutual advantage of all. To maintain the commons the presence of a sense of personal and collective efficacy is needed.

## 15.3 Social capital and environmental change: the question of trust

### 15.3.1 Participation, trust and sustainable community

In recent years, the dimension of social capital has been widely associated with the sustainable development of local or small-scale societies. Scholars of agricultural development have been in the forefront of understanding the potential of social capital to manage the natural resources of local communities in a sustainable manner. They argue that social and human capital, embedded in participatory groups within rural communities, has been central to equitable and sustainable solutions to local development problems (Pretty and Ward 2001). Local sustainable development and social capital often affect each other. Where sustainable agriculture, and the associations that endorse it, are diffuse, a pattern of problem-solving guided by norms of mutual trust and reciprocity is encouraged. It means that, where sustainable practices are at work, an increasing level of social capital affects the local rural community, improving its quality of life (Flora 1995). In a nutshell, social capital has a large impact on the community's capacity to take up social and economic opportunities and to manage change (Falk and Kilpatrick 2000; Kilpatrick and Abbott-Chapman 2005; OECD 2001; World Bank 2001).

This approach can be applied to the improvement of urban neighbourhood quality of life, encouraging sustainable consumption patterns, as in our research. The establishment of wide associational activities produces interpersonal and systemic trust. These collective activities—such as worker solidarity, goods and information exchanges, credit and consumption groups, civil engagements, neighbourhood mutual help and voluntary activity—are embedded within social networks and are sustained by confidence in the motives of others. They increase citizens' participation in public affairs and civic life and, at the same time, affirm interpersonal trust. Furthermore, civic engagement and generalised trust, and the dynamic that sustains them, have important consequences for citizens' confidence in political institutions. In general terms, people who trust others have greater confidence in political institutions and a wider engagement in local public life (Brehm and Rahn 1997; Claibourn and Martin 2000; Gould 1993; Kwak et al. 2004). In short, networks, participation and trust (at different levels) undoubtedly give to community members a sense of social and physical well-being, a sense of a good and decent life and provide the right milieu to lead and manage change in the field of environmental and consumption sustainability.

### 15.3.2 Systemic and interpersonal trust

Trust 'is a by-product of behaviour towards others based on the norm of reciprocity and networks of civic engagements, which can be facilitated by the nature of governmental institutions and the level of socioeconomic development' (Misztal 1996: 199). There are two types of trust: 'institutional or systemic trust' and 'interpersonal trust' (Luhmann 2002; Mutti 1998). By institutional trust we mean the trust that citizens have in social institutions, as long as these institutions continue to offer concrete responses to their demands, to take decisions and to put such decisions into practice so as to guarantee a certain degree of existential stability. A lack of institutional or systemic trust reveals instability in the natural or social order in which subjects are involved and implies a certain degree of uncertainty about the routine events of daily life. Interpersonal trust corresponds to the expectation that *Alter* (the other) will not manipulate communication, will give a real, and not biased, representation of his or her own role and behaviour and of his or her real identity. Basically, the expectations of *Ego* concern the sincerity and credibility of *Alter*: that is, expectations of transparency and abstention from lying, fraud and trickery (Goffman 1971).

Trust, as Harold Garfinkel suggests, is 'a person's compliance with the expectancies of attitude of daily life as a morality' (Garfinkel 1967: 50). Trust ensures that the routine elements of a situation are confirmed, which permits the 'rational action' of the actor. If there is no trust, the uncertainties of the rules of the social game are disclosed, putting the regularity and the stability of the world with which the actor daily interacts in crisis. Thus the lack of systemic trust weakens the elements of our daily life that we 'take for granted', creating confusion, anomie and aggressive and discriminatory attitudes (Garfinkel 1967: 50-51, 173; 1963: 219).

Trust in institutions has been examined in depth in both the sociological and the political literature, and all agree that the part played by trust is crucial for the effective functioning of institutions. Social organisation is more efficient when there is trust, when there are norms that regulate communal living and when there are networks of associations (Putnam 1993). These sources of social capital are 'moral resources' that cut down on the betrayals and behaviours of 'free-riders'. Consequently, trust in the institutions of social life is essential. This is because these institutions not only forge the politics but also the identity of citizens, their public and private behaviour and the norms of communal living. Furthermore, institutions are directly responsible for many aspects of social and public life. Institutions regulate and control economic life, the environment, technological innovation, general culture, and education and training. These institutions are perhaps the main public good on which actors depend in order to delineate their life plans (Donolo 1997).

The rise of systemic distrust is connected to the development of an asymmetry of power relations in society. The perception of environmental risks highlights the fact that corporate actors (firms, governments, corporations) are largely responsible for the ecological problems that threaten people's daily life. To be conscious of ecological risks means underlining the responsibility of big corporate actors in creating many of the dangers faced by 'natural persons', by single individuals (Coleman 1982).

## 15.4 Methodological and normative aspects

### 15.4.1 Methodology

The research used a 'focus group' methodology. In both of the areas studied, the people who were invited to become part of the focus groups were those who had previously been identified as 'nodes' in the local relationship network. However, although it was relatively simple to identify the 'key' figures it was much harder to discern and decipher the relationships between these figures, the 'configurations' of which they were a part. With regard to the method used for the focus groups, we tried to use the scenario-backcasting approach as described by Karl Dreborg (1996). The main interest here is not what will happen in the future, but rather which direction people want to move in. In this case the scenarios may show either the possible path, or the desirable path, that social development should take. In short, because the scenario approach is clearly in the utopian tradition it can offer different options for the future development of society.

First, participants described their situation by focusing on the most unpleasant aspects of their neighbourhood. Subsequently, we developed scenarios that were suggested by the participants. They showed how important social problems could be coped with and solved. Starting with desirable futures, we tried to outline ways of achieving these goals, usually through political and institutional measures, but also by means of a shift in the quality of collective action.

### 15.4.2 The features of the neighbourhoods studied

The urban areas where we set up the focus groups are very different one from the other. The urban area includes the housing complex of Via Maroncelli, where 120 families live, and its neighbourhood, Pio X, which contains a large settlement, mainly occupied by foreign immigrants. The area around Via Maroncelli is a semi-suburban area, near an area of large commercial structures, shopping centres and offices, which have little to do with the everyday life of the zone; it is subjected daily to heavy traffic, many different social classes mingle and there are few essential services.

The urban area around Via Pinelli is very different. This is on the southern outskirts of the city but it is a long way both from major roads and from shopping centres and other facilities. About 250 families live in the area distributed among a group of buildings constructed in various periods. The oldest buildings date from the 1970s; others were built in the early 1990s, and the most recent were put up a few years ago and were designed and built to meet environment-friendly criteria. The zone is largely cut off from the city itself and from other urban settlements; thus there was neither much pollution nor noise. However, such isolation also means that there will be a lack of accessible urban services near at hand.

## 15.5 Social services, social networks and the quality of life

### 15.5.1 Social services and housing in the two areas

In this step we examined the quality of life in the two areas in relation to the existence, or absence, of urban social services. Most of the participants living in via Maroncelli focused on the absence of such services or, more specifically, on the lack of five types of services:

- Structural social services, such as a chemist's, a post office or more efficient public transport

- Services concerned with looking after and caring for elderly people who are often alone or dependent on care from outside or from their family

- Services designed to improve the quality of urban life, such as green areas, parks, traffic reduction, street cleaning and general maintenance of public places and areas

- Services concerned with law and order, including the city police, who are not able to control antisocial behaviour in the area

- Spaces for local residents to socialise in. Whether there are or are not meeting places available in a neighbourhood is very important for the quality of life of the people living there. Ideally, housing complex design should help people to become less 'strange' to each other and to help them to trust each other. Luckily, in this public housing complex there is a meeting place that works to bring residents together, but outside the settlement there is nothing like it at all

Unlike the residents in via Maroncelli, residents in Via Pinelli feel that the overall quality of life in their neighbourhood is relatively good, mainly thanks to the low density of housing in the area, the presence of green spaces and very little traffic pollution. People in Via Pinelli are willing to go without the convenience of having certain services and facilities close by (shops, schools, offices, means of transport) in order to benefit from 'silence', 'clean air' and a traffic-free life. The main problem that residents in Via Pinelli are becoming concerned about is the increasingly fast and unregulated process of urbanisation they see around them. Moreover, there are also problems of organising municipal services (grass cutting, rubbish collection and so on) and of managing public spaces and areas.

### 15.5.2 Social networks and reciprocity

What is interesting in the outcomes from the focus group of Via Maroncelli is that there are no prevalent social networks in the neighbourhood. Weak and strong connections are at work there. Regarding weak connections, we can identify 'bonding' relations among members of local groups or between residents in the housing complex and 'bridging' relations among neighbours and distant friends, associates and colleagues. What is weak is the 'linking social capital' or, in other words, the capacity of individuals and the community to control resources, ideas and information from formal institutions beyond the immediate area of the community. This weakness is the result of a low level of participation in common local issues. When discussing the quality of the social relationships in the area, participants showed three distinct types of attitudes:

- First, they complained about the limited nature of their social bonds, revealing a sense of spatial isolation which may well be partly due to the way the area was initially laid out: it is bounded, cut off, on at least three sides, by busy roads and by big commercial and industrial buildings. Thus it would seem that there is the desire to set up a large network of people who would be able to develop joint, communal, actions for the good of the local inhabitants

- Second, the group members had noted how weak links sometimes create awkward situations and social distance for some categories of people, especially the elderly and children. Usually, distance in relationships allows for only a limited degree of reciprocity, an accumulation of 'chits' that the actors hope will permit such reciprocity. But this type of relation, based on a 'utilitarian' exchange, was not enough for our witnesses, who wanted bonds based on deeper shared meanings and values. It is easy to identify the desire to exchange freely with one another without forming 'debts'. In this case, the favour does not have to be repaid immediately or the debt instantly paid off. We could call this behaviour as offering 'something for nothing' (Gouldner 1975)

- The third, and perhaps most commonly held, attitude that emerged during the focus group meeting concerns participants' overall satisfaction with their existing social relations. This satisfaction would suggest that the general, widespread, reciprocity described above does, to some extent, already exist

In Via Pinelli the relationship model at work is different from that in Via Maroncelli. In Via Pinelli there are two groups of tenants, each of which are based on rather different relations, bonds of friendship and collaboration. The two groups have formed around the key persons on the estate whom we identified and then invited to the focus groups. The two groups are also physically separated in two different sets of buildings. If one were to describe the style adopted by the two groups one could say that one group is characterised by the spontaneity of its relations, 'Rabelaisian-type' friendships, whereas the other is marked by a far more rational—or, better, 'reasonable'—conception of rights and duties, as in the reciprocity found in common action. In Weberian terms one could say that the first group privileges values of friendship and solidarity, reinforced by specific and declared class membership, whereas the second group is closer to the rationality of the aims, where the aims and the results are more important than the shared values.

Given the differences we found, it is not hard to see why, in Via Pinelli, there is no residents' committee as found in Via Maroncelli: setting up a committee means electing representatives who are able to mediate and find ways of getting residents to agree. The fact that in Via Pinelli there are a variety of different points of view, which we have simplified into two main models, makes such unity impossible, or, rather, it is not seen a priority by the residents concerned.

### 15.5.3 Confidence and distrust

Another aspect that emerged from the discussion in both of the neighbourhoods is the issue of trust and diffidence. The existence of trust in an area is a basic requirement for a better quality of life, but it can, for a while, be broken. The type of situation described by the focus group is not unusual in public housing complexes because there is always the possibility that a new occupant will arrive who will upset the existing social balance in the neighbourhood. Distrust and diffidence develop when people begin to fear others, who try to dominate or who are violent. Such people, who do not fit in, often do not accept, indeed even resist, the existing tacitly shared norms of the previous residents. From this point of view, one could say that 'all actions as perceived events may have a constitutive structure, and that perhaps it is the threat to the normative order of events as such, that is the critical variable in evoking indignation, and not the breach of the "sacredness" of the rules' (Garfinkel 1963: 198).

As people begin to feel more and more abandoned by institutions, so distrust and diffidence gradually develop between them. Residents from Via Pinelli said much the same about the degree of trust they felt in institutions. Residents, even though they are willing to do voluntary social work, reap no benefits from their efforts or, worse, meet with indifference and disinterest on the part of the official organs that should be able to offer clear answers to legitimate demands. Thus self-organisation is necessary, not so that social subjects can cooperate better with the institutions; rather, it becomes the only way in which these same actors can find meaning in some segments of collective, everyday life. However, the poor performance of institutions does not help to reduce local and neighbourhood conflicts.

The attitudes and opinions expressed during the course of the focus group made us reflect on the impact such ideas will have on the way in which public goods are consumed. The degree of attention and responsibility residents show with regard to public goods is, interestingly, reflected in the level of responsibility, of awareness, they show in relation to the consumption of energy resources. The same care and attention they pay to cultural resources, to public spaces and to their sociality is reflected in their responsible management of energy and other resources, such as heating, water consumption, rubbish recycling and use of electricity in stairwells and entrances.

## 15.6 Environmental challenges and social capital

### 15.6.1 Introduction

In this research we have looked in particular at the way in which people and local society deal with environmental risks. The way in which they take up the problem depends, obviously, on a complex set of inextricably linked factors, but what is certain is that the quality of social relationships, the level of cultural awareness and the ability to mobilise and put political pressure on policy-makers radically affect behaviour in the face of risks (Bush *et al.* 2002).

Eivind Stø, Harald Throne-Holst and Gunnar Vitterso have suggested that, in this postmodern period, individuals play an important part in the measures and goals of environmental policy, whether it be in their role as citizens, employees or consumers. Governments can make laws and emit directives, they can tax production and consumption of some products and subsidise others, but it is the individuals themselves who have the last word. It is an individual, personal matter whether to start or stop jogging, drinking alcohol or make a consumer choice in the food market (e.g. in order to reduce individual consumption of fat). In short, consumers can choose between individual and collective strategies, and individuals can act strictly as a consumer in the market or also tackle the questions that go beyond markets, into politics (Stø et al. 2001).

We broadly agree with this argument even though the problem of institutions is, in part, ignored. On the basis of the data gathered through the focus groups, it would seem that if individuals are to modify their attitudes in such a way as to promote environmental sustainability then they must be put into a position where they will be able to act in a suitable manner. One could say that different types of sustainable behaviour are guided by different actors. In some cases certain attitudes depend, to a large extent, on institutional actions; in other cases, they depend on the quality of relationships and on social position; in yet others, they are the result of more personal, individual choices. But such processes emerge only when there is a relationship of trust between the actors involved and it is possible to overcome the resistance, the power, of natural inertia, as in the case of traffic, where social inertia and everyday collective behaviour are taken for granted. When trust in institutions, or in others, is limited, or when people are too closed and cut off from each other, then decisions about what to consume and how to consume it are often seen as unimportant and thus neglected. When, however, individual action is perceived as part of a broader collective action that decides on the quality and the way in which 'the commons' will be used, any decision taken will have a moral force that makes it efficacious and a source of inspiration to others.

### 15.6.2 Recycling of waste

The practice of waste management is the institutional response to an environmental threat in the context of a risk society. Waste management regulates the excesses of the consumerist society, containing that risk and threat. The institutionalisation of such practices acts as a neutralising agency and assists in maintaining the status quo of consumption and production. But, often, this management and local government fail to maintain unity in the public sphere, causing breakdown, tensions and conflicts both in the institutional and in the social fields (Broderick 1997; Ost 2002; Viale 1999).

The literature concerning social determinants of waste management is nowadays relatively broad, often focusing on a social capital perspective. Research carried out mainly in developing countries such as India, Bangladesh and Pakistan showed that social capital plays a role in the community-based provision of a public good such as waste collection (Gupta 2004; Pargal et al. 1999; Sekher 2004). Social capital and networks are vitally important because waste collection involves positive externalities leading to limited incentives for individual action. Also, waste collection is an activity wherein one individual's actions alone will not have much impact, so collective and institutional action is necessary, even more so in the case of differentiated waste collection. A participatory solid waste management system handled at the level of the local community can undoubtedly help to form a long-term plan for sustainability of waste management.

We think that social capital is a crucial determinant of such collective action: only when there is good interaction between citizens, stakeholders and public authorities—interaction based on trust, shared norms, reciprocity and participation—will strategies of waste collection be successful (Osti 2002; Taylor and Todd 1995). The social capital both of the family and of the neighbourhood helps to transform an initially difficult and unlikely action into an everyday habit that, as time goes by, becomes easier and easier to do, habit that also depends on the positive influence exerted by networks in which people are involved (Kitts 1999).

The participants in the focus group of Via Maroncelli all showed that they were well aware of the positive impact waste recycling has on the environment. Their comments revealed a high degree of willingness to take positive action in order to reduce one of the causes of current environmental problems. Moreover, they also mentioned the ethical question the economy faces with regard to society and the environment. Their criticism of the economic rationale that underlies waste recycling is very important because it takes up the problem of the social distribution of the benefits of collective action and solidarity.

Our witnesses are well aware that as long as differentiated waste collection is successful then there will be a trust relation between public institutions and citizens. In short, a relationship based on trust in public institutions is crucial even for an apparently simple problem such as differentiated waste collection. Obviously, lack of trust in institutions is not simply a question of people's prejudices regarding institutions; rather, it stems from the real perceived shortfalls and deficiencies of the institutions themselves. Furthermore, trust is not encouraged when the ordinary citizens involved, who agree to collaborate as a collectivity in order to make a public service function or to improve a public good, feel that they are being excluded from the benefits that their cooperation produces.

In Via Pinelli, most of the focus group participants were already separating their refuse even before it became official city policy. They now hope that as separation is official it will become easier because of the provision of new special public bins. Basically, people are not satisfied with the service; their willingness to cooperate and to separate refuse is often being discouraged by institutional attitudes. There is also a pressing need for individuals' efforts and commitment to separate refuse for collection to be recognised by the authorities. The best incentive would be tax cuts or discounts or, at the very least, some improvement in the quality of the service offered to citizens.

### 15.6.3 Consumption of organic products

The question of organic and wholefood products is inextricably linked to the question of trust in the food market. A variety of research has been carried out on this topic. At European level the Trustinfood Project conducted surveys in many European countries, showing that levels of consumer trust in food differ greatly among countries, moving from the highest levels of trust in northern Europe to the lowest levels in the south. What is very dissimilar is consumer trust in public institutions, market actors and family and personal networks. It means that these different dimensions are marked by different performances; in brief, each society is characterised by different levels of social capital (Poppe and Kjærnes 2003). Food acquisition and consumption requires a wide range of trusting social relations to be drawn on. If the importance of trust and networks in food consumption is clear, the importance of these is far more crucial for organic food consumption, because it implies changes in habit and taste.

All the participants in the focus groups were rather sceptical about the quality (and veracity) of the true quality of organic products (be they vegetables or meat), because they did not trust eco-labels, the certification system or the declarations made about the quality and benefits of the goods themselves. This distrust has long been studied by economists and sociologists. It would seem that consumers' past disappointments with goods, because of pollution, toxins or other risks, have encouraged many of them to change their purchasing patterns and attitudes to everyday products. After the 'mad cow disease' scandal[1] many Italian consumers started to buy only organically produced meat or even, in some cases, stopped eating meat altogether. This obviously greatly befitted the producers and distributors of organic foods. However, the momentary panic was not sufficient to convert people to organic products, not only because there are still not enough producers to satisfy potential demand but also because people always develop preferences during their interactions on the market: they look at prices, quality, brand, ecological standards; they have their habits. It is also true that the new organisations that put eco-labels on the market are more vulnerable because of the lack of trust between suppliers and customers, a trust that develops only over time.

At the level of daily life and habit, consumers are probably affected by other mechanisms that have nothing to do with trust in the market or in its ability to react to shifts in consumer preferences; these can be correlated with the social networks the consumers themselves are caught up in. As Ronald Burt (2000) observes, consumers must have information on the goods, sellers, buyers and prices available when they are selecting the best deal. This is the point at which network mechanisms enter the analysis. The structure of previous relations among people and organisations in a market can affect, or replace, information. Replacement happens when market information is so ambiguous that people use network structures as the best available source of information. It is precisely what happens when people approach a new market, such as the organic products market, where there is a marked lack of information. Such assumptions underlie 'network contagion', in which 'contagion' ensures the transmission of beliefs and practices more readily between certain people (Burt 2000). In this case, people with insufficient information about new organic products prefer to trust in the old beliefs transmitted and confirmed by social contagion and imitation within the social network.

Consumption of organic products is limited not only by the generalised distrust of their quality; there are other reasons too. To sum up, focus group members consume very few organic products for three main reasons:

- They do not trust the declarations of quality made regarding such products, given that pollution is so widespread
- Such products are far too expensive for them to use every day
- The distribution network of organic foods is poorly organised and there is little readily available information about them

---

1 In this case, cattle carried bovine spongiform encephalopathy (BSE), leading to the fatal Creutzfeldt–Jakob disease (CJD) in humans.

### 15.6.4 Means of transport and shopping

The policy of creating large shopping centres is highly developed in the city studied. One direct consequence of these large centres is an increase in traffic and, consequently, in pollution. The proliferation of such centres is creating a new type of consumer, who probably has a lot of free time, and has a new hobby: when the weekend arrives, he or she takes the car and takes a trip round all the big shopping centres in the area. This represents a real change in way of life and, also, increases traffic and pollution. This type of consumer either does not realise or ignores the fact that, among other things, lower prices are cancelled out by the cost of travelling and petrol. The other more serious consequence is that local shops are being forced to close for lack of customers. This is obviously causing huge problems for the elderly, or for those without cars or other means of transport, in getting to and from the big shopping centres. This phenomenon is also connected with the role of social capital in facilitating sustainable behaviour that I am exploring here. Investment in social capital reduces consumption, because time devoted to social interaction reduces time spent travelling and shopping.

Our subjects do their shopping in a different way from that described above, just because they are involved in gripping networks of relations. As many other families, they tend to take their car and do a big bulk shop in a supermarket once a week, buying the basics, which they then top up on a daily basis in the food shops in their local area. In neighbourhoods where there are small shops, consumers often enjoy going there, because they build up a trust relationship with traders with regard to the products they buy. Some still prefer to do their shopping in the old city centre where there is a far wider choice and often better quality than elsewhere and where shopping becomes an enjoyable and social task.

Thus, most people use the car once a week to do the shopping for basic goods (detergents, coffee, pasta, etc.) in large supermarkets or discounts stores because it is cheaper, but they seem to prefer buying foodstuffs in the small, local, shops. None of the participants seemed to be particularly enthusiastic about the large shopping centres, which they tend to see as being places where there are so many special offers that shoppers just get confused and end up spending more than they would have done elsewhere because they fall into the trap of buying things that are useless, or superfluous, to their real requirements.

## 15.7 Conclusions

The main results of this study concern the fact that social actors have a complex perception of quality of life. This derives from three social dimensions that greatly influence the lives of individuals:

- The first dimension is that of the physical and natural urban environment that surrounds the local community. It is concerned with the physical objects that populate the social space of actors. The quality and the functionality of the constructed environment that surrounds them influence the quality of life of people in a local area. The proximity of shops and primary services, the efficiency and adequacy of means of transport, schools and hospitals, the quality of the environment—for example, the air we breathe, the food we eat—are all crucial indicators when defining a person's quality of life

- A second, fundamental, dimension is related to the broader system of social relations in which the actors are involved. A large part of people's daily activities depend on the institutional system, activities that are taken for granted because they are guaranteed and regulated by what is often an impersonal set of decisions and actions supported and promoted by institutional actors

- A third dimension that emerged is the structure of the existing relationships, especially local, within which individuals are embedded. At this level we have considered the individual *Ego*'s ability and potential to mobilise resources through the network in order to achieve certain goals, to obtain some help, to shape his or her own lifestyle and consumption patterns

Trust is the fundamental resource that influences each individual's level of satisfaction in relation to these three dimensions. This study has shown that interpersonal and institutional trust are crucial for the actors involved, both with regard to their perception of the quality of their lives and in relation to the decisions that they will take concerning environmental friendly consumption and action.

Interpersonal trust is well established in the two neighbourhoods. Even though there had been episodes when the implicit rules of social exchange were not entirely respected, it is, in general, well established and reciprocity is the rule. In Via Pinelli, the existing tensions could be blamed on the different cultures and social origins of the inhabitants, but this does not mean that the rules of reciprocity are not respected at the level of everyday exchanges.

When considering institutional trust the question becomes more complicated. A lack of systemic trust weakens those elements of our daily lives that we take for granted, creating confusion and anomie. Our subjects said that they had little very trust in either the political or the institutional system. This is important for our study. All our interviewees were well aware both of the power institutions have in regulating social transactions and of the fact that institutions are also responsible for strategies for developing urban sustainability. The results of our study have shown that trust in institutions is indispensable if some attitudes are to prevail. For example, the householders' act of separating rubbish depends directly on the level of trust they have in the organisation that manages separated rubbish collections. Whether or not to consume organic eco-labelled products also depends on trust in market institutions. When there is little trust in the ability to regulate and control the market, consumers find little reason to embark on new consumption patterns. In this case the inertia that comes from habit and, also, social imitation prevails. Here the intermediary networks of actors do have some impact on consumption models. Separated rubbish collection for recycling also depends on the moral pressure that the members of the network in which actors are embedded are able to exert, Indeed, the decision to buy organic eco-labelled products also depends on the social 'contagion' exercised by an individual's network of reference; if these networks tend, for some reason or another, towards encouraging sustainable consumption then they are able to 'contaminate' members in a radical, convincing way.

# References

Bourdieu, P. (1979) 'Les trois états du capital culturel', *Actes de la recherche en sciences socials* 30: 3-6.
Brehm, J., and W. Rahn (1997) 'Individual-Level Evidence for the Causes and Consequences of Social Capital', *American Journal of Political Science* 41.3: 999-1,023.
Broderick, S. (1997) 'Waste Not, Want Not: A Case Study of Ambivalence in Modernity', unpublished paper.
Burt, R.S. (1992) *Structural Holes* (Cambridge, MA/London: Harvard University Press).
—— (2000) 'The Network Structure of Social Capital', in R.I. Sutton and M.B. Staw (eds.), *Research in Organisational Behaviour* (Greenwich, CT: JAI Press): 345-423.
Bush, J., S. Moffatt and C.E. Dunn (2002) 'Contextualisation of Local and Global Environmental Issues in North-east England: Implications for Debates on Globalisation and the "Risk Society" ', *Local Environment* 7.2: 119-33.
Claibourn, M.P., and P.S. Martin (2000) 'Trusting and Joining? An Empirical Test of the Reciprocal Nature of Social Capital', *Political Behaviour* 22.4: 267-91.
Coleman, J. (1982) *The Asymmetric Society* (Syracuse, NY: Syracuse University Press).
—— (1988) 'Social Capital in the Creation of Human Capital', *The American Journal of Sociology* 94S: 95-120.
—— (1990) *The Foundations of Social Theory* (Cambridge, MA: Harvard University Press).
Donolo, C. (1997) *L'intelligenza delle istituzioni* (Milan: Feltrinelli).
Dreborg, K. (1996) 'Essence of Backcasting', *Futures* 9: 813-28.
Falk, J., and S. Kilpatrick (2000) 'What is Social Capital? A Study of Interaction in a Rural Community', *Sociologia Ruralis* 40.1: 87-110.
Field, J. (2003) *Social Capital* (London/New York: Routledge).
Flora, C.B. (1995) 'Social Capital and Sustainability: Agriculture and Communities in the Great Plains and Corn Belt', *Research in Rural Sociology and Development* 6: 227-46.
Garfinkel, H. (1963) 'A Conception of, and Experiments with, "Trust" as a Condition of Stable Concerted Action', in O.J. Harvey (ed.), *Motivation and Social Interaction* (New York: Ronald Press): 187-238.
—— (1967) *Studies in Ethnomethodology* (Englewood Cliffs, NJ: Prentice Hall).
Goffman, E. (1971) *Relations in Public* (New York: Basic Books).
Goldman, M. (ed.) (1998) *Privatising Nature: Political Struggles for the Global Commons* (London: Pluto Press).
Gould, R.V. (1993) 'Collective Action and Network Structure', *American Sociological Review* 52.2: 182-96.
Gouldner, W.A. (1975) 'The Importance of Something for Nothing', in idem, *For Sociology* (Harmondsworth, UK: Penguin Books): 260-99.
Granovetter, M. (1973) 'The Strength of Weak Ties', *American Journal of Sociology* 78.6: 1,360-80.
Gupta, S.K. (2004) 'Rethinking Waste Management in India', *Humanscape* 11.4.
Kilpatrick, S., and J. Abbott-Chapman (2005) 'Community Efficacy and Social Capital', paper presented at the *Future of Australia's Country Towns Conference*, Bendigo, Australia, 11–13 July 2005.
Kitts, J.A. (1999) 'Not in Our Backyard: Solidarity, Social Networks and the Ecology of Environmental Mobilisation', *Social Inquiry* 69.4: 551-74.
Kwak, N., D.V. Shah and R.L. Holber (2004) 'Connecting, Trusting and Participating', *Political Research Quarterly* 57.4: 643-52.
Luhmann, N. (2000) *Vertrauen: Ein Mechanismus der Reduktion sozialer Komplexität* (Stuttgart, Germany: Lucius & Lucius).
Misztal, B.A. (1996) *Trust in Modern Societies*. (Oxford UK: Polity Press).
Mutti, A. (1998) *Capitale sociale e sviluppo* (Bologna, Italy: Il Mulino).
OECD (Organisation for Economic Cooperation and Development) (2001) *The Well-being of Nations: The Role of Human and Social Capital* (Paris: OECD).
Osti, G. (2002) *Il coinvolgimento dei cittadini nella gestione dei rifiuti* (Milan: Franco Angeli).
Ostrom, E. (1990) *Governing the Commons* (Cambridge, UK/New York: Cambridge University Press).
Pargal, S., M. Huq and D. Gilligan (1999) *Social Capital in Solid Waste Management: Evidence from Dhaka, Bangladesh* (WP-16; World Bank Social Capital Initiative; Washington, DC: World Bank).

Poppe, C., and U. Kjærnes (2003) *Trust in Food in Europe. A Comparative Analysis* (Oslo: National Institute of Consumer Research).
Portes, A., and P. Landolt (1996) 'The Downside of Social Capital', *The American Prospect* 26: 18-21.
—— and J. Sensenbrenner (1993) 'Embeddedness and Immigration: Notes on the Social Determinants of Economic Action', *American Journal of Sociology* 98.6: 1,320-50.
Pretty, J., and H. Ward (2001) 'Social Capital and the Environment', *World Development* 29.2: 209-27.
Putnam, R.D. (1993) *Making Democracy at Work* (Princeton, NJ: Princeton University Press).
—— (2000) *Bowling Alone: The Collapse and Revival of American Community* (New York: Simon & Schuster).
Sekher, M. (2004) 'Keeping Our Cities Clean: Urban Solid Waste Management in Karnataka', *Journal of Social and Economic Development* 6.2: 159-75.
Stø, E., H. Throne-Holst and G. Vitterssø (2001) 'The Role of Consumers in Environmental Successes', paper presented at *the International Workshop on Participation and Consumption: Social Action Patterns for Urban Sustainability*, Marsala, Sicily, Italy, 26 May–3 June 2001.
Taylor, S., and P. Todd (1995) 'An Integrated Model of Waste Management Behaviour: A Test of Household Recycling and Composting Intentions', *Environment and Behavior* 27.5: 603-30.
Viale, G. (1999) *Governare i rifiuti* (Turin, Italy: Bollati Boringhieri).
Woolcock, M. (1999) 'Social Capital: The State of the Notio', paper presented at a Multidisciplinary Seminar on *Social Capital: Global and Local Perspectives*, Helsinki, Finland, 15 April 1999.
—— (2001) 'The Place of Social Capital in Understanding Social and Economic Outcomes', in J.F. Helliwell (ed.), *The Contribution of Human and Social Capital to Sustained Economic Growth and Wellbeing: International Symposium Report* (Ottawa: Human Resources Development Canada and OECD).
World Bank (2001) *World Development Report 2000/2001: Attacking Poverty* (Washington, DC: World Bank).

# 16
# Linking sustainable consumption to everyday life
A social-ecological approach
to consumption research

*Irmgard Schultz and Immanuel Stieß*
Institute for Social-Ecological Research, Germany

## 16.1 Introduction

The way people consume is shaped by attitudes, routines and the organisation of everyday life. Thus, strategies towards sustainable consumption must start from a consumer perspective and take different everyday life practices and situations into account. In recent years, the Institute for Social-Ecological Research (ISOE) in Frankfurt am Main has developed a social-ecological approach to sustainable consumption, linking social-ecological lifestyle analysis to the study of everyday behaviour and environmental impact assessment. Within this framework, different fields of consumption (nutrition, mobility, tourism) have been investigated. In some surveys the environmental impact has been calculated, identifying behavioural patterns, motivational backgrounds and the environmental impacts of different social groups. These studies have great potential for the promotion of sustainable consumption behaviour as well as for the design of innovative products and services.

In this chapter, we will first introduce the conceptual and methodological framework of the social-ecological research approach to sustainable consumption (Section 16.2). We will then demonstrate the approach more deeply, drawing on the example of nutrition styles (Section 16.3). We sum up the findings in Section 16.4, bringing in a differentiated understanding of consumers.

## 16.2 A social-ecological approach to sustainable consumption

Since the UN Conference on Environment and Development in Rio de Janeiro in 1992 there has been a wide range of research and policy efforts to promote sustainable consumption patterns. Somehow, these efforts do not seem to be reaching consumers or affecting them enough to lead to fundamental changes in consumer behaviour. Furthermore, increased and growing consumption counteracts the eco-efficiencies gained by 'green' products (the so-called rebound effect). Against this background, more research is needed to understand consumption behaviour with regard to sustainable consumption patterns. Having this in mind, the social-ecological approach to sustainable consumption focuses on the role of orientations and motivational backgrounds for consumption behaviour. Within this approach it stresses the logic of the organisation of everyday life in a multi-dimensional framework.

### 16.2.1 Consumption and the organisation of everyday life

The way people eat or move is a result of individual and collective habits that are woven into the fabric everyday life (Shove 2004). Consumption is linked to everyday activities that can be described as social practices (Giddens 1984). Social practices are more or less institutionalised collective phenomena that are reproduced through everyday action and are governed more by habits and routines than by deliberate and rational choice. Routines are incorporated chains of action characterised by a low degree of reflexivity. Nevertheless, they rely on incorporated competences and on preconscious specific knowledge. Routines allow for immediate action and quick and easy decision-making by reducing the complexity of everyday situations. At the same time, they are invested with emotions and attitudes.

Consumption has very much to do with the way in which individuals organise their daily lives (Cogoy 1999). Nevertheless, it is not an individual act. 'Ordinary consumption' (Gronow and Warde 2001) is framed by cultural and social contexts of households and family life, which are themselves characterised by structures of gendered division of labour, the societal organisation of intimacy and gendered power relations (Schultz 2006). Everyone's consumption is characterised, among other things, by acquiescence to external pressures, routinisation, normalised expectations, vicarious acquisition, personalised appropriation, the dictates of convention and public interests. To express this specific perspective of consumption as being anchored in the needs, interests, fears and expectations of socially embedded consumers we speak about everyday life (Stieß and Hayn 2006).

Everyday life embraces a broad variety of differing activities, such as cooking, shopping, caring work and so on. These activities do not happen in isolation. They have to be connected to form a coherent whole. Individuals have to balance different and sometimes conflicting demands of work life, household organisation, family management and their own needs, taking their own preferences as well as the wishes of other family or household members into account. Thus, everyday life turns out to be a complex mix of practices that has to be actively constructed, maintained and modified (Voss 1995). The organisation of everyday life is a complex task requiring planning, organisational skills and time.

## 16.2.2 Lifestyle research and sustainable consumption

Owing to the societal change that has taken place in Western European countries since the late 1960s, the erosion of traditional institutions, individualisation and differentiation of society implied that traditional sociological categories, such as income, professional position and education, were not sufficient to analyse new social differences. Thus, approaches that stress 'subjective factors' and motivational backgrounds in addition to structural factors have been developed in sociology. These additional considerations have been subsumed under the term *lifestyle*, a term that was already in use by Max Weber (1948). Segmentation of society by lifestyle, as suggested by the newly emerging lifestyle research, represented a means to take these developments into account. Lifestyle concepts emphasise the importance of sociocultural criteria and attitudes for the classification of social segments (cf. Schultz and Weller 1997). By subdividing societies not only vertically (i.e. into upper, middle and lower classes) but also horizontally, according to sociocultural criteria, a picture of different milieus in the social sphere is created, each milieu being distinguished by its own lifestyle. Elements of lifestyles characterise the self-identity of a group, but they also distinguish the group *vis-à-vis* other groups. Thus, lifestyle research is able to map the pluralisation of societies. Furthermore, lifestyle concepts represented an explanation of why societies were not bound to disintegrate by the growing tendency of individualisation. Lifestyles can be understood as a mode of social integration that gives individuals the opportunity to be socially integrated while living in disparate and particularised surroundings (cf. Götz 2001).

The most important impulses for the development of the lifestyle concept in sociology came from Pierre Bourdieu (1979). He was the first to present a multi-dimensional model of social space, with socioeconomic and socio-demographic variables (e.g. income, professional position, education and number of children) along the vertical dimension and preferences and tastes (e.g. on the media, art, literature, products, leisure and politics) along the horizontal dimension. However, he still called attention to the importance of the vertical dimensions for the development of differences in taste, and he did not name concrete groups or lifestyles. Today, sociological lifestyle concepts are considered to be an extension and fulfilment of social structure analysis (cf. Müller 1992).

Although market research does not focus on the analysis of society, the development of lifestyle concepts in market research was motivated by similar reasons as in sociology: owing to societal change, demographic variables and other classical segmentation variables had become less useful for predicting and explaining consumer behaviour. Moreover, in order to improve advertising and marketing, there was an increasing need to know the consumer better.

In recent years, the advantages of lifestyle approaches for the sustainability discourse have been widely recognised. If one stresses the subjective dimension (such as values, orientations and attitudes), consumption is not conceived as the rational choice of a *homo œconomicus* but rather is understood as being guided by motivation (cf. Stieß and Götz 2002). Drawing on lifestyle-based typologies, one can operationalise influential factors that are generally considered to be 'incalculable': attitudes, orientations, emotions and motivational backgrounds. The influence of these 'soft factors' on a person's behaviour can even be measured in terms of actual consumption behaviour. Since these so-called 'soft factors' have proven to be, in fact, the truly 'hard factors' when it comes to changing consumption pattern, typologies of different target groups contain valuable knowledge that can be used to develop strategies to foster more sustainable consumption patterns (cf. Stieß and Götz 2002).

The multi-dimensionality of the lifestyle concept makes it promising for the formulation of sustainable consumption policies (cf. Schultz and Götz 2006). A group-specific analysis can identify different social segments and draw a holistic picture of the respective target group that helps in the development of group-specific information campaigns or product marketing. In the field of ecological research, lifestyle concepts were adapted to develop strategies differentiated according to the motivational backgrounds and needs of different target groups (Reusswig 1994)

### 16.2.3 The social-ecological lifestyle approach

General objectives of lifestyle research in sociology have been to describe societal change. In market research the predominant aim is to promote the selling of products and services. This narrow focus on consumers' purchasing decisions, however, appears hardly suitable for research into sustainable consumption Conventional lifestyle approaches fail to consider crucial environmental and social aspects of consumption patterns because they neglect the complex entanglement of consumption practices in everyday life. Mostly, they do not account for how consumers use and remove products and, in general, they do not account for environmental impacts.

To adapt the lifestyle concept to research into sustainable consumption patterns, ISOE has developed a social-ecological approach to lifestyles that also takes into account motivational factors behind behaviour seen throughout the entire consumption process, including at the planning stage and during buying, use and disposal. This approach analyses three dimensions and their relationships:

- The social background of households (their social situation and household context)
- Orientations in relation to:
    - Lifestyles: that is, general values and orientations with regard to work, leisure, consumption, the environment, health and so on
    - Specific fields of consumption: such as attitudes, emotions, preferences, dislikes and so on
- Indicators of actual consumption behaviour as practised daily

The social-ecological typology evolves from group-specific patterns of orientations, behaviour and social situation and can be used as a target-group model. We label the typology according to the specific research interest or need area in question; we thus have typologies for 'consumption styles', 'mobility styles', 'nutrition styles' and so on. The identified lifestyles can be analysed with regard to the three dimensions listed above and can be used as target groups for the development of socially differentiated information, marketing and consumer advice offers (cf. Empacher and Götz 2004).

This model implies some important advantages for use in the field of sustainability. First, it takes into account the household context and the organisation of time, work and leisure. It thereby acknowledges the embeddedness of individuals in a concrete social context, a family or household. Thus, it allows us to measure gender-specific time-use patterns for different consumption tasks and other patterns of the gendered division of labour (cf. Littig 2001; Empacher *et al.* 2002b). Conventional lifestyle research sees individuals as embedded only in a lifestyle group and abstracts them from their immediate social surroundings. However, these surroundings have considerable influence on individual decisions, because such decisions usually rely on bargaining among different household members. As a consequence, a change of behaviour towards a more sustainable behaviour pattern also depends on other household members.

Second, an important advantage of this approach is the inclusion of the behavioural dimension. This allows for the testing of hypotheses to what extent orientations and motivational factors influence actual behaviour. Moreover, it enables a consideration of the material dimension of consumption and thus allows this approach to link up with natural scientific studies of metabolic processes and their impact on the environment. Therefore, it is possible to analyse environmental impacts of different lifestyles and to estimate the potential to reduce such impacts within different lifestyles.

Last, strategies can be targeted more effectively if the environmental impacts of consumption habits and the amount of equipment or goods used by the different target groups are known.

## 16.3 Social-ecological consumption research

In recent years, ISOE has carried out several studies, focusing on sustainable consumption patterns from the perspective of everyday life. Seminal work was undertaken in a study on behalf of the German Federal Environment Agency (Umweltbundesamt [UBA]), entitled 'Household Exploration of the Conditions, Opportunities and Limitations Pertaining to Sustainable Consumption Behaviour' (Empacher *et al.* 2002b). The study was based on a qualitative empirical survey of 100 German households. The households were carefully preselected from demographic, geographical and lifestyle-related criteria to establish typical consumption styles within German households. The lifestyle and orientations of respondents were ascertained with use of open interviews. In addition, a standardised questionnaire was used to record selected indicators relating both to equipment owned (consumer goods) and to consumer behaviour in key areas of household consumption (nutrition, clothing, mobility and so on; cf. Empacher *et al.* 2002a).

Other research focused on an in-depth examination of consumption practices in different fields of consumption, such as nutrition, mobility and tourism (Götz *et al.* 2002; Schubert 2004). From these investigations, the typology of nutrition styles will be presented in this section.

### 16.3.1 Nutrition in everyday life

Nutrition is a key area of consumption. Although the share of household food expenditure has declined steadily over the past few decades, food consumption continues to cause large negative environmental effects. Most important are the indirect effects of food production and processing in Europe and other regions of the world. Effects include emissions to water, soil and air from livestock, agriculture, industry and transport as well as waste (Kristensen 2004).

In the context of a joint research project on 'Food Change: Strategies for Social-Ecological Transformations in the Field Environment–Food–Health' the social-ecological lifestyle approach has been applied to the field of nutrition.[1] The main objective of the transdisciplinary, cooperative, project was to identify, scientifically, starting points to foster a more healthy and sustainable means of nutrition. Within this research project, ISOE examined the way in which nutrition is embedded in consumers' everyday lives. The established typology of nutrition styles gives advice on the development of strategies towards a 'food change' from a consumer perspective. The typology links the study of consumers' motivations to nutrition-related behaviour and to their environmental impact (determined by an analysis of material flows).

The nutrition-style approach studies the interplay between food-related orientations and the way people manage their diet in everyday life. Based on this approach, an empirical survey with a dual—qualitative and quantitative—methodology was carried out. The quantitative survey investigated the attitudes of a representative group of 2,039 adults towards nutrition and health as well as their behaviour. Furthermore, data on social situation and life context as well as detailed information on their occupations were collected. Their attitudes on nutrition and health as well as some of their socio-structural characteristics were clustered, and seven distinct nutritional styles were identified.

### 16.3.2 A typology of nutrition styles

#### 16.3.2.1 Introduction

The typology illustrates the diversity of current nutrition styles in Germany, ranging from uninterested fast-fooders to those with highly health-oriented nutrition styles. Thus, these typologies provide a better understanding of the cognitive, motivational and structural barriers that prevent people from following a more sustainable diet in the context of everyday life. Moreover, the typology helps to identify the potential and starting points for people to make a 'food change'.

Figure 16.1 provides a synopsis of the seven nutrition styles. The nutrition styles are arranged according to life situation (horizontal axis) and interest in issues related to food and nutrition (vertical axis). The different nutrition styles are sketched in the following sections.

---

1 The research project 'Food Change: Strategies for Social-Ecological Transformations in the Field Environment–Food–Health' was carried out by a consortium of several institutions: the Institute for Applied Ecology (the Öko-Institut e.V. [ÖI]; coordinator), the Institute for Social-Ecological Research (ISOE), the Institute for Ecological Economy Research (Institut für ökologische Wirtschaftsforschung [IÖW]), the KATALYSE Institute for Applied Environmental Research, and the Austrian Institute for Applied Ecology (Österreichisches Ökologie-Institut [ÖÖI]). The project was funded by the German Federal Ministry of Education and Research (Bundesministerium für Bildung und Forschung [BMBF]) within the Social-Ecological Research Programme.

FIGURE 16.1 Nutrition styles and life situation: position of nutrition styles according the phase-of-life model posited by the Institute for Social-Ecological Research (ISOE), Germany

Source: Stieß and Hayn 2005: 34

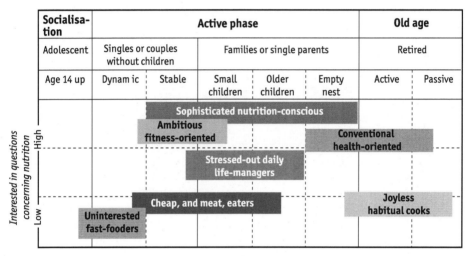

*Phase of life or life situation*

### 16.3.2.2 Uninterested fast-fooders

The 'uninterested fast-fooders' are indifferent to nutrition and health-related issues. Regular eating habits as well as cooking are not part of their daily routine. The 'uninterested fast-fooders' like to eat out, especially in their leisure time, in the company of their peer group. This nutrition style is widespread among younger singles and couples; men are over-represented in this group.

### 16.3.2.3 Cheap, and meat, eaters

For the 'cheap, and meat, eaters' food has to be inexpensive and its preparation simple and not time-consuming. Convenience products are therefore highly appreciated. Meat is considered an ideal meal, as its preparation offers a wide range of easy and creative meals. The 'cheap, and meat, eaters' break with rigid nutrition routines, and shared meals have lost their importance. This nutrition style can be found primarily among young and middle-aged singles, couples and families.

### 16.3.2.4 Joyless habitual cooks

The 'joyless habitual cooks' have very little awareness of nutrition issues. Deeply rooted nutrition routines structure their day and provide guidelines. Eating has the character of a duty; it is rarely connected with enjoyment and pleasure. This nutrition style is found mostly among retired singles and couples.

### 16.3.2.5 Ambitious fitness-oriented

The 'ambitious fitness-oriented' prefer high-quality food and follow a very disciplined diet in order to increase their achievement potential and physical fitness. They balance job-related and private demands with a healthy diet. Therefore, high-value products and food perceived to be healthy (e.g. organic food) as well as functional food play an important role. This nutrition style can be found primarily among couples and families in their child-rearing phase, frequently among freelancers and the self-employed and among households where both partners work.

### 16.3.2.6 Stressed-out daily-life managers

The 'stressed-out daily-life-managers' have a strong interest in questions of nutrition, especially in order to provide a balanced diet to their children. The double burden of career and family and a lack of available support from other family members make it difficult to realise this demand: getting groceries and preparing them becomes a tedious task. This nutrition style is widespread among women in the child-rearing phase.

### 16.3.2.7 Sophisticated nutrition-conscious

The 'sophisticated nutrition-conscious' have a strong interest in nutrition issues and show great sensibility for the integral meaning of nutrition in connection with health. They pay great attention to quality, freshness and the origin of the products. Organic food is considered a benefit to body and soul; synthetic additives are strictly rejected. This nutrition style is bound neither to a specific phase of life nor to a specific age.

### 16.3.2.8 Conventional health-oriented

The 'conventional health-oriented' highly value good food and have a strong interest in nutrition. Cooking, shopping and eating in a communicative atmosphere are highly appreciated. They prefer regional and seasonal products. Their desire to enjoy food collides with the wish to battle weight and health problems. This nutrition style is found among households in their after-family phase.

### 16.3.2.9 Conclusions

The typology of nutrition styles demonstrates that the way nutrition is organised in the context of everyday life varies broadly. The survey shows that most meals are consumed at home. However, the 'uninterested fast-fooders', the 'cheap, and-meat, eaters' as well as the 'ambitious fitness-oriented' eat more frequently away from home than do the other groups. At the same time, the 'ambitious fitness-oriented' manage to reduce the complexity of everyday life quite successfully. They combine a relatively high degree of delegation of nutrition-related work with the establishment of new routines that allow for an easier combination of work and family life. In contrast, the 'stressed-out daily-life managers' have few resources for delegation. They suffer from the burden of everyday life and have a strong desire to have some of the load taken off them.

At the same time, the typology of nutrition styles makes it clear that nutrition-related behaviour is shaped to a large extent by attitudes and orientations towards food and nutrition. For the 'sophisticated nutrition-conscious' and the 'conventional health-oriented' eating is a matter of pleasure and taste. For them, as well as for the 'stressed-out daily-life managers', nutrition is closely associated with issues of health or fitness ('ambitious fitness-oriented'). In contrast, the 'uninterested fast-fooders', the 'cheap, and meat, eaters' or the 'joyless habitual cooks' show only little awareness of nutrition. However, keeping in mind that the 'uninterested fast-fooders' are a very young group, this nutrition style can be considered to have a rather transient character. Many in this group are likely to change their attitudes towards nutrition and related behaviour when they start families.

Thus, the typology shows that nutrition styles are not distributed equally among the population. Most of them, such as the 'uninterested fast fooder' or the 'conventional health-oriented' have a distinct correlation to specific life phases or life situations (see Fig. 16.1).

### 16.3.3 Environmental impact of nutrition styles

Assessment of environmental impacts usually is calculated on the basis of highly aggregated data. Measurements refer either to the level of the entire population of a country or region or to the inhabitants of a smaller territorial unit (e.g. a city). The environmental impacts of individual behaviour can be calculated only on a per capita basis. This approach, however, does not allow for distinguishing the environmental impacts of different social groups.

In order to obtain a more realistic picture, the analysis of nutrition styles was linked to environmental indicators using the GEMIS methodology. GEMIS is a life-cycle analysis programme and database for energy, material and transport systems that been developed by the German Öko-Institut.[2] In GEMIS, environmental impacts are accounted for in terms of carbon dioxide ($CO_2$) and sulphur dioxide ($SO_2$) equivalents.

Drawing on the GEMIS methodology, the Öko-Institut calculated the environmental impacts related to the production, processing, distribution, transport, storage and preparation of food (Wiegmann et al. 2005). Linking life-cycle assessment to key indicators taken from the empirical survey—such as the composition of meals (especially whether the meal contains meat, organic food or chilled food), the means of preparation, the place of eating (at home or away from the home) and so on—the environmental impacts of the different nutrition styles can be calculated, as shown in Figure 16.2.

The results show that the nutrition styles vary with respect to their environmental impacts. In particular, the place of consumption (consumption in or away from the home) as well as the frequency of meat consumption account for differing environmental impacts. Unsurprisingly, perhaps, the nutrition of the 'uninterested fast-fooders'

2 www.oeko.de/service/gemis

FIGURE 16.2 **Nutrition styles and environmental impact**
Source: Eberle et al. 2006: 81

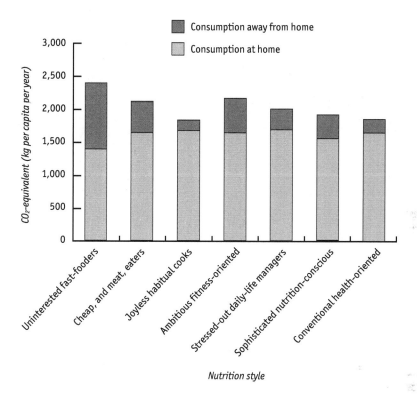

and of the 'cheap, and meat, eaters' have particularly high environmental impacts. In contrast, the relatively high $CO_2$ impact of the 'ambitious fitness-oriented' is more surprising. This impact is due to their relatively high propensity to eat out. In this respect, the typology offers a clear link to supply-side-related strategies. To reduce environmental pressures relating to chilled food, it is more promising to provide more eco-efficient provision systems than to promote the cooking of meals with fresh ingredients at home. Having a closer look at the high environmental impacts of out-of-home consumption the 'food change' project developed the proposal to establish a 'sustainability label' for eco-efficient out-of-home offers.

## 16.4 Overall conclusions

The social-ecological approach to sustainable consumption provides a fruitful and multidimensional framework for the study of sustainable consumption, linking consumption and environmental research. On an analytical level, the approach offers a better understanding of the complex interplay between motivational backgrounds, consumption patterns and their embeddedness in the routines and demands of everyday life. Moreover, it allows for a more appropriate assessment of the environmental impacts of consumption behaviour.

Focusing on everyday life from a bottom-up perspective (i.e. from a consumer perspective), it has great potential for the development of sustainability strategies that meet the various demands and needs of different and socially embedded consumers.

At the same time, the approach provides valuable insights for the creation of strategies to change consumption to a more sustainable pattern. The potential for changes to sustainability can be made visible by identifying starting points and barriers ('emotional anchors') from which a shift to a more sustainable consumption behaviour and to more sustainable provision systems may depart. The focus on consumer orientations and attitudes as well as on the needs and demands of everyday life provides a knowledge base for the design of socially differentiated information services and consumer advice as well as the development of sustainable products and services that meet consumer wishes and needs and 'fit' into everyday life.

Furthermore, the focus on everyday life and consumer perspectives is essential in developing strategies for radical changes from a long-term perspective. As Eivind Stø *et al.* argue (see Chapter 13), radical changes towards sustainable consumption need a combination of micro and macro strategies in the fields of politics and markets. To meet the different needs and demands of consumers in their role as citizens, voters and political consumers it is essential to gain a holistic picture of them by understanding them as well as everyday life actors in the context of different social-ecological lifestyles.

Thus, the social-ecological approach to sustainable consumption helps to overcome the shortcomings of approaches focused only on production or markets. It allows, in particular, for a shift in sustainable consumption research towards a consumer perspective. This perspective has great potential not only to improve consumer information and communication but also to inform sustainable product design, innovation strategies and sustainability policies.

An important result of the social-ecological transdisciplinary project on 'food change' is the dialogue with different stakeholders in the field of nutrition, health and environment: that is, with food producers and retailers, health agencies, consumer organisations and so on. For all of these, the insights gained into the everyday demands and motivation of the different nutrition styles have been very illuminating. The typology is an important tool in gaining a better understanding of consumers' demands and behaviour. It provides a valuable knowledge base on which actor networks can draw to design and promote more sustainable products and services that work under the conditions of everyday life (Shove 2004). In a more comprehensive sense, a consumer perspective (i.e. a *KonsumentInnenperspektive*; see Hayn *et al.* 2005) can support accountable and responsible strategies, making sustainable nutrition an issue of public concern.

# References

Bourdieu, P. (1979) *La distinction: Critique social du jugement* (Paris: Les éditions de minuit).
Cogoy, M. (1999) 'The Consumer as a Social and Environmental Actor', *Ecological Economics* 28.3: 385-98.
Empacher, C., and K. Götz (2004) 'Lifestyle Approaches as a Sustainable Consumption Policy: A German Example', in L.A. Reisch and I. Røpke (eds.), *The Ecological Economics of Consumption* (Cheltenham, UK: Edward Elgar): 190-206.
——, K. Götz, I. Schultz and B. Birzle-Harder (2002a) 'Die Zielgruppenanalyse des Instituts für sozial-ökologische Forschung', in Umweltbundesamt (Federal Environmental Agency) (ed.), *Nachhaltige Konsummuster: Ein neues umweltpolitisches Handlungsfeld als Herausforderung für die Umweltkommunikation* (Berlin: Erich Schmidt): 87-181.
——, I. Schultz, D. Hayn and S. Schubert (2002b) 'Die Bedeutung des Geschlechtsrollenswandels', in Umweltbundesamt (Federal Environmental Agency) (ed.), *Nachhaltige Konsummuster: Ein neues umweltpolitisches Handlungsfeld als Herausforderung für die Umweltkommunikation* (Berlin: Erich Schmidt): 182-214.
Giddens, A. (1984) *The Constitution of Society* (Cambridge, UK: Polity Press).
Götz, K. (2001) 'Sozial-ökologische Typologisierung zwischen Zielgruppensegmentation und Sozialstrukturanalyse', in G. de Haan, E.D. Lantermann, V. Linneweber and F. Reusswig (eds.), *Typenbildung in der sozialwissenschaftlichen Umweltforschung* (Opladen, Germany: Leske & Budrich): 127-38.
——, W. Loose, M. Schmied and S. Schubert (2002) *Mobility Styles in Leisure Time: Final Report for the Project 'Reduction of Environmental Damage Caused by Leisure and Tourism Traffic'* (short version; commissioned by the Federal Environment Office; Frankfurt am Main: Federal Environment Agency [Umweltbundesamt, UBA]).
Gronow, J., and A. Warde (eds.) (2001) *Ordinary Consumption* (London: Routledge).
——, U. Eberle, R. Rehaag, U. Simshäuser and G.Scholl (2005) *KonsumentInnenperspektive: Ein integrativer Forschungsansatz für sozial-ökologische Ernährungsforschung* (DP-8 Frankfurt am Main/Hamburg/Cologne/Heidelberg/Berlin: Institut für Sozial-ökologische Forschung, Öko-Institut e.V., Katalyse-Institut e.V., Institut für ökologische Wirtschaftsforschung; www.ernährungswende.de/pdf/DP8_KP_2005_final.pdf [accessed 18 January 2008]).
——, U. Eberle, I. Stieß and K. Hünecke (2006) 'Ernährung im Alltag', in U. Eberle, D. Hayn, R. Rehaag and U. Simshäuser (eds.), *Ernährungswende: Eine Herausforderung für Politik, Unternehmen und Gesellschaft* (Munich: oekom): 73-84.
Kristensen, P. (2004) *Household Consumption of Food and Drinks* (background paper for the European Environment Agency [EEA] Report on Household Consumption and the Environment; Roskilde, Denmark: Danish National Environmental Research Institute [NERI]).
Littig, B. (2001) *Feminist Perspectives on Environment and Society* (Harlow, UK: Pearson Education).
Müller, H.-P. (1992) *Sozialstruktur und Lebensstile: Der neuere theoretische Diskurs über soziale Ungleichheit* (Frankfurt am Main: Suhrkamp).
Reusswig, F. (1994) *Lebensstile und Ökologie: Gesellschaftliche Pluralisierung und alltagsökologische Entwicklung unter besonderer Berücksichtigung des Energiebereichs* (Frankfurt am Main: IKO-Verlag für interkulturelle Kommunikation).
Schubert, S. (2004) 'Mobility Styles in Leisure Time: A Lifestyle Approach for a Better Understanding and Shaping of Leisure Mobility', paper presented. at the *OECD Est! Workshop on Leisure Travel, Tourism Travel, and the Environment*, Berlin, 4–5 November 2004, www.isoe.de/ftp/tagungen/schu_oecd04.pdf [accessed 25 August 2007].
Schultz, I. (2006) 'The Natural World and the Nature of Gender', in K. Davis, M. Evans and J. Lorber (eds.), *Handbook of Gender and Women's Studies* (London/Thousand Oaks, CA/New Delhi: Sage Publications): 376-96.
—— and K. Götz (2006) 'Konsum', in E. Becker and T. Jahn (eds.), *Soziale Ökologie: Grundzüge einer Wissenschaft von den gesellschaftlichen Naturverhältnissen* (Frankfurt am Main: Campus): 360-70.

—— and I. Weller (1997) *Nachhaltige Konsummuster und postmaterielle Lebensstile: Eine Vorstudie im Auftrag des Umweltbundesamtes* (Publication 30/97; Berlin: Federal Environment Agency [Umweltbundesamt, UBA]).

Shove, E. (2004) 'Changing Human Behaviour and Lifestyle', in L.A. Reisch and I. Røpke (eds.), *The Ecological Economics of Consumption* (Cheltenham, UK: Edward Elgar): 111-31.

Stieß, I., and K. Götz (2002) 'Nachhaltigere Lebensstile durch zielgruppenbezogenes Marketing?', in D. Rink (ed.), *Lebensstile und Nachhaltigkeit. Konzepte, Befunde und Potentiale* (Opladen, Germany: Leske & Budrich): 247-63.

—— and D. Hayn (2005) *Ernährungsstile im Alltag: Ergebnisse einer repräsentativen Untersuchung* (DP-5; Frankfurt am Main: Institut für sozial-ökologische Forschung; www.ernaehrungswende.de/pdf/dp5_ernaehrungsstile.pdf [accessed 18 January 2008]).

—— and D. Hayn (2006) 'Alltag', in E. Becker and T. Jahn (eds.), *Soziale Ökologie: Grundzüge einer Wissenschaft von den gesellschaftlichen Naturverhältnissen* (Frankfurt am Main: Campus): 211-23.

Voß, G.G. (1995) 'Entwicklung und Eckpunkte des theoretischen Konzepts', in Projektgruppe Alltägliche Lebensführung (ed.), *Alltägliche Lebensführung: Arrangements zwischen Traditionalität und Modernisierung* (Opladen, Germany: Westdeutscher Verlag): 23-44.

Weber, M. (1948) *The Protestant Ethic and the Spirit of Capitalism* (New York: Scribner).

Wiegmann, K., U. Eberle, U.R. Fritsche and K. Hünecke (2005) 'Umweltauswirkungen von Ernährung: Stoffstromanalysen und Szenarien' (DP-7; Darmstadt/Hamburg, Germany: Ernährungswende; www.ernaehrungswende.de/pdf/DP7_Szenarien_2005_final.pdf [accessed 25 August 2007]).

# 17
# Emerging sustainable consumption patterns in Central Eastern Europe, with a specific focus on Hungary

*Edina Vadovics*
Central European University, Hungary

## 17.1 Introduction

It is becoming increasingly accepted that in order to reach sustainable development, or, in other words, to use the vocabulary of ecological footprinting, to reduce 'humanity's overshoot', consumption also needs to become more sustainable (e.g. see EEA 2005; Jackson 2004). However, despite the growing number of studies relating to sustainable consumption, there is no generally agreed and accepted definition available for the term. For the purposes of the present chapter, sustainable consumption is defined 'as the use of goods and services that respond to basic needs and bring a better quality of life, while minimising the use of natural resources, toxic materials and emissions of waste and pollutants over the life-cycle, so as not to jeopardise the needs of future generations' (Norwegian Ministry of Environment 1995: 9). Accordingly, it encompasses economic, environmental and social dimensions, all of which need to be changed or adjusted in order to achieve consumption patterns and levels that are indeed sustainable.

The direction of change towards more sustainable consumption may be both top-down and bottom-up, but so far most of the attention seems to have been focused on top-down processes, achieving rather good results in creating more sustainable production patterns. However, as humanity's growing ecological footprint indicates, this has not resulted in a decrease in aggregate consumption and associated environmental as well as social impacts (WWF 2006). In Central Eastern European (CEE) countries, especially, far less consideration has been given to achieving more sustainable patterns and levels of consumption, either top-down or bottom-up. There are several reasons for this; however, by far the most important one is that it is strongly believed that increased levels of consumption will facilitate the development of the region, as is believed for any region in the world (Durning 1992; Jacobs 1997).

Another very important reason is that consumption was constrained in the not so distant socialist past of the region, and, as a result, consuming more and achieving levels of material consumption similar to those in the Western part of Europe is the principal motivation for most of the population. Nevertheless, one can find a variety of bottom-up, voluntary, consumer citizen-driven initiatives in the region that aim at creating a more equitable society with less environmental impact (i.e. they promote more sustainable patterns and levels of consumption). The initiatives are either new lifestyle patterns reflecting a need to cater for needs not satisfied by mainstream society or are patterns that are preserved from the past and have been there for a long time.

The aim of an EU-funded research project called EMUDE (Emerging User Demands for Sustainable Solutions) was precisely to find and describe bottom-up community initiatives in countries of the European Union, focusing mostly on Western Europe but also collecting cases in the new member states, as well as to consider the wider-scale applicability of these more sustainable everyday life patterns.[1] In the EMUDE project such consumer-driven sustainable consumption initiatives were termed as 'creative communities'. Creative communities are defined as groups of innovative citizens who organise themselves to solve a problem or to invent new ways of carrying out everyday activities and in doing so move towards social and environmental sustainability. Thus, creative communities are what Seyfang (2005) terms 'grass-roots initiatives for sustainable development', Georg (1999) calls 'citizen initiatives' and Jackson (2004) 'community-based initiatives'. Since this chapter is based on the EMUDE research project, the term 'creative communities' is adopted.

The role of creative communities, or, in other words, bottom-up, voluntary initiatives, in achieving sustainable consumption is vital, because the effect of top-down efforts, or policies on consumer behaviour change, is uncertain (Stø et al. 2006). Thus, study of how the desired changes come about voluntarily, and how the processes leading to them could be motivated to occur on a wider scale instead of mostly on the margins of society, is of great importance.

In what follows, creative communities from Central Eastern Europe are introduced. However, in order to contribute to a better understanding of such communities, I will first provide some background information on the region, both in terms of the relatively recent transition to a market economy and democracy and the impact of that transition on welfare and consumption trends.

---

1 EMUDE was supported by the EU 6th Framework Programme. More information on the project can be found at www.sustainable-everyday.net/EMUDE [accessed 25 August 2007].

## 17.2 Some background on transition and consumption in Central Eastern Europe

Central Eastern Europe is a region that can be defined in different ways. In this chapter, the term refers to the Czech Republic, Estonia, Hungary, Latvia, Lithuania, Poland, Slovakia and Slovenia, 8 of the 10 countries that joined the European Union in 2004. It is important to note here that these countries represent diverse cultural traditions. The objective of this chapter; however, is not to attempt a comprehensive analysis taking into consideration all the diversity of these countries but to point out general trends characteristic of the region.

### 17.2.1 The transition to a market economy

The countries of the CEE region all belonged to the so-called socialist bloc, which means that they all had centrally planned economic, political as well as welfare systems. Out of the three actors usually considered in the provision of welfare—the market, the family and the state—in the socialist countries most of the responsibility rested on the state (Esping-Andersen 1990). The state, with the involvement of and heavily relying on state-owned companies, rather exceptionally (Wagener 2002):

- Assured complete employment (achieved through over-staffed state companies)
- Provided free education for all, including nursery and kindergarten care and education to allow for the employment of women
- Provided for a social as well as a healthcare system that was freely available to everyone
- Tried to ensure that everyone was included in society and had access to the provisions of the welfare system
- Through subsidies, kept the prices of products and services low, making them available to the majority of the population

At the same time, Wagener (2002: 156) calls attention to the fact that in the socialist system 'workers stood under the protective guarantee of a paternalistic state, and were free of any responsibility'. This has significant implications for the level of participation in democratic processes as well as in voluntary community activities in the decades following the regime change. People who were not called on to express their views or take part in the decision-making processes at local, regional and national levels will need time to adjust their expectations as well as learn the skills needed for successful and meaningful participation.

It also necessary to note that, although the political and economic system was meant to be egalitarian and non-discriminatory, of course one could find examples of exclusion. There were also differences between the rich and the poor, but, compared with the differences existing in capitalist societies today, they were much less pronounced.

Furthermore, in the background, there was mounting pressure on the environment as it was viewed as a resource freely available for exploitation, and neither the material nor the energy efficiency of production were priority issues (Tellegen 1996). Additionally, full employment, free education for all, well-planned and cheap public transport systems as well as a good-quality social and healthcare system were very expensive to maintain. These, together with the increasing number of people demanding more political freedom, the freedom of speech and the freedom to self-organise, necessitated a change that took place in the late 1980s and early 1990s (Ferge 2001; Frigyesi and Kapolyi 2005).

Although it was greatly needed and anticipated, the transition to a market economy, capitalism and democracy, bringing with it the move to decentralisation, came as a great shock to a large part of the population in CEE countries. The most important reason for this is that the state had to discontinue its paternalistic role of providing for everyone. Thus, CEE societies had to face:

- The closing-down of factories, resulting in unemployment, as well as the closing of factory-supported nurseries and kindergartens which, to a large extent, had facilitated the employment of women

- An enlarged the gap between the rich and the poor measured in terms of income inequality and the Gini coefficient (Svejnar 2002; Soubbotina 2004)

- A steep decrease in the purchasing power of salaries

- The privatisation of welfare services

(For more detail, see EC 2003; Ferge 2001; Frigyesi and Kapolyi 2005; Haney 1999; Svejnar 2002.)

Despite these negative trends, according to a survey conducted by the Public Opinion Research Centre in Poland in 1999 (reported in Svejnar 2002), most people feel that the transition was, overall, a positive process. This can be claimed even though in CEE countries more people feel that the process of regime change has resulted in more losses than gains in terms of material conditions for living.

These changes in the society and economy of CEE countries explain why there was a sudden need for creative, civil society solutions to attempt to close the gap in the provision of social services and the satisfaction of people's needs in different areas of life.

### 17.2.2 Consumption and consumption trends in Central Eastern Europe

Household consumption was constrained in the past regime—so much so that some authors label this period as 'the economy of permanent shortage' (Buchowski 1996: 85). In contrast, the efficiency of production was not an issue as there was no real competition in the market. The political and economic transition process, however, resulted in significant changes that affected both consumption and production.

On the one hand, inefficient companies needed to be closed down, and competition picked up. On the other hand, gross domestic product (GDP) as well as household consumption, though first stagnating or decreasing in some of the CEE countries, started rising during the second part of the 1990s (Figs. 17.1 and 17.2).

The political and economic changes and their impact on consumption and production patterns as well as on consumption levels are mirrored very well by the rather sudden

FIGURE 17.1 Gross domestic product (GDP) in selected Central Eastern Europe countries between 1989 and 2003, taking 1989 as 100%

Source: adapted from: Kollányi 2004: 46

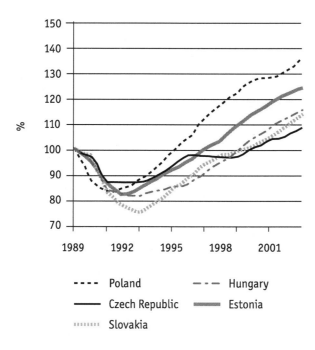

FIGURE 17.2 Gross domestic product (GDP) and household consumption in Hungary between 1989 and 2003, taking 1989 as 100%

Source: adapted from: Kollányi 2004: 47

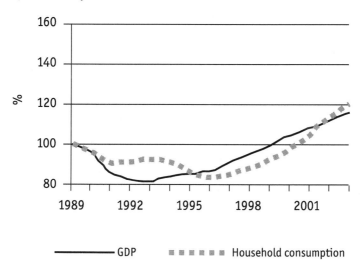

fall in per capita 'ecological footprints' at the beginning of the 1990s, as shown by the examples of the Polish and Hungarian cases (Fig. 17.3). The drop in the size of per capita footprints also indicates that in CEE countries, at least for the initial period after the regime change, pressure on the environment decreased with decrease in emissions, mostly as a result of the closing down of the most polluting companies (Archibald *et al.* 2004; Svejnar 2002). More recently, increased levels of consumption are expected to result in increased pressure on the environment and in ecological footprints that start growing again.

FIGURE 17.3 Change in the per capita footprint in [a] Poland and [b] Hungary since 1961

Source: Global Footprint Network 2006

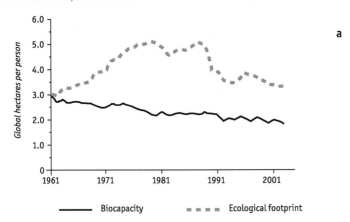

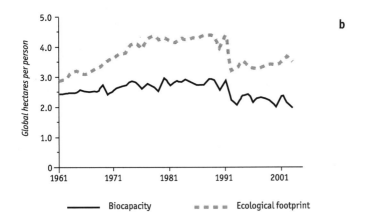

Data available from sources such as UNEP's GEO Data Portal[2] and the European Environment Agency (EEA 2005) also show that overall consumption trends have been quickly counteracting the initially positive environmental impact of reduced volumes and increasingly efficient production. After a period of stagnation, and at times decrease, household spending has started increasing again, as illustrated in Figure 17.4 through the example of Hungary, and in Figure 17.5. Despite this trend, spending is still well below Western European levels, although the average per capita ecological footprint of CEE citizens is about three-quarters that of the average Western European, owing mostly to inefficiencies in production (WWF/GFN/NC-IUCN 2005).

FIGURE 17.4 **Total household final consumption expenditure (in constant 2000 US$ millions) in Hungary between 1986 and 2003**

Source: GEO Data Portal, geodata.grid.unep.ch [accessed 25 August 2007]

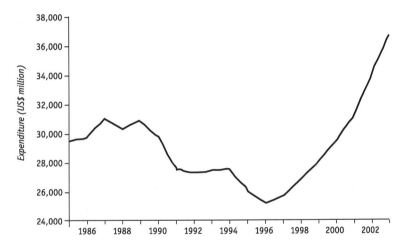

In summary, it can be said that, although levels of consumption in the new member states are currently about one-fifth of those in old member states, consumption and its associated environmental impact are rising rapidly, sometimes even faster than in the EU-15 countries, or, in other words, quicker than in the old member states or the European Union (SEI-T 2005). Furthermore, as growth in consumption is considered to be beneficial for economic development and is closely associated with increased levels of welfare, increasing it is the most important goal of new member states. Thus, it is likely to grow further unless a rather drastic change in thinking about the real relationship between GDP, levels of consumption and the quality of life is effected. It will be necessary for policy-makers and consumers to consider the fact that, based on a growing number of studies, increases in GDP and consumption do not necessarily result in higher levels of well-being (NEF 2004; Venetoulis and Cobb 2004; Worldwatch Institute 2004).

2 UNEP online database, at geodata.grid.unep.ch [accessed 25 August 2007].

FIGURE 17.5 Household expenditure per capita (in constant 1995 euros) in [a] the EU-15 countries and [b] five new member states (note the difference in scale in the expenditure axes)

Source: EEA 2005: 17-18

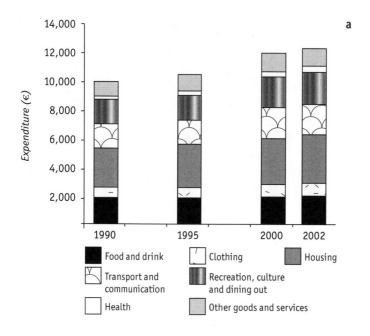

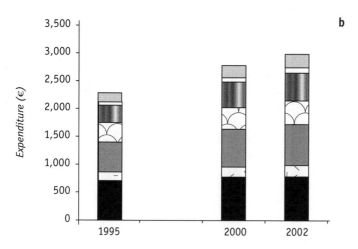

## 17.3 Creative communities in Central Eastern Europe

Creative communities have existed, and more are emerging, in the CEE region just as in Western Europe. The reasons for people to join forces and search for novel and more sustainable ways of performing everyday activities are just as diverse in CEE countries as they are in Western Europe. The most important motivations identified in Central Eastern Europe during the EMUDE project are:

- To find local and customised responses to environmental as well as to social problems, needs and wants that are fulfilled by neither the state nor the market
- To solve pressing social problems such as unemployment and the depopulation of villages
- To ensure better protection of the environment and a more efficient and environmentally friendly use of resources
- To find alternatives to capitalism and the market economy—the only solutions presented after the change of regime at the end of the 1980s and beginning of the 1990s

These motivations were all found to be playing an important role in the region (EMUDE 2006; Vadovics 2007), and it would be difficult to say which one of them is the most powerful. It must be emphasised, however, that in the majority of cases creative communities, although they may be focusing on the solution of a particular problem, often provide answers to a variety of issues. Thus, in a large proportion of cases, both social and environmental sustainability are positively affected. Below, some of the motivations are examined, grouped into two sections. Section 17.3.1 introduces creative communities that emerged in response to the need to provide services that the state or state-owned companies catered for in the past regime and that are now not satisfied. Section 17.3.2 focuses on initiatives that outline a development model alternative to the capitalist model, the only one considered by mainstream politics and economics. Finally, in Section 17.4 specific characteristics of CEE countries are described, characteristics that could be taken advantage of in promoting creative communities and, through them, more sustainable consumption patterns.

### 17.3.1 'Filling the gap': providing services no longer granted by the state

#### 17.3.1.1 Childcare

In Section 17.2.1 above it was outlined that the transition process, although welcomed by the majority of society, resulted in certain negative impacts from the point of view of the well-being of people. It was specifically mentioned that inefficient companies needed to be closed down, resulting, among other things, in a shortage of nurseries and kindergartens. Thus, a shortage in nurseries and kindergartens resulted and the remaining childcare institutes had to work with much larger groups of children, their staff became overburdened and a number of children could not find places in any schools or nurseries. This situation was not conducive to the development of children; in addition, some mothers were unable to take on jobs. Thus the well-being of some families decreased.

In response to this, and also in response to the want on the part of parents to provide good-quality education to their children, CEE countries now have small, often family-run nurseries, kindergartens, day-care services or 'mini kindergartens' (detailed examples of cases can be found in Meroni 2007). These kindergartens contribute to local social sustainability by hiring teachers and staff who were previously unemployed, establishing links and cooperation between the stakeholders in nursery education, such as parents, the local government and teachers. Environmental sustainability is enhanced by the location of nurseries close to where the children live (thus reducing transportation needs), using toys made locally and of natural materials (often crafted by the parents). Food for the children is also purchased locally, often from local growers. Furthermore, these small establishments allow for a more flexible approach to a variety of issues, such as: parents being able to choose whether they want their children to stay, for the whole day, for part of the day or only for a couple of hours. In these schools children learn more sustainable everyday life patterns by being part of such a lifestyle and it is hoped that this will have an impact on their adult life.

### 17.3.1.2 Care for the elderly

Care for the elderly was also shaken during the transition years. As the state is no longer able to provide fully for their care, some care homes had to be closed down. This, and the rather sharp decrease in the purchasing power of pensions, necessitated that people get together and find solutions to help those elderly who do not have families willing or able to care for them, still an important dimension of taking care of the elderly in most of the CEE region. Thus, there has been a renewed interest in starting or enlivening senior clubs (originally promoted by the communist state), which provide space for the elderly to get together, to engage in creative crafts and activities, to sell the products they make in their own shop, or to dine together cheaply. Such clubs thus help the elderly to engage actively in society, to keep local traditions, arts and crafts alive and to teach these skills to the younger generation. This also promotes the use of locally available resources and materials, making an important contribution to environmental sustainability.

### 17.3.1.3 Unemployment

An additional very important challenge in the region is the combined effect of unemployment, resulting from the closing-down of companies, and the lack of employment in rural areas, leading to urbanisation and the depopulation of villages. These factors motivated a number of very devoted people and communities to initiate so-called eco-village projects. Although this is closely linked to the solutions discussed in this section, it also highlights another important motivation for initiating creative communities, described in the next section.

## 17.3.2 Finding alternative ways of development

Eco-villages in the region create an alternative development model for settlements. They can be viewed as projects providing complex solutions to a number of problems existing in contemporary CEE society. The EMUDE research project identified numerous cases in the region; in Hungary alone there are more than a dozen eco-village initiatives. What is common to them all is that they promote an environmentally friendly lifestyle—including housing, work and food—while attempting to overcome social problems. They offer complex solutions to various challenges present in the countryside: environmental degradation (addressed in the eco-villages by the use of organic agriculture and the utilisation of alternative energy sources), depopulation of villages (addressed by the creation of local jobs and attracting people back from the cities) and a lack of local jobs (addressed by reviving local crafts and traditions). Thus, as well as promoting more sustainable consumption patterns, they may also serve as models for more sustainable local development.

The Model Eco-friendly Hamlet in Poland (Meroni 2007), for example, strengthened the local social fabric by involving people in various projects in order to save the local school from closing.[3] Teachers, parents and pupils decided to establish an ecological education centre in the school that organises various courses as well as ecological awareness-raising and education events. Projects are also developed to improve the image of the village and at the same time to act against its depopulation by providing an attractive living space for inhabitants. Eco-tourism, the revival of local crafts and renewable energy generation are all part of the new development. As more local people get involved, more projects are conceived of and implemented with the coordination of a local not-for-profit organisation. As a consequence, the social structure of the village is strengthened as people unite and cooperate in creating a better place to live. The resulting environmental advantages are, for example, a raised level of awareness of environmental issues, the use of renewable energy sources and the saving of energy.

An example from Hungary is the Gömörszőlős Sustainable Village Project in the north of the country.[4] Its aim is the revitalisation of a small village inhabited by a couple of hundred people, mostly elderly and the solution of social and (un)employment problems based on the principles of sustainable development. The project is managed by the Ecological Institute for Sustainable Development, a not-for-profit organisation that drew up a comprehensive development plan for the village; the plan is being piloted in that village but can later be used in other villages of the region. It includes local institutional capacity-building and training for people; this is provided in a demonstration and training centre in the village, in a building that was renovated using traditional and environmentally friendly technologies.

An important feature of the project is that, similar to the Polish Model Eco-Friendly Hamlet, it does not create a completely new village, as is frequently the case with eco-village initiatives; it instead sets out to develop an already existing one with the involvement and education of the local people.

---

3 More information on this creative community can be found at www.przyslop.zawoja.pl [accessed 25 August 2007].
4 Source: leaflet of the Gömörszőlős Sustainable Village Project produced by the Ecological Institute for Sustainable Development in Miskolc, Hungary. More information is available at www.ecolinst.hu/index3.html [accessed 24 January 2008].

A different case from Hungary is the Open Garden Foundation, a not-for-profit organisation in central Hungary.[5] It runs a community-supported agriculture project through the operation of a demonstration organic market garden and a vegetable box scheme, both of which employ local people as well as people with disabilities. Its objective is to promote the consumption of locally grown seasonal food, to connect producers and consumers directly and, by doing so, contribute to the creation of a healthy community. The most important motivation for initiating the project was to spread the idea and practice of environmentally friendly growing and agriculture, to provide local people with fresh, healthy and seasonal produce and to reconnect them (i.e. the consumers) as much as possible to nature as well as to natural processes by involving them in planting and harvesting activities. These obviously also have positive community-building impacts and help to reconnect producers with consumers as well as with different groups of consumers, or, in other words, with local society.

It needs to be mentioned here that in the majority of cases the present economic and infrastructural setting is often not conducive and supportive to creative community initiatives. Novel ways of employment, education, energy generation, waste-water treatment or food production are often not accepted by authorities issuing the various permits necessary for practising them. In this sense, some Western European countries are more advanced (e.g. The Netherlands, which offers people a choice of which type of energy generation to support; Spaargaren 2003). However, the EMUDE project concluded that all EU countries could do a great deal to improve the infrastructural and legal setting to support creative communities (EMUDE 2006). As a result of these challenges, creative community initiatives very often struggle to 'stay alive', or, in other words, to achieve economic sustainability in the currently unsustainable circumstances, and, consequently, rely heavily on devoted 'heroes', national and EU funding and, to a great extent, on volunteers.

## 17.4 Central Eastern European characteristics that can be built on in promoting creative communities and sustainable consumption

### 17.4.1 Experience in being part of a group

In the past regime, ruling communist parties in CEE countries organised a wide variety of group activities and, in fact, greatly encouraged participation in various communal events *within* the realms of the party. As a result, people were involved in youth groups, belonged to the local housekeepers' association, took part in local clean-ups, tree-plantings and party festivals, helped their children to collect waste separately, attended clubs for seniors, participated in neighbourhood watch schemes and took part in numerous other collective activities. Thus, it can be said that the communist party tried its best to

---

5 More information on the project can be found at www.nyitottkert.hu [accessed 25 August 2007] and Vadovics and Hayes 2007.

include everyone in some kind of a group, each of which had quite a full and varied schedule of activities (Buchowski 1996).

This has had both a negative and a positive impact on the formation of creative communities in the region. To begin with the former, it needs to be mentioned that, after the transition, groups and communities began to be associated with the old regime; thus the act of getting rid of them was considered to be a sign of development. At the same time, a large number of people had a very positive experience of belonging to and being active in groups, which they began to miss after the transition. This experience and the positive memories could be harnessed in presenting newly emerging creative communities to CEE societies.

Furthermore, it needs to be borne in mind that, although independent and politically active civil society was largely missing from life under the communist rule, people were active in extended kin groups and informal interest groups. These were a necessary part of survival in 'the economy of permanent shortage' (Buchowski 1996). In these groups, people often shared cars and tools and assisted one another in house-building, gardening or animal husbandry—practices that still form a natural part of life for a great number of people and can be seen to be utilised in creative community activities.

### 17.4.2 Consuming local products

Consuming local or at least regional products was encouraged in the past regime to reduce dependence on imports from outside the socialist bloc. This movement was kept in the region but has gained new momentum in a globalising world. Its aim now is the protection of local jobs and local products by buying these products instead of imported ones. This can even be said to be a mainstream movement, with a specific logo in Hungary (e.g. for products made in that country).

Another characteristic is the tradition of buying local fruit and vegetables at a local market. This practice is weakening with the increasing popularity of hypermarkets. However, in the fairly recent past in Hungary, for example, each village and town held a market at least once a week on Saturday, where local people went to sell their produce and thus supplement their income. Village as well as town people purchased their weekly supply of fresh local fruit and vegetables at these local markets.

However, with the change of economic system, the increasing popularity of supermarket shopping and more stringent EU legislation for markets (regarding food safety, hygiene and market operation), many markets needed to be closed down. In this way, small growers were forced out of the market or now need to travel long distances to be able to sell their produce.

As this tradition is still fairly strong in the CEE region there is great potential for strengthening and improving existing initiatives with ideas from Western European cases such as those collected and studied in the EMUDE research project (Meroni 2007).

### 17.4.3 Culture of re-use and recycling

Recycling and re-use on the household scale has a long and strong tradition in Central and Eastern Europe, partly because of to 'the economy of permanent shortage' before the transition, which has already been mentioned and discussed above. People are reluctant to throw their things away; they can always find a new use for old objects, give them to their neighbours or relatives or get rid of such objects at the annual junk clearance organised by municipal governments. On junk-clearance days, people put their unwanted objects, furniture, clothes, bicycles, etc. on the street and, before the municipal government collects the items, anyone can come and take them away for free. This practice and the strong tradition of re-use may explain why initiatives teaching people new ways of re-use and recycling were found to be widespread and popular in the region in the EMUDE project. Examples include the Mööblikom (a furniture redesigning studio) and Materjalid.net, both of them in Tallinn, Estonia. Mööblikom invites people to bring their unwanted pieces of furniture to the organisation, where the furniture will be repaired, renewed and sold to new owners.[6] Materjalid.net is a project to recycle used construction material; it educates people about how they can utilise discarded construction materials as well as collecting such material from them or providing them with materials (Meroni forthcoming).[7]

### 17.4.4 Existing infrastructure

The most significant type of infrastructure that needs to be mentioned concerns the good public transport networks in CEE countries that were designed to reach even the remotest villages. With most of the attention, including financial resources (EEA 2003), now drawn to road and motorway building, they are quickly deteriorating, but their maintenance and upgrading could greatly contribute to social inclusion by making it possible for those in remote villages who cannot afford to own cars to travel to schools and jobs in their local town. Although such a system may not reduce the need for bicycle associations, mobile bicycle repair people, car sharing or 'walking bus' initiatives found overwhelmingly in urban areas in Western European countries, from the point of view of reducing pressure on the environment, public transport networks should be kept in the focus of mobility-oriented solutions.

## 17.5 Conclusions and considerations for further research

This chapter has introduced numerous issues relating to sustainable consumption in the new EU member states. Its aim, however, was not to provide a comprehensive picture but to draw attention to promising trends and initiatives that could facilitate the transition to a more sustainable society. It also aimed at pointing out factors that hinder the transition process and thus sustainable consumption and that consequently need to be changed if current unsustainable trends are to be altered.

The first important conclusion to draw is that in the new CEE member states of the EU there is a great amount of experience and variety of skills to build on in a move to promote creative communities, a participatory welfare society and sustainable local devel-

---

6 More information on this creative community can be found at www.mooblikom.com [accessed 25 August 2007].
7 More information on the initiative is available at materjalid.net [accessed 25 August 2007].

opment. At the same time, as this region is moving rapidly towards the mainstream Western European way of living and consuming as well as becoming globalised, these experiences and skills are becoming lost and forgotten. Thus, there is an urgent need to strengthen what still exists and to provide people with positive examples from the West that could be easily adapted to CEE circumstances and local needs. The fact that CEE countries still have, or at least remember, the patterns of life that Western societies are relearning and rediscovering is something that should be valued and protected.

Meanwhile, new patterns of consumption such as car sharing, solidarity purchasing and shared housing could be introduced and presented as alternatives to the consumer society. In doing so, it needs to be borne in mind that both parts of the EU have valuable experiences, production and consumption patterns as well as skills to offer one another, and these could be taken advantage of in better ways.

A second important conclusion is that a new kind of welfare—an 'active welfare' (EMUDE 2006; Stø et al. 2006)—seems to be being created. Active welfare is where people are mobilised in their roles as consumers and citizens and, as a result, they assume more responsibility for the provision of welfare. Such a welfare system is characterised by more voluntary action and activity, especially to provide for the needs and/or wants that neither the state nor the market can satisfy in today's rapidly changing and restructuring society. At the moment, it appears that creative communities make a more sustainable society possible in that they provide increased levels of welfare (social sustainability), are less demanding on the environment and are, or can become, more sustainable economically as they rely on local resources and local people serving local needs and wants. A growing body of research suggests that the increased levels of networking among people that can be observed in creative communities, and the resulting trust and reciprocity, enhance economic performance and has been found to be important for the economic success of countries (Briceno and Stagl 2006; Fukuyama 2002; Rudd 2000; Worldwatch Institute 2004). Thus, the new forms of livelihood represented by creative communities could be utilised in sustainable local development both in the short run and in the long run.

From this point of view, it is important to note that in CEE countries the welfare state has gone through changes and is being restructured, although the importance of this is second, at best, to economic development. However, it appears that real and long-term economic development is not possible without a well-functioning active welfare state and strong communities. So, the encouragement and support of creative communities should indeed become long-term sustainable development priorities for countries.

Finally, it needs to be remembered that these new forms of welfare (or creative communities), although numerous examples exist, represent only a rather minor section of society, and the people working in and for them have to overcome a variety of difficulties to succeed in creating not only socially and environmentally, but also economically sustainable solutions. As Sanne (2002) observed, consumers are often locked into inflexible systems in their everyday activities and lives (such as in eating and moving and in their housing) that not only limit the choices available to them but also often make it impossible to implement more sustainable, creative community options. Additionally, choices are limited further by funding and investment priorities, as well as loan and mortgage policies that do not take sustainable consumption options and alternative solutions into consideration. Thus, in order to create a more sustainably consuming society, systemic changes will need to be made.

# References

Archibald, S.O., L.E. Banu and Z. Bochinarz (2004) 'Market Liberalisation and Sustainability in Transition: Turning Points and Trends in Central and Eastern Europe', *Environmental Politics* 13.1: 266-89.

Briceno, T., and S. Stagl (2006) 'The Role of Social Processes for Sustainable Consumption', *Journal of Cleaner Production* 14.17: 1,541-51.

Buchowski, M. (1996) 'The Shifting Meanings of Civil and Civic Society in Poland', in C. Hann and E. Dunn (eds.), *Civil Society: Challenging Western Models* (London: Routledge): 79-99.

Durning, A.T. (1992) *How Much Is Enough? The Consumer Society and the Future of the Earth* (London: Earthscan Publications).

EC (European Commission) (2003) 'Social Protection in the 13 Candidate Countries: A Comparative Analysis', European Commission, Directorate-General for Employment and Social Affairs; europa.eu.int/comm/employment_social/publications [accessed 25 August 2007].

EEA (European Environment Agency) (2003) *Europe's Environment: The Third Assessment* (Copenhagen: EEA).

—— (2005) *Household Consumption and the Environment* (Copenhagen: EEA).

EMUDE (Emerging User Demands for Sustainable Solutions) (2006) *Creative Communities: Towards Active Welfare and a Distributed Economy: Final Results* (EMUDE Project Consortium).

Esping-Andersen, G. (1990) *The Three Worlds of Welfare Capitalism* (Cambridge, UK: Polity Press).

—— (2002) 'Towards the Good Society, Once Again?', in G. Esping-Andersen, D. Gallie, A. Hemerijck and J. Myles (eds.), *Why We Need a New Welfare State?* (New York: Oxford University Press): 1-26.

Ferge, Zs. (2001) 'Welfare and "Ill-fare" Systems in Central-Eastern Europe', in R. Sykes, B. Palier and P.M. Prior (eds.), *Globalisation and European Welfare States: Challenges and Change* (New York: Palgrave): 127-53.

Frigyesi, V., and L. Kapolyi (2005) 'Szociálpolitikat az Europai Unióban' ('Social Policy in the European Union'), *Közgazdasági Szemle (Journal of Economic Literature)*, March 2005: 289-305.

Fukuyama, F. (2002) 'Social Capital and Development: The Coming Agenda', *SAIS Review* 22.1: 23-37.

Georg, S. (1999) 'The Social Shaping of Household Consumption', *Ecological Economics* 28.3: 455-66.

Global Footprint Network (2006) 'National Footprints', www.footprintnetwork.org/gfn_sub.php?content=national_footprints [accessed 25 August 2007].

Haney, L (1999) ' "But We Are Still Mothers": Gender, the State, and the Construction of Need in Postsocialist Hungary', in M. Burawoy and K. Verdey (eds.), *Uncertain Transition: Ethnographies of Change in the Postsocialist World* (Lanham, MD: Rowman & Littlefield): 151-89.

Jackson, T. (2004) 'Negotiating Sustainable Consumption: A Review of the Consumption Debate and its Policy Implications', *Energy and Environment* 15.6: 1,027-51.

Jacobs, M. (1997) 'The Quality of Life: Social Goods and the Politics of Consumption', in M. Jacobs (ed.), *Greening the Millennium? The New Politics of the Environment* (Oxford, UK: Basil Blackwell).

Kollányi, M. (2004) 'A magyar GDP 1989–2003 között' ('The Hungarian GDP between 1989 and 2003'), *Fejlesztés és Finanszírozás (Development and Financing)* 3: 45-52.

Meroni, A. (ed.) (2007) *Creative Communities in Europe: People Inventing Sustainable Ways of Living* (Milan: Edizioni Poli.design).

NEF (New Economics Foundation) (2004) *Chasing Progress: Beyond Measuring Economic Growth* (London: NEF).

Norwegian Ministry of Environment (1995) *Report on the Oslo Ministerial Roundtable: Conference on Sustainable Production and Consumption* (Oslo: Norwegian Ministry of Environment).

Rudd, M.A. (2000) 'Live Long and Prosper: Collective Action, Social Capital and Social Vision', *Ecological Economics* 34.234: 131-44.

Sanne, C. (2002) 'Willing Consumers, Or Locked In? Policies for a Sustainable Consumption', *Ecological Economics* 42: 273-87.

SEI-T (Estonian Institute for Sustainable Development) (2005) *Household Consumption Trends in New Member States* (background report prepared for the European Environment Agency; Tallinn, Estonia: SEI-Tallinn).

Seyfang, G. (2005) 'Shopping for Sustainability: Can Sustainable Consumption Promote Ecological Citizenship?', *Environmental Politics* 14.2: 290-306.

Soubbotina, T. (2004) *Beyond Economic Growth: An Introduction to Sustainable Development* (Washington, DC: World Bank; www.worldbank.org/depweb/english/beyond/global/beg-en.html).

Spargaaren, G. (2003) 'Sustainable Consumption: A Theoretical and Environmental Policy Perspective', *Society and Natural Resources* 16: 687-701.

Stø, E., H. Throne-Holst, P. Strandbakken and G. Vitterrsø (2006) 'A Multi-dimensional Approach to the Study of Consumption in Modern Societies and the Potentials for Radical Sustainable Changes', in M.M. Andersen and A. Tukker (eds.), *Proceedings of the Workshop on Perspectives on Radical Changes to Sustainable Consumption and Production* (RISØ/TNO; available from www.score-network.org).

Svejnar, J. (2002) 'Transition Economies: Performance and Challenges', *Journal of Economic Perspectives* 16.1: 3-28.

Tellegen, E. (1996) 'Environmental Conflicts in Transforming Economies: Central and Eastern Europe', in P.B. Sloep and A. Blower, (eds.), *Environmental Problems as Conflicts of Interest* (London: Edward Arnold): 67-97.

Vadovics, E. (2007) 'Emerging Creative and Sustainable Solutions in Central Eastern Europe', in A. Meroni (ed.), *Creative Communities in Europe: People Inventing Sustainable Ways of Living* (Milan: Edizioni Poli.design): 151-53.

—— and M. Hayes (2007) 'Open Garden: A Local Organic Producer–Consumer Network in Hungary: Going through Different Levels of System Innovation', in *Proceedings of the Workshop Cases in Sustainable Consumption and Production: Workshop of the Sustainable Consumption Research Exchange (SCORE!) Network*; available from www.score-network.org.

Venetoulis, J., and C. Cobb (2004) *The Genuine Progress Indicator. Measuring the Real State of the Economy* (Oakland, CA: Redefining Progress).

Wagener, H.-J. (2002) 'The Welfare State in Transition Economies and Accession to the EU', *West European Politics* 25.2: 152-74.

Worldwatch Institute (2004) *State of the World 2004: Special Issue on Consumption* (Washington, DC: Worldwatch Institute).

WWF/GFN (Global Footprint Network)/NC-IUCN (Netherlands Committee for International Union for Conservation) (2005) *Europe 2005: The Ecological Footprint* (Brussels: WWF European Policy Office).

——/ZSL (Zoological Society of London)/GFN (Global Footprint Network) (2006) *Living Planet Report 2006* (Gland, Switzerland: WWF).

# Part 5
# System innovation policy perspective

# 18
# Review: system transition processes for realising sustainable consumption and production

*Maj Munch Andersen*
Oe-DTU, Denmark

## 18.1 Introduction

This part on system transitions seeks to highlight systemic aspects of sustainable consumption and production (SCP). That is, to set a focus on eco-innovations that are so radical and systemic in character that they involve complementary changes in production and consumption patterns, often involving considerable institutional change. It is a logical next step in the evolution of the environmental policy agenda, which has shifted from neighbour regulation and end-of-pipe solutions towards process, products, functions and, more recently, systems of consumption and production. With the SCP agenda developing very much internationally, global development issues (rather than a narrow Western European approach) are in focus, and the very different conditions in developed, transition and developing countries must be considered. The term 'system innovation' is by now a well-established reference concept for such radical, systemic change.

Another important policy and research trend related to the SCP agenda is the rising attention to 'eco-innovation' since the 1990s (e.g. see Andersen 1999; den Hond 1996; Fukasako 1999; Fussler and James 1996; Rennings 2000; WBCSD 2000).[1] Analytically, it

---

1 Eco-innovations are defined as 'innovations which are able to attract green rents on the market' (Andersen 1999: 19); that is, the concept is closely related to competitiveness. The focus of eco-innovation research is on the dynamics of innovation (rather than on the environmental effects), seeking

puts emphasis on 'green' competitiveness; policy-wise, it seeks to forward greater synergy between environmental and innovation policy. It has gained momentum in recent years, via the process of the EU Environmental Technologies Action Plan (Andersen 2004a, 2004b, 2004c; European Commission 2003; Foxon 2003, 2004; Foxon *et al.* 2005b) and the Dutch Presidency of the EU in 2004 (Kemp and Andersen 2004; Kemp *et al.* 2004). The EU (2006) is starting to integrate eco-innovation more systematically into innovation policy, and individual countries are developing a stronger innovation approach to environmental policy (e.g. Japan, Sweden, the UK, Finland, Denmark, China and the USA). Compare, for example, the 'Green Growth' strategy of UNESCAP, China's strategy for the 'Circular Economy', Japan's resource-efficiency goals, and the US aim for renewable energy production and efficiency. Conceptually, eco-innovation is grounded in evolutionary economic theory, in which the 'innovation system' approach plays a key role. The innovation concept by definition covers the entire innovation process (from idea generation, prototypes, early production and upscaling, to marketing and value creation on the market) and hence links production and demand. Although demand is only a part of the wider consumption process, eco-innovation research can contribute a much-needed bridge between the consumption and production parts of SCP.

The 'system innovation' approach, generally speaking, relates to environmental policy and the 'innovation system' approach relates to innovation policy; thus the two approaches chosen represent each side of the policy areas involved in the recently emerging tendencies. These approaches hence take centre stage in this chapter, which is structured as follows. Section 18.2 reviews the two groups of theories for understanding systemic change. Section 18.3 looks at related governance models. Section 18.4 discusses possible syntheses between the different approaches and Section 18.5 briefly touches on their limitations. Section 18.6 brings together some conclusions.

## 18.2 State of the art: understanding system behaviour

### 18.2.1 Introduction

Both the 'system innovation' and 'innovation system' approach draw on basic understandings of complex systems theory. This theory focuses on systems as being self-organised and regulated by feedback loops and matches and mismatches of sub-elements in the system (e.g. O'Connor and McDermott 1997). Such general system theories go beyond this chapter. We focus here on literature oriented towards innovation in one way or other. In this, the ambition of the chapter is to shed light on how we fundamentally understand respectively *systems* and *innovation* related to the SCP agenda.

The introduction has already indicated that each approach has rather different roots and applications. **System innovation** research is not yet fully consolidated; it is oriented

to capture the degree to which environmental issues are becoming integrated into the economic process.

at environmental issues and is undertaken by a differentiated research community of, for example, policy analysts (emphasising the policy regime), sociologists (emphasising the socio-institutional framework) and evolutionary (innovation) economists (emphasising innovation processes). The term 'innovation' refers here more broadly to transition processes, where innovation economics uses it strictly in the sense of *novelty*, which leads to value creation on the market and hence connects it to competitiveness. The **innovation system** concept, in contrast, is today a well-established framework for innovation economic analysis at the national and regional level (Archibugi *et al.* 1999; Edquist 2001; Edquist and Hommen 2006; European Commission 2002, 2006; Lundvall 1992, 2005; OECD 1999, 2000, 2002a). This concept and the substantial research lying behind it is only beginning to be applied to environmental analysis (Andersen 1999, 2002, 2004a, 2004b, 2004c; Andersen and Rasmussen 2006; Foxon 2003, 2004; Foxon *et al.* 2004, 2005b; Hübner *et al.* 2000; Kemp and Andersen 2004; Midttun and Koefoed 2005; Rand Europe 2000a, 2000b; Rennings 2000; Saviotti 2005; Schienstock 2005; Segura-Bonilla 1999; Smith 2002; Weber and Hemmelskamp 2005). Therefore, this chapter elaborates how this line of thinking could contribute to SCP research and policy-making.

The above reflects confusion and lack of clarity over how the two concepts of 'system innovations' and 'innovation systems' relate to each other (see also Chapter 19, by Weber *et al.*). The lack of a bridge illustrates the weak connection between innovation research and environmental research as well as between the two related policy areas (Andersen 2004c). This represents a problem to the overarching SCP agenda. This section aims to limit this gap by discussing the two concepts in more detail,

### 18.2.2 The system innovation approach

#### 18.2.2.1 Introduction

The core concern of the system innovation approach, also known as 'transition management', is normative: how to stimulate major system or regime shifts for sustainability. It operates, in other words, at a fairly high level of aggregation (the system level) and is somewhat policy-oriented. The system innovation concept is used in many ways, so it does not represent one theoretical approach, but there are some unifying factors.

Although those working on the core transition research situate the system innovation concept within a mix of evolutionary economic theory and institutional or politico-logical theory, the concept is also used by a diverse group of researchers who have a more technical, design or mixed background. In such research, the social side and system transition dynamics may receive less attention. System innovation may too easily be seen as the best way to achieve SCP, only because, by definition, it entails simultaneous change in technology and infrastructure and in wider institutions. Framed like this, it becomes a planning or policy concept rather than an analytical concept, with clear limitations from an evolutionary perspective.

The first limitation is the claim that system innovations are by definition good and are the only (or main) solutions to environmental problems. But system innovations cannot be said per se to have positive, if any, sustainability impacts. There is, fundamentally, insufficient correlation between type of innovation and level of eco-efficiency. Also, the system innovation concept may be used to imply that a radical 'saltationist' change model is the only way to obtain systemic change. This perspective neglects the evolutionary and cumulative nature of the innovation process, well proven within general innovation research (e.g. see Dosi 1982; Nelson and Winter 1982). Rather, we should pay more attention to the interdependences between different types of innovations (incremental and radical, generic and specialised, systemic and stand-alone) and their various roles in the different stages of the innovation cycle (Perez 2000). Furthermore, at times in this SCP research the system (society) is ill-defined, with references simply to system innovations rather than specifying the consumption and production processes and the actors and institutions involved. This may be harmful when it leads to an overly simplistic representation of the innovation process.

A final limitation is the rather absolute definition of system innovations. The system innovation concept implies that innovations are either systemic or not. But there is no dichotomy between systemic and in-house or stand-alone innovations; rather, there is a continuum. Most innovations are systemic to some degree and call for complementary innovations in various parts of the innovation system to succeed (e.g. in other firms, knowledge institutions and infrastructure and in other planning practices, regulation and consumption practices; Langlois and Robertson 1995; Perez 2000; Teece 1986, 2000).

Such issues reflect more generally a fundamental problem of the SCP agenda: the quest to address very complex problems, such as radical system transitions, while still seeking clear and simple policy solutions, such as a cry for system innovations. A move forward could be to develop more specific definitions of innovations and system transitions. There is a need for careful analyses of the systemic nature of each eco-innovation pursued, so that the necessary specified complementary changes may be addressed.

The following sections present two more mature transition policy frameworks of an evolutionary character for the SCP agenda: transition management (Section 18.2.2.2) and sociotechnical regime transformation (Section 18.2.2.3).

### 18.2.2.2 Transition management for sustainable consumption and production

This group of research is well-defined and mature. The core representatives consist of the Dutch transition management school, in which, for example, Kemp (see Chapter 20) Rotmans, Grin, Elzen, Geels, Schot, Hoogma, Hofman and Green are key figures. The empirical focus is on 'system innovations' rather than SCP as such. Initially, it was developed for large-scale sociotechnical infrastructure systems, such as energy and water supply and transport, but it has been taken up and applied to other system innovations as well. Kemp's chapter in this book contributes important new insights on how to address the SCP agenda from this perspective (Chapter 20).

The Dutch transition management school builds mainly on evolutionary economic theory but draws also on more sociological and politicological studies. The theoretical basis may vary somewhat between different research groups as the framework is looked at by a broader audience. The school seeks to set up a long-term strategic framework to facilitate collective transition processes and regime shifts towards specific system innovations (Hoogma *et al.* 2002; Kemp 1994; Kemp and Rotmans 2005; Kemp *et al.* 1998, 2001; Rotmans *et al.* 2001; Schot *et al.* 1994).

A sub-part of the school focuses on analysing past system innovations, with the ambition to build a transition theory of long-term system innovation (e.g. Elzen *et al.* 2004;

Geels 2002a, 2002b, 2002c, 2004; Schot 1998; Schot *et al.* 1994). These analyses mix sociology and institutional theory while only partly referring to evolutionary innovation research on long-term systemic innovations (the literature on innovation cycles such as Perez 2000). They have developed their own conceptual model of major system transitions on the basis of historical examples, such as a transition from a coal-based to a gas-based energy-supply system (e.g. see Geels 2002a, 2002b, 2002c, 2004). In their emphasis on the role of users in these transition processes as well as on institutional and political factors in system transition processes, they fit well into an SCP agenda. There is generally more emphasis on external drivers than on the innovation processes in the market compared with general innovation research. These analyses aim to feed into transition management recommendations, and the governance part is very similar to that of other transition management work (we will return to this in Section 18.3.2, on policy).

A multi-level model seeks to capture an evolutionary understanding of the system and of how systems (or rather 'sociotechnical regimes', as the system is termed) may change (see Fig. 18.1). The model has been developed not only to analyse but also, primarily, to further long-term system innovations (as we shall see in Section 18.3.2). It consists of:

- The micro level of 'technological niches', denoting protected spaces in which radical innovations can be developed

- The meso level of 'sociotechnical regimes', carried by a variety of societal actors (e.g. companies, public authorities and users associated with a specific technology or functional area; see Fig. 18.2)

- The macro level of 'sociotechnical landscape', or the 'external environment' that encompasses, for example, cultural norms or dominant economic or governance regimes that can be influenced by the actors in the regime under analysis only with great difficulty

The technological niches make up a key element in the understanding of systemic change from this school, where they are seen as necessary confined spaces for radical innovations to evolve and reach commercial maturity. The landscape provides a 'hard

FIGURE 18.1 Niches, regime and landscape

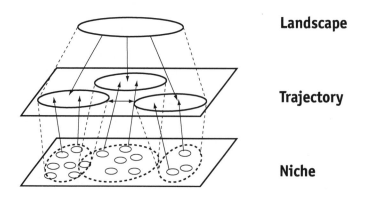

## FIGURE 18.2 Regime or sociotechnical configuration for the example of transport

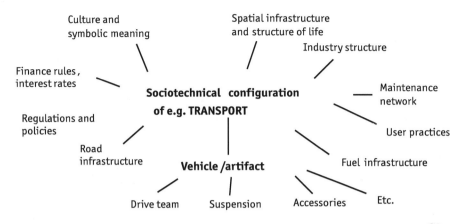

context' for developments in the regime and niches. The degrees of freedom of the development of the regime are limited by the mutual dependence of the different elements (technical practices, user practices, institutions, cultural meaning and so on). This line of work therefore places a strong emphasis on co-evolutionary processes as part of system transitions (i.e. the co-evolution of technology and society) and feedback between subsystems and the system as a whole (e.g. see Kemp and Rotmans 2005; see also Chapter 20). This co-evolution is at a very general (macro or societal) level whereas micro co-evolutionary processes on the market (the coordination of production and learning between competing firms and the co-evolution of science, technology and industrial organisation as part of the economic process) are given limited attention in this discipline compared with in general innovation economic research.

### 18.2.2.3 Sociotechnical regime transformation

Expanding on the Dutch transition model, Smith, Stirling and Berkhout have build what they term a 'quasi-evolutionary model of regime transformation' based on a combination of political science and evolutionary economic theory aiming to come up with policy suggestions (Berkhout et al. 2004; Smith et al. 2005). Criticising more descriptive and structural approaches they seek in their transition model to put greater emphasis on agency. The system they also term social-technical regimes, but they seek to go further in identifying the key functions that contribute to the reproduction of such sociotechnical regimes.

An important contribution of their approach is that they identify two aspects essential for regime transitions: (1) the degree to which the selection pressures on a regime is articulated towards a particular problem or direction and (2) the adaptive capacity of the regime, which depends on the degree to which the resources (factor endowments and capabilities) required for regime transformation is present and coordinated within the regime.

On the basis of this they seek to differentiate system transition processes, identifying four different types of transitions. These depend on the contexts in which these transitions are taking place, which are (a) the level of coordination to purposively transform a regime (high or low) and (b) on the location of resources (internal or external to the regime). The framework thus gives rise to four types of transitions termed:

- Endogenous renewal (high level of coordination, internal to the regime)
- Reorientation of trajectories (low level of coordination, internal to the regime)
- Emergent transformation (low level of coordination, external to the regime)
- Purposive transition (high level of coordination, external to the regime)

This typology, then, highlights the sources of agency for various system transitions, stressing that these vary considerably.

The research builds to some degree on evolutionary economic sectoral innovation system analysis (Jacobsson and Bergek 2004; Jacobsson and Johnson 2000; see also Section 18.2.3), but the focus here is mainly on the macro policy level, as it is in the discussion of articulation and coordination, with the (micro) innovation dynamics (e.g. interfirm coordination) being given little attention. The research is important for trying to differentiate different types of system transitions, making the system innovation concept and regime shifts more operational. However, there are natural limitations to the explanatory power of such a typology given the very complex nature of such major systemic transitions. The more explicit policy recommendations central to the debate will be dealt with in Section 18.3.

## 18.2.3 The innovation system approach

### 18.2.3.1 Introduction

The innovation system approach differs considerably from the approaches hitherto described, in several ways:

- It was not developed for or oriented towards sustainability or SCP specifically but towards innovation more generally
- It is directed towards all kinds of innovation, not only to major systemic innovation
- It represents, first of all, an analytical framework that also, but not primarily, serves as a framework for governance
- It is a more pure evolutionary *economic* approach, with a strong emphasis on the market and competitiveness. Research focuses on any innovation (technical, service, marketing or organisational) that, by definition, leads to value creation in the market (European Commission 2006).

The innovation system framework is nowadays well consolidated, making up the main basis for innovation policy at the international level (in OECD countries and in the EU) and in many countries (European Commission 2002, 2006; OECD 2000, 2001a, 2001b). The innovation system research has evolved for the past ten years based firmly on evolutionary economic theory. By now there are also quite well-defined empirical frameworks and methods for innovation system analysis (Edquist 1997a, 1997b; Edquist and Hommen 1999; Freeman 1995; Lundvall 1985, 1988, 1992, 2005; Nelson 1993). It has been further operationalised by the OECD and the European Commission (European Commission 2002, 2006; OECD 2000, 2001a, 2001b). This resulted in an alternative framework to the mainstream neoclassical economic one to inform economic policy, arguing that competitiveness depends on knowledge and learning rather than on traditional labour productivity (Lundvall 2005). The notion of the 'knowledge-based economy' has now become widely accepted in national and international science, innovation and economic policy and even, hesitantly, among mainstream economists (European Commission 2002, 2006; OECD 2000, 2002a). For the SCP discussion the rising recognition of 'marketing innovation' as an important source of competitiveness is relevant, putting increasing emphasis on branding, values, design and marketing as key competitive factors. This points at the potential of 'green' profiling and branding as a competitive factor (European Commission 2003, 2006; Kemp and Andersen 2004).

The innovation systems framework is primarily applied at the national level (national innovation system [NIS] analysis). The argument is that, despite a globalising economy, learning is still very localised and is a major part of the national institutional setting, noticeably regarding policy but also with respect to culture and to various other institutions (Maskell and Malmberg 1997). Increasingly the framework is also applied to wider regions such as the EU, treating, for example, the EU as an innovation system that can be compared with that of, for example, the USA, Japan or China. Sectoral innovation system (SIS) analyses are emerging as a new research field (Jacobsson and Bergek 2004; Breschi and Malerba 1997; Malerba 2002, 2005). This research tries to link up in-depth analyses of sector-specific innovation patterns with wider national innovation system analyses (see also the functional analysis in Section 18.2.3.3).

Innovation system analysis may be divided into two main strands—the organisational and the functional—as we will discuss below. These draw on the same theoretical basis—that is, they share the same understanding of innovation, but differ in their definition of the (innovation) system.

### 18.2.3.2 Innovation systems: the organisational strand

The organisational strand of innovation system analysis, currently the dominant strand in research and policy, focuses on the key organisations and institutions that influence the innovation process. Core proponents are Freeman (1995), Nelson (1993) and, noticeably, Lundvall (1985, 1988, 1992, 2005). An innovation system is defined as 'those elements and relations, which interact in the production, diffusion and use of new and economic useful knowledge' (Lundvall 1992: 2). The main emphasis is on the market, company innovation and competitiveness in a larger institutional setup. The innovation system consists of three main elements (European Commission 2002; cf. Lundvall 2005):

- The **innovation dynamo**. This consists of key knowledge producers and users: that is, firms and knowledge institutions and the interaction between them
- The **transfer factors**. These are interactions and flows of knowledge, funding and the influence between them

- The **wider institutional setting**. This influences innovation, and, in particular, policy conditions

The concept of 'interactive learning' is central, underlining the key role of the interaction between users and producers for the innovation process (Lundvall 1985, 1988, 1992, 2005). The market is therefore fundamentally seen as being 'organised' into network-like relations rather than being coordinated by anonymous price coordination (Lundvall 1988).

Focus here is, however, primarily on firms and other professional users. Technological communities, informal groups and innovative networks are recognised as important actors in the innovation process (Lundvall 2005; OECD 2001b). Consumers are not considered part of the innovation dynamo, being users with little bargaining power and normally entering (too) late or indirectly in the innovation process. There are exceptions where expert consumers play important roles as 'lead users' ahead of and actively shaping new market trends (e.g. mountain bikers, wind surfers, computer experts and so on; Jeppesen and Molin 2003; von Hippel 1988). Generally, however, in innovation economic research there is a historical neglect of the demand side of innovation (Edquist and Hommen 1999). Also, market assessments related to product innovation tend to be poorly handled by companies (Leonard 1995).

The empirical (comparative) analysis of different innovation systems allows for an understanding of their specific innovation patterns, structural characteristics and development over time. Such regular analyses of given innovation systems are important if we are to understand system dynamics and hence guide innovation policy towards *specific* system failures in different innovation systems. Such studies show that innovation patterns vary widely between different countries (Edquist and Hommen 2006; Nelson 1993).

The concept is used mainly to benchmark the performance of innovation systems (e.g. innovation rates and competitiveness) rather than how they form and transform (Andersen 2006; Lundvall 2005). Such dynamics are seen as a (path-dependent) function of the *co-evolution* of science, technology, organisations (companies) and institutions—in a largely unplanned manner (Edquist 1997b; Freeman and Louçã 2001; Freeman 1995; Lundvall 1988, 1992, 2005; Malerba 2002, 2005; Perez 2000). The changing organisation of production and learning *across* a variety of actors in the global knowledge economy is at the centre of analysis. The focus is on these ongoing co-evolutionary processes rather than on (orchestrated) system transitions, of interest to SCP.

The discussion of major radical systemic change is seldom coupled directly to innovation system analysis but forms a part of the underlying evolutionary economic theory. Systemic transitions are seen as part of business or innovation cycles (i.e. the rise of new technologies or industrial sectors). Major system transitions are termed 'technological revolutions', 'techno-economic paradigm change', 'technological paradigms' and 'technological–natural trajectories' (see Dosi 1982; Freeman and Louçã 2001; Freeman and Perez 1988; Nelson and Winter 1982; Perez 2000, 2002). This research emphasises a more cognitive or paradigmatic explanation of systemic change than the ones we have discussed so far. The basic argument is, inspired by Kuhn 1970, that technology development, parallel to scientific work, follows certain heuristics and is characterised by path dependence.

Innovation cycle theory, supported by empirical analyses, seeks to identify patterns in technology evolution and industry formation. The focus is on gestation times and the role of various actors—for example, incumbents or start-ups, big or small firms, and knowledge institutions—in various stages of the innovation cycle. Two major phases are referred to: the fluid **formative** stage and the **consolidating** or **market expansion** phase, characterised by stabilisation processes around a dominating design (Abernathy and Utterback 1978; Dosi 1982; Freeman and Perez 1988; Perez 2000). The innovation analyses show how radical innovations are always accompanied by a series of incremental innovations, and that departure from existing trajectories may take place at any given time in the technology development process (Perez 2000, 2002). Institution formation is seen as an inherent part of (radical) innovation, including factors such as standard-setting and intellectual property rights (Freeman and Perez 1988; Perez 2000, 2002). Naturally, for the SCP research the formative phase is of key concern; but it is in the inclusion also of the stabilisation phase that production and consumption elements may be linked, which is greatly needed for an understanding of the overall system transition processes in the long run.

Traditionally, major system transitions are tied to the emergence of more horizontal or enabling technologies such as information and communications technology (ICT), biotechnology and the upcoming nanotechnology. The rise and pervasive impact of these on the economy and society are among the most studied objects in evolutionary economic theory (Bresnahan and Trajtenberg 1995; Freeman and Louçã 2001; Helpman 1998); certainly, the role of ICT for the functioning of (national) innovation systems has been given much attention (OECD 2002a).

There is, generally speaking, a strong technology perspective in the understanding of system transitions from this research perspective. This is changing somewhat lately with the broadening of the innovation concept and increasing focus on service innovation, marketing and organisational innovation in innovation economic research.

### 18.2.3.3 Innovation systems: the functional approach

The functional approach in innovation system analysis is more recent and is not as influential. A core proponent is Edquist (2005), who argues that rather than defining the innovation system as constituted by organisations it should be defined by specifying key functions. He suggests ten such functions:

- The provision of research and development (R&D)
- Competence-building
- The formation of new product markets
- The articulation of user needs
- The creation and changing of organisations for the development of new fields of innovation
- Networking around knowledge
- The creation and changing of institutions (e.g. IPR laws, environmental regulation)

- The carrying-out of incubation activities
- The financing of innovation
- The provision of consultancy services for innovation

This approach is being criticised by the organisational approach for lacking consistency: the functions are heterogeneous in character and many other functions could be pointed to. Furthermore, all functions are treated as equally important and are not linked to each other. This neglects what is known about innovation and hence seems a step back in theory formation on innovation system dynamics (Lundvall 2005).

More relevant to the SCP agenda is the functionalist sectoral innovation system analysis, which relates back to technology systems analysis (e.g. see Carlsson and Stankiewicz 1991). This sectoral innovation system analysis seeks to take on a stronger problem-oriented approach to innovation than does the national innovation system school: i.e. it looks at not only how to analyse but also how to promote a specific sector or technology area. A quite well-known analytical framework is presented by Jacobsson and Bergek (2004) and by Jacobsson and Johnson (2000), pointing to five key functions of these sectoral innovation systems:

- The creation and diffusion of 'new' knowledge
- The guidance of the direction of search among users and suppliers of technology
- The supply of resources such as capital and competences
- The creation of positive external economies, both market- and non-market-mediated
- The formation of markets

This line of thinking resembles more the transition research discussed above (Section 18.2.2) in clearly aiming at identifying key factors to be influenced by policy. The emphasis is, however, more strictly on the evolutionary economic, with a focus on micro dynamics (e.g. at the level of firms and knowledge processes) and relating these to innovation cycle dynamics.

For the SCP agenda the framework is interesting because it includes supply and demand aspects. The formation of markets is emphasised but, interestingly, so is the more direct guidance of the direction of search among users and suppliers of technology (unusual within innovation economics). It has a main focus on (dedicated) research funding for directing search, which seems a somewhat narrow perspective that could be further developed. Also, this perspective has a weaker emphasis on agency than the other functional approaches. Nevertheless, in a SCP context the framework could contribute to more work on how to build and transform innovation systems, be they national, regional or sectoral. Not surprisingly, it has been elaborated, in a sustainability context, as a framework for analysing and promoting renewable energy technologies. It is quickly being picked up by environmental researchers (see Foxon 2003, 2004; Foxon *et al.* 2005a; Mackard and Truffer 2006), though so far with a main focus on energy research.

## 18.3 Governance of system transitions to sustainable consumption and production

### 18.3.1 Introduction

Overall, the innovation system approach presents a strong framework for analysing system characteristics, innovation dynamics and system evolution, but with some limitations towards the demand side and particularly the consumer role (for a general critique of the innovation system framework, see Miettinen 2002). The innovation system framework could form a basis for in-depth empirical analysis of the ongoing transformation of national, regional and sectoral innovation systems and the role of eco-innovation in this. It could also contribute to specific studies of major radical eco-innovations. However, so far such studies are very few in number (for descriptive studies of the 'greening' of national innovation systems, see Hübner *et al.* 2000; Rand Europe 2000a, 2000b; for case studies of [radical] eco-innovation in an innovation system framework, see Andersen 2005; Andersen and Rasmussen 2006; Foxon 2003; Foxon *et al.* 2005a; Midttun and Koefoed 2005; Segura-Bonilla 1999; see also Chapter 21, by Mont and Emtairah).

Yet the differences in understanding of systemic change between the approaches discussed also give rise to some important differences in policy recommendation. In this section we will shortly look at their key policy suggestions relevant for furthering a transition to SCP patterns.

### 18.3.2 The system innovation approach and policy

The system innovation approach overall has a strong policy focus, seeking to facilitate regime shifts and major system innovations.

The Dutch transition management school presents a well-developed framework for policy recommendations. Specific collective management strategies are presented to initiate long-term multi-level transition processes. It is interesting because it has been put into policy practice both in The Netherlands and in Belgium. According to Kemp (Chapter 20) the intention is to create an innovative governance context that enables processes of co-evolution: joint problem perception, vision, agenda, instruments, experiments and monitoring through a process of social learning about radical innovations and new system innovations. A four-step model is suggested to establish transition paths for system innovations:

- Define a vision for the long-term future of an industry or sector
- Agree on strategic goals for the medium term
- Set out transition paths for how these might be achieved
- Agree on the steps or 'learning experiments' along these paths

The central policy suggestion is, in short, to identify and further 'experiments'. The key policy suggestion here is the establishment of confined technological niches that may allow the 'breeding' of new technologies. The ambition is to enable an open process that ensures that a variety of options are being pursued for as long as possible. The more historic transition research refers to the same transition management model, but adds in elements for achieving linkages between niches and pays attention to broad breakthroughs resulting from such linkages and to leapfrog dynamics (Elzen *et al.* 2004; Geels 2002a, 2002b, 2002c, 2004; Schot 1998).

The transition management model is impressive since it has shown to be an operational way to capture evolutionary transition processes and a feasible way to direct policy attention to radical (eco-)innovation. It is, however, important to recognise the limitations in such a staged transition management process. A range of actors and processes important for obtaining radical innovation will necessarily be outside the domain of these processes, particularly factors related to globalisation. There is a need for further action beyond positive experiments to achieve consolidation and market expansion. Furthermore, we need greater insight into how the model would work in other settings, such as in larger countries (e.g. China and the USA) where such staged processes are obviously more complicated, as well as developing countries and transition economies facing quite different conditions, especially institutionally.

The sociotechnical regime transformation model shares the basic recommendations of transition management, with the emphasis on creating shared visions and expectations about future innovations. However, rather than pointing to niche formation it includes the idea that governance should seek to sustain or alter different transition contexts that are a function of the availability of resources and to influence how they are coordinated (as discussed in Section 18.2.3.3). A limitation of this approach is that the four transition contexts make up a complicated policy message that is not easily translated into concrete policy measures. The basic message of bringing attention to the availability and coordination of resources as a precondition for obtaining system transitions is very valid, though.

The other important policy message from this group is the emphasis of attention to agency and power in regime transformation. The focus here is on power relations in the institutional setting, whereas power struggles and competition in the market in system transition processes are given little attention.

### 18.3.3 Innovation systems and policy

The innovation system framework differs considerably in its policy messages as it did in its understanding of systemic change to the transition research described above. It is a core principle in innovation as well as research policy not to be too prescriptive. The core goal of innovation policy is to promote competitiveness through high innovative capacity, not to pursue wider societal goals. The intention is to enable free research and development to allow for as much creativity and variety as possible. Science and technology choices made by public authorities ('picking winners') are considered to be inferior to those of the market under competition. This is only beginning to change somewhat in relation to the EU Lissabon process; compare, for example, the greater emphasis placed on eco-innovation aspects (European Commission 2006).

The general principle of free research and development does not mean a *laissez-faire* approach to the market. The innovation system perspective has represented a shift in policy rationale in research and innovation policy from 'simply addressing market failures that lead to underinvestment in R&D towards one which focuses on ensuring the agents and links in the innovation system work effectively as a whole, and removing blockages in the innovation system that hinder the effective networking of its components' (European Commission 2002). In relation, the innovation system concept has supported a shift from linear to interactive thinking of innovation (cf. Lundvall and Borras 2005). By setting up the right framework conditions policy may seed and channel innovation processes. It concerns, for example, access to risk capital, incubation environments, support for collaboration between knowledge institutions and firms or start-ups, knowledge diffusion, securing intellectual property rights, demonstration activities, strengthening the absorptive capacity of users and foresight exercises that 'wire up the national innovation system' by raising new visions and prioritising research and innovation policy (Kaivo-oja et al. 2002; Martin and Johnston 1999). Regulation is also a part of this, but there is no attempt to 'manage' the innovation process directly. The approach is hence fundamentally more 'humble' towards the market than the transition governance approaches discussed earlier.

Innovation policy is generally directed towards small and medium-sized enterprises (SMEs), immature and radical innovations and start-ups, which are considered most subject to system failures (which incumbents and mature technologies are not). Radical innovation is thus in focus, though major system transitions are generally not pursued as such, partly because these may be seen as obstructive to national competitiveness (by being too disruptive) and, partly, as stated, because of an unwillingness to pick the winners. Exceptions include the broad interest for shifts to an ICT society and policy attention on securing a future energy supply, which has created interest in the advancement of renewable technologies as well as more energy-efficient consumption patterns.

Innovation policy tends to be concerned with the supply side rather than the demand side (Georghiou 2006). This has been changing somewhat recently. A recent communication from the European Commission to the Council places emphasis on the creation of 'lead markets' as a core initiative (European Commission 2006). The lead market concept is increasingly seen as an important instrument for furthering eco-innovation at the sectoral level and forms part of the ongoing EU Innova (sectoral innovation) policy project. The discussion on lead markets for sustainability needs, though, to be linked more explicitly to the SCP agenda (on the discussion on lead markets and sustainability, see Beise and Rennings 2003; Jänicke and Jacob 2004). The same applies to the link between innovation policy and eco-innovation and the SCP agenda

Key elements in an innovation system transition model towards SCP could be to wire up the innovation systems for eco-innovation more generally and turn environmental issues into a competitive advantage (e.g. Andersen 2004a, 2004b, 2004c; Foxon 2004; Foxon et al. 2005b). Four key elements would be to:

- Develop well-functioning markets for 'green' products, via 'green' public procurement, lead market initiatives and fiscal incentives to promote 'green' products

- Promote organisational change, noticeable proactive environmental strategies in companies and financial institutes; also *Leitbilder* (visioning) to show the benefits of eco-innovations can play a role (cf. Schienstock 2005)

- Facilitate and make efficient 'green' knowledge production and learning across various actors in national and sectoral innovation systems (this includes a destruction of the existing 'non-green' well-established knowledge and search practices and the establishment of a strong 'green' knowledge infrastructure)
- Promote 'green' entrepreneurship (Andersen 2004a, 2004b)

This approach, in fact, entails a shift in policy attention in current SCP practices from promoting specific system innovations towards a 'greening' of the overall national, regional and sectoral innovation systems, seeking to influence the 'innovation dynamo', the transfer factors and the wider institutional setting, with 'green' competitiveness as a core driver of change. This means that consumers are given somewhat lesser emphasis. A core argument related to SCP is that it is necessary to recognise and address the *general* friction to eco-innovation embedded in current innovation systems to various degrees caused by the widespread but uneven lock-in to non-sustainable production and consumption practices—for example, in the financial sector, in the many companies with a low direct environmental impact and in households (Andersen 1999, 2004a).

Additionally, extraordinary systemic policy efforts are needed, much in line with the transition research thinking, to achieve breakthroughs of *specific* systemic eco-innovations. The sectoral innovation system framework could be used here. This entails establishing a broad package of soft, regulatory and fiscal policy measures that may allow new eco-innovations to go through both the formative and the consolidation stage. Emphasis should be particularly on fostering a variety of novel solutions; support for spin-offs and start-ups is essential.

Overall, the innovation system concept is potentially helpful to the SCP policy area in shedding light on the system dynamics and failures of specific national and sectoral innovation systems that need to be addressed to achieve radical system transitions. This is especially important when considering the global orientation of SCP and hence the need to discuss policies for SCP in very different specific (national or regional) contexts.

More importantly the assumptions regarding innovation and system dynamics developed within this framework could guide policy development for SCP in important ways, leading to a stronger knowledge-based and market-focused approach. The innovation system concept is today much used but also abused by not taking the core assumptions seriously (e.g. in the edited book of Weber and Hemmelskamp, *Towards Environmental Innovation Systems* [2005] many of the contributions fail to build on the theoretical foundations of innovation systems theory, though referring to the concept). If properly used the system approach could present a framework for promoting SCP policies targeted at different innovation systems around the globe.

## 18.4 Synthesis: models of change

The discussion of the different approaches to system transition for SCP reveals that there is currently no great synthesis emerging. There are quite substantial differences both in the understanding of systemic change and in policy suggestions to further these despite the fact that evolutionary economic theory feeds into many of the theories in various ways. Three major differences will be pointed to: the role of markets, the emphasis on analysis rather than policy, and policy measures.

The first, and perhaps most important, is the representation and role of *market*: that is, whether change is sought via the market or by staging processes outside the market. There are strong ideological elements in this difference but also analytical ones, depending on the perception of the market, which differs greatly between neoclassical (and hence much resource-based and environmental economics) and evolutionary or institutional economic thinking. The neoclassical market perception has for a long time had a strong influence on environmental policy-making and research, leading to a significant emphasis on external drivers as a means of 'greening' the economy (since, by definition, rationality and competitiveness cannot change).

The system innovation approaches suggest, to various degrees, the need to stage transition processes outside the market, whereas the innovation system approach suggests the need for transitions achieved by moulding the market. The strong focus on the market in this approach may to many SCP researchers and policy-makers appear cynical, conservative or even naive. But the argument here is that to neglect the power of market forces is naïve rather than helpful in achieving system transitions. Any company that wishes to pursue sustainability strategies needs simultaneously to pursue competitive goals to stay in business. Some SCP research and policy initiatives currently tend to neglect the incentive structures of business and innovation characteristics in innovation systems, though these are central in order to redirect the system . Emphasis is often placed on isolated consumption behaviour rather than on seeing such behaviour as part of the (global) economy. This naturally makes it difficult to address major system transitions.

The emerging eco-innovation trend in research and policy may potentially place a stronger focus on the market, innovation dynamics and competitiveness when or if it is applied to the SCP agenda—particularly if it is being utilised in innovation system analysis to address the more systemic and overarching agendas so central to the SCP agenda. This is especially important in order to address the interconnectedness of consumption and production. The upcoming 'sectoral innovation system' analyses also represent possibly new opportunities for aligning competitiveness with the SCP agenda. This framework links the overall innovation dynamics of the globalising knowledge economy with the eco-innovation processes at the specific technological, sectoral or functional level.

There are as yet only very few attempts to reconcile these different views on the market with SCP. Some are present in this part of the book. For example, in Chapter 19 Weber *et al.* suggest 'adaptive foresight' methods be used to expand the transition management model with sectoral innovation system and foresight analyses; this approach places greater emphasis on agency and micro-innovation dynamics and adaptive policy processes. And in Chapter 21 Mont and Emtairah attempt to combine the innovation system framework with product-service system (PSS) analysis and consumption policy in a novel attempt to link innovation systems directly to consumption research and the SCP agenda.

The second fundamental difference to be pointed to lies in the emphasis on policy rather than analysis. Most approaches have transition governance for SCP as a starting point in building their model for change. This easily leads to a less rigorous analytical framework or a less thorough empirical analysis of dynamics, trends and barriers to the 'greening' of production and consumption in specific contexts. This hampers the possibility for targeted policy action to remove these barriers, and SCP policy may be undertaken blindly, as mere trial and error.

In the transition management approach (e.g. Kemp, Chapter 20) and the sociotechnical regime approach (of Berkhout *et al.* 2004) analyse the transition context more rigorously and are helpful here in pointing to specified models of change. It is a question, though, whether these analyses capture the specific innovation dynamics and trends in given contexts (i.e. in regional, national and sectoral innovation systems). The innovation system change approach stresses that we need to know these before we can change them efficiently.

The innovation system perspective offers deep insights into the micro processes of systemic transitions. However, this perspective, although applying a system perspective, still has a very strong firm and market focus in the analysis as well as a tendency to rest explanations on technology. This may be insufficient for some types of major system transitions, particularly in connection with infrastructural and wider institutional and planning changes. Here, the more policy and institutional approaches are more suitable. Generally, the role of the consumer, but not the professional user, is neglected. Changes in consumption patterns by consumers and changes in user behaviour rather than transaction behavior also lie outside this field.

The third important difference in the approaches mentioned lies in the policy measures. Noticeably, the emphasis is on a planning–dialogue approach (negotiations, visions, experiments, regulations), as represented in particular by the transition research approaches, as opposed to a knowledge–learning approach, as represented mainly by the innovation system approach (key elements are seen to be the need to build up a 'green' knowledge base in the innovation system and to secure efficient knowledge production and learning across various actors in national as well as sectoral innovation systems). There is a difference in the emphasis, respectively, in forcing through system transitions and the more long-term (and possibly slower) creating incentives for systemic change. These differences relate back to the various understandings of the market and innovation and the belief in changes via *or* outside the market. This also refers to assumptions on sustainability: whether eco-innovation and SCP are or could be primarily promoted by external requirements (i.e. primarily through regulation, which is the traditional and dominating perspective in environmental research and policy-making) or whether eco-innovation and SCP can be promoted endogenously (i.e. via the market, as presented by the emerging eco-innovation trend, which seeks to integrate environmental issues in the economic process).

Together, these three elements mean that the models of change in the system transition approaches discussed are in fact quite different. There is currently no grand theory for system transition processes towards SCP. At the same time, there are some signs of increasing synergies represented both by SCP policies and eco-innovation policies, on the one hand, and the new market-led trends in innovation policy, on the other. Against this background, Georghiou (2006) made an interesting analysis of and contribution to the novel demand-side trend in innovation policy, a discussion of much relevance to SCP policies. This new trend in innovation policy provides new opportunities for the SCP agenda. Georghiou's recommendations manage to pull in important aspects both from the transition research and from the innovation system framework and seems, with its focus on supply–demand coordination, to be a promising avenue for future SCP policy work. We still need, though, to see more of the actual implementation of these new ideas.

## 18.5 Limitations of system transition theories towards realising sustainable consumption and production

The system transition theories discussed in this chapter contribute in important ways to the SCP agenda in setting the focus on radical and systemic change processes. However, much (most) of the research still needs to address the SCP agenda more specifically. None of these approaches manages to link up changes in production and consumption sufficiently, despite the claimed systemic perspective; in particular, consumption aspects are neglected. We need more insight into the linkage between changes in production and changes in consumption patterns in systemic transition processes, both when it comes to understanding the dynamics and when it comes to working out policy measures towards these dynamics.

The reasons for this are that, analytically as well as policy-wise, consumption and production are traditionally treated as separate domains. No single research field covers the SCP theme satisfactory. And perhaps it is asking too much. The SCP agenda is very broad whereas research needs to specialise. Efforts to close the gap between consumption and production is currently not helped by the fact that production tends to be addressed mainly by economists whereas consumption tends to be addressed by sociologists and political scientists. However, innovation research—economic *and* sociological—does tell us that consumption and production are central and inherently interdependent parts of innovation dynamics. The evolutionary economic thinking that lies behind the transition management, regime transition and innovation system approaches discussed here, to varying degrees, forwards a much more interactive and co-evolutionary perspective on the innovation process and hence on the relation between consumption and production. These approaches all need, however, to address the SCP agenda in more depth both theoretically and empirically, before a framework for system transitions for SCP may be developed. Newer research on user-driven innovation, on the one hand, and firms as central market makers and shapers, on the other, are promising avenues towards understanding the very complex consumption–production dynamics and how these change over time and in different contexts (Lundvall 2005).

However, the complexity of consumption–production dynamics should be taken very seriously if SCP research is to be more rigorous. Consequently, we need to break down the SCP agenda into specific sub-problems, transition processes and contexts, each characterised by different kinds of processes and actors. The current transition theories discussed here have not managed to do so far. This is not surprising, since they have not been developed to address the SCP agenda specifically; rather, their aim was to address more the general sustainability or innovation agenda. Worse, the oversimplified innovation representation related to the system innovation concept that exists in part of the SCP systems transitions research may do more harm than good to the SCP agenda.

The upcoming eco-innovation research agenda could contribute in important ways to the SCP area, mainly in bringing in stronger insights from general innovation economic research on innovation dynamics and hence a focus on supply–demand coordination, which has hitherto not been given the attention it needs within SCP research. This could inform us more on system transition processes and how consumption and production are interlinked, particularly concerning issues of market development and (global) competitiveness. There are, however, still major limitations to this approach towards addressing changes in consumption patterns, though there are promising signs of greater attention being given to the demand side. Sociological consumption research is needed here, which is strong on consumption patterns but weak on linking this up to production patterns and firm strategising; politicological research is strong on changes in policy regimes and utility systems but weak on consumption and production processes.

SCP research needs to include all three areas to address system transition processes. There is, in other words, a need to pull in more theoretical approaches to address the full SCP theme. This does not necessarily call for more transdisciplinary research as this tends to lead to less rigour on the theoretical side. But there is a need for these different strands to undertake more empirical studies targeted at SCP issues, which might lead to fruitful theoretical developments within the different fields as well as within the specific SCP area.

## 18.6 Conclusions

We are still far from an overall framework for system transition towards SCP but there are promising avenues in some of the newer research areas as well as in policy tendencies. A main conclusion from the review of system transition approaches towards SCP given in this chapter is a need for a more differentiated approach to SCP system transitions rather than aiming for one, unified, grand theory. System transitions are complex and many-faceted, making it necessary to break down the SCP agenda into specific problems and to situate these in specific contexts. Much of the system transition research has not so far been particularly oriented towards the specific SCP agenda. There is hence work to be done in developing approaches to SCP.

A key role for the system transition research towards SCP, as opposed to the other approaches presented in this book, is to contribute to a better understanding of supply–demand coordination—which ought to make up the very core of SCP research and policy-making. It remains, however, a fundamental problem that consumption and production are poorly linked, and this is the case both analytically and policy-wise. Analytically, they belong to separate research domains, with sociology being strong on consumption, innovation economic research being strong on production and politicological research being strong on policy analysis and in part utility systems relevant for both consumption and production.

Policy-wise, the SCP agenda touches on a number of separate policy areas (environment, science and innovation, urban planning, transport and so on), meaning that policy efforts are fundamentally fragmentised and build on very different rationales (Andersen 2004c). There are, however, promising signs of a more holistic and demand-side approach in innovation policy-making which may open up new exciting opportunities for supply–demand coordinated policy action, a trend the SCP area should seek to exploit.

For the SCP research and policy-making to do so, much more attention needs to be placed on the firm as a central actor in the innovation process and on competitiveness—elements that currently are not at the core of the SCP agenda. It is therefore a question whether such a bridge between general innovation policy and SCP will be built. The newer trends in sustainability research and policy-making emphasising eco-innovation and 'green' market development (perhaps more than SCP as such) are very interesting. The recent trends in policy are, overall, more promising (more advanced) than the trends in research in seeking to build systemic policies for the demand side, including for eco-innovation. This trend is currently, however, stronger at the EU level than at the national level. Possibly, this could lead to the EU becoming a major actor for SCP in the future.

If synergies could be created between these trends the SCP agenda could possible gain wider acceptance and momentum, analytically as well as policy-wise. Embracing science and innovation research and policy-making could be an important step to allow the SCP agenda to move beyond minor experiments to become mainstream and an integrated element of economic and wider social processes and policy areas. But this is an agenda that raises important challenges to SCP research and policy as hitherto practised.

Overall, the review has shown that how we understand systems and innovation matters considerably for how we seek to transform these. Greater care should be made in clarifying the underlying assumptions of SCP to allow theory to provide better guidance for SCP policy-making in the future.

## References

Abernathy, W.J., and J.M. Utterback (1978) 'Patterns of Industrial Innovation', *Technology Review* 80.7: 40-47.
Andersen, E.S., and B.-Å. Lundvall (1988) 'Small National Innovation Systems Facing Technological Revolutions: An Analytical Framework', in C. Freeman and B.-Å. Lundvall (eds.), *Small Countries Facing the Technological Revolution* (London: Frances Pinter): 9-36.
Andersen, M.M. (1999) *Trajectory Change through Interorganisational Learning: On the Economic Organisation of the Greening of Industry* (PhD thesis; PhD Series 8.99; Copenhagen: Copenhagen Business School).
—— (2002) 'Organising Interfirm Learning: As the Market Begins to Turn Green', in T.J.N.M. de Bruijn and A. Tukker (eds.), *Partnership and Leadership: Building Alliances for a Sustainable Future* (Dordrecht, Netherlands: Kluwer Academic): 103-19.
—— (2004a) 'Innovation System Dynamics and Sustainable Development: Challenges for Policy', paper presented at the *Innovation, Sustainability and Policy Conference*, Kloster Seeon, Germany, 23-25 May 2004.
—— (2004b) 'Partnership for Green Competitiveness: An Innovation System Approach', paper presented at the *12th International Conference of Greening of Industry Network*, Hong Kong, 7-10 November 2004.
—— (2004c) 'An Innovation System Approach to Eco-innovation: Aligning Policy Rationales', paper presented at *The Greening of Policies: Interlinkages and Policy Integration Conference*, Berlin, Germany, 3-4 December 2004.
—— (2005) 'Path Creation in the Making: The Case of Nanotechnology', paper presented at the *DRUID Conference*, Copenhagen, Denmark, 27-29 June 2005.

—— (2006) 'Boundary Spanning, Knowledge Dynamics and Emerging Innovation Systems: Early Lessons from Nanotechnology', paper presented at the *DIME Workshop*, Lund, Sweden, 26–27 April 2006.

—— and B. Rasmussen (2006) *Nanotechnology Development: Environmental Opportunities and Risks* (Report EN-1550; Roskilde: RISØ).

Archibugi, D., J. Howells and J. Mitchie (eds.) (1999) *Innovation Policy in a Global Economy* (Cambridge, UK: Cambridge University Press).

Beise, M., and K. Rennings (2003) *Lead Markets of Environmental Innovations: A Framework for Innovation and Environmental Economics* (DP-03-01; Mannheim, Germany: ZEW).

Berkhout, F., and J. Hertin (2001) *Impacts of Information and Communications Technologies on Environmental Sustainability: Speculations and Evidence* (Report to the OECD: Brighton, UK: University of Sussex).

——, A. Smith and A. Stirling (2004) 'Sociotechnical Regimes and Transition Contexts', in B. Elzen, F.W. Geels and K. Green (eds.), *System Innovation and the Transition to Sustainability: Theory, Evidence and Policy* (Cheltenham, UK: Edward Elgar): 48-75.

Breschi, S., and F. Malerba (1997) 'Sectoral Innovation Systems', in C. Edquist (ed.), *Systems of Innovation: Technologies, Institutions and Organisations* (London: Frances Pinter): 130-56.

Bresnahan, T.F., and M. Trajtenberg (1995) 'General Purpose Technologies: Engines of Growth', *Journal of Econometrics* 65: 83-108.

Carlsson, B., and R. Stankiewicz (1991) 'On the Nature, Function and Composition of Technological Systems', *Evolutionary Economics* 1: 93-118.

David, P.A. (1985) 'Clio and the Economics of Qwerty', *American Economic Review* 75.2: 332-37.

Den Hond, F. (1996) *In Search of a Useful Theory of Environmental Strategy: A Case Study on the Recycling of End-of-life Vehicles from the Capabilities Perspective* (PhD thesis; Amsterdam: Vrije Universiteit).

Dosi, G. (1982) 'Technological Paradigms and Technological Trajectories: A Suggested Interpretation of the Determinants and Directions of Technological Change', *Research Policy* 11 :147-62.

——, C. Freeman, R. Nelson, G. Silverberg and L. Soete (eds.) (1988) *Technical Change and Economic Theory* (London: Frances Pinter).

Ecotec (2002) *Analysis of the EU Eco industries, their Employment and Export Potential* (report for DG Environment, European Commission; Birmingham, UK: Ecotec Research & Consulting).

Edquist, C. (1997a) 'Systems of Innovation Approaches: Their Emergence and Characteristics', in idem (ed.), *Systems of Innovation: Technologies, Institutions and Organisations* (London: Frances Pinter/Cassell Academic): 1-35.

—— (ed.) (1997b) *Systems of Innovation: Technologies, Institutions and Organisations* (London: Frances Pinter/Cassell Academic).

—— (2001) 'Innovation Policy: A Systemic Approach', in D. Archibugi and B. Lundvall (eds.), *The Globalising Learning Economy* (Oxford, UK: Oxford University Press): 219-38.

—— (2005) 'Systems of Innovation: Perspectives and Challenges', in J. Fagerberg, D. Mowery and R.R. Nelson (eds.), *The Oxford Handbook of Innovation* (Oxford, UK: Oxford University Press): 15-31.

—— and L. Hommen (1999) 'Systems of Innovation: Theory and Policy for the Demand Side', *Technology in Society* 21.1: 63-79.

—— and L. Hommen (eds.) (2006) *Small Economy Innovation Systems: Comparing Globalisation, Change and Policy in Asia and Europe* (Cheltenham, UK: Edward Elgar).

Elzen, B., F.W. Geels and K. Green (eds.) (2004) *System Innovation and the Transition to Sustainability: Theory, Evidence and Policy* (Cheltenham, UK: Edward Elgar).

European Commission (2002) *Report on Research and Development* (EPC/ECFIN/01/777-EN, final; Brussels: EC Economic Policy Committee Working group on R&D, January 2002).

—— (2003) 'Developing an Action Plan for Environmental Technology', europa.eu.int/comm/environment/etap [accessed 25 August 2007].

—— (2006) *Putting Knowledge into Practice: A Broad-Based Innovation Strategy for the EU* (COM 2006-502, final; Brussels: European Commission).

Foxon, T.J. (2003) *Inducing Innovation for a Low-Carbon Future: Drivers, Barriers and Policies* (London: The Carbon Trust).

—— (2004) 'Innovation Systems as the Setting for the Co-evolution of Technologies and Institutions', in *Proceedings of 'Organisations, Innovation and Complexity: New Perspectives on the Knowledge Economy' Conference*, Manchester, UK, 9–10 September 2004.

——, R. Gross, A. Chase, J. Howes, A. Arnall and D. Anderson (2005a) 'The UK Innovation Systems for New and Renewable Energy Technologies', *Energy Policy* 33.16: 2,123-37.

——, P. Pearson, Z. Makuch, and M. Mata (2005b) *Transforming Policy Processes to Promote Sustainable Innovation: Some Guiding Principles* (report for policy-makers; London: Imperial College, March 2005).

Freeman, C. (1988) 'Japan: A New National System of Innovation?', in G. Dosi, C. Freeman, R. Nelson, G. Silverberg and L. Soete (eds.), *Technical Change and Economic Theory* (London: Frances Pinter): 330-48.

—— (1992) *The Economics of Hope* (London: Frances Pinter).

—— (1995) 'The "National System of Innovation" in Historical Perspective', *Cambridge Journal of Economics* 19: 5-24.

—— and F. Louçã (2001) *As Time Goes By: From the Industrial Revolutions to the Information Revolution* (Oxford, UK: Oxford University Press).

—— and C. Perez (1988) 'Structural Crisis of Adjustment, Business Cycles and Investment Behaviour', in G. Dosi, C. Freeman, R. Nelson, G. Silverberg and L. Soete (eds.), *Technical Change and Economic Theory* (London: Frances Pinter): 38-66.

Fukasaku, Y. (1999) 'Stimulating Environmental Innovation', in OECD (ed.), *STI Review 25: Special issue on Sustainable Development* (Paris: OECD): 47-64.

Fussler, C., and P. James (1996) *Driving Eco-innovation* (London: Pitman).

Geels, F.W. (2002a) 'Technological Transitions as Evolutionary Reconfiguration Processes: A Multi-level Perspective and a Case Study', *Research Policy* 31: 1,257-74.

—— (2002b) *Understanding the Dynamics of Technological Transitions* (PhD thesis; Enschede, Netherlands: University of Twente).

—— (2002c) 'Towards Sociotechnical Scenarios and Reflexive Anticipation: Using Patterns and Regularities in Technology Dynamics', in R. Williams and K.H. Sorensen (eds.), *Shaping Technology, Guiding Policy: Concepts, Spaces and Tools* (Cheltenham, UK: Edward Elgar): 355-81.

—— (2004) 'From Sectoral Systems of Innovation to Sociotechnical Systems: Insights about Dynamics and Change from Sociology and Institutional Theory', *Research Policy* 33: 897-920.

Georghiou, L. (2006) *Effective Innovation Policies for Europe: The Missing Demand Side* (Strategy paper for the Economic Council of Finland during Finland's EU Presidency; Helsinki: Prime Minister's Office, 20 September 2006).

—— and M. Keenan (2006) 'Evaluation of National Foresight Activities: Assessing Rationale, Process and Impact', *Technological Forecasting and Social Change* 73: 761-77.

Green, K., and P. Vergragt, (2002) 'Towards Sustainable Households: A Methodology for Developing Sustainable Technological and Social Innovations', *Futures* 34: 381-400.

Helpman, E. (1998) *General Purpose Technologies and Economic Growth* (Cambridge, MA: MIT Press).

Hoogma, R., R. Kemp, J. Schot and B. Truffer, (2002) *Experimenting for Sustainable Transport: The Approach of Strategic Niche Management* (London/New York: E&FN Spon).

Hübner, K., J. Nill and C. Rickert (2000) *Greening of the Innovation System? Opportunities and Obstacles for a Path Change towards Sustainability: The Case of Germany* (WP-47/00; Berlin: Institute for Ecological Economy Research).

Jacobsson, S., and A. Bergek (2004) 'Transforming the Energy Sector: The Evolution of Technological Systems in Renewable Energy Technology', *Industrial and Corporate Change* 13.5: 815-49.

—— and A. Johnson (2000) 'The Diffusion of Renewable Energy Technology: An Analytical Framework and Key Issues for Research', *Energy Policy* 28.9: 625-40.

Jänicke, M., and K. Jacob (2004) 'Lead Markets for Environmental Innovations: A New Role for the Nation State' *Global Environmental Politics* 4.1: 29-46.

Jeppesen, B., and M. Molin (2003) 'Consumers as Co-developers: Learning and Innovation Outside the Firm', *Technology Analysis and Strategic Management* 15.3: 363-83.

Kemp, R. (1994) 'Technology and the Transition to Environmental Sustainability: The Problem of Technological Regime Shifts', *Futures* 26.10: 1,023-46.
—— (2002) *Synthesis Report of 1st Blueprint Workshop on 'Environmental Innovation Systems'* (Brussels: Blueprint Network).
—— and M.M. Andersen (2004) 'Strategies for Eco-efficiency Innovation', strategy paper for the *EU Informal Environmental Council Meeting*, Maastricht, Netherlands, 16–18 July 2004 (The Hague: VROM).
—— and J. Rotmans (2005) 'The Management of the Co-evolution of Technical, Environmental and Social Systems', in M. Weber and J. Hemmelskamp (eds.), *Towards Environmental Innovation Systems* (Berlin: Springer): 33-56.
——, J. Schot and R. Hoogma (1998) 'Regime Shifts to Sustainability through Processes of Niche Formation: The Approach of Strategic Niche Management', *Technology Analysis and Strategic Management* 10: 175-96.
——, A. Rip and J. Schot (2001) 'Constructing Transition Paths through the Management of Niches', in R. Garud and P. Karnoe (eds.), *Path Dependence and Creation* (Mahwah, NJ: Lawrence Erlbaum Associates): 269-99.
——, M.M. Andersen and M. Butter (2004) 'Background Report about Strategies for Eco-innovation', background report for the *EU Informal Environmental Council Meeting*, Maastricht, Netherlands, 16–18 July 2004 (The Hague: VROM).
Kaivo-oja, J., J. Marttinen and J. Varelius (2002) 'Basic Conceptions and Visions of the Regional Foresight System in Finland', *Foresight* 4.6: 34-45.
Kuhn, T. (1970) *The Structure of Scientific Revolutions* (Chicago: University of Chicago Press, 2nd edn).
Langlois, R.N., and P.L. Robertson (1995) *Firms, Markets and Economic Change: A Dynamic Theory of Business Institutions* (London: Routledge).
Leonard, D. (1995) *Wellsprings of Knowledge* (Boston, MA: Harvard Business School Press).
Lundvall, B.Å. (1985) *Product Innovation and User–Producer Interaction* (Ålborg, Denmark: Ålborg Universitetsforlag).
—— (1988) 'Innovation as an Interactive Process: from User–Producer Interaction to the National System of Innovation', in G. Dosi, C. Freeman, R. Nelson, G. Silverberg and L. Soete (eds.), *Technical Change and Economic Theory* (London: Frances Pinter).
—— (ed.) (1992) *National Systems of Innovation* (London: Frances Pinter).
—— (2005) 'National Innovation Systems: Analytical Concept and Development Tool', paper presented at the *DRUID Conference*, Copenhagen, Denmark, 27–29 June 2005.
—— and S. Borras (2005) 'Science, Technology, Innovation and Knowledge Policy', in J. Fagerberg, D. Mowery and R.R. Nelson (eds.), *The Oxford Handbook of Innovation* (Oxford, UK: Oxford University Press): 599-631.
—— and B. Johnson (1994) 'The Learning Economy', *Journal of Industry Studies* 1.2: 23-42.
——, B. Johnson, E.S. Andersen and B. Dalum (2002) 'National Systems of Production, Innovation and Competence Building', *Research Policy* 31.2: 213-31.
Mackard, J., and B. Truffer (2006) 'Innovation Processes in Large Technical Systems', *Research Policy* 35: 609-25.
Malerba, F. (2002) 'Sectoral Systems of Innovation and Production', *Research Policy* 31.2: 247-64.
—— (2005) 'Sectoral Systems of Innovation: A Framework for Linking Innovation to the Knowledge Base, Structure and Dynamics of Sectors', *Economics of Innovation and New Technology* 14.1–2: 63-82.
Martin, B.R., and R. Johnston (1999) 'Technology Foresight for Wiring Up the National Innovation System', *Technological Forecasting and Social Change* 60: 37-54.
Maskell, P., and A. Malmberg (1997) 'Towards an Explanation of Regional Specialisation and Industry Agglomeration', *European Planning Studies* 5.1: 25-41.
Midttun, A., and A.L. Koefoed (2005) 'Green Innovation in Nordic Energy Industry: Systemic Contexts and Dynamic Trajectories', in M. Weber and J. Hemmelskamp (eds.), *Towards Environmental Innovation Systems* (Berlin: Springer Verlag): 115-36.
Miettinen, R. (2002) *National Innovation System: Scientific Concept or Political Rhetoric?* (Helsinki: Sitra Publications).

Muchie, M., P. Gammeltoft and B.-Å. Lundvall (eds.) (2003) *Putting Africa First: The Making of African Innovation Systems* (Ålborg, Denmark: Ålborg University Press).

Nelson, R. (1993) *National Systems of Innovation: A Comparative Analysis* (Oxford, UK: Oxford University Press).

—— and S. Winter (1982) *An Evolutionary Theory of Economic Change* (Cambridge, MA: Harvard University Press).

O'Connor, J., and I. McDermott (1997) *The Art of Systems Thinking: Essential Skills for Creativity and Problem-solving* (London: Thorsons/HarperCollins).

OECD (Organisation for Economic Cooperation and Development) (1999) *Managing National Innovation Systems* (Paris: OECD).

—— (2000) *Knowledge Management in the Learning Society* (Paris: OECD).

—— (2001a) *Innovative Clusters: Drivers of National Innovation Systems* (Paris: OECD).

—— (2001b) *Innovative Networks: Cooperation in National Innovation Systems* (Paris: OECD).

—— (2001c) *Cities and Regions in the New Learning Economy* (Paris: OECD).

—— (2002a) *Dynamising National Innovation Systems* (Paris: OECD).

—— (2002b) *Policies to Promote Sustainable Consumption: An Overview* (policy case-studies series; Paris: OECD).

Perez, C. (2000) 'Technological Revolutions, Paradigm Shifts and Socio-institutional Change', in E. Reinert (ed.), *Evolutionary Economics and Income Equality* (Aldershot, UK: Edward Elgar).

—— (2002) *Technological Revolutions and Financial Capital* (Cheltenham, UK: Edward Elgar).

Rand Europe (2000a) *Stimulating Industrial Innovation for Sustainability: An International Analysis* (report for the Dutch Ministry of Housing, Spatial Planning and the Environment; Leiden, Netherlands: Rand Europe).

—— (2000b) *Stimulating Industrial Innovation for Sustainability: An International Analysis* (nine country reports; Leiden, Netherlands: Rand Europe).

Rennings, K. (2000) Redefining Innovation: Eco-innovation Research and the Contribution from Ecological Economics', *Ecological Economics* 32: 319-22.

Rotmans, J., R. Kemp and M. van Asselt (2001) 'More Evolution than Revolution: Transition Management in Public Policy', *Foresight* 3: 15-31.

Saviotti, P. (2005) 'On the Co-evolution of Technologies and Institutions', in M. Weber and J. Hemmelskamp (eds.), *Towards Environmental Innovation Systems* (Berlin: Springer): 9-32.

Schienstock, G.(2005) 'Sustainable Development and the Regional Dimension of the Innovation System', in M. Weber and J. Hemmelskamp (eds.), *Towards Environmental Innovation Systems* (Berlin: Springer): 97-114.

Schot, J. (1998) 'The Usefulness of Evolutionary Models for Explaining Innovation: The Case of the Netherlands in the 19th Century', *History and Technology* 14: 173-200.

——, R. Hoogma and B. Elzen (1994) 'Strategies for Shifting Technological Systems: The Case of the Automobile System', *Futures* 26.10: 1,060-76.

Segura-Bonilla, O. (1999) *Sustainable Systems of Innovation: The Forest Sector in Central America* (SUDESCA Research Paper 24; PhD thesis; Ålborg, Denmark: Department of Business Studies, Ålborg University).

Smith, A., A. Stirling and F. Berkhout (2005) 'The Governance of Sustainable Sociotechnical Transitions', *Research Policy* 34: 1,491-1,510.

Smith, K. (2002) 'Environmental Innovation in a Systems Framework', paper presented at the *1st Blueprint Workshop 'Environmental Innovation Systems'*, Brussels, Belgium, January 2002.

Teece, D. (1986) 'Profiting from Technological Innovation: Implications for Integration, Collaboration, Licensing and Public Policy', *Research Policy* 15: 27-44.

—— (2000) 'Strategies for Managing Knowledge Assets: The Role of Firm Structure and Industrial Context', *Long Range Planning* 33: 35-45.

Van der Meulen, B. (1999) 'The Impact of Foresight on Environmental Science and Technology Policy in the Netherlands', *Futures* 31.1: 7-23.

Von Hippel, E. (1988) *The Sources of Innovation* (New York: Oxford University Press).

WBCSD (World Business Council for Sustainable Development) (2000) *Eco-Efficiency: Creating More Value with Less Impact* (Conches/Geneva: WBCSD).

Weber, M., and J. Hemmelskamp (eds.) (2005) *Towards Environmental Innovation Systems* (Berlin: Springer).

# 19
# System innovations in innovation systems
Conceptual foundations and experiences with Adaptive Foresight in Austria

*K. Matthias Weber and Klaus Kubeczko*
ARC systems research, Austria

*Harald Rohracher*
IFZ–Inter-University Research Centre, Austria

## 19.1 Introduction

The need for system innovations towards sustainability has frequently been underlined in policy as well as in scientific debates. Different approaches and inroads have been suggested for how to guide long-term transformation processes of sociotechnical systems in line with the principles of sustainable development. In this chapter we aim to draw together some of our recent experiences with attempts to capture such processes conceptually and to tackle them by means of policy, with the emphasis being put on the role of research, technology and innovation (RTI) policy.

At the basis of our argument lies the assumption that radical changes (such as a transition to sustainable production systems) cannot be reduced to isolated technical innovations but require innovation at a systemic level—that is, the interplay of changes in technology, actor constellations and institutions.

Adapting earlier definitions of environmental system innovations (Butter 2002; Rennings *et al.* 2003), one can define system innovations for sustainability as a set of innovations that provides a service in a novel way or offers new services, involving a new logic (guiding principle) and new types of practice and giving rise to a step change in several of the dimensions of sustainability.

System innovations tend to imply changes both at the level of the components and at the level of the architecture of a technology (Henderson and Clark 1990) and equally at the level of social and institutional arrangements, such as mechanisms of coordination (regulation, governance) or patterns of interaction at the supplier and the user side of innovation. Time also matters for the realisation of system innovations. They tend to be realised only at a slow pace, especially when pre-existing technologies, structures and institutions need to be changed. On the one hand, this implies that system innovations should be initiated early on to become effective in time and in order to avoid painful and fast adaptations in moments of crisis. On the other hand, they tend to be initiated only once the improvement potential of less complex innovation options has been exhausted.

With these considerations in mind, the concept of transition management (and other similar concepts) has been introduced (Rotmans *et al.* 2001). In essence, it deals with collective strategies to initiate smooth and long-term multi-level transformation processes. Initially being developed for large-scale sociotechnical infrastructure systems such as energy supply, transport and water supply, its basic notions are now being taken up and adapted to other technological areas as well.

The transition concept is based on a multi-level model of technological change. It distinguishes the 'breeding' of new technologies in confined technological niches from a meso level of sociotechnical regimes (e.g. the system of mobility) and a broader context of the sociotechnical landscape, which encompasses cultural norms, values or dominant economic or governance regimes. The term 'sociotechnical regime' means a rule set or grammar that structures the sociotechnical co-evolution process. The way such a regime evolves 'is structured by the accumulated knowledge, engineering practices, value of past investments, interests of firms, established product requirements and meanings, intra- and inter-organisational relationships [and] government policies' (Kemp *et al.* 2001: 273). The creation of novel technologies is thus shaped by the interaction at three levels: the micro level of interacting users, firms and households, the meso level of technological regimes and the macro level of sociotechnical landscapes. These levels change simultaneously in a co-evolutionary process, though at different time-scales. The value of such a concept is that it reflects the multi-dimensional character of processes of sociotechnical change and the embeddedness of local practices and niches in historically grown contexts with their specific dynamics.

The dynamics of innovation are also addressed in other fields of innovation literature, where a systems language has become the dominant approach over the past ten years. While the notion of sociotechnical regimes puts an institutional perspective into the foreground (norms, practices, 'grammar'), the system concept puts more emphasis on actors and their relations, the boundaries between system and environment, emergent patterns in the dynamics of systems and so on. Innovation-system research is currently being enriched by insights from complexity science in order to understand better the underlying dynamics of the transformation of innovation systems: for instance, to tackle the dynamics of research and development (R&D) collaboration networks (Barber *et al.* 2006).

With the functional approach, a new inroad has been opened for strengthening the normative value of the innovation systems thinking, an issue for which innovation systems research has been criticised in recent years. Moreover, regimes and systems can be seen as complementary concepts, each emphasising different analytical perspectives and specifically contributing to a better understanding of transition dynamics and the design of policy strategies.

With regard to research, technology and innovation policy, forward-looking approaches have been widely applied in the past ten years as a tool to underpin strategy formation and priority-setting both in government policy and in firms. They are also regarded as an essential element in devising strategies for system innovations, but it is still a major challenge to move from foresight to the formulation of concrete decision options.

In this chapter, we revisit the debates about system innovation and the management of long-term processes for realising system innovations, underpinned by examples of production–consumption systems and research, technology and innovation policy. We reposition these debates in a conceptual framework that will be outlined in Section 19.2. For that purpose, we will organise the discussion around four main dimensions of system innovations: (1) the descriptive dimension, (2) the analytical dimension, (3) the normative dimension and (4) the strategic dimension.

These four dimensions need to be clarified in order to build a solid conceptual foundation for devising a methodology that could help guide collective processes of managing transitions as well as individual processes of decision-making. While different conceptual frameworks are possible in principle, we believe that a systems language is best suited for our purposes because it allows us to use and embed different important concepts within a coherent language. On that basis, a methodology is introduced in Section 19.3, which we call 'Adaptive Foresight'. It will be underpinned by examples and experiences from recent research projects and strategy processes in RTI policy that dealt either specifically with processes of long-term transitions to sustainable production–consumption systems or with broader RTI policy strategies. Finally, in Section 19.4 we will draw some conclusions on the experiences with Adaptive Foresight in a policy context.

## 19.2 Building blocks of a conceptual framework: transition fields as complex, sociotechnical innovation systems

In the following section we will introduce a perspective on sociotechnical transformations that takes up elements from existing transition thinking but merges them with a systems approach to innovation, which puts a stronger and explicit emphasis on the role of actors. More recent variants of innovation systems approaches that focus on the sectoral or on the functional dimension will be used to strengthen the normative and the technology-specific dimensions of transition analysis. These elements are regarded as important to make the approach useful as a basis for policy-making and policy advice in the area of production, where agency and normative considerations regarding the direction of technological change and the coherence of policies are crucial.

### 19.2.1 The descriptive dimension: the patterns of transition processes

As a first approximation to conceive and understand sociotechnical transformation processes, descriptive models that capture the aggregate evolutionary path of system innovations can be used. The notion of transitions has been introduced in recent years as a framework that is useful to describe the essential patterns of change of system innovations (Rotmans *et al.* 2001).

Based on historical examples, past cases of transitions have been reconstructed (e.g. see Geels 2004) and generalised into a model of how transitions take place. Several examples of transitions can be identified. For instance, the move from a coal-based energy-supply system to an oil-based system or from an industrial to a service economy can be interpreted as transition phenomena. Usually, transitions are long-term processes that can stretch over several decades,[1] and they are characterised by a co-evolution of institutional settings, markets, technologies, cultures and behavioural patterns.

At an aggregate level, transitions are described in a three-level model of niches, regime(s) and (sociotechnical) landscape. The core of the model was developed initially on the basis of the innovation systems framework, but it also gives ample room to technological aspects as embedded in the practices of users, suppliers and other actors as well as to institutional settings, which are captured by different types of sub-regimes (e.g. user and market regimes, policy regimes) (Geels 2004). One of the strengths of the transition framework is thus the explicit consideration of sociotechnical arrangements rather than just innovations.

Smith *et al.* (2005) have differentiated the initial transition model by introducing four types of transition, which depend on the contexts in which these transitions are taking place. This context is determined by (a) the level of coordination to purposively transform a regime (high or low) and (b) on the location of resources (internal or external to the regime). This framework gives rise to four types of transitions, which the authors call endogenous renewal (high, internal), reorientation of trajectories (low, internal), emergent transformation (low, external) and purposive transition (high, external). The role of the sociotechnical landscape or, generally speaking, of the (system) environment is thus taken into account explicitly as a source of resources as well as of disturbances.

The initial transition model has been developed for large sociotechnical and infrastructure systems or functional subsystems of society (mobility, energy provision). However, such a broad delineation does not work so well for other subsystems, such as production, which are more heterogeneous in terms of technologies, actors and institutional and regulatory settings. In these cases, effective transition strategies must address more specific levels of products, services or policy fields, and thus a more fine-grained and pragmatic delimitation of subsystems is needed.

Currently, research and technology development programmes aiming to induce changes in the production system often concentrate on the funding of individual projects and technologies or are focused on individual industries (e.g. see the analysis of the Austrian RTD programme, 'Factory of Tomorrow', in Weber *et al.* 2003). These approaches build on technology-specific or sectoral perspectives on innovation systems.

---

1 Transition processes can thus be interpreted as comparatively smooth transformations, whereas ruptures, revolutions or breaks imply a rather fast and abrupt type of transformation process.

However, there are instances where it might be more appropriate to delineate systems with a focus on the demand side or areas of needs. This has been demonstrated, for instance, for the household area by the EU-funded SusHouse project (Vergragt 2000). What seems to be lacking is a more flexible and problem-specific operationalisation of innovation systems at an intermediate level of analysis—that is, a description that is sufficiently flexible and concrete to relate easily to individual technologies or specific needs but at the same time sufficiently general to address systemic characteristics and to link up with more abstract visions of the future. Unlike in some specific fields, such as software development, where the issue of user-led innovations is a major issue, user orientation and the involvement of users are generally not regarded as important elements of innovation systems.[2]

An integration of the user view into the systemic view seems to be an obvious necessity, especially when dealing with sustainability issues. We are therefore using the term 'production and consumption system' to delineate the core of the system under transition. Hence, it includes the actors along the value chain as well as the user side.

Thus, when trying to apply the approach of transition management to more specific domains within the broader system of production–consumption, a finer 'granularity' in defining the system or transition domains is needed. One approach to address changes and innovations within the production–consumption system at an intermediate level is to specify more limited 'transition fields' that can be interpreted as emerging systemic entities (i.e. can be demarcated from an environment and develop some degree of internal coordination and logic) but are more homogeneous and easier to address by policy measures than conventional broadly defined systems in transition, such as water or energy supply. Such a transition field can be specified as a segment of a production–consumption system that is sufficiently coherent and limited in scope to be addressed by an operational and well-defined transition process. It is critical in the sense that it exerts a strong influence on adjacent segments of the system. To address such transformations, transition fields should link the level of technologies with the level of society (e.g. with the areas of need to be fulfilled).

In more detail, we mean the following by these two terms:

- Areas of need. As a starting point, we take the observation that production is meant to serve certain 'needs'. The fulfilment of these needs is ensured by systems of production and consumption—that is, by sociotechnical systems that range from the provision of raw materials (including recycling) to service concepts and products for final consumption. The closer you get to the needs end in this production–consumption chain, the more prominent are the social practices and routines of the users for defining this stage. By concentrating on satisfying these final needs, it is possible to overcome the conventional product orientation and think about alternative ways of meeting needs by providing the necessary services and/or products. For instance, the final need of 'mobility' can be met by individual car ownership, but it can also be met by means of new service approaches such as car sharing, door-to-door mobility services and

2 In spite of Lundvall's (1992) pioneering work on user–supplier relationships in national innovation systems, the role of users tends to be under-represented in most innovation systems research (Rohracher 2005).

so on. Often these alternative concepts are embedded in wider guiding visions that capture both these new services and the changes in social, regulatory or organisational practices required. 'Sustainable mobility', 'sustainable household', 'factory of the future' or 'green chemistry' are examples, and are often expressed even in terms of concrete performance targets

- Individual technologies. The practice of research programmes shows that individual projects often focus on specific technologies, without paying particular attention to final needs or social context. In other words, they tend to be conducted in a rather isolated way and need to be integrated and adapted at the level of the transition fields. That is, these specific technical solutions need to be contextualised at the level of transition fields in order to their enable wider uptake. In principle, an individual technology can be applied in different domains to fulfil certain functionalities or satisfy needs, and functionalities can be realised by using different individual technologies. For instance, travel information can be provided in real time to your mobile phone with use of GSM (Global System for Mobile Communications), but it can also be transmitted by means of local information terminals or by phone services. Integrated intermodal services can be achieved by standardised integrated ticketing (as in urban transport networks) or by means of an electronic bidding service, where prices may vary according to demand and so on

Transition fields draw on areas of needs as well as on the supply of technological opportunities. At the same time, transition fields represent an arena that gives space to mediation efforts between them. They can be interpreted as arenas where new systemic solutions for ensuring the provision of functionalities are shaped. As such, they provide the ground for alternative sociotechnical innovation systems to emerge, shaped by the interplay of technological opportunities, actor strategies, user needs and institutional frameworks. In other words, it is in these transition fields, where specific technologies are tied together with and embedded in social and organisational practices in order to offer alternative solutions. Transition fields are thus an intermediate systemic level, at which the integration of individual technology projects into the provision of needs-oriented products and services takes place, guided by an orientation towards long-term sustainability.

For instance, in the case of sustainable mobility, the introduction of an intermodal and integrated mobility service is an example of a transition field. Real-time travel information could be another example, as well as online or mobile booking. As an example of an intermediate transition field, vehicle maintenance services could be mentioned.

In principle, transition fields can be categorised in several different ways. They can be described in terms of the different functionalities they address, in terms of policy arenas or in terms of networks or industrial branches. The definition of 'transition fields', we suggest, thus introduces more flexibility into the delimitation of subsystems to be addressed by the intervention of policy actors, market actors or actors from civil society. Especially in the case of manufacturing, a sectoral differentiation can be useful (at least in some cases) if this is in line with the production–consumption system under study. 'Printing' is an example of a functionality that can be represented roughly by specific industry branches whereas 'dyeing' is scattered over many industrial branches and policy arenas. In order to reduce the complexity of subsequent scenario-development processes it is certainly helpful if a transition field is fairly homogenous, does not consist of too many actor constituencies and relates to a limited set of areas of needs and technology fields.

## 19.2.2 The processes driving transition dynamics: from a quasi-evolutionary model to a complex model

Apart from the aggregate description and patterns of transition processes and the delimitation of what we have called transition fields, there is a second important aspect of analysis to be considered: namely, the analysis of the processes driving transition dynamics at the micro level. Agency and interactions must be key elements of such a model of transition dynamics. This is particularly important because the conceptual framework is supposed to serve as a foundation for political (or other) interventions aiming to influence the behaviour and decisions of other actors.

The transition models developed so far are presented as quasi-evolutionary and stress the co-evolution of research, technologies, markets, regimes and organisations. They essentially use the classical notions of variation, selection and stabilisation or alignment, with regimes playing an important role as coordinating mechanisms (e.g. see Geels 2005; Smith *et al.* 2005). We suggest that models that put the notion of self-organisation and complexity at the centre are better equipped to capture the role of actors, interactions and decision-making processes. They can build on the body of knowledge that has been accumulated in complex systems research over past years.

For the analysis of transition fields the principles of self-organisation will thus be applied to capture transition dynamics. From this perspective, transition processes at an aggregate level can be interpreted as the result of underlying self-organising processes.

Central to our perspective is the consideration of interactions between different types of **actors** playing a role in shaping a transition field: technology suppliers and developers, policy-makers, administrative bodies at different levels, stakeholder and intermediary organisations.

In the end, all these actors take **decisions** that exert an influence on the evolution of a transition field. As a first simple approximation, it may suffice to distinguish between innovation decisions, adoption decisions and policy decisions. These decisions can result in the flow of money in a market, in the exchange of services, products or other technological artefacts or in the exchange of information in the implementation of a legal act. Policy decisions comprise, for instance, the setting-up of a research programme, the introduction of a new regulatory framework or the creation of a new intermediary agency that serves as a carrier organisation for a new technology or transition field. Decisions are shaped by several spheres of influence, including interactions with other actors, future expectations, preferences and objectives, resources, established sets of rules and beliefs and so on. From the perspective of agent-based modelling, one could speak of the decision rules that characterise an agent.

Obviously, although decisions are finally taken by individual actors, decisions are influenced and prepared—to a varying degree—by **interactions** with other actors, embedded in hierarchies, networks or markets. We speak of decision arenas where such decisions are prepared. Further important elements in our conceptual model are **expectations** and **visions**, and related to these are the **objectives** an actor may pursue.

At the level of the transition field, it is possible to identify certain aggregate features that characterise the transition field. These are the **structural properties** of the transition field, which comprise the rules and institutions that guide interactions within the transition.

These building blocks can now be used to analyse complex self-organising mechanisms and the way in which they determine the observable aggregate patterns of dynamics of transition fields. In this context, the central role of the concept of **circular causalities** (i.e. reinforcing, delaying or stabilising feedback mechanisms) for our interpretation of innovation and transition dynamics must be highlighted. The notion of circular causalities has been used to explain the origin of emergent properties in self-organising systems. This concept can be applied to the three levels of sociotechnical innovation systems: that is, both between the levels of the transition field (e.g. vertical relations between structural properties and the actor level) and within each of the levels (e.g. horizontal relations between actors).

Such emergent properties can be organisational structures, diffusion patterns or regulatory frameworks. They result from the decisions of different types of actors and have been prepared in interactions and negotiations. Once realised, these emerging properties are also perceived by the actors and translated—in conjunction with their respective expectations, preferences and objectives—into a next series of decisions, thus giving rise to self-reinforcing or self-defeating feedback loops. Obviously, this is not a closed process that takes place solely within a transition field but has implications for the wider context, just as developments in this wider context are taken into account in the formation of expectations and objectives that co-determine decisions.

Circular causalities could in principle also be interpreted in evolutionary terms—that is, as being at the origin of variation, selection and stability. The joint operation of reinforcing and delaying mechanisms, applied to different technology options, gives rise to a selection process. For example, a circular causality of the learning-by-experimenting type strengthens, if successful, the possibilities for a technology to be perceived as superior to others. This favours its selection by users, manufacturers and policy-makers. If, in parallel, the competing options are not favoured, for example, as a result of unsuccessful experiments, a selection pattern emerges as a result of the combination of these circular causalities. Similar arguments can be made with respect to variation and stabilisation, which have been elaborated elsewhere (Weber 2005). They show that the conceptual model building on principles of complexity and self-organisation as well as on the explicit consideration of actors and their behaviour provides an underlying rationale of the (quasi-)evolutionary model that has been used so far to explain transition processes.

### 19.2.3 Orientating transitions: functions of innovation systems and sustainability

The third dimension to be investigated concerns the normative dimension of transition processes: that is, the question of the direction of change. Usually, the broad notion of sustainable development is used as the *Leitbild*, or vision, but it tends to remain vague because it is contingent on future framework scenarios and thus needs to be translated into certain guiding principles, objectives or assessment dimensions. At this more concrete level, there is also a need for a continuous adjustment of objectives while maintaining the general principles and vision. Innovation is usually seen as a crucial element when it comes to defining objectives to be pursued for achieving sustainability. Unfortunately, the innovation systems approach as the dominant approach in innovation research, although being useful for descriptive purposes and as a heuristic device, has been rather weak when it comes to providing a theoretically founded normative guidance for how to shape and manage innovation systems.

Recent discussions in the innovation systems literature emphasise different 'functions' of innovation systems (e.g. Bergek et al. 2005; Chaminade and Edquist 2006; Hekkert et al. 2006; Kubeczko et al. 2006a, 2006b). These functional approaches, although diverse in terms of the sub-functions, in most cases include the need to orient or guide innovation processes in a certain direction. This is particularly well documented in cases where innovation systems are based on alternative energy resources (e.g. Jacobsson and Bergek 2004; Jacobsson and Johnson 2000). We thus suggest combining the functions of innovation systems framework with sustainability objectives.

In general, we suggest three types of functions of innovation systems need to be distinguished:

- **Structure**. Innovation systems provide structures for innovation activities which support activities in the innovation system itself. This can be achieved by the introduction of actors, institutions (rules, norms, etc.), networks or artefacts

- **Orientation**. Orientation can be provided by means of *Leitbilder* (visions) or other 'open methods of coordination', or by more concrete means such as information flows or financial incentives

- **Adaptability**. Adaptability[3] is a prerequisite for a system to maintain its other functions over time. Possible ways to maintain this function are through strategic intelligence or the involvement of users in the innovation process

The first function is in line with the identified need to overcome systems failure and support cooperation between different actors in the innovation system (e.g. between science and industry). It relates to any kind of innovation system and is not specific to system innovations.

The second function, of orientation, is crucial in the context of system innovation because it provides the possibility to combine normative arguments about the direction of sociotechnical change within an innovation systems framework.

The third function is crucial for addressing the dynamic properties of innovation systems. It stresses a key characteristic of complex, self-organising systems: namely, the ability to adapt to changes in the environment, either in a reactive way (i.e. in response to external developments) or in a proactive way (i.e. by anticipating external developments).

---

3 The term *adaptability* may be interpreted in a purely passive sense, reflecting simply reactive behaviour. In our interpretation, the term *adaptability* can equally imply proactive behaviour.

## 19.2.4 Strategies and pathways towards the future

The key challenge of transition management is to ensure the adjustment between open and collective learning and innovation processes with respect to societal goals implicit in the transition concept. Conventional transition management strategies generally consist of the following elements: (1) development of long-term sustainability visions and overarching joint strategic agendas; (2) organisation of a multi-actor network, mobilisation of actors and execution of projects and experiments; and, finally, (3) monitoring and evaluation processes as feedback loops in the collective learning process.

Transition management is a very optimistic approach in terms of seeing a collective ability to shape the future. It assumes that these three types of processes allow sociotechnical systems to be initiated and steered in a direction that is compatible with societal objectives such as sustainability. However, in reality, the future evolution of many, if not most, sociotechnical systems is beyond the control even of well-concerted efforts of the key actors in an individual country. The room for manoeuvre is highly constrained by global technological and economic developments, by EU policy and by societal developments that evolve independently of the sociotechnical system in question. In a globalised knowledge society, external factors tend to exert a much stronger influence on the future pathway of a sociotechnical system than those factors that can be influenced even by a collective effort at the national or sectoral level.

The fact that many of the determinants of future transition processes are beyond our influence calls for an adaptive strategy for managing transition processes. Support for the development of policy strategies by way of what we call 'Adaptive Foresight' thus draws extensively on recent advances in adaptive and strategic planning.

The classical approach to adaptive planning is based on a staged development process with key decision milestones to confirm or reject a development trajectory depending on the progress made and changes in context. More recently, adaptive planning approaches have emphasised the need to combine scenario development (as a means to capture a wide range of possible future contexts) with the maintenance of a set or portfolio of real options (as a basis for making choices). It has thus developed into a much more flexible tool for supporting strategic decision-making, especially when it is combined with processes of continuous monitoring and evaluation (Eriksson 2003). The portfolio-based approach is particularly well suited to dealing with situations where there are numerous strategically acting individuals and organisations.

Sociotechnical scenarios represent a first important building block for operationalising the principles of adaptive planning. They serve to capture the range of possible futures in which the addressee might be operating and which he or she can partly influence. In other words, scenario development serves in the first instance to make different possible future contexts for action explicit, in scientific–technological and economic as well as in sociopolitical and institutional terms. For the purpose of adaptive planning, sociotechnical scenarios should be neither purely exploratory nor purely normative. In fact, it is important that they represent a mixture of desirable and undesirable elements in order to be able to arrive at non-trivial conclusions.

Usually, exploratory scenarios are the core element in adaptive planning exercises that are aimed at optimising the potential economic benefits of a firm. From a policy perspective, however, exploratory scenarios alone are not sufficient as a basis for strategy development. They are useful to describe how different future worlds might look, but the normative dimension plays a much more significant role in public policy-making than in private firms. There is a need to tackle explicitly a range of political objectives as a basis for assessing and selecting different decision options that will allow these, largely exogenously driven, exploratory scenarios to be influenced in a direction that could be regarded as desirable from a societal and/or political point of view. At the same time, it needs to be recognised that the actual influence of government decisions on the future course of sociotechnical transitions tends to be quite limited. Even a very proactive national government, for instance, is by no means in a position to determine the future development path on its own; its room for manoeuvre is restricted by the extent to which the exploratory scenarios developed are driven by exogenous factors and by other actors. From a societal and governmental perspective, the exploratory scenario approach thus needs to be combined with strong normative elements in order to inform future-oriented policy strategies that are in line with major societal objectives (e.g. sustainable development).

The notion of portfolios of real options is the second main building block considered here.[4] Initially, the notion of portfolios of options was introduced in finance, describing a set of financial options that allows investors to minimise (i.e. to hedge) financial risks. Sophisticated models have been developed to deal with financial portfolio optimisation. In the world of research and innovation, modelling approaches have also been tested but, in contrast to financial options, real options are much more difficult to capture in quantitative models (Schauer 2007).

The key argument behind maintaining a portfolio of real options can best be explained by a simple question: 'what preparatory and precautionary actions can be taken today in order to make sure that an actor can adapt to unexpected adverse conditions and exploit unexpected upcoming opportunities tomorrow?' This approach admits that our possibilities of shaping the future are constrained to a large extent by developments beyond our influence and that we must be able to adapt to these. It implies that a good portfolio of real options is characterised by two main features. First of all, there are options that are helpful under the conditions of all scenarios considered. These 'robust' options represent the quick gains from a portfolio-oriented scenario analysis. In other words, robust policies are those that would allow one to do reasonably well under the conditions of several scenarios considered. Second, there are policy options that enable adaptability—that is, they are meant to maintain the ability to rapidly exploit upcoming opportunities or to cope swiftly with major uncertainties.

The notion of a robust and adaptive portfolio of options can be applied both to technologies and to policies. Looking first at technologies, this approach is compatible with policy recommendations to develop competences in generic technologies that are likely to be applied in a wide range of sectors, the relative importance of which may differ across scenarios. Obviously, although generic technologies are characterised by a certain degree of robustness and adaptability, they are usually not sufficient to ensure the level of adaptability needed. Advancing additional technological options may thus be

4 The concept of real options is usually applied with respect to a specific actor or addressee who has the choice between different options. It is important to be very clear about the addressee of an adaptive planning activity and about the scope of his or her decision options because this determines what is to be regarded as an endogenous or an exogenous variable—that is, as a factor he or she can or cannot influence. In most policy-oriented exercises, this addressee can be society at large, but in more operational terms it is usually a public agent such as a ministry or a regional government.

necessary. The history of renewable and energy-saving technologies may serve as an example of the advantages of an adaptive technology portfolio approach. After the first oil crisis the introduction of alternative and efficient energy technologies was promoted, but it took several years to make some of the options in question (e.g. heat pumps, combined heat and power generation and solar collectors) technologically and economically viable. In those countries where these technologies had already been developed early on, their uptake took place much more quickly and thus facilitated the adaptation to the scarcity and high price of fossil fuels.

In terms of research and technology policy, adaptability can be achieved by keeping a fairly broad range of research programmes and activities in order to be able to build on the acquired competences if a research field 'takes off' or if changing circumstances require a course to be adapted or changed swiftly. Rather than focusing on a very restricted set of research and technology priorities (which is currently the dominant line of argumentation to strengthen technological competitiveness), a portfolio-based and risk-conscious research and technology policy strategy would aim at advancing a broader set of options in a kind of 'wait-state' and at supporting technologies that are inherently adaptive. Obviously, as it is not possible to keep all options open, it is still necessary to prioritise, but the notion of portfolios of real options should be taken into account as an additional building block for designing strategies to cope with the uncertainties, risks and opportunities inherent in different future scenarios.

The timing of exerting policy options is another important aspect of a portfolio-based policy approach because of the criticality of the timing for the impact of a decision. We know from technology and innovation policy studies that there are only small 'windows of opportunity' for initiating an action (for instance, starting a major R&D programme to give a new, emerging, technology area a boost) that is supposed to effectively influence a technological development path. By choosing the right moment in time, the operation of self-reinforcing mechanisms can be exploited and thus the financial and political effort can be minimised (Erdmann 2005).

## 19.3 Adaptive Foresight as a methodology for policy support

### 19.3.1 Introduction

The conceptual framework outlined in Section 19.2 reflects our understanding of how transformations of sociotechnical systems 'work', and what basic implications this has for policy strategies aiming to influence these transformations. It has been formulated in such a way to stress those elements that we regard as key for an innovation systems approach that is intended to capture the characteristics of system transitions towards sustainability.

In line with our interest in making constructive use of theoretical and conceptual insights in policy advice, we have developed and tested a process methodology for guiding and supporting sociotechnical transitions towards sustainability, which we describe below. It takes the four key elements—differentiation of transition patterns, complex dynamics, orientation and strategic behaviour—into consideration in the process design, particularly when it comes to developing policy options. The move from sociotechnical scenarios towards strategic behaviour and policy portfolios probably represents the most innovative feature of our conceptual and methodological approach, at least with respect to research on sociotechnical transition that has been mainly inspired by innovation system analysis and social studies of technology.

Based on the conceptual framework outlined, we have developed and implemented different projects that were set up either as broader participatory exercises involving different actors or, more specifically, to support decision-making processes of individual policy actors (e.g. ministries, programme agencies, etc.). The methodology of Adaptive Foresight as suggested below links the perspectives of transition management and sociotechnical innovation systems with the experiences gained in adaptive planning and foresight processes (for a framework that shares some features of Adaptive Foresight, see Markard 2005).

### 19.3.2 Adaptive Foresight: a methodology for scenario and policy portfolio development

Adaptive Foresight aims at making foresight more effective for policy-making by (a) having a more realistic level of aspiration with regard to the ability to shape the future and put stronger emphasis on the necessity to adapt also to future developments, by (b) stressing the need for iterative monitoring and learning, and by (c) using a flexible and adaptive methodological approach that may well differ over the different phases of the policy cycle it aims to support (e.g. in terms of the participatory character, or the balance between analytical and prospective methods).

### 19.3.3 Methodological foundations

Nowadays, more is expected from policy research than retrospective analysis and the creation of general hypotheses as inputs to policy-making. Instead, more attention needs to be paid to providing forward-looking knowledge and transparent methodologies to support policy strategy development. In this section, such a methodology will be presented under the title of 'Adaptive Foresight'. It is based on insight gained from past experiences with foresight methods and sociotechnical scenarios, but it also reflects principles of adaptive planning and reflexive governance (on sociotechnical scenarios and their application to the evolution of different large technical systems, see also Elzen et al. 2004; however, Elzen et al. applied mainly the scenario method, predominantly for experimental purposes; their method was not linked to actual policy-making processes). Conceptually, the Adaptive Foresight methodology builds, as outlined above, on a combination of building blocks—innovation systems research, transition management and science and technology studies—that emphasise the importance of complex mechanisms for the explanation of innovation dynamics. Such a 'sociotechnical innovation systems' perspective, as discussed in Section 19.2, delivers the main requirements that need to be met by the methodology. The provision of an explicit theoretical foundation for the methodology has several advantages: it facilitates the interaction process with participants in the scenario workshops and it allows an explicit interpretation of the findings of the process and thus contributes to the overall understanding of potential future innovation dynamics and the role of policy interventions in these dynamics .

### 19.3.4 Looking back: problem definition, system delimitation and analysis

Before looking ahead as part of a policy strategy development process it is necessary to conduct a solid retrospective analysis of the field of investigation—that is, of recent and current developments along the lines of the conceptual framework of complex innovation systems. And, even before starting such analytical work, two initial clarifications need to be achieved. First of all, it must be clear what the focal issue of the strategy development exercise is—that is, the main question or problem that the addressee of an exercise would like to have tackled. In government-led exercises, these focal issues are usually related to societal goals and the (policy) strategies to pursue in order to achieve them. They can be at highly aggregate levels (e.g. 'how can we achieve an overarching transition towards a more sustainable energy-supply system?') or more specific to individual sectors and policies (e.g. 'what innovation and technology policy strategies should be pursued in order to contribute to the joint objectives of sustainable mobility and enhancement of the competitiveness of domestic transport technology industries?').

Second, the analytical boundaries of the innovation system that determines the evolution of the focal issue need to be clarified. This implies a distinction between the inside and outside of the system under study—that is, a distinction between the aspects that can be influenced by, for instance, national policy, and those that are outside its scope and thus are to be regarded as exogenous factors. This delimitation of the innovation system will be further refined in the course of the scenario-building exercise.

There are several typical elements and aspects that need to be addressed in the course of an innovation system analysis:

- Actors (i.e. the entities that exert an influence on processes of innovation, either directly or indirectly). These entities do not necessarily have to be individuals or organisations but can be based on looser forms of coordinated behaviour such as networks. The set of actors that makes up an innovation system can obviously change over the course of time

- Decision-making processes and interactions. These are the objectives, rationales and behaviour of the actors identified and the processes of interaction by which these decisions are influenced

- Structures (i.e. an analysis of the structural characteristics of the system and their transformation). These include institutional and organisational changes such as, for instance, a liberalisation of relevant markets or a financial crisis as well as major technological developments or changes of sociotechnical regime

- Sociotechnical knowledge base (i.e. the entirety of the distributed knowledge that is available to the different actors). Apart from describing the capabilities and competences that exist in an innovation system and the knowledge base, there is also information about the performance and potential of new technologies and innovations. Moreover, such knowledge may be used to anticipate impacts and generate future visions or *Leitbilder*

Apart from describing these changing patterns in actors, interactions, structures and knowledge of an innovation system it is also important to understand the underlying system dynamics in terms of interdependences between the different levels of analysis (e.g. between actors and structures). Currently, mechanisms from complex systems research increasingly are tested to capture basic principles of innovation system dynamics (Fischer and Fröhlich 2001), but it has to be admitted that this is an area in which much research work still remains to be done.

In methodological terms, this first phase is usually based on a combination of analytical desk research to collect important background information and interactive workshops to define the focal issue and interpret it with respect to key mechanisms determining system change.

**PROJECT EXAMPLE 19.1 The impact of information and communications technology on transport and mobility: the development of a portfolio of robust and adaptive policy options**

The impact of information and communications technology on transport and mobility (ICTRANS) is a scenario development exercise that aims to explore, in a qualitative way, policy options at the European level in order to enhance the impact that the use of information and communication technology (ICT) could have on the volume, modal choices and efficiency of mobility service provision in the realms of living, working and producing (Wagner *et al.* 2003). As the potential impacts of ICT on mobility are subject to a high degree of uncertainty and are dependent on the future of the information society in general, the scenarios developed had to span a wide spectrum of possible futures. Moreover, potential impact chains of ICT use on transport and mobility had to be identified before the actual exploration of future impacts could be implemented. The role of different policy options for shifting impacts in a more sustainable direction could then be explored under the conditions of the different scenarios. Some options clearly turned out to be 'robust' in the sense of being helpful in all scenarios, whereas others could be identified as being 'adaptive' in the sense of being needed to prevent negative impacts in one specific scenario or to maintain the ability to exploit upcoming opportunities specific to that scenario.

### 19.3.5 Looking ahead: combining exploratory and normative elements

The forward-looking step consists of three main parts: namely, an exploratory part, a specification part and a normative part. The exploratory part aims to define different possible future scenarios that serve as a framework for further specification and also for adjusting the scenarios according to the normative objectives and goals.

Both endogenous and exogenous developments (i.e. the ones from within the system and from the outside) are taken into account as possible factors of influence that are likely to shape the future evolution of the system under study.

After collecting the range of factors of influence, an analysis of current trends and possible breaks in trend can be performed in order to differentiate between factors that are likely to be stable and others that are still open to evolve in qualitatively new ways. The second type of factor is particularly useful in differentiating one scenario from another. Such trends and breaks in trend can be of a technological as well as of a socio-economic nature.

Trends and trend-breaks are the basic material from which to develop scenarios. Here, a variety of methods is available, ranging from 'bottom-up' methods of constructing and clustering storylines into scenarios (as e.g. in Project example 19.1) to 'top-down' methods that concentrate on selected dominant factors of influence, possibly supported by cross-impact analysis, Delphi methods or simulations to strengthen the credibility of results. However, most of these methods rely, in the end, on expert or stakeholder judgement, thus reflecting these actors' implicit theories about impact chains.

What is essential for these scenarios is that they are plausible, multifaceted and challenging. They must be regarded as tools to stimulate thinking about potential future developments rather than as predictions of the future. Moreover, the distinction between endogenous and exogenous factors of influence helps identify where the opportunities and limitations for action are for the main actors addressed (e.g. for a national government). Trends and trend-breaks also point implicitly to important decision options for the different actors (i.e. to the strategic moves they can make). Simulation games (*Planspiele*) are tools by which these implicit strategic options can be made explicit.

When the basic scenarios are defined, they simply represent different frameworks and still leave much room for further specification. In fact, they simply define a corridor of actions that needs to be refined and related to the focal issue under investigation. As guidance for how to describe such framework scenarios, the main types of factors of influence can be taken as a starting point. Typically, categories such as technologies, producers, users, policy and so on are defined by the participants (i.e. categories that can easily be related to actors' strategies).

The normative dimension is addressed in the subsequent stage. At this point, goals and values of the different actors come into play. A possible starting point is therefore a debate on visions related to the focal issue of the exercise. This is useful in order to clarify shared (or diverging) policy and/or societal goals, ambitions and the underlying values of the actors and stakeholders involved. A second element is related to the potential future risks and opportunities tied to the focal issue of the exercise.

Both elements serve, in the end, to reach agreement on the dimensions along which the exploratory scenarios are to be assessed. In many current cases, sustainable development is taken as a guiding vision that needs to be operationalised in terms of different assessment dimensions: for instance, as in the case of Project example 19.2, along the lines of the pattern developed by the German Helmholtz Society (Coenen and Grunwald 2003).

When assessing the scenarios along these main dimensions, it would be unrealistic to expect clear-cut statements on which is the better or the worse scenario from the perspective of sustainable development. First of all, the scenarios are designed in a way to avoid simply 'good' or 'bad' scenarios; on the contrary, multifaceted scenarios are regarded as the most productive. Moreover, in most cases, the level of uncertainty of impacts is such that it will, at best, allow the identification of 'critical issues' that require the attention of policy-makers or that potentially could have significant impacts on the focal issue. These tend to be issues where additional research needs to be conducted in order to better understand potential impacts in the future. In other words, the assessment serves to identify potential but uncertain levers for shaping future development in a more desirable direction.

The basic argument behind this approach is that the framework scenarios determine to a large extent the future evolution of the focal issue, but, within a corridor defined by the scenarios, there is still room for manoeuvre and adjustment in a direction that converges with the negotiated goals and objectives.

The assessment and subsequent identification of critical issues opens up the opportunity to modify the scenarios developed towards the best possible variants (e.g. 'what could be best achieved within each of the framework scenarios?'), but it is not a necessary condition for the subsequent backcasting and portfolio analysis.

Methodologically, this step is based largely on interactive and creative methods but can, at least in principle, be supported by computer tools, forecasts and modelling.

PROJECT EXAMPLE 19.2 **Wood–plastic composites, fibre composites and biopolymers from renewable resources: scenario development for adaptive policies**

In order to support the medium-term to long-term strategy development of a research programme on 'The Factory of Tomorrow', a scenario-building exercise has been implemented for the area of wood–plastic composites, fibre composites and biopolymers from renewable resources (see Weber *et al.* 2005). The emphasis was put on the policy options on the basis of which Austrian technology and innovation policy could influence the future evolution of this transition field towards more sustainable production systems. As the ability to shape the evolution of this field from within a rather small policy area in a small country such as Austria are comparatively limited, policy options were explored within the context of different framework scenarios that were driven largely by developments external to the reach of Austrian technology and innovation policy. By using this approach, the limits to the political shaping of the future were explicitly taken into account, and the emphasis put rather on the need to adapt to and prepare for scenarios that are driven by external factors, not the least international and European developments.

## 19.3.6 Pathways towards the future: multiple backcasting

Conventional backcasting approaches take a single and desirable image of the future as their starting point (e.g. see Vergragt 2000). While this is a valid method for clarifying necessary steps towards a desirable future, it underrates the limitations imposed on the ability to shape a desired future, and in particular the dependence on decisions outside of one's own control. In order to make use of backcasting in a more realistic, strategy-oriented manner, its application to each of the scenarios developed is regarded as more helpful ('multiple backcasting').

In essence, backcasting looks at the steps necessary to achieve a specific scenario. This requires the analysis of barriers and incompatibilities with which the realisation of a scenario may be confronted in the course of time. Key decision and bifurcation points need to be identified, and the compatibility of, for instance, technologies, values and actors' interests to be assessed as a first consistency check.

By staging the pathways leading to the realisation of different scenarios, it is possible to identify needs for action and intervention that can serve as an important input to the subsequent portfolio analysis. In principle, this stepwise backcasting of the scenario pathways allows for the discussion of the appropriate timing for policy and other measures: for instance, in terms of 'windows of opportunity' for introducing a new technology or for starting a policy initiative. By developing consistent pathways, the backcasting exercise represents a second level for testing the credibility of a scenario.

Methodologically, backcasting tends to rely on qualitative methods in order to capture the full range of aspects that can potentially come into play in the course of a scenario pathway, but consistency checks, in particular, can also be supported by means of quantitative tools.

### 19.3.7 Portfolio analysis: robust and adaptive policy options

So far, individual scenarios have been developed, refined and analysed. Each of the scenarios and pathways can be characterised in terms of technologies and policies that have been realised. The options delivered by the scenarios have also been assessed with respect to the focal issue.

From today's perspective, portfolio analysis then looks across different scenarios in order to assess and select those technology options and corresponding policies that promise to be either robust or adaptive (or both). In other words, robust options are fairly easy to identify because they are positively assessed in all or most scenarios. Adaptive options have been identified as part of the normative stage of scenario development when possibilities were sought to move the basic scenario in a more desirable direction. Adaptive options are thus either crucial for avoiding major negative impacts or for exploiting specific opportunities in a single scenario. These kinds of insights should then serve as an input for today's policy-makers to prioritise, for instance, emerging technologies and to design corresponding policies.

Of particular interest are technologies that embody characteristics of both robustness and adaptability (i.e. in general terms they are beneficial in all scenarios, but their specific 'shape' depends on the conditions of the respective scenarios). In fact, many technologies tend to have a double-edged character, because they can be beneficial under certain circumstances and detrimental under others. The type of impact they will have will depend on the context of use and on often politically defined framework conditions and objectives (see Project example 19.3). Embedded systems, to take an example from the ICT context, are expected to have a very positive impact within an optimistic information society scenario. The same technology, however, can be abused in a 'big brother'-type scenario when used for invading the privacy of individuals.

The matter is further complicated by the fact that policy options can have an impact at different levels. For instance, they can refer to the promotion of individual technologies (e.g. an R&D programme) as well as to the structural settings of the innovation systems (e.g. the liberalisation of energy supply). Moreover, the impact of policies depends on their time of implementation.

These examples show that the analysis of portfolios of real options is far from easy and that technological options and policy options are closely intertwined. However, the basic principle still holds that promising technology and policy portfolios are composed of options that promise to have at least the potential to help improve the issue in focus in all scenarios (robust options) and either avoid major problems or help exploit opportunities in some selected scenarios (adaptive options).

Methodologically, interactive methods can be used to discuss different options from a range of viewpoints. In order to come up with new and fresh ideas for policy options, comparative analysis of other countries' practices can be instructive. Finally, while being by far less sophisticated than is the case in financial portfolio analysis, modelling tools are being developed that promise to be applicable to real options in quantitative terms: for instance, for analysing research and technology portfolios (Schauer 2007).

**PROJECT EXAMPLE 19.3 The future of freight transport in Austria: overcoming barriers between political administrations**

In a project dealing with the future of freight transport in Austria, several different, and sometimes also contradictory, policy objectives had to be taken into account in order to derive insights into opportunities for defining the future strategy for transport technology and innovation policy. Transport policy objectives as well as environmental and industrial policy considerations had to be brought to bear for the conceptualisation of future policy initiatives and strategies. Although, for instance, transport and technology policy were hosted in the same ministry, the level of cooperation between the respective directorates general was only weak. It was thus crucial to clarify these objectives to incorporate them in suggestions for new technology policy initiatives as well as for the success of the entire exercise (Seibt *et al.* 2002; Weber *et al.* 2002; Whitelegg 2004).

## 19.3.8 Policy strategies: from open participation to closed processes

Especially when it comes to defining policy strategies, where the fundamental orientations and guidelines for policy need to be discussed as a basis for triggering and framing more specific initiatives, it is extremely difficult if not impossible to involve a broader audience. Hidden agendas and political bargaining positions cannot be discussed in an open or even public setting. This holds for private firms as well as for government bodies. However, such debates are essential for consolidating forward-looking insights and making them effective in policy-making. For the sake of bringing foresight to bear on policy-making, we argue that the foresight community needs to accept supporting policy-preparing exercises that are of a rather closed nature as a necessary and complementary element to public and participatory foresight exercises (cf. Project example 19.4).

In the early phases of opinion-building, open consultation and participation are necessary to exchange information, to improve our ability to sense and assess future developments along their socioeconomic and technological dimensions. In the later stages, however, when individual actors need to make up their minds about their strategies and to make concrete decisions, such forward-looking consultations need to be kept internal to an organisation in order to protect its knowledge and to improve its ability to act. Equally, these internal processes need the support of foresight specialists and should thus be considered more explicitly by the corresponding communities (Project example 19.4).

**PROJECT EXAMPLE 19.4 The development of research, technology and innovation policy strategy at the ministerial level**

As part of a strategy development process in one of Austria's ministries in charge of issues relating to research, technology and innovation (RTI), an internal foresight and strategy development process was launched in 2005. It serves to redefine the strategic guidelines for RTI policy within the confines of the ministry's responsibilities. It builds extensively on available strategic and prospective intelligence that had been generated in several projects over the previous year. Many of these projects were of an analytical nature; others were set up as consultative, forward-looking exercises. The strategy process itself has been organised as a fully internal process without external participation other than from some policy advisors and process consultants. Owing to the sensitivity of issues raised in the process (e.g. relating to the positioning *vis-à-vis* other ministries, agencies and the Austrian Council for Science and Technology Development), a closed and purely internal process was the only possibility for discussing these matters thoroughly. Building extensively on foresight knowledge and applying foresight-type methodologies in a closed setting, this process has been a highly effective way of bringing foresight knowledge to bear in the development of policy strategy.

### 19.3.9 Policy implementation and learning: monitoring, shaping and adjusting to the future

The development of policy options and portfolios, and even policy strategies, is just input for actual policy design and implementation. In other words, so far we have been discussing mainly the early phases of the policy process. The actual design and implementation of specific policies, and the learning processes that take place in the course of the stage of the policy cycle, from design to implementation, represent the wider context in which Adaptive Foresight processes are supposed to contribute.

If the principles behind Adaptive Foresight are to be effective, they will thus need to be closely tied not only to policy design but also to policy implementation and learning, at a strategic as well as at a local level. More specifically, the experiences gained in the course of local implementation need to be monitored and fed back to strategy development.

In other words, Adaptive Foresight should be interpreted as part of a broader continuous learning process that comprises the implementation and evaluation of specific policy measures as well as a monitoring of relevant developments of policy as a whole.

Thus, strategy development, policy design, implementation and learning should not be understood as distinctly separate phases but rather as a continuous process of mutual adjustment. This adjustment refers to goals and objectives, to the identification of new sociotechnical options, to the growing knowledge and understanding of their impacts, to the design of new types of policy options and to their integration into portfolios. However, in the course of this process, the degree of participation may change from being an open participatory process to being a closed internal strategy and policy-design processes.

Within such a comprehensive setting, the impact of guiding policy strategies should not be underestimated because, in particular, public policy strategies fulfil an orienting function for many private actors as well, and, in the best case, play an implicit coordinating function for their decision-making.

One of the main difficulties of this continuous process consists of the fact that all actors involved can resort to strategic and adaptive (and thus interdependent) behaviour. This is why issues of policy coordination—both between different policy areas and between public and private actors—have started to play such an important role in policy-making.

Nevertheless, in practice, processes of scenario development and portfolio analysis will rarely be conducted on a continuous basis. They will rather be repeated every few years, coinciding with an update of the overall technology and innovation policy strategy. The practical tools and methods are available, based on many years of experience with foresight, adaptive planning, evaluation and monitoring. What is still missing is the integration of these methods in a continuous and long-term strategy-development process.

## 19.4 Conclusions and assessment

The conceptual framework and methodology presented have been developed and tested in the course of a number of forward-looking projects in Austria and for the EU that have been aimed either at underpinning or at directly supporting processes of shaping future development paths in a sustainable direction. In all cases, this implied exploring opportunities and conditions for system innovations and, in most cases, these analytical and forward-looking activities were closely tied to the concrete development or adaptation of policy strategies or research programmes.

In this policy context, the conceptual framework turned out to be particularly useful for a number of reasons:

- It clearly puts actors, their decisions and interactions as well as new and existing policy institutions or organisations at centre stage

- It provides a rationale for sustainability-oriented RTI policy, but set within the context of the—at least in Austria—dominant paradigm of innovation systems that inspired most recent initiatives in RTI policy
- It delivers 'realistic' policy advice, in the sense that both clear-cut ('robust') options as well as defensive and offensive ('adaptive') ones are identified. It avoids overambitious options by stressing the limitations to the collective ability to shape the future

From a methodological point of view, the Adaptive Foresight approach has proven to be very useful in three respects:

- By accepting the need for foresight in the context of internal strategy support, preceding research and participatory processes can be effectively brought to bear on policy-making and policy strategy development
- Moreover, the methodology is realistic in terms of the level of aspiration with respect to the political shaping of change; a purely normative *Leitbild* type of approach may be useful to develop orientations, but it is less helpful when it comes to devising concrete strategies that take into account the limited possibilities for shaping future development paths
- Although the cases presented here are based on experiences gained in a public policy environment, the methodology could also be applied to strategy processes for firms or non-governmental organisations because it is geared towards providing a public forum for forward-looking debates and for then targeting the results of these debates to decision-making for individual actors

# References

Barber, M., A. Krueger, T. Krueger and T. Roediger-Schluga (2006) 'Network of European Union-Funded Collaborative Research and Development Projects', *Physical Review E* 73.036132.

Bergek, A., S. Jacobsson, B. Carlsson, S. Lindmark and A. Rickne (2005) 'Analysing the Dynamics and Functionality of Sectoral Innovation Systems: A Manual', paper presented at the *DRUID Conference*, Copenhagen, Denmark, 27–29 June 2005.

Berkhout, F., A. Smith and A. Stirling (2004) 'Sociotechnical Regimes and Transition Contexts', in B. Elzen, F.W. Geels and K. Green (eds.), *System Innovation and the Transition to Sustainability: Theory, Evidence and Policy* (Cheltenham, UK: Edward Elgar): 48-75.

Butter, M. (2002) 'A Three Layer Policy Approach for System Innovations', paper presented at the *1st Blueprint Workshop*, Brussels, Belgium, January 2002.

Chaminade, C., and C. Edquist (2006) 'From Theory to Practice: The Use of Systems of Innovation Approach', in J. Hage and M. Meevi (eds.), *Innovation, Science and Institutional Change* (Oxford, UK: Oxford University Press).

Coenen, R., and A. Grunwald (eds.) (2003) *Nachhaltigkeitsprobleme in Deutschland: Analyse und Strategien* (Berlin: Sigma).

Elzen, B., F.W. Geels, P.S. Hofman and K. Green (2004) 'Sociotechnical Scenarios as a Tool for Transition Policy: An Example from the Traffic and Transport Domain', in B. Elzen, F.W. Geels and K. Green (eds.), *System Innovation and the Transition to Sustainability* (Cheltenham, UK: Edward Elgar): 251-81.

Erdmann, G. (2005) 'Innovation, Time and Sustainability', in M. Weber and J. Hemmelskamp (eds.), *Towards Environmental Innovation Systems* (Springer: Berlin): 195-208.

Eriksson, E.A. (2003) 'Scenario-Based Methodologies for Strategy Development and Management of Change', in M.O. Olsson and G. Sjöstedt (eds.), *Systems Approaches and their Application: Examples from Sweden* (Dordrecht, Netherlands: Kluwer).

Fischer, M.M., and J. Fröhlich (eds.) (2001) *Knowledge, Complexity and Innovation Systems* (Berlin: Springer).

Geels, F. (2002) 'Technological Transitions as Evolutionary Reconfiguration Processes: A Multi-level Perspective and a Case-study', *Research Policy* 8.9: 1,257-74.

—— (2004) 'From Sectoral Systems of Innovation to Sociotechnical Systems: Insights about Dynamics and Change from Sociology and Institutional Theory', *Research Policy* 33: 897-920.

—— (2005) 'Processes and Patterns in Transitions and System Innovation: Refining the Co-evolutionary Multi-level Perspective', *Technological Forecasting and Social Change* 72.6: 681-96.

Green, K., and P.J. Vergragt (2002) 'Towards Sustainable Households: A Methodology for Developing Sustainable Technological and Social Innovations', *Futures* 34: 381-400.

Hekkert, M.P., R. Suurs, S. Negro, S. Kuhlmann and R.E.H.M. Smits (2006) 'Functions of Innovation Systems: A New Approach for Analysing Technological Change', mimeograph, Copernicus Institute for Sustainable Development and Innovation, Department of Innovation Studies, Utrecht University, and Fraunhofer Institute for Systems and Innovation Research, Karlsruhe.

Henderson, R., and K. Clark (1990) 'Architectural Innovation: The Reconfiguration of Existing Product Technologies and the Failures of the Established Firm', *Administrative Science Quarterly* 35: 9-30.

Jacobsson, S., and A. Bergek (2004) 'Transforming the Energy Sector: The Evolution of Technological Systems in Renewable Energy Technology', *Industrial and Corporate Change* 13.5: 815-49.

—— and A. Johnson (2000) 'The Diffusion of Renewable Energy Technology: An Analytical Framework and Key Issues for Research', *Energy Policy* 28.9: 625-40.

Kemp, R., A. Rip and J. Schot (2001) 'Constructing Transition Paths through the Management of Niches', in R. Garud and P. Karnoe (eds.), *Path Dependence and Creation* (Mahwah, NJ: Lawrence Erlbaum Associates): 269-99.

Kubeczko, K., A. Kaufmann and K.M. Weber (2006a) *Urbanes Innovationssystem: Konzeptionelle Grundlagen und Politikansätze* (working paper; Vienna: ARC Systems Research, February 2006).

——, E. Rametsteiner and K. Weiss (2006b) 'The Role of Sectoral and Regional Innovation Systems in Supporting Innovations in Forestry', *Forest Policy and Economics* 8.7: 704-19.

Lundvall, B.-Å. (ed.) (1992) *National Systems of Innovation. Towards a Theory of Interactive Learning* (London: Frances Pinter).

Markard, J. (2005) 'Investigating Radical Innovation Processes: Innovation System Analysis', paper presented at the *ARC Systems Research Seminars*, Vienna, Austria, 4 November 2005.

Rennings, K., R. Kemp, M. Bartolomeo, J. Hemmelskamp and D. Hitchens (2003) *Blueprint for an Integration of Science, Technology and Environmental Policy: Final Report* (Brussels: European Commission).

Rohracher, H. (ed.) (2005) *User Involvement in Innovation Processes: Strategies and Limitations from a Socio-technical Perspective* (Munich: Profil-Verlag).

Rotmans, J., R. Kemp and M. van Asselt (2001) 'More Evolution than Revolution: Transition Management in Public Policy', *Foresight* 3: 15-31.

Schauer, B. (2007) *Portfolio Selection Considering Risk and Project Interrelations* (PhD thesis; Vienna: TU Vienna).

Seibt, C., D. Schartinger, P. Wagner and M. Weber (2002) *Technologien für die zukünftige Entwicklung des Güterverkehrs in Österreich* (working paper; Seibersdorf, Austria: ARC Seibersdorf Research).

Smith, A., A. Stirling and F. Berkhout (2005) 'The Governance of Sustainable Sociotechnical Transitions', *Research Policy* 34: 1,491-1,510.

Vergragt, P.J. (2000) *Strategies towards the Sustainable Household: Final Report of the SusHouse Project* (Delft, Netherlands: Delft University of Technology).

Wagner, P., D. Banister, K. Dreborg, E.A. Eriksson, D. Stead and K.M. Weber (2003) *The Impact of ICT on Transport and Mobility* (ESTO research report; Seville, Spain: IPTS; Vienna: ARC Systems Research).

Weber, K.M. (2005) 'What Role for Politics in the Governance of Complex Innovation Systems? New Concepts, Requirements and Processes of an Interactive Technology Policy for Sustainability', in J.N. Rosenau, E.U. von Weizsäcker and U. Petschow (eds.), *Governance and Sustainability: New Challenges for States, Companies and Civil Society* (Sheffield, UK: Greenleaf Publishing): 100-18.

——, A. Geyer, D. Schartinger and P. Wagner (2002) *Zukunft der Mobilität in Österreich: Konsequenzen für die Technologiepolitik* (research report; Seibersdorf, Austria: ARC Seibersdorf Research).

——, K.-H. Leitner, K. Whitelegg, I. Oehme, H. Rohracher and P. Späth (2003) 'Middle-range Transitions in Production–Consumption Systems: The Role of Research Programmes for Shaping Transition Processes towards Sustainability', in *Proceedings of the Conference on the Human Dimension of Global Environmental Change*, Berlin, Germany, 5–6 December 2003.

——, K. Kubeczko, K.-H. Leitner, I. Oehme, H. Rohracher, P. Späth and K. Whitelegg (2005) *Transition zu nachhaltigen Produktionssystemen* (final report; Vienna: ARC Systems Research).

Whitelegg, K. (2004) *Patchwork Policy-making: Linking Innovation and Transport Policy in Austria* (case study report for the OECD NIS MONIT Project; Seibersdorf, Austria: ARC Systems Research).

# 20
# Transition management for sustainable consumption and production

*René Kemp*
UNU-MERIT, ICIS, Drift and TNO, Denmark

## 20.1 Introduction

Consumption constitutes a big problem for sustainable development. For low-income countries, sustainable development means *more* consumption. For rich countries it means *different* consumption, not reductions in material consumption. Governments and people in rich countries are just as committed to growth as people in low-income countries, perhaps even more. For a consumer there is no such thing as enough consumption. This makes sustainable consumption almost an oxymoron because how can we achieve lower environmental impact if we are consuming more? 'Green' products help to reduce environmental impacts. To achieve even greater reductions together with other sustainability benefits, more is needed than product changes: we need system innovations: that is, transformations and changes in systems of provision and behaviour (Rotmans *et al*. 2000; Kemp and Rotmans 2004; Smith *et al* 2005; Weaver *et al*. 2000).

Such new systems are unlikely to emerge through the normal operation of markets or through business sustainability strategies. System innovation is inexorably linked with institutional change, guiding images and joined-up efforts (Kemp and Rotmans 2005), different social practices and a new type of normality (Shove 2004), all of which are feeding on one another. The process of change is non-linear, structured by a patchwork of sociotechnical regimes, shaped by macro factors. In thinking about system innovation, the notion of transition is a useful concept, bringing into focus the time aspects and issues of path dependence, emergence, co-evolution and resistance to change, which make the steering of such a process very difficult. In this chapter I offer a discussion of transformations, sustainability transitions and the possibilities for managing societal transformations towards sustainability goals through transition management, a steering concept that I developed with Jan Rotmans and Derk Loorbach, which is currently used in The Netherlands by the national government as a model for sustainability policy.

## 20.2 Societal transformations as transitions

Transformations are the outcome of historical processes. They are not caused by a single factor. Technical change is an inherent part of the transformations that occurred in the process of modernisation as shown by the landmark studies of Mumford (1934) and Landes (1969), and by more recent studies such as those by Rosenberg (1982), Freeman and Louçã (2001) and Geels (2002). Technology is both endogenous to these transformations and at the same time a driver for change.

A transformation is a process of structural change. Transitions are change processes from an old to a new equilibrium. An example is the demographic transition. In a transition there is a change in form, with the new form constituting the basis for development. An example is the transition from weeding to pesticide-based agriculture or the shift from public transport to car-based transport. A transition denotes a change in dynamic equilibrium, being the outcome of interaction processes. In transitions there is destabilisation (of an old regime) and stabilisation. The new dynamic equilibrium may be superseded by a new equilibrium. The term *transition* is often used in connection with former state-controlled economies that have changed to market economies. In this literature the term is used simply as a general term for a widely disparate phenomenon. A distinction is made between approaches for managing the transition, such as shock therapy and gradualist reforms, but, as far as we know, there is no distinction between types of transition.

To us, the concept of transition is a useful organiser for theorising about transformations. In this chapter I distinguish two types of transition:[1]

- Evolutionary transitions, where the outcome is not planned in an important way
- Goal-oriented (teleological) transitions, when a (diffuse) goal or vision of the end-state is guiding decision-makers, orienting strategic decisions of private decision-makers

Most transitions appear to be of the first type. An example of the first is the transition from the use of sailing ships to steam boats in the 19th century and the shift from horse

---

1 Other typologies are offered by Geels (2002) and Berkhout et al. (2004). I do not speak of planned transitions because transitions cannot be planned. Only projects within transitions can be planned, not the transitions themselves, being non-linear processes of change.

and carriage to automobiles in the first half of the 20th century (described in Geels 2002, 2005) and the whole process of mechanisation. An example of the second is the development of centralised electricity systems (described in Hughes 1989) and the transition from piston engine aircraft to jetliners. Another example of a goal-oriented transition is the transition towards the single European market. It should be noted that also in a goal-oriented transition there is an evolutionary process of a variation, selection and retention (reproduction). It is not a linear or deterministic process but a cumulative process in which there are surprises and setbacks.

Most transitions in the sociotechnical realm are evolutionary transitions, driven by short-term gains in a changing landscape rather than by a long-term goal. The generative element in societal change processes is well described by Paul David in his discussion of technical change as a myopic, cumulative, process:

> Because technological 'learning' depends on the accumulation of actual production experience, short-sighted choices about what to produce, and especially about how to produce it using presently known methods, also in effect govern what subsequently comes to be learned. Choices of technique become the link through which prevailing economic conditions may influence the future dimensions of technological knowledge. This is not the only link imaginable. But it may be far more important historically than the rational, forward-looking responses of optimising inventors and innovators which economists have been inclined to depict as responsible for the appearance of market- or demand-induced changes in the state of technology (David 1975: 4).

In the literature on technological paradigms (Dosi 1988) and regimes (Georghiou *et al.* 1986; Nelson and Winter 1982; Van de Poel 2002) the structured nature of technical change is recognised and emphasised. Both perspectives agree that technical change is not perfectly responsive to changes in demand and cost conditions but occurs along relatively well-identifiable trajectories of change.[2]

A first attempt at theorising about transitions was made by Rotmans *et al.* (2000, 2001). It is said that the process of change in a transition is non-linear; slow change is followed by rapid change when concurrent developments reinforce each other, which again is followed by slow change in the stabilisation stage. There are multiple shapes a transition can take, but the common shape is that of a sigmoidic curve such as that of a logistic (Rotmans *et al.* 2000, 2001).

Transition processes of societal development are believed to be composed of four distinctive phases (Rotmans *et al.* 2001; see also Fig. 20.1):

- The **predevelopment** phase, where there is very little visible change but a great deal of experimentation
- The **take-off** phase, where the process of change gets under way and the state of the system begins to shift

---

2 The reasons for this are discussed in Kemp 1994. It is not a matter of learning-curve effects and network externalities alone. There are also institutional sources behind such patterns, notably (a) interests vested in the continuation of a trajectory, (b) the self-assumed roles of the actors, (c) interpretative frameworks and beliefs and (d) endogenous preferences and habits.

## FIGURE 20.1 Four phases of transition

Sources: adapted from Rotmans et al. 2000, 2001

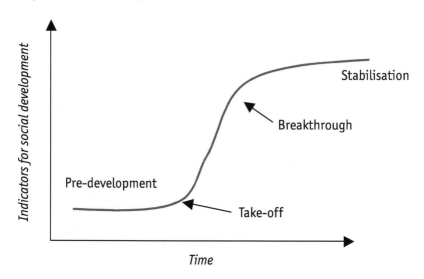

- The **breakthrough** or acceleration phase, in which structural changes occur in a visible way through an accumulation of sociocultural, economic, ecological and institutional changes that react to each other; during this phase, there are collective learning processes, diffusion and embedding processes[3]
- The **stabilisation** phase, where the speed of societal change decreases and a new dynamic equilibrium is reached

A transition can be accelerated by one-off events, such as a war, a large accident (e.g. Chernobyl) or a crisis (such as the oil crisis) but it cannot be caused by such events. In transitions there is multicausality (Geels 2002; Rotmans et al. 2000, 2001). Transitions are the result of endogenous and exogenous developments in regimes and the macro landscape: there are crossover effects and autonomous developments. Technical change interacts with economic change (changes in cost and demand conditions), social change and cultural change, which means that one should look for process explanations such as virtuous cycles of reinforcement (positive feedback) and vicious cycles of balancing (negative feedback).

Transitions are the result of system innovations and other changes. This is visualised in Figure 20.2.

Transformations involving system innovation are interesting from the viewpoint of sustainability because they offer the prospect of large-magnitude environmental benefits, alongside wider social benefits through the development of systems that are inherently more environmentally benign. Examples of system innovation are: the hydrogen

---

3 In Rotman's earlier publications this phase is called the acceleration phase. 'Breakthrough phase' is probably a better term because it refers to the *absolute* amount of change, which is greatest at this stage, rather than the relative amount of change (the speed).

FIGURE 20.2 **A transition is the result of system innovations and other innovations and changes**

Source: Butter *et al.* 2002

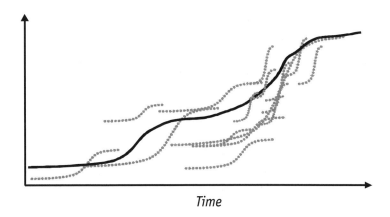

economy, industrial ecology and customised mobility (other examples of system innovation are biomass-based chemistry, multiple sustainable land-use [the integration of the agricultural function with other functions in rural areas] and flexible, modular, manufactured construction; Ashford *et al.* 2001).

The idea of 'green' benefits from system innovation is not new. Freeman (1992) discussed the idea of a 'green' techno-economic paradigm in his book *The Economics of Hope*, and the idea was further investigated in the project 'Technology and the Transition to Environmental Stability' (Kemp 1994). In the early work on transitions the difficulties of a transition from an old to a new system received most attention. It was said that a transition (regime shift) was difficult to achieve because of the adaptation of the selection environment to the old trajectories and because of the dynamic learning and scale effects the old solutions had benefited from, creating institutional and economic barriers to the development of new trajectories (Kemp 1994; Kemp and Soete 1992).

This raises the question how a transition is ever possible. One answer offered (Hoogma *et al.* 2002; Kemp *et al.* 1998, 2001; Rip and Kemp 1998) is that niches for new technologies are an important element. It is said that a new solution needs an application, which could be found in the existing system or outside it (e.g. in the military). Within such niches learning processes can occur that are important for the wider diffusion process. Such learning processes help to create virtuous cycles that allow a new technology to escape lock-in—by helping the technology to overcome the initial barriers of high costs, the non-availability (or high cost) of complementary technologies, and misfits between the new technology and the external environment during the infancy period of a new technology, when it has not yet benefited from dynamic scale and learning economies. Niches are thus believed to be important for learning, investment and alignment processes and are seen as an important element in processes of co-evolution. Niches act as incubators for the new technology, helping the new technology and companies survive the selection pressures that are especially harsh for a new, fledgling, technology in the early period of development.

Drawing on the multi-level model of Rip, Geels (2002) studied several transitions and found that radical new technologies are first used in niches, which are instrumental to processes of learning and adaptation. Transitions are understood from the interplay between niches, regimes and the sociotechnical landscape. The landscape consists of the social values, policy beliefs, world-views, political coalitions, built environment (factories, etc.), prices and costs, trade patterns and incomes. The term 'landscape' refers to the sociotechnical 'lay of the land' (e.g. its gradients), making certain advances easier to follow (see Geels 2002; Geels and Kemp 2000; Rip and Kemp 1998).

Sociotechnical **regimes** are at the heart of Geels's scheme of technological transitions. The term 'regime' refers to rules and institutions, by which we mean the dominant practices, search heuristics, mental models and ensuing logic of appropriateness that pertain in a domain (either a policy domain or a technological domain), giving it stability and orientation, and guiding decision-making. Faced with sustainability problems, regime actors will opt for change that is non-disruptive from the industry point of view, which leads them to focus their attention on system improvement instead of system innovation.

Over time, this may change. Under the threat of losing their market (or losing market share) incumbent companies may decide to change course and commit their resources and powers to the development of a new system. Some companies, however, never make the transition and become extinct. The history of technology is full of companies that failed to make a transition. Also, the pioneers may also pass away. A well-known example is DeHavilland, the pioneer of the first passenger turbojet. The distinction between niches, regimes and sociotechnical landscape helps us to understand processes of structural change that could be seen as the outcome of the interaction of multi-level processes. A common mechanism works through landscape factors that put pressure on a regime of production in which practices and technologies are challenged by new solutions pioneered in niches, with regime actors initially fighting and resisting alternative solutions, focusing their attention and money on improving existing technologies, but, over time, changing course by investing in radical solutions. This is happening now in the case of fuel cells, with DaimlerChrysler and Ford, long resisting alternative types of propulsion, joining forces with Ballard, a Canadian-based manufacturer of fuel cells. Another example is BP, the UK oil giant, which has moved into the production of renewable energy. When this happens—that is, when the belief systems (world-views) and strategies of key actors change—new developments gain momentum and a regime shift may occur.

In the transition scheme, governments are not autonomous actors. Policy is seen as *part* of transitions or transformations instead of an external force. Policy reacts to the problems associated with technology use and is deeply influenced by the interests, values, beliefs and mental models within these systems (through the representation of the interests in the political process) and by the values and beliefs of society at large. Quite often there is a clash between values and beliefs within sectors and those within society, with government acting as a mediator. Government cannot be regarded as a completely dependent variable: government agencies have their own programmes and programmatic views and are important institutions in their own right, quite often resistant to change. Within sectors there are, of course, conflicts too, between workers and management and between incumbent companies and newcomers.

## 20.3 Possibilities for managing transitions

What are the possibilities for managing transitions? Can transitions be brought about through steering activities? From what I have said it may be apparent that transitions cannot be managed in a controlling sense, for the simple reason that transitions are the result of the interplay of many unlike processes, several of which are beyond the scope of management, such as cultural change, which can be considered as a sort of autonomous process. What one can do, however, is influence the direction and speed of change through various types of steering.

Transitions defy control but they can be influenced: both the rate of change and nature of change can be influenced. The management of transitions can be done through the (direct and indirect) use of three coordination mechanisms: markets, hierarchy, and structure or institutions.[4] Market coordination is when prices coordinate economic decisions. In the second, hierarchical, case of planning (either state planning or company planning) economic activities are centrally coordinated through a plan or through a set of goals. The third type of coordination is through structure or institutions.[5] By this we mean the coordination that is achieved through standard practices, trust, collective norms, networks and shared expectations and beliefs. Institutions play a coordinating role by limiting the choice-set, providing orientation and reducing uncertainty. Without them the world would be rather unpredictable and actors would be without orientation. One important institution is the self-assumed role of companies (what kind of company they are and want to be). Networks too are an important institution, the importance of which is increasingly being recognised (Powell 1990). Institutional change is an element of a transition, and policy should be concerned with it by facilitating processes of institutionalisation.

The idea of institutions as collective properties shaping further change is very important for thinking about transitions and transition management. It brings into focus possibilities for intervention at a different level: the level of collective structures and matrices of institutions. Here we are thinking about policies oriented at (1) market structures and networks and (2) actors' views and beliefs. Examples of the first are policies of market liberalisation and privatisation. Examples of the second are policies aimed at altering the engineering consensus and assumptions; cluster policies aimed at creating new clusters and product constituencies. What these policies have in common is that they are oriented towards creating a structuralist element under which micro behaviour will proceed.

---

4 The term 'management' is used here in the sense of shaping, not in the sense of control (Palmer and Dunford 2002).
5 The notion of 'institution' used here is a broad one, which includes interpretative frameworks and belief systems that colour problem definitions and includes engineering consensus about the relevant problems and appropriate approaches for solving problems in a technical domain (cf. Parto 2003; Rip and Kemp 1998).

## 20.4 Sustainable transitions

Sociotechnical transitions are interesting from a sustainable consumption and production point of view because they may offer sustainability benefits in the form of lower emissions, a shift away from depletable resources and employment opportunities. Examples are: industrial ecology (the closing of material chains), the hydrogen economy, integrated water management, and customised mobility.

So far, sustainability goals have been pursued through environmental policy, laying down specific requirements for products and processes, and through subsidy policies for the use and development of environmental technologies. Past policies led to a considerable 'greening', but progress is often viewed to be insufficient from a sustainability point of view (NMP4 2001; Weaver *et al.* 2000). The possibilities for gradual improvement should be further exploited, but one should also explore the possibilities of system changes that may lead to greater benefits. Support for the latter type of change is warranted because the time-scale of system innovation is 25 years or more and beyond the mutual coordination possibilities of individual actors who have a short-term orientation. Economic change and technological progress in a market economy is driven by short-term economic benefits rather than long-term optimality (Kemp and Soete 1992). System innovation meets several barriers, and the environmental problems may be viewed as problems of system coordination as well as problems stemming from the non-internalisation of external costs (Smith 2002). We are often trapped into certain ways of doing things.[6]

Given the institutional barriers to system innovation, policy interventions should be oriented not just towards conditions in the economic framework (through the use of taxes and regulations) but also towards beliefs, people's outlook on things, expectations, institutional frameworks and arrangements. Indicative planning through the setting of goals, and the creation of networks and constituencies for alternative systems, are ways to do so. These regime policies should complement policies that change the cost structure. Furthermore, apart from changes to policies, we need changes in politics, which should be oriented more towards the long term and towards sustainability goals. The policy regime has to change too.

The power of the market in efficiently allocating the decisions of millions of actors should be utilised to the greatest possible extent—for instance, through the use of market-based instruments such as emissions trading—but it should be combined with the intelligence of people in terms of generating ideas about alternative systems and institutions. Bottom-up initiatives such as experiments with new technologies should be encouraged and exploited. Transition management is not limited to bottom-up initiatives but also has top-down elements. Examples are the visioning of long-term goals, control policies, the establishment and maintenance of portfolios of options and industrial policies. What we need is for people's desire for a better society for themselves and for later generations (which is ill-served by free markets) to be institutionalised in the political system and be used as a guide for policy and economic decision-making. Sustainability has to be discovered (created in the act of discovery), which is why we need a great deal of variation. It is not for the government to pick solutions, and support should be temporary. In my view, sustainability is best worked towards in a flexible, for-

---

6 Linscheidt (1999) talks about development traps.

ward-looking manner, using all three coordination mechanisms to manage transitions: markets, hierarchy and structure.

## 20.5 Transition management

This section offers a model for managing transitions to sustainability. The model has been developed by the author with Jan Rotmans and others for the 4th National Environmental Policy Plan of The Netherlands (NEPP4; in Dutch: NMP4). Transition management consists of a deliberate attempt to work towards a transition into what is believed to be a more sustainable direction. There are different ways of trying to achieve such a transition. One can opt for the use of economic incentives or rely on a planning and implementation approach or some combination of the two: for example, by using market-based indicative planning based on sustainability visions. We opted for the last option, which allows us to combine the best of both worlds: the reliance on markets helps to safeguard user benefits and promote efficiency, whereas the use of targets informed by long-term visions of sustainability helps to orient sociotechnical dynamics towards sustainability goals. We thus hope to achieve efficiency, flexibility and long-term welfare benefits.

The basic steering philosophy is that of **modulation**, not dictatorship, or planning and control. Transition management joins in with ongoing dynamics and builds on bottom-up initiatives. Ongoing developments are exploited strategically. Transition management for sustainability tries to orient dynamics to sustainability goals. The long-term goals for functional systems are chosen by society either through the political process or in a more direct way through a consultative process. The goals can be quantitative or qualitative. The goals may refer to the use of a particular solution (fuel cell vehicles, road pricing or multimodal transport) but, in preference, refer to performance, such as non-congested transport that is safe, accessible and that minimises nuisance.

The goals and policies to further the goals are not 'set in stone' but are constantly assessed and periodically adjusted in development rounds. Existing and possible policy actions are evaluated against two criteria: first, the immediate contribution to policy goals (e.g. in terms of kilotonnes of reduction of carbon dioxide [$CO_2$] and reduced vulnerability through climate-change adaptation measures) and, second, the contribution of the policies to the overall transition process. Policies thus have a **content goal** and a **process goal**. Learning, maintaining variety and institutional change are important policy aims, and policy goals are used as means to achieve the desired end. The use of development rounds brings flexibility to the process, without losing the long-term focus. Transition management is oriented towards achieving structural change in a stepwise manner. A schematic view of transition management is given in Figure 20.3.

Whereas existing policy is for the most part based on short-term goals, transition management is oriented towards the realisation of long-term goals with use of visions of sustainability. The short-term goals are based on the long-term goals and comprise learning goals. Sustainability visions are explored with use of small steps. Transition management is based on a two-pronged strategy. It is oriented towards both system improve-

## FIGURE 20.3 [a] Current policy compared with [b] the transition management approach

Source: Kemp and Loorbach 2003

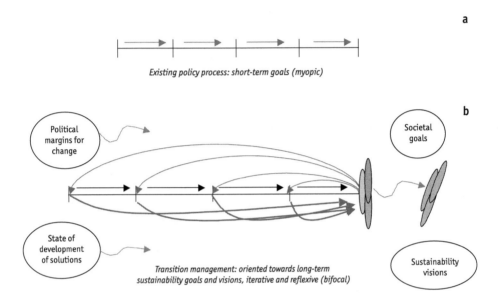

ment (improvement of an existing trajectory) and system innovation (representing a new trajectory of development or transformation).

The role of government varies in each transition phase. For example, in the predevelopment stage there is a great need for social experimentation and for the development of visions. In the breakthrough phase there is a special need to control the side-effects of the large-scale application of new technologies. Throughout the entire transition the external costs of technologies (old and new) should be reflected in prices. This is not easy. Taxes are disliked by any person who has to pay them. Perhaps it helps if taxes are introduced as part of a politically accepted transition endeavour, with the revenue generated used to fund the development of alternatives.

Transition management breaks with the old planning-and-implementation model aimed at achieving particular outcomes. It is based on a different, more process-oriented philosophy. This helps to deal with complexity and uncertainty in a constructive way. Transition management is a form of process management to achieve a set of goals determined by society; in this process the problem-solving capabilities of that society are mobilised and translated into a transition programme, which is legitimised through the political process. Transition management does not aim to realise a particular path at any cost. It engages in the exploration of some paths, with exit strategies. It does not consist of a strategy of forced development, going 'against the grain', but uses bottom-up initiatives and business ideas of alternative systems offering sustainability benefits in addition to user benefits.

Key elements of transition management are:

- The development of sustainability visions and the setting of transition goals
- The use of transition agendas
- The establishment, organisation and development of a transition arena
- The use of transition experiments and programmes for system innovation
- The monitoring and evaluation of the transition process
- The creation and maintenance of public support
- A cycle of learning and adaptation

Each of these is discussed in turn below.

### 20.5.1 The development of sustainability visions and the setting of transition goals

Transition management is based on sustainability visions and the use of transition goals and agendas. The act of organising an envisioning process aimed at sustainable development is cumbersome. It requires the ability to set aside one's own preferences and concomitant everyday 'noise'. It also requires insight and imagination to look ahead one or two generations. Last but not least it requires some sort of minimal agreement on what sustainability means for a specific transition theme in a situation where opinions usually diverge. Many sustainability visions are still imposed by the government on other parties in a top-down manner or originate from a select group of experts who are far from representative of the broad social setting needed.

The long-term visions of sustainability should be used as a guide to formulate programmes and policies and to set short-term and long-term objectives. To adumbrate transitional pathways, the transition visions must be appealing and imaginative so as to be supported by a broad range of actors. Inspirational final visions are useful for mobilising social actors, although they should also be realistic about innovation levels within the functional subsystem in question.

There will usually be different sustainability visions and different pathways for achieving them. This is visualised in Figure 20.4 in the form of a basket of images with different paths leading to the images. Over time, the transition visions should be adjusted as a result of what has been learned by the players in the various transition experiments. Based on a process of variation and selection new visions emerge, others die out and existing visions will be adjusted. Only during the course of the transition process will the most innovative, promising and feasible transition visions and images be chosen. This evolutionary goal-seeking process means a radical break with current practice in environmental policy, where quantitative standards are set on the basis of studies of social risk and adjusted for political expediency. Risk-based target-setting is doomed to fail when many issues are at stake and when the associated risks cannot eas-

FIGURE 20.4 **Transition as a goal-seeking process, with multiple transition images and goals**

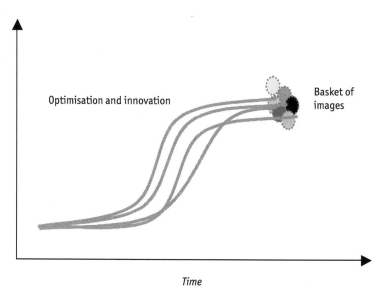

ily be expressed in fixed, purely quantitative, objectives. This holds true not only for climate change but also for sustainable transport.

Transition management thus differs from so-called 'blueprint' thinking, which operates from a fixed notion of final goals and corresponding visions. Transition management uses goals too, but these are derived from the long-term objectives (through so-called 'backcasting') and contain qualitative as well as semi-quantitative measures. Apart from **content** goals or objectives (which at the start can look like the current policy objectives, but later will increasingly appear to be different), transition management uses **process** objectives (concerning the speed and quality of the transition process) and **learning** objectives (concerning what has been learned from the experiments carried out, what is blocking progress and the identification of things that we want to know). Learning, therefore, is a policy objective in its own right.

## 20.5.2 The use of transition agendas

Based on the common problem perception and the shared sustainability vision(s) a joint transition agenda can be designed. This is important because all participants take their own agenda into the transition arena, whereas a joint transition agenda contains common problem perceptions, goals, action points, projects and instruments. The means for realising an effective execution of the proposed plans are important in order to resolve the problems on the transition agenda as adequately as possible. Actually, a transition agenda is a joint action programme for initiating or furthering transitions. It is important to register which party is responsible for which type of activity, project or instrument to be developed or applied. Government would be responsible for certain policies, and industry for certain innovation activities. The monitoring of this joint action programme is important to guarantee that the transition agenda is complied with as far as possible.

An adequate transition agenda forms a binding element in the transition process, in which participants need each other. It coordinates action between mutually dependent actors. Coordination is thus achieved not only through markets but also through collective choice and new institutions. The transition agenda requires a kind of balance between structure and flexibility. Structure is needed to work at a scale appropriate to the issue in question and to frame the issue in terms of themes and subthemes. Coherence between the various subthemes and the level or scale at which they are examined is a separate, important point on the transition agenda. Structuring the transition agenda is time-consuming, but this cost is paid back in the form of increased quality in the transition management process (Dirven *et al.* 2002). Flexibility is needed because the transition agenda is dynamic and changes over time. The transition agenda helps to translate long-term thinking into short-term action. Agenda-setting is an iterative and cyclical process and is a learning process in itself.

### 20.5.3 The establishment, organisation and development of a transition arena

A novel and important aspect of transition management is the establishment and organisation of a transition arena: a platform at which innovators and visionary people meet. It would operate in addition to (and partly independent from) the normal policy-making networks dominated by incumbent companies having an interest in the status quo. The selection of participants for this transition arena is of vital importance. These participants need to have some basic competences at their disposal: they need to be visionaries, to be forerunners, to be able to look beyond their own domain or working area and to be open-minded thinkers. They must function quite autonomously within their organisations but also have the ability to convey the developed vision(s) and set it (them) out within their own organisations. They need to be willing to invest a substantial amount of time and energy to play an active role in the transition-arena process.

Government has a task not just in the setting up of a transition arena but also in the facilitation of interactions within the transition arena, not simply in process terms but also in terms of substance. A continuous process of feeding the participants in the arena with background information and detailed knowledge on a particular topic is necessary, enabling a process of co-production of knowledge among the participants. This is of vital importance, because experience shows that, in most cases, arena participants have insufficient time, specific knowledge or sense of overview to immerse themselves in complex problems. The arena is a novel institution for visionary (out-of-the-box) thinking, feeding into innovation decisions of organisations willing to innovate. The goal is not necessarily to strive for consensus but to discuss problem perceptions, long-term goals and transition paths.

### 20.5.4 The use of transition experiments and programmes for system innovation

Programmes for system innovation are a key element of transition management. Here, one should think of a programme for intermodal transport and for decentralised electricity systems. The programmes should be time-limited and be adapted in the light of experience. An important element of these programmes is the use of transition experiments (i.e. strategic experiments designed to learn about system innovation and transition visions). The crucial point is to measure to what extent these experiments and projects contribute to the overall sustainability system goals and to measure in what way a particular experiment reinforces another experiment. Are there specific niches for experiments that can be identified? What is the attitude of the regime actors towards these niche experiments? The aim is to create a portfolio of transition experiments that reinforce each other and whose contribution to the sustainability objectives is significant and measurable.

Preferably, these experiments need to link up with ongoing innovation projects and experiments in such a way that the existing effort put into these innovation experiments can be used as much as possible. Often, many experiments already exist but these are not set up and executed in a systematic manner, as a result of which the required cohesion is lacking.

The experiments are best undertaken as part of a portfolio approach. Because transition processes are beset with structural uncertainties of different kinds it is important to keep a number of options open and to explore the nature of these uncertainties through the transition experiments. Through learning experiences with transition experiments the estimation of these uncertainties will change over the course of the transition process. This may in turn lead to an adjustment of the transition visions, images and goals. In this search and learning process, scenarios can play an important role, particularly explorative scenarios (see Elzen *et al.* 2004) that attempt to explore future possibilities without too many decision-making constraints. Explorative scenarios allow for an investigation of which options and experiments are most promising and feasible, and which ones to drop out. This leads to a necessary variation of options, taking account of possible sustainable futures.

### 20.5.5 The monitoring and evaluation of the transition process

Transition management involves monitoring and evaluation as a regular and continuous activity. Two different processes should be monitored: the transition process itself and the cycle of transition management. The monitoring of the transition process itself consists of the monitoring of macro developments, niche developments and regime developments.

The monitoring of the transition management cycle consists of (1) the monitoring of actors within and 'outside' the transition arena, including their behaviour, networking activities, alliance forming and responsibilities with regard to activities, projects and instruments; (2) the monitoring of the transition agenda, including the actions, goals, projects and instruments agreed on; and (3) the monitoring of the actions themselves, including the barriers, prospects and points to be improved.

The overall learning philosophy is that of 'learning by doing'. The monitoring of learning processes, however, is easier said than done. The phenomenon of 'learning' is for many still an abstract notion that cannot easily be translated into components for monitoring. It is therefore important to formulate explicit learning goals for transition experiments that can be monitored.

The evaluation of learning processes is in itself a learning process and may lead to an adjustment of the developed transition vision(s), transition agenda and transition management process within the transition arena. The set interim objectives are evaluated to see whether they have been achieved; if this is not the case, they are analysed to see why not. Have there been any unexpected social developments or external factors that were not taken into account? Have the actors involved not complied with the agreements that were made?

Following this, a new transition management cycle starts, which takes another few years. In the second round the proliferation of the required knowledge and insights is central, which requires a specific strategy for initiating a broad learning process.

### 20.5.6 The creation and maintenance of public support

Because these transition management cycles take several years within a longer-term context of 25–50 years, the creation and maintenance of public support is a continuous concern. When quick results do not materialise and setbacks are encountered it is important to keep the transition process going and avoid a backlash. One way to achieve this is through participatory decision-making and the societal choice of goals. But societal support can also be created in a bottom-up manner, by engaging in experiences with technologies in areas in which there is local support. The experience may remove fears elsewhere and give proponents a weapon against the forces of conservatism. With time, solutions may be found for the problems that limit wider application. Education, too, can allay fears, but real experience is probably a more effective strategy.

### 20.5.7 The cycle of learning and adaptation

Transition management is a cyclical and iterative process involving adaptation (cf. Rammel and van der Bergh 2003). Each cycle consists of four main activities: establishing and further developing a transition arena for a specific transition theme; the development of long-term visions for sustainable development and of a common transition agenda; the initiation and execution of transition experiments; and the monitoring and evaluation of the transition process. One such transition cycle may take between two and five years, depending on the practical context within which one has to operate.

### 20.5.8 Instruments

Instruments for transition management are in a certain sense endogenous to the process. Transition management does not call for a great upheaval in policy instruments but says that different policy fields should be better coordinated. Existing policies could be improved and extended as follows (see Smits and Kuhlmann 2004):

- **Science policy**: carry out sustainability assessments of system innovations, roadmap the transition path, study past and ongoing transitions, and focus on the role of policy and the usefulness of various governance models
- **Innovation policy**: create innovation alliances, institute R&D programmes for sustainable technologies, use transition experiments, and align innovation policies with transition goals
- **Sector policy**: create niche policies (through procurement, regulations or the use of economic incentives), remove of barriers to the development of system innovations, and formulate long-term goals and visions to give direction to research and innovation

The advantages of mission-based strategies are combined with those of 'technology-blind' NSI (National System Innovation) policies aimed at entrepreneurship, collaboration and market dynamism.

## 20.6 Transition policies in The Netherlands

Transitions to sustainability are the focus of attention in The Netherlands. There is a widely shared view that the current trajectories of fossil-fuel-based energy, intensive farming and car-based transport are not sustainable environmentally or economically. Attention is being given to alternative systems in addition to possibilities for making the trajectories more sustainable. Examples of such alternative systems are: customised mobility based on different transport modes, energy systems based on renewable sources, and precision farming.

The interest in system innovations is motivated by environmental and economic reasons. The alternative systems should be attractive not only from an environmental point of view but also from an economic point of view (in terms of providing jobs, generating income and providing superior products and services to individual end-users).

Given uncertainty about what systems are best, the Dutch government has opted for the simultaneous exploration of multiple options through the use of adaptive policies, based on iterative and interactive decision-making. New systems are being 'grown' in a gradual manner, relying on feedback and decentralised decision-making. The following transitions are explored in The Netherlands:

- Transition to sustainable energy:
  - Goal: to develop a system of energy supply that is reliable, efficient and low in emissions

- Transition to biodiversity and sustainable use of natural resources:
  - Goals: to maintain biodiversity, which is essential for food supply, fertility of soils and climate; to prevent the over-use of natural resources; and to promote the re-use and recycling of those natural resources

- Transition to sustainable agriculture:
  - Goals: to realise an agricultural system that has a minimal impact on the environment and that does not impair human health, landscape qualities or animal well-being
- Transition to sustainable mobility:
  - Goal: to create a transport system that produces low emissions and little nuisance from noise while maintaining high levels of accessibility, safety and spatial value

Attention is given to the interplay between various transitions. For instance, in the energy transition attention is given to energy crops and to transport.

Following discussions about various system innovations, 26 transition paths have been chosen for sustainable energy within six transition themes. For each team, a platform is being established, typically chaired by a person from business.

Through the transition paths (see Table 20.1) it is hoped to achieve an additional reduction in $CO_2$ emissions of 180 million tonnes (see Fig. 20.5). It is too early to tell whether such reductions will be achieved and whether the model of transition management is a good model for achieving this. Expectations about transition management are rather high, even when historical transitions research (Geels 2005) suggests that transitions in sociotechnical systems defy control and effective steering. Policy can do little more than increase the *chance* for a transition to occur and to shape the features of it.

At the time of writing this chapter (2007), the approach of transition management is broadly supported in The Netherlands by government, business, politicians and the main government advisory councils. Reasons for this are that it fits the Dutch tradition of cooperation between societal actors and that the policies did not cause great pain. Thus far, existing systems are not really threatened by the new systems. Non-sustain-

FIGURE 20.5 **Time path for carbon dioxide ($CO_2$) emissions in The Netherlands**

Source: Presentation Hugo Brouwer in London, 2005; published in Taskforce energietransitie, 2006

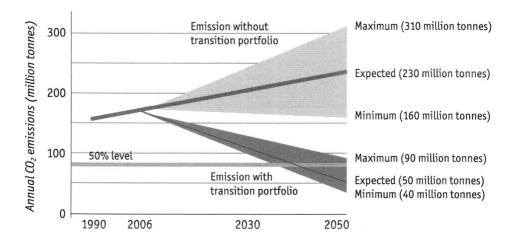

TABLE 20.1 Themes, goals and transition paths in the Dutch energy transition

| Theme | Goal | Transition path |
|---|---|---|
| New gas | To become the most sustainable gas country in Europe | • Decentralised electricity generation<br>• Energy-efficient greenhouses<br>• Hydrogen fuel<br>• Clean fossil fuels<br>• Built environment |
| Sustainable mobility | Factor 2 reduction of GHG emissions for new vehicles in 2015 and Factor 3 reduction for all vehicles in 2030 | • Hybrid propulsion<br>• Biofuels<br>• Hydrogen vehicles<br>• Intelligent transport systems |
| 'Green' resources | Substitution of 30% of resources for energy by 'green' resources | • Biomass production in The Netherlands<br>• Chains for biomass import<br>• Wise biomass co-production<br>• Synthetic natural gas<br>• Sustainable chemistry |
| Chain efficiency | 20–30% extra improvement of product chains by 2030 | • Optimising the waste chain<br>• Precision farming<br>• Process intensification<br>• Multimodal transport<br>• Clearing house for swapping bulk products (to avoid unnecessary transport)<br>• Symbiosis (closing material loops)<br>• Micro co-generation<br>• Energy-efficient paper production |
| Sustainable electricity supply | To make electricity supply more sustainable | • Renewable energy sources<br>• Decarbonisation and co-generation<br>• Electric infrastructure<br>• Electricity use |

able technologies are not being phased out or seriously penalised. It is hoped that the commitment to sustainability transitions will help to make tough political choices about control policies, which will be necessary to move to alternative systems, but whether this will happen is far from certain. Transition management is not an instrument but a framework for policy-making and politics (governance).

## 20.7 Conclusions

In discussions about sustainable development in the United Nations Environment Programme (UNEP) and other fora, the idea of sustainable consumption emerged as an umbrella term that brings together a number of key issues, such as meeting needs, enhancing the quality of life, improving efficiency, minimising waste, taking a life-cycle perspective and taking into account the equity dimension: integrating these component parts in the central question of how to provide the same or better services to meet the basic requirements of life and the aspiration for improvement, for both current and future generations, while continually reducing environmental damage and the risk to human health (UNEP 2002). Achieving this requires the commitment of business, government actors and citizens to sustainability goals.

Thus far, sustainable consumption has been pursued through ecologically improved products, not through comprehensive change in the form of system innovations—that is, alternative systems for fulfilling societal functions such as transportation, communication, housing and food production (Geels 2004). To stimulate system innovations, the Dutch model of transition management appears to be useful. It helps to bring together many actors (technologists, designers, governments, business and citizens) to work on sustainability transitions. Transition management takes on board the criticism of sociologists that ecological modernisation is too supply-oriented (Shove 2004; Spaargaren 2003), neglecting issues of lifestyle and normality. Transition management sees supply-side changes and demand-side changes as interlinked and opts for change from both sides rather than from one just side.

Transition management helps to steer processes of co-evolution in a reflexive manner, inserting feedback into societal decision processes and investment decisions (Kemp and Loorbach 2006). Transition management could be viewed as 'evolutionary governance' as it is concerned with the functioning of the variation–selection–reproduction process: creating variety in terms of technology and systems of provision informed by visions of sustainability; shaping new paths; and reflexively adapting existing institutional frameworks and regimes. It is a model for escaping lock-in and moving towards solutions offering multiple benefits for society as a whole, not just for consumers. It is *not* a megalomaniac attempt to control the future but an attempt to insert normative goals into evolutionary processes in a reflexive manner. Learning, maintaining variety (through portfolio management) and institutional change are important policy aims. Evolutionary change is believed to bring greater benefits than revolutionary change. Evolution is believed to be 'the greatest designer of all'. Of course, evolution must be guided by ideas of what constitutes progress. Concepts of ecodesign and industrial ecology will remain useful but are to be combined with specific visions for sustainable transport and energy.

# References

Ashford, N., W. Hafkamp, F. Prakke and P. Vergragt (2001) 'Pathways to Sustainable Industrial Transformations: Co-optimising Competitiveness, Employment, and Environment', Final Report to the Ministry of Environment and Spatial Planning, Government of The Netherlands, 30 June 2001.

Berkhout, F., A. Smith and A. Stirling (2004) 'Technological Regimes, Transition Contexts and the Environment', in B. Elzen, F. Geels and K. Green (eds.), *System Innovation and the Transition to Sustainability: Theory, Evidence and Policy* (Cheltenham, UK: Edward Elgar): 48-75.

Butter, M. (2002) 'A Three Layer Policy Approach for System Innovations', paper presented at the *1st BLUEPRINT workshop, 'Environmental Innovation Systems'*, Brussels, Belgium, January 2002.

David, P. (1975) *Technical Choice, Innovation and Economic Growth: Essays on American and British Experience in the Nineteenth Century* (Cambridge, MA: Cambridge University Press).

Dosi, G. (1988a) 'The Nature of the Innovation Process', in G. Dosi, C. Freeman, R. Nelson, G. Silverberg and L. Soete (eds.), *Technical Change and Economic Theory* (London: Frances Pinter): 221-38.

Dirven, J., J. Rotmans and A.-Pieter Verkaik (2002) 'Samenleving in Transitie: Een vernieuwend gezichtspunt' ('Society in Transition: A New Perspective'), *LNV, ICIS en Innovatienetwerk Groene Ruimte en Agrocluster*, April 2002.

Elzen, E., F. Geels and K.Green (2004) (eds.) *System Innovation and the Transition to Sustainability: Theory, Evidence and Policy* (Cheltenham, UK: Edgar Elzar): 137-67.

Freeman, C. (1992) *The Economics of Hope* (London: Frances Pinter).

—— and F. Louçã (2001) *As Times Goes By: From the Industrial Revolutions to the Information Revolution* (Oxford, UK: Oxford University Press).

Geels, F.W. (2002) 'Technological Transitions as Evolutionary Reconfiguration Processes: A Multi-level Perspective and a Case Study', *Research Policy* 31.8-9: 1,257-74.

—— (2004) 'From Sectoral Systems of Innovation to Socio-technical Systems: Insights about Dynamics and Change from Sociology and Institutional Theory', *Research Policy* 33.6-7: 897-920.

—— (2005) *Technological Transitions: A Co-evolutionary and Sociotechnical Analysis* (Cheltenham, UK: Edward Elgar).

—— and R. Kemp (2000) *Transities vanuit sociotechnisch perspectief (Transitions from a Sociotechnical Perspective)* (background report for the study 'Transitions and Transition Management' of ICIS and MERIT for NMP-4; Enschede and Maastricht, Netherlands, November 2000).

Georghiou, L., J.S. Metcalfe, M. Gibbons, T. Ray and J. Evans (1986) *Post-innovation Performance: Technological Development and Competition* (London: Macmillan).

Hoogma, R., R. Kemp, J. Schot and B. Truffer (2002) *Experimenting for Sustainable Transport: The Approach of Strategic Niche Management* (London: EF&N Spon).

Hughes, T.P. (1989) 'The Evolution of Large Technological Systems', in W.E. Bijker, T.P. Hughes and T.J. Pinch (eds.), *The Social Construction of Technological Systems: New Directions in the Sociology and History of Technology* (Cambridge, MA: MIT Press): 51-82.

Kemp, R. (1994) 'Technology and the Transition to Environmental Sustainability: The Problem of Technological Regime Shifts', *Futures* 26.10: 1,023-46.

—— and D. Loorbach (2003) 'Governance for Sustainability through Transition Management', paper presented at the *EAEPE 2003 Conference*, Maastricht, Netherlands, 7–10 November 2003.

—— and D. Loorbach (2006) 'Transition Management: A Reflexive Governance Approach', in J.-P. Voss, D. Bauknecht and R. Kemp (eds.), *Reflexive Governance for Sustainable Development* (Cheltenham, UK: Edward Elgar): 103-30.

—— and J. Rotmans (2004) 'Managing the Transition to Sustainable Mobility', in B. Elzen, F. Geels and K. Green (eds.), *System Innovation and the Transition to Sustainability: Theory, Evidence and Policy* (Cheltenham, UK: Edward Elgar): 137-67.

—— and J. Rotmans (2005) 'The Management of the Co-evolution of Technical, Environmental and Social Systems', in M. Weber and J. Hemmelskamp (eds.), *Towards Environmental Innovation Systems* (Heidelberg/New York: Springer): 33-55.

—— and L. Soete (1992) 'The Greening of Technological Progress: An Evolutionary Perspective', *Futures* 24.5: 437-57.

——, J. Schot and R. Hoogma (1998) 'Regime Shifts to Sustainability through Processes of Niche Formation: The Approach of Strategic Niche Management', *Technology Analysis and Strategic Management* 10.2: 175-95.

——, A. Rip and J. Schot (2001) 'Constructing Transition Paths through the Management of Niches', in R. Garud and P. Karnoe (eds.), *Path Dependence and Creation* (Mahwah, NJ: Lawrence Erlbaum Associates): 269-99.

Landes, D. (1969) *The Unbound Prometheus: Technological and Industrial Development in Western Europe from 1750 to the Present* (Cambridge, MA: Cambridge University Press).

Linscheidt, B. (1999) *Nachhaltiger technologischer Wandel aus Sicht der evolutorischen Ökonomik: staatliche Steuerung zwischen Anmaßung von Wissen und drohender Entwicklungsfalle* (Umweltökonomische Discussion Paper 99-1; Cologne: FiFo Institute).

Mumford, L. (1934) *Technics and Civilization* (New York: Harcourt, Brace).

Nelson, R., and S. Winter (1982) *An Evolutionary Theory of Economic Change* (Cambridge, MA: Belknap Press/Harvard University Press).

NMP4 (4th National Environmental Policy Plan of the Netherlands) (2000) *Een wereld en een wil: Werken aan duurzaamheid (A World and a Will: Working towards Sustainability)* (The Hague: NMP4).

NRLO (1999) *Innovating with Ambition, Opportunities for Agri-business, Rural Areas and the Fishing Industry* (The Hague: NRLO).

Palmer, I., and R. Dunford (2002) 'Who Says Change Can Be Managed? Positions, Perspectives and Problematics', *Strategic Change* 110: 243-52.

Parto, S. (2003) 'Transitions: An Institutionalist Perspective', MERIT research Memorandum 2003-019.

Powell, W.W. (1990) 'Neither Market nor Hierarchy: Network Forms of Organization', in B.M. Staw and L. Cummings (eds.), *Research in Organizational Behavior* (Greenwich, CT/London: JAI Press): 295-336.

Rip, A., and R. Kemp (1998) 'Technological Change', in S. Rayner and E. Malone (eds.), *Human Choices and Climate Change*, 2 (Columbus, OH: Battelle Press).

Rosenberg, N. (1982) *Inside the Black Box: Technology and Economics* (Cambridge, UK: Cambridge University Press).

Rotmans, J., R. Kemp and M. Van Asselt (2001) 'More Evolution than Revolution: Transition Management in Public Policy', *Foresight* 3.1: 15-31.

——, R. Kemp, M. Van Asselt, F. Geels, G. Verbong and K. Molendijk (2000) *Transities & Transitiemanagement: De casus van een emissiearme energievoorziening* (final report of the study on 'Transitions and Transition management' for the 4th National Environmental Policy Plan [NMP4] of the Netherlands; Maastricht: ICIS/MERIT, October 2000).

Shove, E (2004) 'Sustainability, System Innovation and the Laundry', in B. Elzen, F. Geels and K. Green (eds.), *System Innovation and the Transition to Sustainability: Theory, Evidence and Policy* (Cheltenham, UK: Edward Elgar): 76-94.

Smith, A., A. Stirling and F. Berkhout (2005) 'Governance of Sustainable Sociotechnical Transitions', *Research Policy* 34: 491-550.

Smith, K. (2002) 'Environmental Innovations in a Systems Framework', paper presented at the 1st BLUEPRINT workshop, 'Environmental Innovation Systems', Brussels, Belgium, January 2002.

Smits, R., and S. Kuhlmann (2004) 'The Rise of Systemic Instruments in Innovation Policy', *International Journal of Foresight and Innovation Policy* 1. 1-2: 4-32.

Spaargaren, G. (2003) 'Sustainable Consumption: A Theoretical and Environmental Policy Perspective', *Society and Natural Resources* 16: 687-701.

UNEP (United Nations Environment Programme) (2002) *Product-Service Systems and Sustainability: Opportunities for Sustainable Solutions* (Paris: UNEP).

Van de Poel, I. (2002) 'The Transformation of Technological Regimes', *Research Policy* 32.1: 49-68.

Weaver, P., L. Jansen, G. Van Grootveld, E. Van Spiegel and P. Vergragt (2000) *Sustainable Technology Development* (Sheffield, UK: Greenleaf Publishing).

# 21
# Systemic changes and sustainable consumption and production
Cases from product-service systems

*Oksana Mont and Tareq Emtairah*
International Institute for Industrial Environmental Economics, Sweden

## 21.1 The nature of the challenge

Central to the sustainable production and consumption (SCP) agenda is the need for radical changes not only in the ways we produce but also in the ways we consume. The SCP agenda emerged within the understanding that it is not possible to reach the necessary reductions in environmental impact and resource consumption purely by technical solutions directed at improving the efficiency of production processes and 'greening' products. Research demonstrates that aggregate environmental impact continues to rise because of an increasing population and increasing levels of affluence. It was hoped that technological improvements could compensate for increases in these factors. However, to keep within the limits of environmental impact of the year 1990, some commentators argue that a Factor 10, 20 or higher improvement in material and energy efficiency is needed by 2025 (Jensen 1993; Schmidt-Bleek 1995).

Is it possible to reach this goal? What kinds of changes are needed to achieve Factor 10 or 20 improvements? These questions are not new within the environmental debate. Historically, environmental problems were addressed at different levels, including production processes and products. For example, experiences with cleaner production projects and eco-efficiency initiatives demonstrate that it is economically feasible to reach Factor 2 improvements, after which, on average, the efforts become more expensive than the expected savings (IVA 1998). The best cases of product improvement with product redesign provide us with maximum Factor 4 improvements and, on average, Factor 2 (Brezet and van Hemel 1997; Rathenau Institute 1996).

Lately, however, attention has been focused on the potential benefits from more systemic changes, encompassing changes in both production and consumption and in the relationships linking the act of producing and consuming. An approximate exemplification of such changes can be found at the function innovation level, which includes systems for shared use, pay-per-use offers and functional sales, which can be summarised under the name of product-service systems (PSSs). At this level, on average, Factor 2–3 improvements have been reported in various examples, including: washing centres (Hirschl et al. 2003; Weaver et al. 2000), car sharing (Sperling et al. 2000), ski rental services (Hirschl et al. 2001) and integrated pest management (Schmidt-Bleek and Lehner 1998).

Summing up the impact of these changes, it seems to be possible to reach Factor 6–7 improvements in environmental performance, but the challenge of Factor 10–20 still remains. Some proponents argue that higher levels of environmental improvements can be found at the so-called system level, which includes demand-side strategies and initiatives towards sustainable consumption. However, there are a number of conceptual and analytical challenges with system-level changes.

The discussion on system-level changes with regard to SCP has so far been mainly at a conceptual level, with, at best, a focus on the specification of requirements of sociotechnical changes (Geels 2002; Kemp et al. 1998). There are very few cases in the real world to rely on in exploring the dynamics of change, particularly with regard to desired changes towards SCP patterns. We do not, in fact, know what novelties and discoveries can lead to desired changes, nor have we been good at anticipating these changes. A discovery taking place somewhere on the margin might actually, through uncoordinated processes, lead to fundamental changes in the structure of dominant industries, creating a new consumption paradigm and perhaps new consumers and users.

However, from a long history of innovation studies, we do have various ideas about the dynamics of sociotechnical change and the influencing factors. In one conceptualisation, innovation is viewed as a dynamic phenomenon embedded in techno-economic systems and sociocultural systems. Kemp and Rotmans (2001) have taken this line to illustrate how system-level changes and the transition to a more sustainable society entails time-moderated shifts in the different sociocultural, political and techno-economic subsystems. Further, the emergence of the concept of innovation systems—linking the innovating firms, their interactions with other firms and organisations and the institutional environment in spatial and sectoral contexts—illustrates the broader scope for analysis of innovation processes and the dynamics of change (Asheim and Gertler 2004; Edquist and Hommen 1999; Lundvall et al. 2002; Malerba 2005). The innovation system approach may bring more structured discussion on the various factors influencing the development, diffusion and use of innovations. However, the emphasis is still on how innovations come about and diffuse in the market, with the innovating firm taking a central role. Still, even within this view, the central concern is the rate of innovation and not necessarily the direction of innovation (Edquist 2005). The call for system-level changes is directional in one sense and ideological in another, as it prescribes the desired actions of actors in producing and consuming. So what lies underneath such system-level change? What are the nature and dynamics of change involving transformation both in production patterns and in consumer behaviour? And what role do institutions play in this transformation?

In this chapter we look at experiences with function innovation and the concept of PSS as a starting point for theoretical and conceptual exploration of the above questions in the framework of SCP. We draw on existing cases of different levels of maturity within PSSs. However, we do not claim that these cases represent system-level changes as described by proponents of the system innovation idea. The cases are interesting because they approximate desired changes for system-level improvement, or what we have termed systemic changes, affecting the act of producing, the conception of utility and function and the act of consuming. In the next section we describe and position PSS cases within the system innovation idea. In Section 21.3 we discuss novel changes that are assumed to occur in PSS cases, leading to a discussion, in Section 21.4, of the dynamics and transformations needed in PSSs to attain SCP. It is important to point out that the novelties described are not necessarily novel to the world but rather to the actors involved in the innovation process. Some conclusions are presented in Section 21.4.

## 21.2 Systemic changes and product-service system cases

### 21.2.1 Levels of systemic changes towards sustainable consumption and production

The central thesis of the PSS concept is that by redesigning systems that deliver the function of product by offering an alternative to private ownership additional environmental gains can be made. The concept builds on the need to stimulate innovation at the function level, which is considered to be at the high end on a scale of changes promoted within the idea of systems innovation. Tukker and Tischner (2006) distinguish three levels of change towards SCP:

- **System optimisation**. This level includes improvement to existing systems of production and consumption with use of existing tools, such as the ecodesign of products and services, and providing information to consumers about environmentally sound alternatives and consumption patterns. There is no change in the structure of the production–consumption system. Incentives for change are a mixture of hard policy instruments (primarily for producers) and soft policy instruments (for producers and consumers). The typical sustainability improvement at this level is 20–30%

- **System redesign**. At this level, the system of provision is redesigned to become less environmentally burdensome and the change is often based on function innovation. For example, instead of fulfilling the need for mobility by private car ownership, an integrated mobility system is offered in which people use public transport where feasible, with car-sharing systems as a backup option. At this level, the typical sustainability gains are 50% or more. Together with tools that optimise the system, the structure of the production–consumption interaction changes and new forms of function provision with new products and services or new combinations of existing products and services are found, albeit still shaped within the existing context and market framework

- **System innovation.** At this level not only products, services and production systems are optimised and new ways of satisfying consumption needs are found within existing institutional frameworks and infrastructures, but new infrastructures, spatial planning and incentive systems are developed and implemented that promote more sustainable lifestyles. One example is provided by community-based washing centres in Sweden; these have become part of urban planning, increase previous standards for energy and water efficiency and provide the washing function to millions of households

In this categorisation, we can say that PSS cases can be found both at the system redesign level and at the system innovation level. Therefore we can expect different dynamics in the change processes leading to PSS, depending on the level of change. The starting point for our discussion is function innovation, which has been discussed in PSS literature in two ways. Some authors equate leasing and renting with function innovation. If we take the same starting point, then the system optimisation level may also include function innovation. However, research demonstrates that the environmental outcomes of this kind of function innovation may not lead to significant improvements and may even increase the environmental impact of providing the function. For example, comparison of various scenarios for providing the function of do-it-yourself tools, including ownership, community-based sharing schemes and rental companies, demonstrated that the distances households had to drive to rent the tools made rental companies the worst alternative (Mont 2004). Thus, the function innovation that is understood as only leasing or renting does not guarantee environmental improvement and does not represent the system optimisation level.

However, if function innovation is understood as the starting point for developing products, services and actor networks that are suited for providing function, then environmental improvements have a better chance of occurring. For example, instead of designing a car with environmental criteria in mind (ecodesign), a car manufacturer may ask: 'what is the function that I would like to sell to the customers?' The answer would probably be 'mobility'. The next step for the manufacturer would be to organise a system for mobility provision that would include establishing contacts with other mobility providers, such as public transportation, car-sharing schemes and rental companies, designing cars for multiple use, and facilitating services that substitute the need for transportation (e.g. videoconferencing). Reduction of environmental impacts would then be expected from better-designed cars (such as the development of electric or hybrid vehicles) for car sharing and renting, the adjustment of train and bus schedules and the support of municipalities in providing designated parking places for shared cars—all steps aligned to provide mobility function in the least environmentally burdensome way.

To sum up, in our understanding of PSSs, function innovation can be the starting point for transition; however, the extent of systemic change depends on the degree of changes in the entire product-service value chain and not merely on function delivery. In the following section we elaborate further on levels of change in specific cases from business-to-business (B2B) and business-to-consumer (B2C) markets.

## 21.2.2 Levels of changes in product-service system cases

We make a distinction between two types of cases. In the B2C market two interesting cases to examine are the car-sharing and communal washing centres in Sweden. In the B2B market we look at the two of the most cited examples, of Xerox and Interface Inc.

### 21.2.2.1 Car sharing

The number of car-sharing systems is slowly growing in many countries. It is an alternative system for satisfying the mobility needs of people. Although it is still a niche market, the car-sharing idea is popular among people who do not use a car often or among families who want to avoid buying a second car. Members of car-sharing organisations typically have access to several types of cars, which they can book through a telephone or Internet booking system and pay for by a monthly bill.

The majority of existing car-sharing systems are provided by actors other than car manufacturers and therefore, usually, no design changes to cars occur. Car sharing is a niche market and there are largely no supporting normative or regulatory institutions. Moreover, the idea of sharing cars contradicts the established norm of car ownership and the perception of a car as a status symbol. Car-sharing organisations are working at improving the image of shared cars as a status good. For example, Mobility, a Swiss-based car-sharing organisation, sacrificed placing its logo on shared cars in order to satisfy the needs of its members for status. Another interesting way to embed car sharing into every day life is practiced by StadtAuto, in Bremen, and StattAuto, in Berlin, which offer full mobility services by combining public transport and car sharing into a single mobility solution (Glotz-Richter 2001). At the regulatory level, car sharing has recently paved its way into the list of possible solutions to transportation-related problems in, for example, the EU white paper on *European Transport Policy 2010: A Time to Decide* (European Commission 2001). In addition to including car sharing as a part of the Dutch Policy Plan on the Environment and the Economy (Meijkamp 1999), the Dutch government established a foundation for the promotion of car sharing (Meijkamp 2000). Summarising the features of car sharing outlined above, we can categorise car-sharing systems as being at the system redesign level but not at the system innovation level.

### 21.2.2.2 Communal washing centres in Sweden

Communal washing centres in Sweden are an example of system-level innovation. The communal washing centres provide the function of clean clothes to households by providing a system of facilities that have become an integral part of urban development. The idea was initiated by a real-estate company, HSB, in the 1920s and was picked up in the 1930s–'40s by the Swedish Housewives' Association and other women's organisations (Hagberg 1986) as a measure to help women with housework. In the mid-1940s, the decision to integrate women into the labour market was taken and the issue of assisting women in their washing activities received regulatory support. For example, during 1939–46 direct financial support was given to cooperative washing centres in the countryside (Kjellman 1989). An official report from 1947 advocated self-service washing centres and washing services external to washing at home (SOU 1947). At the beginning of the 1960s households in the countryside started using private washing machines, but in cities the use of communal washing centres was spreading (Henriksson 1999). First, in the 1950s, real estate companies started regularly to equip newly built houses with washing rooms (Mitchell 1993) and, by the end of the 1960s, 80% of the population had access to small washing centres, equipped with automatic washing machines. Of these, less than half of the population owned a washing machine at home.

Later, Swedish regulators developed guidelines for building and housing companies on the location of communal washing centres and the baseline equipment to be provided. In addition, the Association of Tenants and Society of Tenant–Owners provided recommendations on the accessibility and availability of communal washing centres for newly built and existing multi-flat houses. In the mid-1990s, the energy authority and other organisations devised guidelines and advocated the instalment of energy-efficient equipment in communal washing centres. Therefore, nowadays, communal washing centres use semi-professional washing machines and drying equipment, which is more efficient than individual washing equipment in terms of both water and energy use. Since this equipment is used more intensively by many households it is worn out faster and is updated to new equipment more often. Thus, the function of clean clothes is delivered through a collective system that has a lower environmental impact than the system where the function is provided through individual ownership of washing machines. Unlike the car-sharing example, communal washing centres are not a niche market but are very widely used in Swedish towns and cities.

### 21.2.2.3 Xerox Corporation

Xerox Corporation started as a photocopier manufacturer but over the past decade has become a document company, focusing on the entire commercial documentation process. Xerox's Asset Management Programme is one of the most elaborated PSSs in the B2B market. Products are leased or sold under a multi-year contract, which guarantees customer satisfaction through functioning machines and payment of a fixed price per copy. As was mentioned before, leasing does not necessarily lead to environmental improvements. In the B2B market, leasing practices are widespread and companies lease for a variety of reasons. For example, Xerox was leasing copiers as early as the 1960s, because of users' unfamiliarity with the new process and for tax purposes. However, a system design was needed in order to keep the leasing strategy successful. This means that products and processes had to be designed for remanufacturing. The product design process was enhanced by applying the commonality principle to component design. A new Asset Recycling Programme was set up as soon as products started to come back from the market.

These programmes demand more time in the design phase of products and components. However, the possible losses resulting from an increased time to market and loss of economy of scale through regional (re)manufacturing are offset by considerable savings in the procurement of raw materials and in waste disposal. A historical bias exists against remanufactured products, but Xerox has taken steps to overcome this by promoting the remanufacturing products as 'proven workhorses' and by giving a three-year replacement guarantee on all products. By the end of the 1990s, in Europe, the demand for remanufactured Xerox machines exceeded supply by 50% (Ayres *et al.* 1997; Ferrer and Ayres 2000). To support the new business model, Xerox also provides quality training to its suppliers and special training to its service personnel. After the success of Xerox, many large manufacturers of office-equipment (e.g. Ricoh and Océ) shifted to leasing and remanufacturing.

### 21.2.2.4 Interface Inc.

Another case from the B2B market is Interface Inc., which operates a carpet-leasing programme for its range of commercial carpet tiles, called Evergreen Lease, and based on the Evergreen Service Contract. The Evergreen Lease is based on a modular system of flooring. Interface Inc. maintains control of all stages of the product's life, from production, through use and maintenance, to recycling and disposal. The system leads to a number of environmental benefits, such as a greater life expectancy for the flooring and its refurbishment, a closed-loop recycling system and the production of new carpets containing a percentage of recycled materials, resulting in a lower rate of consumption of raw materials. For the customer, the system has lower costs.

When introducing the system, the company faced several problems, such as the general reluctance of customers to embrace the concept of a continual lease arrangement and difficulties with developing technology for recycling old carpets. The company also encountered significant economic problems with operating closed-loop systems (Fishbein *et al.* 2000). Difficulties were associated with the relatively inexpensive raw materials for carpet production, which made carpet recycling a more costly enterprise. In addition, Interface learned that it needed to help its customers with contractual law, specific in each country of operation, and with preparing leasing contracts, which were more complex than the standard sales contracts. Another problem was that the Evergreen Lease Programme that was planned as an operational lease did not meet a number of requirements of the Financial Accounting Standards Board. This made financial institutions unwilling to finance the leasing arrangements. It was also unclear how Interface and its clients should handle this lease within the traditional accounting system and what consequences this might have for tax treatment. After these problems, the company offered take-back and recycling options in addition to its standard capital leases. This example demonstrates that making the economics of closed-loop system work might be a challenging task, the fulfilment of which not depends not only on the company's performance but also on existing institutional frameworks.

Despite these difficulties, Interface is persevering with the concept with the aim of developing a sustainable floor-covering business that is able to reclaim and recycle its products and 'never have to take another drop of oil from the Earth' (Interface Inc. 1998). The leasing concept extends the company responsibility for the entire life-cycle of a carpet and contributes to the carpet-reclamation initiative of Interface—called ReEntry—within which more than 85 million pounds weight of material have been diverted from landfills since 1995 (Interface Inc. 2006).

The idea of providing a service to customers is not new in itself, but acceptance of environmental responsibility and turning that into a profit centre is novel. In addition, changes are required not only in company culture but also among its customers and existing financial and regulatory frameworks. Similar to Interface Inc., a number of other companies are selling the function of flooring (e.g. MilliCare, DuPont, BASF, Monsanto, Collins & Aikman and Milliken Carpet).

## 21.3 Innovative processes in product-service systems towards sustainable consumption and production

The two aforementioned B2C cases of car sharing and communal washing centres are both considered to be successful cases in the PSS literature, but our brief evaluation demonstrated that they are quite different in terms of the societal changes involved in their dissemination, including the diversity of actors involved in promoting them and the regulatory and normative institutions that are being modified in the transformation process. The other two cases, from the B2B market—Xerox and Interface Inc.—are cases of incumbent actors leading the transformation in the business model within existing regulatory and normative institutions.

In general terms, PSS cases illustrate a number of changes that can be characterised as novel to the constellation of actors involved in the production–consumption process. This implies new learning and 'delearning' among actors, new interactions and relationships and, perhaps, new 'rules of the game'. The critical question is how these changes co-evolve for a successful shift to take place. In the following section we elaborate on key transformational processes in PSSs.

### 21.3.1 Shift in ownership

One of the main features of PSSs is the changing nature of relationships between producer and user. Whereas in traditional sales, value is directly associated with 'the amount buyers are willing to pay for what a firm provides them' (Porter 1985: 38), in PSSs value is co-produced and recreated and cannot be reduced to single monetary metrics (Hampden-Turner 1990). In traditional sales, the profit centre and the point of transaction is tied to a material product, whereas in many PSS cases profit is tied to the number of functional units the material product delivers. Thus in the PSS case consumers do not have to own a product to be able to use it. They pay per use: that is, per unit of function that the consumers themselves or that service providers extract from products; they do not pay for the product.

In PSSs, incentives for providers are created to increase the number of functional units that the product delivers—because functional units are the price carriers—and consequently to treat products as capital assets that are worth maintaining, just like production equipment (Braungart 1991). Thus, the shift in ownership structure provides incentives for increasing product durability, to design products for easy upgrading, re-use and remanufacturing. In addition, if PSS providers are responsible for the operation of products in the use phase, incentives are also created to reduce the costs associated with the use phase, including costs for consumables and auxiliary products as well as costs for maintenance and for upgrading services. For customers, incentives are created to learn how much they use the product, and thus the cost of the use phase becomes transparent and creates incentives for users to reduce product use. In that way, possibilities for shared use of products is created.

The shift in ownership in the B2C markets may sometimes be associated with higher barriers than in B2B cases. In B2C markets the shift typically occurs in the form of switching from owning a product to leasing, renting or sharing systems. The success of these systems depends, to a large extent, on the type of product to be shared or leased and on how customers feel about owning the product. A distinction between so-called functional and emotional products may explain some of the differences. In the case of functional products, customers are not emotionally attached to these products and are interested mainly in the function delivered. For example, utility companies provide 'functional products' (i.e. energy, water) with which customers do not have an emotional attachment. Unlike functional products, as well as providing function emotional products also create an image, secure a certain status and make customers feel emotionally attached to them. Therefore, a somewhat different incentive structure for customers needs to be developed if emotional products are to be substituted with systems where the ownership of products is abandoned. It is not so easy, as many studies have shown, to replace car ownership with mobility services provided by car-sharing and car-pooling schemes. Purely financial incentives, which work better in the case of functional products, need to be substantiated by image-making and/or status-building incentives in the case of emotional products.

### 21.3.2 Shift from product to offer

As the name suggests, in PSSs what is being sold is not only a product but also an offer, comprising products and services. At first glance, this might not seem to be a big change. However, taking into consideration that the majority of companies are used to developing and selling products, the shift to providing offers becomes a significant challenge that entails many changes within and outside the product manufacturing company. First, being part of a PSS, services need to be designed with similar considerations as products. The issue of service design has not received similar attention to product design (Bullinger *et al.* 2003) and has been discussed mainly within the context of service marketing. However, service design should include a much broader perspective than simply marketing. For example, services need to be adapted to the current systems of provision. In addition, service design needs to include environmental considerations to ensure that the offer does have a lower environmental impact than traditional business models of selling products. Second, services also include special 'sales techniques' that need to be changed from the existing practice of 'product-transfer' sales into 'functional sales', where consumers can experience the functional qualities of products and understand the service component in the communication process with personnel. In PSSs, the point-of-sale becomes the point of service. This entails changes in the design of the 'sale' stage, which should include different techniques of selling product use.

### 21.3.3 Shift from supply-chain to value-chain actor networks

The shift to a PSS may have a considerable impact on the structure of the supply chain. New actors may play a role in the existing product chains (Halme and Antonnen 2004; Mont 2004), providing services that add value to products or that close the product lifecycle. There may be a need to consider not only the product chain in the design phase but also networks of companies that may jointly fulfil customer needs in a less environmentally burdensome way (Mont 2004). Sometimes, non-traditional actors deliver utility to customers in the most efficient and effective way (e.g. actors who are usually placed outside the traditional supply chain, such as real estate companies and non-commercial cooperatives).

In addition to these considerations, the shift to actor networks is necessitated by the need to improve the product-service offer and the life-cycles involved. For example, in the case of the sale of washing machines, production processes can be improved and resource use optimised throughout the life-cycle of the washing machine. In the case when a function of clean clothes is sold, optimisation should take place not only at this level but also throughout the life-cycle of the detergents, electricity and water used and with regard to maintenance and to the end-of-life of the washing machine. Thus, the complexity of actor networks increases in comparison with the sale of products when a PSS is delivered or optimised.

## 21.4 Discussion on dynamics and transformation

In general terms, specific innovations are the result both of intentional and of unintentional choices and actions taken by individuals or organisational actors. However, a growing body of literature has come to recognise the systemic nature of innovation and the importance of learning and institutions in innovation activities. Thus actors, interactions and institutions are important components in the conceptualisation of innovation processes (Malerba 2005). The innovation system approach brings these components into a system framework to conceptualise the activities enabling and constraining these processes (Edquist 2005). The system boundaries can be defined on a spatial scale, such as the national or regional, or on a sectoral level. The dynamics and transformation of innovation systems are the result of several different processes. These processes may involve the co-evolution of various elements, such as technology, knowledge bases, learning, demand, and firms, non-firm organisations and institutions (Malerba 2005). Similarly the transformation towards a PSS entails the co-evolution of various elements.

As illustrated previously, transformation to a PSS entails intricate shifts, including a shift in the structure of ownership and in the structure of the supply chain. Clearly, we see also that these processes can take place in different product-service markets and social contexts. The concept of a sectoral system of innovation provides an interesting and relevant entry point for the conceptualisation of change processes towards PSSs. This is particularly relevant to PSS cases because the sectoral system of innovation is at the level where societal functions are fulfilled.[1] Malerba (2005) showed clear differences in the behaviour of different sectoral systems such as the role of knowledge, actors and institutions. This has implications for our discussion. First, the stimulation of a transformation to a PSS needs to consider these sectoral differences. However, in the case of a PSS when starting from function innovation, broader systemic changes for SCP need to take place. For example, new agents might be involved beyond the current boundaries of the sector. Therefore, the focus should be on the dynamics and transformation of particular societal functions (e.g. on mobility rather than cars).

---

1 Geels (2004) refers to examples such as transport, communications, materials and housing. However, in the case of PSSs we extend this to the utility from the perspective of the user, such as mobility, shelter and so on.

Looking back at the innovative processes in PSS cases, the shifts occur in the presence of incentives (economic or regulatory) for existing actors (i.e. those involved in the production, diffusion and use of a particular societal function) to change and adapt, or for new agents to enter into the chain. The dynamics were different in PSS cases between B2B and B2C markets. Even within B2C cases there were differences in relation to the meaning given to the act of consuming, such as the distinction made between extracting function (electricity for lighting) versus emotional utility (image and status).

A shift in ownership structure occurs in diverse user contexts. In the B2B cases, the consumers are generally organisations. The concerns were mainly about costs, the tax regime and trust in the new business model. In the B2C cases, consumers are generally individuals and households. The critical issues here were meanings, norms and routines in owning products and the very act of consuming. The dynamics here were much more complex and path-dependent. This is perhaps more visible in the case of communal washing centres in Sweden. The context of sharing evolved through a long history of normative rules and values within the fabric of the society. Several historical reasons contributed to the formation of the prevalent norms in Swedish society on which the existence of collective sharing schemes rests. Examples of this could be the concept of *lagom* ('just enough', or 'with moderation'), the acceptance of the notion of collective good, and the tradition of non-conflict. The concept of a common good originates from early religious postulates which implied that it was sinful to strive for more than the satisfaction of individual needs. In modern times, 70 years of rule of social democrats in Sweden has also contributed to the institutionalisation of the solidarity tradition. These historical decisions have greatly contributed to the evolution and acceptance of community-based forms of living representative of contemporary Swedish society. If similar values are present in a society, there is perhaps a better chance that collective sharing systems or business models based on leasing arrangements will be easier to embed into everyday choices and will receive grater acceptance than in cultures with values based on possessive individualism.

A point to make from this is perhaps that an understanding of the sociocultural and even historical context in which consumption–production activities take place is needed to provide useful insights into how sociotechnical innovations can be institutionalised and embedded into everyday life.

A shift from product to product-service offers has implications for systems of provision and relationships among the actors involved. This perhaps explains why car sharing has not evolved into a fully fledged systemic change and has remained a niche market. The dominant actors in the car business—auto manufacturers—perceive themselves to be in the business of car making, not in the business of mobility. Therefore, mobility provision through means other than private car ownership is perceived by them as a threat. They are typical actors of the prevailing regime and are locked into old ways of thinking. Their main focus is on improving existing technologies and optimising existing systems and not on innovating new systems. Therefore the driving forces behind car sharing are external agents. Car-sharing organisations, in contrast, operate in a niche market in which new ideas for mobility provision are being developed, often in close collaboration with other mobility actors, such as public transportation companies and car rental and leasing firms. Car-sharing organisations thus establish new actor networks for the provision of mobility. However, in order for car sharing to become a mainstream option for mobility, other changes are needed that would shape social values, the built environment and infrastructure as well as economic incentives. In this respect, the development of collective structures suited for and facilitating car sharing is of particular importance. It is perhaps impossible to devise policies that specifically promote car sharing per se. However, by altering the incentive structures (restrictive policies) for the use of cars, particularly in urban centres, the opportunities for shared systems could be enhanced.

Communal washing centres, though, were never a niche market in Sweden. Their development went hand in hand with and, at some periods of time, rivalled the technological development of washing machines for individual use. In this sense, communal washing centres were radical innovations, combining innovation at both the technical and the organisational levels. From the first examples of communal washing centres provided by the housing company HSB in the early 1920s, the journey took a few decades until a significant number of communal washing centres were built in Sweden in the 1950s.

The B2B cases of Xerox and Interface Inc. demonstrate that, although a shift in the business model from the sale of a product to the sale of a function may encounter problems with existing regulatory, normative and financial frameworks, the idea is viable and is spreading in various sectors. The change to the sale of function instead of products leads to a change in the structure ownership and responsibility. Depending on the size of operations and the market, many companies have to find collaborative networks and new agents to help in the provision of the function. In case of Xerox, Interface Inc. and washing centres, service organisations are involved in servicing the products and in ensuring the function is provided.

## 21.5 Concluding remarks

What the PSS cases demonstrate is that changing the ownership structure and shifting from the sale of products to the sale of offers do not on their own guarantee an improvement in economic, environmental or social sustainability. Companies do not have to drastically change the way they do business in order to optimise existing production processes, products and services. However, in order to move to the next level of systemic changes towards SCP a shift in the ownership structure may become necessary. Even here, on its own, changing the ownership structure along the product life-cycle may not guarantee any sustainability improvements, as traditional cases of leasing and renting demonstrate. What are needed are systemic changes in the structure of consumption–production interactions and new ways of delivering function. The ability to do this fundamentally depends on the creation of new institutions or on adjusting existing regulatory and normative frameworks.

The dynamics and transformation in PSS cases provide an insight into some of the challenges provided by systemic changes to SCP. Institutions play a central role in the transformation process. In line with Geels (2004) we recognise the conceptual problems with the sectoral system of innovation, particularly in relation to the role of institutions. One of the largest gaps seen both in institutional frameworks and in innovation systems is a lack of attention and hence understanding of how changes can be instigated in the realm of consumption.

# References

Asheim, B.T., and M.S. Gertler (2004) 'The Geography of Innovation: Regional Innovation Systems', in J. Fagerberg, D. Mowery and R. Nelson (eds.), *The Oxford Handbook of Innovation* (Oxford, UK: Oxford University Press): 291-317.

Ayres, R., G. Ferrer and T. van Leynseele (1997) 'Eco-efficiency, Asset Recovery and Remanufacturing', *European Management Journal* 15.5: 557-74.

Braungart, M. (1991) 'Die Leasinggesellschaft: Unverkäufliche Produkte, Gebrauchsgüter, Verbrauchsgüter' ('The Leasing Business: Unsaleable Products, Consumer Durables, Consumer Goods'), in BAUM (Bundesdeutsche Arbeitskreis für Umweltbewußtes Management [German Environmental Management Association]) (ed.), *Mit Umweltschutz zum Gewinn für die Wirtschaft und die Umwelt (Using Environmental Protection for the Benefit of the Economy and the Environment)* (Hamburg: Kongre-Reader): 109-22.

Brezet, H., and C. van Hemel (1997) *Ecodesign: A Promising Approach to Sustainable Production and Consumption* (Paris: United Nations Environment Programme).

Bullinger, H.-J., K.-P. Fähnrich and T. Meiren (2003) 'Service Engineering: Methodical Development of New Service Products', *International Journal of Production Economics* 85: 275-87.

Edquist, C. (2005) 'Systems of Innovation: Perspectives and Challenges', in J. Fagerberg, D. Mowery and R. Nelson (eds.), *The Oxford Handbook of Innovation* (Oxford, UK: Oxford University Press): 15-31.

—— and L. Hommen (1999) 'Systems of Innovation: Theory and Policy for the Demand Side', *Technology in Society* 21.1: 63-79.

European Commission (2001) *European Transport Policy 2010: A Time to Decide* (Brussels: European Commission).

Ferrer, G., and R.U. Ayres (2000) 'The Impact of Remanufacturing in the Economy', *Ecological Economics* 32.3: 413-29.

Fishbein, B., L.S. McGarry and P.S. Dillon (2000) *Leasing: A Step toward Producer Responsibility* (New York: INFORM/Duke University/Nicholas School of the Environment/Tufts University/The Gordon Institute).

Geels, F. (2002) *Understanding the Dynamics of Technological Transitions: A Co-evolutionary and Sociotechnical Analysis* (Enschede, Netherlands: Twente University Press).

—— (2004) 'From Sectoral Systems of Innovation to Socio-technical Systems: Insights about Dynamics and Change from Sociology and Institutional Theory', *Research Policy* 33.6-7: 897-920.

Glotz-Richter, M. (2001) 'CityCarClub/Car-Sharing: Experience of a Municipality with an Innovative Mobility Scheme as a Strategic Move towards Sustainable Development', in P. Allen, C. Bonazzi and D. Gee (eds.), *Metaphors for Change: Partnerships, Tools and Civic Action for Sustainability* (Sheffield, UK: Greenleaf Publishing): 306-13.

Hagberg, J.-E. (1986) *Tekniken i kvinnornas händer: Hushållsarbete och hushållsteknik under tjugo- och trettiotalen (Technology in Women's Hands: Household Work and Household Technology during the 1920s and 1930s)* (Malmö, Sweden: Liber Förlag).

Halme, M., and M. Antonnen (2004) 'Sustainable Homeservices? Win–Win Possibilities for Housing Providers and Real Estate Services?', paper presented at the *10th European Real Estate Society (ERES) Conference*, 11-13 June 2004, Helsinki, Finland.

Hampden-Turner, C. (1990) *Charting the Corporate Mind* (New York: The Free Press).

Henriksson, G. (1999) *Organisationsformer för hushållens tvätt i Stockholm under 1900-talet* (Stockholm: The Environmental Strategies Research Group, Stockholm University).

Hirschl, B., W. Konrad and G. Scholl (2001) 'New Concepts in Product Use for Sustainable Consumption', paper presented at the *7th European Roundtable for Cleaner Production*, International Institute for Industrial Environmental Economics (IIIEE), Lund University, Lund, Sweden, 2-4 May 2001.

——, W. Konrad and G. Scholl (2003) 'New Concepts in Product Use for Sustainable Consumption', *Journal of Cleaner Production* 11.8: 873-81.

Interface Inc. (1998) *Sustainability Report* (Atlanta, GA: Interface Inc.).

—— (2006) 'Redesign Commerce', www.interfacesustainability.com [accessed 25 August 2007].

IVA (Ingenjörsvetenskapsakademien [Royal Swedish Academy of Engineering Sciences]) (1998) *Möjligheter och hinder på väg mot Faktor 10 i Sverige (Possibilities and Barriers on the Road toward Factor 10 in Sweden)* (Stockholm: IVA).

Jansen, J.L.A. (1993) *Towards a Sustainable Oikos, en Route with Technology!* (Amsterdam: CLTM [Commissie Lange Termijn Milieubeleid; Policy Committee for Long-Term Environment]).

Kemp, R., and J. Rotmans (2005) 'The Management of the Co-evolution of Technical, Environmental and Social Systems', in M. Weber and J. Hemmelskamp (eds.), *Towards Environmental Innovation Systems* (Heidelberg/New York: Springer Verlag): 33-55.

——, S. Johan and R. Hoogma (1998) 'Regime Shifts to Sustainability through Processes of Niche Formation: The Approach of Strategic Niche Management', *Technology Analysis and Strategic Management* 10.2: 175-95.

Kjellman, G. (1989) *Från bykbalja till tvättmaskin: Folklig arbetstradition och teknisk nyorientering* (Linköping, Sweden: TEMA Teknik och social förändring, Linköpings Universitet).

Lundvall, B.-Å., B. Johnson, E.S. Andersen and B. Dalum (2002) 'National Systems of Production, Innovation and Competence Building', *Research Policy* 31.2: 213-31.

Malerba, F. (2005) 'Sectoral Systems of Innovation: A Framework for Linking Innovation to the Knowledge Base, Structure and Dynamics of Sectors', *Economics of Innovation and New Technology* 14.1–2: 63-82.

Meijkamp, R. (1999) 'Car Sharing in the Netherlands', *Journal of World Transport Policy and Practice* 5.3 (Special Issue: 'Car Sharing 2000: A Hammer for Sustainable Development'; ed. E. Britton): 72-88.

—— (2000) *Changing Consumer Behaviour through Eco-efficient Services: An Empirical Study of Car Sharing in the Netherlands* (Delft, Netherlands: Delft University of Technology).

Mitchell, B.R. (1993) *Den eviga byken: Planering för tvätt i flerbostadshus* (Lund, Sweden: Byggnadsfunktionslära, Arkitektursektionen vid Lunds Universitet).

Mont, O. (2004) 'Reducing Life-cycle Environmental Impacts through Systems of Joint Use', *Greener Management International* 45 (Special Issue on Life-cycle Management): 63-77.

Porter, M.E. (1985) *Competitive Advantage* (New York: The Free Press).

Rathenau Institute (1996) *A Vision on Producer Responsibility and Ecodesign Innovation* (The Hague: Rathenau Institute).

Schmidt-Bleek, F. (1996) *MIPSbook or The Fossil Makers: Factor 10 and More* (Berlin/Boston, MA/Basel: Wuppertal Institute).

—— and F. Lehner (1998) *Das MIPS-Konzept: Weniger Naturverbrauch; mehr Lebensqualität durch Faktor 10 (The MIPS Concept: Less Consumption of Natural Resources; More Quality of Life through Factor 10)* (Munich: Droemer).

SOU (1947) *Kollektiv tvätt: Betänkande med förslag att underlätta hushållens tvättarbete (Collective Washing: A Report with Suggestions on How to Make Household Washing Easier)* (SOU 1947:1; Stockholm: Statens Offentliga Utredningar [State Official Inquiries]).

Sperling, D., S. Shaheen and C. Wagner (2000) *Car Sharing: Niche Market or New Pathway?* (Berkeley, CA: University of California).

Tukker, A., and U. Tischner (eds.) (2006) *New Business for Old Europe: Product Services, Sustainability and Competitiveness* (Sheffield, UK: Greenleaf Publishing).

Weaver, P., L. Jansen, G. van Grootveld, E. van Spiegel and P. Vergragt (2000) *Sustainable Technology Development* (Sheffield, UK: Greenleaf Publishing).

# Part 6
# Conclusions and integration

# 22
# Conclusions: change management for sustainable consumption and production

*Arnold Tukker*
TNO, The Netherlands

## 22.1 Introduction

So, at the end of this book we can now try to wrap up what we have learned about the key question we set out to solve: how can we stimulate, foster or force a change to sustainable consumption and production (SCP) and what is the natural role of different types of actors? To follow an order consistent with the structure of this book, in the following sections we will discuss what we can learn about this from a business, design, consumer and system innovation perspective, challenging, in a thought-provoking manner, some simplistic but abounding myths about how this change process could work. However, it is essential to recapitulate first what the goal of the whole SCP project could or should be and to what agenda this could or should lead.

## 22.2 The goals of sustainable consumption and production and the agenda to be pursued

### 22.2.1 Goals

As discussed in Chapter 2, SCP is not a fully objective notion. Although it is clear that 'Spaceship Earth' has boundaries to its exploitation, the location of these boundaries and the type of interventions needed to allow social and economic development to take place within these boundaries often cannot be determined by looking at 'scientific facts' alone. There are the optimists, who feel that market incentives and human ingenuity will ensure that real sustainability crises will be avoided (e.g. Lomborg 2001). Indeed, the Stone Age did not end for lack of stones but rather developed into the Bronze Age; and Malthus's population ceiling was surpassed thanks to the Industrial Revolution. Then there are those who are more concerned, who feel that such breakthroughs will not come automatically but will require hard and conscious efforts (cf. Gore 2006; Meadows *et al.* 2004; Stern 2006).

Obviously, this lack of scientific certainty is not an argument to 'let go' the discussion on SCP or sustainable development. Most companies or individuals take key decisions even though they cannot predict consequences 'beyond reasonable doubt', as the legal standard in crime law asks. This is true for the couple that considers marrying or having children as much as for the business leader who deliberates over whether or not to penetrate a market on a new continent. What matters is that, on the basis of the best available knowledge, they judge which action to take.[1] Any business requiring full certainty before taking action will probably never move and will be swiftly surpassed by even the weakest competitor. In sustainability matters the situation is no different. An SCP policy that is 'evidence-based' is wise, but an SCP policy that acts only on 'evidence beyond reasonable doubt' would be foolish.

In Chapter 2 we gave some points that, on the basis of simple metrics or commonly accepted ethical standards, would seem to be undeniable elements to pursue on the SCP agenda and that need deliberate action:

- A radical reduction of impact per unit satisfaction should be reached, given the rise in the world population from about 6 billion people now to an expected peak of 9 billion people in the late 21st century, and the tremendous wealth per capita rise that many emerging economies and developing countries still have to go through. This reduction can be realised via two routes: smarter production (dematerialisation of products and production) and smarter consumption. In the *Consumption Opportunities* report he edited for the United Nations Environment Programme (UNEP 2002) Manoochehri uses the following elegant division:
    - Dematerialisation (of production). This strategy implies providing the same final service by seeing much less material input into and much fewer emissions from production and products

---

1 Interestingly, it is not uncommon in business literature to describe business development in probabilistic terms such as 'placing a bet' or 'investing to play'. Of course, such moves are made only when the odds look good, but they also reflect that certainty is absent.

- Optimisation (of consumption). This strategy aims at changing consumption patterns by making smarter consumption options available, by choosing more wisely and consciously and by defining an appropriate level of consumption (categorised, respectively, as different, conscious and appropriate consumption)
- Where the SCP agenda may not be the right place to lead the fight for poverty eradication and equity, it must at least support it. Hence:
    - Compensation for compliance with basic environmental and labour or social standards in supply chains should be ensured
    - The potential for 'leapfrogging' in emerging and developing economies should be investigated and tested
- The reasons for the apparently low efficiency of Western economies for providing high-quality life-years should be investigated and understood and its implications be translated into guidelines for organising consumption and production patterns

## 22.2.2 Agenda

As discussed in Chapter 1, the goals to be pursued play out differently in different types of economy (Hart and Milstein 1999; Tukker 2005):

- Consumer economies (Western Europe, the USA, Japan), with a high level of wealth per capita, and where poverty is all but eradicated (some 1 billion citizens). These economies should focus on reducing their material use per unit satisfaction, ensure that basic environmental and social standards are met in the countries of origin of their imports, and improve their efficiency in providing quality of life per unit of gross domestic product (GDP)
- Emerging economies (e.g. China), that are rapidly changing and developing fast to modern consumer economies (some 1–2 billion citizens). For these economies the main challenge will be to look at how they can 'leapfrog' directly towards SCP structures without first copying the existing problematic Western examples
- Base-of-the-pyramid (BOP) economies, where the vast majority of people survive on a few dollars per day, and which concern consumer markets that are of relatively low importance to the others in the global system (some 3–4 billion citizens). In these economies consumption and production structures have to be fostered that allow people to cover their basic needs and to allow subsequent sustainable growth

In combination with the insight that the final consumption of food, mobility, and energy and housing drives over 70% of the life-cycle impact of consumption (at least in developed economies), all this leads to a structure for approaching the change to SCP patterns along three dimensions:

- By type of economy: consumer, emerging, or BOP
- By experts and expertise involved: business specialists, designers, consumer scientists and innovation policy experts
- By domain: with food, mobility and energy and housing as key areas

SCORE!, as an EU-based project, is inevitably biased towards consumer economies. In the next sections we will summarise and reflect on the analyses of how such change can be fostered from the perspectives central in this book: that is, from the business, design, consumer and system innovation perspectives.

## 22.3 Contributions to change

### 22.3.1 Introduction

Parts 2–5 of this book looked at how changes to SCP could be realised from a business, design, consumer and system innovation perspective. The perspectives are not totally unambiguous. One can interpret them as being perspectives from the scientific fields of business, design and consumer research, but also as perspectives of how business, designers and consumers can contribute to change. We could not totally escape this ambiguity in the discussions before, and we will not try to do so here either. In the following sections we will summarise for each perspective the following issues:

- Basic understanding of roles and drivers
- Change model(s)
- Limitation of the change model(s), policy implications and scattered myths

The first two issues concern mainly the second and third sections of the review chapters of Parts 2–5 of this book (i.e. Chapters 3, 8, 13 and 18). The remaining point tries to bolt down the most important messages for practitioners and policy-makers alike in the form of dilemmas, non-working myths and other implications.

### 22.3.2 Business

#### 22.3.2.1 Basic understanding of roles and drivers

Typically, business is driven by the need to create value for its owners. This is not necessarily primarily a short-term monetary issue (though shareholder-owned firms with quarterly reporting obligations may feel significant pressure in this respect); continuity is usually an important factor. The business literature shows that various strategies can be followed to realise long-term profitability, such as diversification (e.g. branding, corporate reputation and the development of unique competences, client relationships and solutions), cost reduction (including risk reduction) and increase in market share. An interesting specific point for privately owned firms is that they also often create 'owner value' by providing an interesting, independent working place or some other platform for self-realisation.

Businesses operate in a context shaped by 'regulated competition', where the player with unique powers usually captures most value, by financiers, which may have an important influence on strategy, and by 'lock ins', which hinder change.

### 22.3.2.2 Change model(s)

Businesses can contribute significantly to SCP. First, they can 'green' their products and production. Further, they can support basic environmental and social standards in their upstream and downstream business chains (e.g. endorsed via [voluntary] schemes such as those of the Forest Stewardship Council [FSC], the Marine Stewardship Council [MSC], Responsible Care or other self-regulation). Related to this is 'choice editing': retailers preselect sustainable products on their shelves so that the consumer automatically buys sustainable products through existing routines. Furthermore, new business models that focus on dematerialised value creation can be implemented, supported by sustainability marketing. Finally, business usually leads development, with the implementation of radical innovations and new solutions adapted to a changing societal context: for example, Apple revolutionised (and dematerialised) the music industry via iPod and iTunes; Nokia, Motorola and Ericsson 'shaped the future' by envisioning a wirelessly connected world (thus models to develop flexibility to 'compete for the future' [Hamel and Prahalad 1994] and related novel business models are of relevance too [see Chapters 4 and 5]). With radical change seen as key to tackling global sustainability problems (including poverty), the pivotal role that business can play is obvious. Radical technical breakthroughs in energy technology may not only contribute to solve global warming but may even create a society where power generation is highly decentralised, with important behavioural change resulting from this (cf. Rifkin 2002). Drivers for such leadership can be:

- Enlightened self-interest (i.e. it is realised that current practices do not allow for the sector to be sustained in the long term, or new business opportunities are found to address the scarcity of environmental resources)
- Tacit or explicit expectations, norms and values in society[2]
- A desire to make a difference to society (by owners who have secured their independence)

This potential is clearly enhanced if government creates 'a level playing field' with 'green' market framework conditions (e.g. by regulation and tax reform, by eliminating perverse subsidies and by 'green' public procurement [see Holliday and Pepper 2001] and by 'greening' the innovation system [see Section 22.3.5 and Chapter 18]).

---

2 Interesting examples here are companies that succeed in 'hooking onto' a tacit, broadly experienced value held among citizens. See e.g. the Bionade example in Box 22.4 on page 432. A proven tactic of consumer organisations is hence to articulate sustainability and social values and check companies on these in their product-testing schemes—a nice example of indirect consumer–business interaction.

### 22.3.2.3 Limitation of the change model(s), policy implications and scattered myths

Still, businesses cannot neglect the business fundamentals. Sustainability improvements quite often are a (smart) reaction to real sustainability-related problems in the environment of the firm: the increasing cost of resources; the imposition of emission caps and other boundary conditions already set or likely to be set by regulators; tacit or explicit customer expectations about sustainability values; or other boundaries that society at large sooner or later has to deal with. Without such boundaries, it is simply too tempting for companies to make use of any freely available social or environmental 'commons' (see Hardin 1968). Furthermore, some of the 'change models' discussed above are incremental or leave at least the incumbent companies firmly in control. This is not without coincidence. Though businesses can foster radical change (Hamel and Prahalad 1994), many firms are 'locked in', and often cannot survive radical change and hence oppose (plans for) true radical change. Further, most of the efforts described deal with dematerialisation, efficiency improvement and social improvement at the production and product side of SCP, rather than making consumption patterns more efficient (as in the case of iPod and iTunes); consumption patterns can be influenced, but more as a by-product (an increased consumption volume and market share is the supreme reigning driver). At this consumption side, the logic of our current economic system provides powerful incentives to businesses to continuously stimulate more consumption (see Chapter 7). The paradigm of growth, measured in monetary units, rewards not only entrepreneurship and progress but also, in essence, perverse business behaviour. Examples include externalising costs (which then have to be paid by society as a whole), selling hitherto free goods in the market economy (which seemingly enhances gross national product [GNP] and turnover but does not provide any additional well-being), and using aggressive marketing approaches that enlarge the aspiration gap, making people uncomfortable with who and what they are, to stimulate more consumption of products and services.[3] Business cannot be expected to seek to reduce consumption. For all these reasons, the governmental role of supporting sustainable market frameworks is very relevant.[4]

### 22.3.3 Design

#### 22.3.3.1 Basic understanding of roles and drivers

Design is a discipline the traditional purpose of which 'is to establish the multifaceted qualities of objects, processes, services' and hence is rather product-oriented. This role of design, however, is transforming. Traditional product design has been expanded with elements such as communication and branding design, and system and strategic design. Indeed, the added value of such intangible design elements now usually by far surpasses

---

3 As discussed in Chapter 2, authors such as Schumacher (1973, 1979) and Illich (1978) argue that such developments in fact reduce the freedom and capabilities of consumers. Schumacher even went so far as to state that where 'good work' should have as a prime goal helping humans to overcome greed, our current economic system actually is dependent on fostering such 'deadly sins'.

4 Such action may lead to individual losers, but probably will not jeopardise competitiveness in general (see Wagner's contribution on the Porter Hypothesis, Chapter 6).

that of the material artefact. A part of the design research community[5] uses an even broader design concept, defining the discipline as 'a reflective activity aiming to solve problems by developing solutions'. Or, in other words: design must not focus on products but on the problems to which these products and services may perhaps be the solution (i.e. there must be a development towards 'solution-oriented' design).

Some influential designers (and architects and city planners) have traditionally felt a strong ethical responsibility, related to the insight that human behaviour and development is highly influenced by design and required behaviour relating to artefacts—this is akin to the notion of 'lock-in' seen in Part 5 on system innovation. Paraphrasing Kipling (1899), one could almost speak of the 'designer's burden'. With the expanded definition of design, this responsibility is now expanded to the contribution of the development of innovative socioeconomic interaction systems.

### 22.3.3.2 Change model(s)

Design can contribute via a variety of approaches to sustainable development, at different levels and with different degrees of potential sustainability gains:

- Product design. This can promote the use of resources with a low impact and the creation of products with a low life-cycle impact. These activities centre on products and production rather than on consumption patterns

- Design and envisioning of 'satisfaction-fulfilment' systems (sometimes referred to as product-service systems [PSSs]). Here, systems are designed to be eco-efficient as well as to promote social equity and cohesion. This approach faces much wider problems than product design alone, dealing with firm networks and even non-governmental organisations (NGOs), communities and governments. The main avenue for change in this case appears to be to foster locally based and network-structured (distributed) economies and creative communities that can enhance bottom-up actions. Such promising signs of new solutions and change can be improved and made stronger over time, so that they will eventually provide more viable and appropriate solutions to problems than has hitherto been the case in mainstream design[6]

- Communication, and brand and advertisement design. In this approach, the environmental and social sustainability values of sustainable products and services are articulated in a creative and novel way, thereby influencing purchasing behaviour and even consumption patterns in the process

The first model has become fairly well known and is, to a certain extent, implemented in business and education. The second approach is now being experimented with in a limited number of projects. The third approach is illustrated by the example of Bionade (see Box 22.4 on page 432).

---

5 It concerns most notably an informal Learning Network on Sustainability (LENS) of Higher Education Institutions set up by Politecnico di Milano; see Chapter 8.
6 Think, for example, of Schumacher's (1979) 'lifeboat' concept: one should ensure the development of alternative, 'appropriate', technological systems that may be used in niche situations but provide alternative ways of doing things when you really need them.

#### 22.3.3.3 Limitation of the change model(s), policy implications and scattered myths

The development of design for sustainability has been supported by stimulation and education programmes and by legislative projects such as the EU Energy-using Products Directive, the Packaging Directive, the ROHS Directive (Reduction of Hazardous substances Directive) and by similar product-oriented demands. Where such incentives are absent, the implementation of ecodesign is still limited; apparently, reliance only on the economic–environmental win–win is insufficient (Tukker *et al.* 2001). This probably reflects some points already mentioned in the Part 2 on business: change fails to occur because of internal inertia, because the issue is not seen as important enough to warrant substantial management attention and/or because of risk-avoiding behaviour.

The implementation of sustainable (product-service) systems is both supported and hampered by sound business reasons. Particularly if one offers a 'result' rather than a product, a company takes much more responsibility than before, where often it cannot influence all variables that influence the result (Tukker and Tischner 2006). And, finally, where communication or brand design can be used positively for reaching SCP goals, it can also be used negatively. We refer to the final remarks in the business section on the sometimes perverse role of marketing (Section 22.3.2).

With regard to the broader issue of eco-efficiency, there is a need for provision systems that promote equity and cohesion; the researchers working in this field articulate and promote promising examples of sustainable and equitable ways of life, but there is relatively little articulation of what they regard to be the pathways by which such niche systems could become mainstream. This, in fact, means articulating a vision of how to deal with the barriers mentioned in the business, consumer and system innovation policy sections of this chapter. (Note: a final limitation is that many of the more traditional designers have an only limited interest in sustainability. They leave the leading role in this respect to marketers, continuing to design according to their insights and specification; alternatively, they may find sustainability criteria to be 'a boundary condition too far' that prevents them to work out their creativity in wildly beautiful shapes and structures.)

### 22.3.4 Consumers and consumer science

#### 22.3.4.1 Basic understanding of role and drivers

Why do we consume and why do we increase our consumption? Even a brief peek at the insights from the consumer sciences, described in Chapters 13–17, shows a reality that is more complex then many probably realise but that, at the same time, is still simple enough to be comprehensible.

First, there is the dichotomy between needs and wants. Neoclassical economics suggests that the choice of what humans can want is infinite. But authors such as Max-Neef (1991) suggest there are finite, few and classifiable (axiological) needs, such as subsistence, protection, affection, identity, creation and freedom. Needs can be met by different 'satisfiers', including ones with a disproportional ecological footprint (indeed, some satisfiers are just perverse 'destroyers', not fulfilling any axiological need). Closely related is the insight that consumption is not simply about providing a material base for life. Through consumption, people create identity, confirm their position, pursue dreams and, indeed, gain meaning to life (see Baudrillard 1998; Jackson *et al.* 2004)—and marketers know this. Fragrances, designer suits and branded shoes are mainly image products of which the material production costs and even retail costs are just a fraction of the value paid. The 'needs' approach provides the hope that, by smartly choosing satisfiers, needs can be met with limited resources. But what is clear in any case is that it is naïve to ask consumers to 'voluntary downscale' and to give up their desires without offering them alternative dreams.

Second, there is ambiguity in our understanding of how consumers choose. A popular view is that of the rationally acting, sovereign consumer, who via his or her voting power on the market decides which producer will or will not be successful. Certainly, there will be markets and product groups where it works like this, but this distracts from the fact that in many cases a consumer's behaviour is driven by a material and social context that leaves often limited choice. Most of a household's expenditure is on non-flashy 'ordinary consumption'. Think of, for example, mortgage payments, contributions to pension funds, payment of utility bills, health insurance, school fees, the cost of commuting. Other contextual givens are, for example, social norms—to wear clean shirts and have a shower daily, or to choose the type of products available in the supermarket (all produced via a globalised production system). Furthermore, full rational choice is a myth. People cannot but take mental short cuts when making the hundreds of small decisions in daily life. They discount the future, emphasise fresh and extreme experiences, tend to feel loss more keenly than gains, and fear dramatic more than average risks (even if they have equal odds; Holdsworth and Steedman 2005). Semi-conscious routine behaviour, such as choosing a pub, shop or brand, is changed only during disruptive events, crises or during other windows of opportunity (such as when moving place of residence, when marrying, when changing jobs and so on).

Finally, the consumer is 'multifaceted'. He or she is not merely a person with voting power on the market, but is also in the political arena, a worker and part of a local community. In this role of citizen, he or she contributes to shaping the norms and values of the societies in which he or she lives and can be a powerful agent in bottom-up actions for change, even in companies.

### 22.3.4.2 Change model(s)

The description above provides various leverage points for changing consumer behaviour and the creation of more sustainable lifestyles. One-dimensional approaches include:

- Making changes to knowledge about products and processes (e.g. via labelling)
- Making changes to attitudes (via awareness-raising campaigns)
- Making changes to norms, values and opinions (as in anti-smoking and anti-drink-drive campaigns; but see also how Stern [2006] and Gore [2006] used the media to boost public support for climate change policy)

- Making changes to the symbolic meaning of consumption (e.g. making sustainability values such as equity, human rights and care for nature part of such symbolic value)[7]
- Making changes to habits and routines
- Creating windows of opportunity for change (e.g. offering alternative opportunities to fulfil needs)

Such approaches in isolation usually do not 'do the trick', particularly if true behavioural change is needed (see Vlek et al. 1999; however, if one is faced with very simple choices that do not require huge shifts in behaviour, simple price incentives and awareness campaigns can work, as in the example of the shift from the use of leaded to unleaded petrol in the early 1990s; see Holdsworth and Steedman 2005). A careful balance of persuasive and dissuasive (policy) approaches is needed to create lasting change. Much attention is needed to 'unfreeze' existing routines and 'refreeze' them into sustainable routines. Motivational approaches creating a supportive social environment (e.g. via role models, social learning and participation) that reflect a sense of community and credible, fair and legitimate shared goals and values work much better than preaching and appealing to individual altruism. Persuasion must be direct and emotionally appealing, repetitive (e.g. by regular feedback) and enforced by other factors (see Holdsworth and Steedman 2005). But, most importantly, motivational techniques must go hand in hand with the creation of alternative behavioural opportunities for fulfilling needs that provide (almost) equal tangible and intangible quality (the motivational and educational element is sometimes called 'software', and the creation of new opportunities is sometimes referred to as changes in 'hardware').

### 22.3.4.3 Limitation of the change model(s), policy implications and scattered myths

In the above, it seems that citizens, in their role as policy actors, workers, consumers or community members, cannot act alone. This obviously is not true. In particular, one can look at the inspiring accounts in Chapters 8 and 10 of this book of local, citizen-driven social innovation to see what people can do to foster change to SCP. However, some clear limitations must be understood:

- Consumers are not totally sovereign but are 'locked in' in situations limiting their choices (in economic terms, this is represented as an inelastic demand curve)
- Routines and habits may be more important than rational, conscious deliberation in making choices of which products and service to consume (even if there is full information and transparency in the market; see Chapter 16)

---

7 This implies articulation of such values in society, via government, consumer organisations or action. In this respect, it was a perfect move by the Secretary General of the UN to launch the Millennium Goals and to embark, via the Global Compact, on a strategic discussion with industry on how to realise this. But see also the efforts of the Slow Food movement (www.slowfood.com [accessed 25 August 2007]) and the Centre for the New American Dream (www.newdream.org [accessed 25 August 2007]) in promoting quality as lifestyle instead of hedonism.

- Simple approaches, such as relying on informative instruments or on adjusted prices, therefore will not lead to adjusted consumption patterns or behaviour if this implies the need to accept lower quality or less symbolic value or to counter existing routines that are reinforced by 'lock-ins'. The use of a tax rebate to facilitate the introduction of a hybrid car works (e.g. at the introduction of the Toyota Prius); advertising campaigns to stimulate car sharing will not work as long as car owning means superior transport quality and symbolic value (though such campaigns may have some, limited, success)

- Bottom-up initiatives started by consumers usually need backup from policy measures to become mainstream

- Consumers, even as a group, are not well placed to deal with sustainability problems where societal limits have to be set (as may be the case for emissions of carbon dioxide [$CO_2$] and use of energy from fossil fuels)

### 22.3.5 System innovation policy

#### 22.3.5.1 Basic understanding of roles and drivers

A variety of theories take a more overarching view on consumption and production. They look at the systems of consumption and production, their institutional setting and how government and other forms of governance can bring about change.[8] Briefly stated, theorists in the field of 'innovation systems' analyse how the relation between factors such as education, knowledge development, availability of venture capital and entrepreneurship determines how effective a country or sector innovates and enhances competitiveness (e.g. Hekkert *et al.* 2007; Nelson 1993). Scholars dealing with 'system innovations' emphasise the normative change to sustainability and related governance (e.g. Chapter 20). The system innovation approach now generally discerns three specific levels in society:

- A meso level (consisting of sociotechnical regimes). These form a set of interdependent and co-evolving technologies, symbolic meanings, infrastructures, consumer practices, rules and expectations reflecting the mainstream way of doing things in a specific field (e.g. food production and consumption habits)

- A micro level (consisting of niche markets). This consists of radical novelties that are not yet widespread but survive in 'protected' spaces, such as in relatively small markets where specific values are relevant (e.g. Slow Food; see footnote 7, page 415)

- A macro level (the 'landscape'). This consists of very stable factors that will not or are very unlikely to change and thus influence and channel development of the regime. Examples are basic trends (meta-trends), deep-lying beliefs and values (meta-values) and 'hard facts' such as geopolitical realities (meta-structures) such as the location of oil resources

---

8 As indicated in Chapter 18, this book does not discuss in detail complex system theory (apart from some text in Section 2.2.2). Attempts to explain changes to SCP in terms of this theory are still rare but could be an interesting research avenue for the future.

## 22.3.5.2 Change model(s)

The two theories described above—the 'innovation system' and 'system innovation' theories—each lead to a change model:

- The system innovation approach aims to direct change in a sustainable direction. The main concept applied is that of 'transition management'. This uses approaches such as visioning, learning by doing, doing by learning, and adaptive management by groups of front-runners to increase the probability of change to sustainable systems. It does so by supporting the development of promising niches, taking measures that create pressure on the regime, and providing a sense of direction for front-runner actors

- The innovation systems approach is interested mainly in fostering innovation and competitiveness rather than normative change. It also works through the market rather than by changing the market (as in the first approach)

Traditional policy instruments can play a role. It is helpful to put pressure on existing regimes via regulation (e.g. to stop using certain heavy metals, or to ensure minimum conditions for workers are met) or to implement financial reforms such as to abolish perverse subsidies, to set up emission trading schemes, or to tax resources (in an effort to reduce environmental impact) rather than labour (see Holliday and Pepper 2001). Yet such an approach is insufficient. Systemic change is also about changing the structure of the system, the mutual relations and feedback loops, the rules and the goals. A key problem in systemic change is hence to align the actions of different actors, to remove market failures and to clear blockages in the innovation system.

## 22.3.5.3 Limitation of the change model(s), policy implications and scattered myths

Nevertheless, the two approaches still have significant limitations. Despite their claimed systems perspective, neither of them truly manages to link up changes in production and consumption. Innovation theories tend to neglect consumers since they are seen as having too weak an influence on innovation [sic!].[9] The agenda is not yet broken down into specific sub-problems and transition processes, and each SCP domain is characterised by different kinds of processes and actors. Only the 'system innovation' school aims to foster a normative direction of change, but whether its 'transition management' approach can deliver is largely awaiting proof.[10] A differentiated approach to SCP system transitions seems to be needed rather than aiming for one unified grand theory. Sociological consumption research is strong on consumption patterns but weak on linking these up to production patterns and firm strategising; policy research is strong on changes in policy regimes and utility systems but weak on consumption and production.

---

9 This was concluded by Andersen after reviewing more than 90 papers in Chapter 18. This seems sound evidence that it is a myth that 'green consumers' alone can lead the change to the SCP nirvana.

10 Given the long time-horizons and the complexity of the systems at stake, no scholar in this field would suggest that traditional means–end management is possible, but the hope is that by developing strategic intent, in combination with learning by doing approaches, it will be possible to influence the direction of sociotechnical change. Some scholars, however, feel that external factors may dominate change and that influencing change is too difficult (see also Chapter 19).

SCP research needs to include all three areas to address system transition processes. A key role for the system transition research towards SCP is to contribute to a better understanding of supply–demand coordination—which ought to make up the very core of SCP research and policy-making.

## 22.4 Synthesis: integration via a systems perspective

### 22.4.1 Introduction

Table 22.1 summarises the findings of Section 22.3. One can see that the views on change to SCP from a business, design, consumer and system innovation perspective are highly (and surprisingly) complementary:

- The business perspective makes it clear that business can contribute significantly to SCP via a variety of mechanisms. However, it can only partially influence the system of which it is a part, having to obey business fundamentals and the prevailing paradigm of economic growth. This leads to logical, but from a sustainability viewpoint, less desirable behaviour such as the externalisation of costs, bringing hitherto free goods to the market economy and enlarging the aspiration gap to sell consumers more products and services

- The design perspective emphasises local distributed economies and creative communities that, via bottom-up action, change the world. This very much resembles the 'niche experiment' approach in system innovation theory (the design perspective, though, is less specific on how scaling up is to be realised). Another key role of design can be to provide visions for sustainable futures, which again in system innovation approaches play an important role in governance

- The consumer perspective shows that consumers can drive change via their voting power on the market and in their role as political agents, workers and citizens, capable of bottom-up action. But, at the same time, consumers face limits in terms of quality of choice and are driven by routines and other boundary conditions. Micro action has to be followed by macro action to realise a lasting implementation of SCP

- The system innovation policy perspective seems well suited to knit these parts together. It shows that businesses and consumers alone have freedom of action but usually must choose between creating radical new niches or taking less radical action compliant with the 'regime' and 'landscape' of which they are a part

The above simply confirms a basic hypothesis of the SCORE! project (see Chapter 1). A holistic, systemic view is needed in discussing changes to SCP. Putting the burden of realising SCP solely on ('green') business and ('green') consumers is a recipe for failure and would be a policy going counter to the evidence we—and many others—have gathered.[11] Individual actors in the system can take some steps alone, but, usually, their interaction has an inherent logic that keeps a treadmill going—leading to an ever higher material need of society (compare Ayres 1998; see also Box 22.1, page 422). A change in feedback loops in the system is hence needed—typically an action beyond the behaviour of one actor, but which may be realised jointly (this is reflected nicely by the title of a recent report by the UK Sustainable Consumption Roundtable: 'I will if you will' [SCR 2006]; compare also the plea of Andersen in Chapter 18 of this book to 'green' the innovation system as a whole). And, despite all talk of a society where powers shift from nation states to business, governments are still the best placed to initiate and moderate the change process.[12]

### 22.4.2 A perspective on systems and change

#### 22.4.2.1 A stylised representation of consumption–production systems

From the perspectives presented in Part 4 in this book in particular, we feel that the system innovation school provides the most fruitful analytical framework to achieve our goal. Its strength is the analytical distinction between a macro, or landscape, level (to be taken for granted in the short and medium term), the production–consumption regime (a level where actors in the system in can in principle influence and negotiate desired change) and niches (where groups can try out new consumption and production practices). Although this concept has been used mainly to understand change of a technical nature,[13] it encompasses user practices, sociocultural values and so on that clearly reside on the consumer side. Furthermore, the concept has many parallels with the idea of 'social practices' which has been embraced firmly by consumer scientists (see Shove 2004; Spaargaren 1997).

Figure 22.1 revamps the original conceptual model for SCORE! (see Chapter 1) with this multi-level concept. (Note: given the goal of this chapter—to wrap up the main implications for governance of SCP for a broad audience—this figure is a considerable simplification of those presented in Part 4 of this book and in similar work, such as that of Geels 2005.) It positions a simple mainstream production–market–consumption interaction above a niche level in a landscape context consisting of meta-trends, meta-values and meta-structures, and meta-shocks. Meta-shocks are sudden disruptive events such as wars and economic crises and, unlike the other meta factors, are a source

---

11 The neoclassical paradigm postulates that sovereign consumers exercise autonomous, rational choice in a fully competitive and transparent market and are the actors to whom dependent producers react. Here, consumers are the prime actors that should signal sustainability preferences to producers. Section 22.3.4 and 22.3.5.2 show how flawed the evidence for this view is (see Shove 2003; Spaargaren 1997). Indeed, that view is even at odds with work of Nobel Prize winners such as Simon (1957), who worked on the concept of bounded rationalities, and Akerloff, Spence and Stiglitz, who worked on the concept of asymmetric information.
12 See, for instance, the call, on 20 February 2007, from 100 businesses and other organisations for governments of the world to establish clear regulations that would cap carbon dioxide production: www.nysun.com/article/48996 [accessed 16 March 2000]; www.earth.columbia.edu/grocc/grocc4_statement.html [accessed 16 March 2000].
13 Geels (2005) gives example cases such as the introduction of steamships, cars and the jet engine; Elzen *et al*. (2004) include as examples changes in the Swiss agrofood production system and the change from a coal-based to a gas-based energy system in The Netherlands.

**TABLE 22.1 Review of the role of business, design, consumer and (innovation) policy for sustainable consumption and production (SCP)** (continued opposite)

| Aspect | Business | Design | Consumer | System innovation policy |
|---|---|---|---|---|
| Key understanding of role and drivers | Maximising shareholder value by diversification (branding, corporate reputation, developing unique competences, client relationships or solutions), cost reduction (including risk reduction) and market share increase<br><br>Alternative: creating 'owner value' by providing a working place or platform for self-realisation<br><br>Context is shaped by 'controlled competition', where the player with unique powers usually captures most value, where external financiers have an important influence on strategy, and where 'lock-ins' hinder change | Design is a discipline whose purpose traditionally is to establish the multifaceted qualities of objects, processes and services. By adding (intangible) brand and advertisement design it is now transforming itself into 'a reflective activity aiming to solve problems by developing solutions'. Designers are seen as having a responsibility for shaping consumption patterns and society at large | Consumption can be framed as fulfilling needs (in principle limited) versus wants (in principle unlimited)<br><br>Consumption is not only about material need fulfilment but also about symbolic value, the creation of identity, the meaning of life and dreams<br><br>Consumer choice can be framed as rational and sovereign, behaviour being a function of attitude, intentions and behavioural control. But it can also be seen as driven by routines and embedded in social practices ('ordinary consumption')<br><br>Humans are both buyers and consumers and political actors and citizens | There are various relevant system concepts<br><br>The *innovation system* concept is interested in understanding development and diffusion of innovation, and argues that for this the right mix of knowledge infrastructure, entrepreneurship, risk capital, launch markets, etc. must be in place.<br><br>The *system innovation* concept sees a partly locked-in, interdependent mainstream regime of technical artefacts, user practices, infrastructure and values, a niche level with novel practices, and a landscape that moulds the degrees of freedom of the regime. Regimes hence usually change incrementally<br><br>*Complex systems theory* gives a more formalised systems view in the form of stocks, flows and feedback loops, emergent properties, etc. |
| Change model: drivers for change to SCP | Environmental crises and prices forcing change<br><br>Tacit or explicit consumer or citizen expectations about firm behaviour, articulated by action, consumer groups or policy<br><br>External drivers and pressures: green public procurement, regulation, etc. setting minimum standards<br><br>CEOs want to show 'noble leadership'<br><br>Voluntary action: labels (MFC, FSC, energy), CSR, 'choice editing'<br><br>Translated into new business models and radical innovative products and services | Contribute to SCP by<br>1. Acting even beyond selecting resources with low impact and designing products with low impact, towards designing 'satisfaction fulfilment' systems with low environmental impact, and high social equity and cohesion<br>2. Developing alternative symbolic value and related patterns of production and consumption, etc. via new scenario and vision building and its communication<br>3. Facilitating local-based and network-structured initiatives and enterprises ('distributed economies') and other visions of a sustainable society (i.e. by 'creative communities') | Influence—in combination!—the 'software' via media and education campaigns (values, attitudes and knowledge of consumers, the symbolic aspects related to consumption) plus the 'hardware' (often material contexts that shape habits, routines, or force 'obliged' consumption)<br><br>Be sensitive to available windows of opportunity or create them<br><br>Also use the consumer in his or her role as citizen to foster change via local action, articulation of sustainability values and political choice | The different system concepts have different change models. The most relevant are:<br><br>*Innovation systems.* Create an innovation dynamo by putting elements of the innovation system in place. It is not normative but is aimed at improving competitiveness<br><br>*System innovation and transition management.* Involve front-runners in a process of goal-oriented modulation, learning and adaptive management<br><br>Common elements: change is evolutionary, path-dependent and interactive; it goes well beyond addressing market failures and addresses systemic failures, and tries to learn about how to create new and better feedback loops, etc. |

TABLE 22.1 (from previous page)

| Aspect | Business | Design | Consumer | System innovation policy |
|---|---|---|---|---|
| **Limit of the role: drivers for change away from SCP** | Business fundamentals must prevail; certain goals can be met only via framework changes<br><br>The competitive market system also rewards companies that make people dependent via the promotion of greed, fear and addictions, that externalise costs and that draw hitherto freely available non-market goods into a market context. Furthermore, firms actively try to influence governments and influence institutions so that their interest is served | How to scale up the experiments and visions against sometimes formidable barriers posed by consumer and business interests is not well articulated<br><br>Branding can also be used to promote greed and materialistic life styles | Since often the consumer is *not* sovereign, one-dimensional instruments such as education, articulation of values and information or pricing for SCP will work *only* if the alternatives available give the same quality, symbolic value and meaning as the original practices<br><br>Bottom-up action has to be reinforced by framework change for lasting effects.<br><br>Consumer wants are probably not limited | System innovation theory has a high hope that the power of front-runners, or new insights created with powerful actors, is sufficient to foster systemic change. But system change creates losers, and they may promote lock-ins *negative* to SCP if that is in their interest. This reflects that understanding of how to stimulate goal-driven transitions is still weak: not all actors relevant for radical innovation could be involved, alignment of change power may prove impossible, the role of planning or negotiation versus knowledge and learning is not clear |
| **Dilemmas** | 'Tragedies of the commons' and 'prisoners' dilemmas' such as long-term sustainability vision versus shareholder reporting systems that reward short-term profits | Traditional designers do not like to lose freedom through additional design specifications related to sustainability criteria or to design just what marketers ask them to | The world can cover everybody's needs, but not everybody's greed | Change via (moulding) the market ('innovation systems') versus change via staging processes outside the market ('systems innovation') |

CSR = corporate social responsibility; FSC = Forest Stewardship Council; MSC = Marine Stewardship Council

## Box 22.1 A view of the treadmill of (un)sustainable consumption

Fuchs and Lorek (2005) and others (e.g. Schumacher 1973) suggest feedback mechanisms in current society that hinder a strong sustainable consumption and production (SCP) policy. In brief, they state that the realisation of consumption objectives that include status, definition of identity and establishment of belonging will require, in the modern, globalised world, more material items than when local social networks still dominated. Messages about sustainable behaviour are widely overpowered by messages promoting material consumption, among others by media portraying role models that are able to consume beyond the wildest expectations of normal human beings. The desire for material consumption is hence deeply entrenched in consumer behaviour and social practices. Industry, for its part, is all too willing to fulfil these wants—or, indeed, co-created them via powerful branding, advertising and other media influence (see Danziger 2006). More production implies more turnover and more profit—for business. At a macro scale the result is a growing economy (albeit in monetary terms, but probably not in terms of quality of life [Marks et al. 2006]) and more jobs—feats that any politician is keen to realise during his or her term in office. And where, in theory, business models exist where intangible value is created, the globalised economy is characterised by a high level of competition, mass markets and substantial pressure to externalise costs. And, given this lack of consumer and business support for strong SCP measures (e.g. in terms of influencing consumption), it will be difficult to expect too much action from government in this respect. This would go counter to the wishes of voters and important industrial lobbying organisations. Indeed, even in various democratic countries it is not unusual for businesses to support candidates for elections (including presidential elections), obviously expecting a benign attitude in return (e.g. in The Netherlands the US ambassador is a business owner who had no experience in foreign politics but did donate a significant sum to the Bush electoral campaign in 2004; although this can be regarded as a relatively harmless outcome, it does illustrate what can happen). Last, but not least, from intergovernmental organisations (IGOs) precious little can be expected in this respect, since the IGOs responsible for sustainable development have little or no power to sanction or enforce standards.

FIGURE 22.1 The production–consumption regime embedded in a landscape context, showing competing (niche) practices

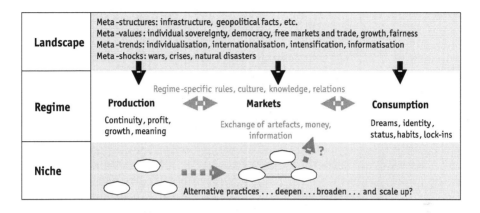

of instability. Meta-structures include existing infrastructure, geopolitical realities and other 'hard facts'. Meta-trends include internationalisation and globalisation, individualisation, informatisation and intensification.[14] Meta-values are widely held beliefs and cultural values such as personal freedom, democracy, equal rights, pursuit of growth, free markets, (unlimited) private ownership and a high level of personal responsibility for personal well-being (Mandelbaum 2002).[15] Further, a relatively high degree of honesty and fairness is expected: for example, with regard to eradicating extreme poverty, providing basic working conditions and other social standards.[16] Such 'meta' factors are usually 'untouchables' for actors in the regime in the short term.

### 22.4.2.2 Consumption–production systems: sources of (in)stability

This multi-level theory explains why intentional radical changes to practices in domains such as food, mobility and housing is often so difficult. First, the landscape poses clear boundaries for regime development. But this regime itself is an interdependent and co-evolving set of technologies, symbolic meanings, services, consumer practices, rules,

---

14 Intensification reflects not only the high degree of interest in change and variation of experiences but also the fact that time pressure on private and professional life is increasing, since more is being done in the same amount of time. All these trends in themselves can have important consequences for (unsustainable) consumption patterns. Examples of such trends are: an increasing number of smaller households and choices for individualised rather than communal services (individualisation), more use of energy-using ICT products (informatisation) and the creation of global relation networks leading to higher transport needs (internationalisation) and greater use of convenience products and services and of time-saving appliances such as dishwashers (intensification).
15 It is not for nothing that 'people who really made it', such as movie stars, business tycoons or sports icons, are seen by many as the ultimate role models. Where this system allows for accumulation of wealth, power and capital, this is seen as acceptable since the overall progress in society ensures that the extra wealth 'trickles down' to those who ended up in the somewhat lower rankings in the race.
16 Although these ideals may not be realised in practice, they have been repeated time and again in formal agreements and hence seem a benchmark for performance (e.g. see the UN Millennium Goals; UN 2006).

financial relations and expectations. It is difficult to change one part without altering the rest. This dynamic equilibrium generally changes only incrementally. A simple example: you cannot put any great number of hydrogen cars on the road without hydrogen gas stations, new safety rules, perhaps even new driving licence standards and so on.

But the theory can also help to find tensions or 'cracks' in the system that can make the stimulation of change easier. Such 'cracks' can be: internal tensions in the production–consumption regime, or misfit between regime and landscape, and can have a normative and operational dimension. Examples include a production structure evidently based on labour exploitation in the Southern Hemisphere (misfit with ethical meta-values), or a sector practising agriculture in greenhouses that because of rising energy prices becomes too expensive (operational misfit).

When promising niches are available that have matured (deepened) and become connected (broadened), and when at the same time 'cracks' develop or 'shocks' in the landscape occur, pressure on the regime may become so high that rapid change may become possible (i.e. niches 'scale up'). The regime breaks down, and niches plus the remnants of the existing regime will develop new structures, which eventually will stabilise and form a new regime (see Geels 2005; Kemp and van den Bosch 2006; see also Box 22.3, page 431).

### 22.4.2.3 Interventions: change in the regime or outside the regime

An intentional (as opposed to the above-described autonomous) regime change is possible usually only if actors align themselves to create a critical mass for change (unless one single crucial 'node' or 'gatekeeper' is present controlling stability in the regime). The power, interest, desires and beliefs, as well as the legitimacy of the position, then determine the position of an actor and the success of that actor in defending it (see Sabatier 1987). Typically, the following situations in such struggles may occur (Guba and Lincoln 1989; Hisschemöller 1993; Schön and Rein 1994):

- There is agreement on (problem) perceptions, and knowledge on how to solve the problem is certain

- There is agreement on (problem) perception, but the situation is too complicated to set out a road for solutions directly

- There is no agreement on the problem perception, and knowledge is uncertain—the problem is 'messy'

In the first two situations the operational or normative tensions and misfits are so obvious that, usually, there is a sufficient sense of urgency to solve the problem and thus legitimacy is created for action. But given the high level of complexity involved in many systems of consumption and production, a substantial proportion of problems encountered will be 'messy': the perspective on the problem can differ, and knowledge about how to reach goals is uncertain. A favourite tactic of potential losers is to articulate such uncertainties and ambiguities in an attempt to de-legitimise intervention. Interests and interpretative frameworks are hopelessly mixed up, and calls on science to arbitrate usually fail because of the 'trans-scientific' nature of the problem (see Weinberg 1972). Prolonged discursive struggles usually then determine the outcome.[17]

If by such processes the critical mass supporting change is lacking, the alternative is to avoid the legitimacy question altogether by developing action at the 'niche' scale: opponents simply do not have influence. Such niches then can satisfy the function of 'lifeboats' (i.e. they are ready for use when needed): the niche is ready to scale up once the context has changed and more support for the new way of doing things is available (see Schumacher 1979).

### 22.4.2.4 Interventions: types and time-horizons

Another perspective on change relates to the type of intervention that can be considered. In one of her many enlightened writings, Donella Meadows, pupil of system analysis guru Jay Forrester and co-author of *Limits to Growth* (Meadows *et al.* 2004), listed 12 places to intervene in a system, ranked in order of influencing power (Meadows 1999). For the sake of simplicity, we pragmatically group them below into four main categories, useful as a kind of 'completeness check' and inspiration for considering SCP policies. These are interventions that:

- Change technical characteristics
- Change in incentives and institutions (affecting feedback loops)
- Create or enhance self-organising capacity in the system
- Adapt goals and paradigms from which such goals have been developed

These levels also give a rough indication of the time-horizon that they may influence. In the short term there is probably no alternative but to stick to interventions that comply with generally accepted norms, values and paradigms in society and with the related aspirations of people. Of course, it is not impossible to change meta-views (and meta-trends) but as these reside at the landscape level this can work only in the long term (to get an indication of the time-horizons at stake, think of the decades it has taken to make prevention of climate change a viable policy proposal, or the time it has taken to embark on substantial measures to discourage smoking and to start to ban advertisements for tobacco products). To make matters worse, as actors are keenly aware that dominant goals and paradigms are so influential in shaping a mandate for measures, we often see

---

17 When knowledge is certain and perspectives cannot or do not differ, scientific analysis can usually underpin impartially what action to take. But in 'messy' problems the frontier between facts and values is blurred, and deliberative processes are needed that may lead to 'informed agreement' on the way forward (see Hisschemöller 1993; Schön and Rein 1994). Such 'messy' problems usually show up as messy when there is conflict of interest. Without that, differences in perspective will not be articulated and the problem will look 'normal' (Collingridge and Reeve 1986). Interpretative frameworks may be loosely defined and have only partially consistent 'value positions' (reflected by 'storylines') to underpin their positions (Hajer 1995). 'Discursive hegemony' then becomes relevant to 'win' a debate. Climate change was, for instance, long contested despite massive scientific research from, for example, the Intergovernmental Panel on Climate Change (IPCC). In 2006, the Stern Report and Gore's Oscar-winning movie *An Inconvenient Truth* were important events in giving a strong position to the discourse that states that climate change has to be acted on. Both have been severely criticised for being simplistic or biased, but that is not relevant: they shifted the image of what is right and wrong within large groups in society (Gore 2006; Stern 2006).

an intensive discursive struggle over these. The creation of self-organising capacity is particularly relevant when there is agreement on the perception of the problem but when the process of change is too complicated to be overseen fully—although, in this case, at least a common perspective on the strategic intent and a basic shared sense of urgency is present. The first two leverage points can be applied when less controversy over problem perception and cause–effect relationships are at stake, in the relatively short term. These leverage points are, compared with the others, also quite specific and hence can usually address production, interaction in the market and consumption.

### 22.4.2.5 Wrap-up

Table 22.2 summarises the findings on interventions discussed in this section (Section 22.4.2). Changes in technical characteristics and incentives can be implemented in the relatively short term, provided there is agreement on the problem and regarding means–end relations. This typically implies that they are regime-compliant and are not fundamental in nature. The main problem left may be that of dealing with rearguard fights of laggards whose interests may be affected by intervention. 'Learning' approaches are appropriate when there is a sense of urgency and a common sense of direction but there is not a clear idea on how this direction can be followed. Finally, adaptation of goals and paradigms usually implies a fundamental struggle between meta-views. Kuhn's (1962) description of paradigm shifts suggests that this will probably require a prolonged, informed, societal debate and that a truly revolutionary, fundamental, impact will be felt if the shift succeeds.

TABLE 22.2 Interventions and related characteristics

| Leverage point | Leverage | Time-horizon | Typical policy context | Main problem |
|---|---|---|---|---|
| Technical and incentive change | Regime-compliant | Short | Agreement on goals and means | Overcoming opposition of 'laggards' |
| Enhanced self-organising capacity and learning | Intermediate | Medium | Agreement on sense of urgency but not on means | Focusing direction and learning about means |
| Adapt goals and paradigms | Fundamental | Long | Disagreement on urgency, goals and means | 'Managing' a mental revolution—in a nice way! |

### 22.4.3 Seven keys to success: implications for policy and action for sustainable consumption and production

#### 22.4.3.1 Introduction

The above analyses suggest that it is not possible to define a single or clearly defined set of models for change to SCP. There are simply too many variables that can differ: the domain (food, mobility, housing), the extent and time-horizon of change (e.g. a single product versus a total system), the level of legitimacy for change and so on. We suggest, therefore, that we take a number of key lessons from the evidence gathered in this book and use these to work out a specific strategy for change to SCP in a specific situation. Four lessons follow directly from the systemic character of the SCP challenge. First, politics must act *with* others rather than delegating action. Second, production, markets *and* consumption must be addressed. Third, it is key to understand what forces work in favour of and what forces work against SCP. Last, we need to look at how the forces that work in favour of SCP can be strengthened and used. Furthermore, Section 22.4.2 made it crystal clear that there are specific approaches for and lessons about change to SCP for the short, medium and longer term. In the following sections we discuss these seven topics.

#### 22.4.3.2 Don't outsource politics: work with ('green') consumers and businesses to create a triangle of change

This will be a short section, but we want to repeat this warning one final time: *don't outsource politics* in the naïve and unjustified belief that 'green' businesses, the market or 'green' consumers will realise SCP on their own. This goes counter to all evidence we have put together in this book and against the evidence that many authors have gathered before us. If you still believe that flawed mantra at this point, go back to the start of this chapter and read it again. Certainly, 'green' consumers and 'green' businesses can 'do their bit', and markets to some extent will reflect scarcity of resources and other sustainability problems providing incentives for some change. But all evidence shows that bottom-up action needs top-down support and framework change (e.g. see Table 22.1, pages 420-21). For things to work, policy-makers have to do their bit and collaborate with business and consumers, creating what the UK Sustainable Consumption Roundtable (SCR 2006) calls a 'triangle of change'. Of course, SCP policy is not an easy subject, but, rather than ducking away for the challenge, the ambition should be to deal with it smartly: a task with which the next six keys may help.

#### 22.4.3.3 Focus on production, markets and consumption, not just the supply side

Fundamentally, the goal of reducing environmental pressure by reducing consumption can be reached via three routes: 'greening' supply, shifting demand to low-impact consumption categories and lowering demand (see Hertwich 2005; Tukker and Tischner

2006; UNEP 2002). Table 22.3 lists these options in more detail, including illustrative impact-reduction factors.[18]

TABLE 22.3 **Fundamental approaches for reducing environmental pressure by consumption**

Source: adapted from UNEP 2002; Tukker and Tischner 2006

| Principle | Approach | Improvement[a] | Parties |
|---|---|---|---|
| Reduce impacts of production and products | Eco-efficiency improvement | Factor 2–4 | Business, government |
| | Radical technical change | Factor X | |
| | Eco-efficient product use | Factor 2 | Consumers |
| Shift consumption expenditure from high- to low-impact products and services | Via 'software': stimulate 'green' consumer choice via education, awareness, etc. | Factor 2[c] | Consumers, government, business |
| | Via 'hardware': providing infrastructure and options that encourage consumers to shift behaviour[b] | | Government, Business |
| Reducing or limiting growth of consumption expenditure | Create win–win situations | Factor 2–4[d] | Society at large |
| | Truly 'downshifting' | p.m. | |

a This is 'static' improvement. Interventions at the consumption side are in fact much more relevant than this table suggests, since they can prevent 'rebounds' in the future.

b The classic example here is to provide a good public transport system and to discourage car travel via taxes and road pricing, the strategies used by Hong Kong and Singapore to avoid the type of 'traffic infarcts' seen in many megacities.

c Suh 2004; Meijkamp 2000, Tukker 2006.

d Marks et al. 2006.

p.m. = Pro memorie

Since society is adverse to interfering with consumer choice and markets, it is not surprising that the SCP debate is often narrowed down to the first principle under banners such as 'dematerialisation' (e.g. the World Business Council for Sustainable Development report, *Sustainability through the Market* [Holliday and Pepper 2001], lists seven keys to success that focus solely on 'greening' business and markets but carefully avoids the consumption issue). However, history has shown that this strategy alone fails because of what has been loosely termed the 'rebound effect': the growth of material consumption.[19] Hence, approaches transforming markets and consumption patterns

18 As this book is focused on the EU, the equity and poverty eradication dimension of SCP inevitably receives less emphasis. However, even in the EU this needs attention in relation to the environmental SCP dimension. Environmental impact is by and large proportional to expenditure, and richer people tend to embark relatively more often on high-impact activities such as air travel (Tukker 2006; SCR 2006).

19 A classic example can be found in mobility, where the trend to more energy-efficient cars was offset by the appetite for buying heavier cars (e.g. sport utility vehicles [SUVs]), higher mileages and higher levels of car ownership. Richard Branson, probably one of the more society-minded entrepreneurs, plans to offer space flights for the masses, which will lead to a next level of impacts from leisure and holiday activities.

must be followed as well. Although 'market transformation' can probably still count on support, it is unlikely that this will ever be the case for explicit 'downshifting' policies that imply a true loss in quality of life. Indeed, most accounts describing downshifting suggest that people experience gains in non-consumption-related quality of life and hence still echo a 'win–win' mantra.[20] At the same time, the good news is that win–win solutions that are not too radical (i.e. that will allow the same quality of life) will, if applied over the full chain of production, probably be sufficient to create radical reductions in environmental impact (see Tukker and Tischner 2006; von Weizsäcker et al. 1997; see also Table 22.3). The 'happy planet index' report, discussed in Chapter 2, provides an important indication that changes to consumption patterns and markets is possible without loss of quality of life (Marks et al. 2006). Long, happy lives can be lived with very different ecological footprints. And it is not difficult to see why. Significant parts of the financial and time expenditure of consumers are on 'duties' such as commuting, business travel and so on or compensate for a low local quality of (social) environment. That changes to such patterns can yield a double dividend is not surprising.

### 22.4.3.4 Understand your system: avoid fighting windmills

When developing an SCP policy it is very important to understand the system at stake, particularly in terms of how the forces in the landscape and production–consumption regime pose barriers—and provide windows of opportunity—for change. Particularly in the short term, meta-trends, meta-values and meta-structures have to be seen as givens. At the regime level, business realities, consumption habits and lock-ins and the desire for status and other dreams pose further boundaries to change. Policies going counter to these would have policy-makers ending up like Don Quixotes fighting against windmills. Understanding these force fields, and using them to foster change, makes them like smart judo players who, despite an apparent lack of strength, time and again manage to defeat much bigger adversaries. Such a system analysis can be done via the following steps:[21]

- Analyse what the relevant meta-trends, meta-values and meta-structures are in the field at stake and what change they legitimate
- Break the regime down into (key) actors (e.g. producers, retailers, consumers), their financial, physical and informative relations, their interests, desires and (policy) beliefs and their power and resources

---

20 For example, Princen (2005) describes how a lack of car access to an island off Toronto has resulted in a community with a high quality of life socially and a low need for material artefacts. Etzioni (1998) argues that downshifting typically is practised by those who have the choice to do so: that is, by economically well-off and secure people. Also, they tend to downshift in such a way that they can still show status: they 'display the objects of poverty in a way that makes it clear [they] are just rolling in dough' (Brooks 1997).
21 Note we describe here a rather static system analysis approach. Interactive and dynamic set-ups may be more appropriate to complex systems (see Rotmans 2003). Nevertheless, simple static schemes can be very useful in devising a policy strategy. The UK National Consumer Council developed, for instance, a simple checklist to develop policy recommendations for change to sustainable consumption based on basic insights into consumer behaviour: 'address consumer priorities', 'tackle specific sustainable consumption beliefs' and 'work with influences on consumer behaviour' (Holdsworth and Steedman 2005).

- See if there are clear 'gatekeepers' that can create change, or see how coalitions that could create a critical mass could be formed
- Check the availability of promising niches

The information in Section 22.4.2.3 shows why this analysis in developing an SCP policy needs so much attention. The use of regulatory and economic instruments often still dominates policy approaches. Harnessing the environmental effects of production and products in this way is one thing, but bluntly trying to use such instruments to influence consumption is easily at odds with, for example, the principle that humans should be free to shape their lives in the way they want, the idea of consumer sovereignty and the concept of free markets. Any policy-maker proposing such measures is unlikely to last long. Box 22.2 shows a smarter approach.

**Box 22.2 How to use (tacit?) consumer expectations to change the coffee chain**

In The Netherlands, coffee certifier Utzkapeh rocketed in just five years from a 0% to a 25% market penetration in 2005. Its approach is to ensure basic social and environmental performance in the coffee production sector, without paying a fixed price premium (as Fair Trade organisations do). Interestingly, the Utzkapeh logo is not actively advertised. Meanwhile, roasters, the most powerful players in the coffee chain, understand very well that consumers see such basic qualities as being inherent to a quality brand—just as a good car does not rust in its first year. Here, we see an interplay of tacit consumer expectations articulated by Utzkapeh being picked up by a dominant industry player, bringing about change. Further, it must be noted that this success was probably greatly supported by earlier, more far-reaching, fair trade initiatives such as the Max Havelaar brand, which is less popular with the roasters, presumably because of the minimum price that comes with Fair Trade and which is seen as market-distorting. Where these initiatives still in themselves tend to have only a small market penetration, they point clearly at problems in the production–consumption system and encourage consumers and producers to act (Tukker 2007).

### 22.4.3.5 Enlist the good forces: articulate positive trends and paradigms and a credible sense of urgency

Our warnings do not imply that consumption-oriented policies will never have a mandate: on the contrary. As stated in Section 22.4.2.3, normative and operational misfits provide a powerful legitimacy and opportunity for action in favour of SCP. The meta-values listed allow citizens, NGOs and policy-makers alike to question and act on, for example, the following points (where smart businesses can use some of them to find new sources of competitive advantage):

- Lack of transparency about the origin of goods and how they are produced
- Poor social and environmental standards in the supply chains of producers of products

- Unfair competition between alternatives: for instance, as a result of perverse subsidies or unaccounted-for externalities
- Monopolies or oligopolies that hinder a proper functioning of the market
- Markets where power balances are so distorted that consumer sovereignty cannot be exercised
- Market systems that seem incapable of solving grinding poverty, or in which economic growth does not lead to increase in quality of life
- Consumption activities that go counter to important values in society

Highlighting meta-values favourable to SCP, formalising them, enlisting support (see Box 22.3), as well as 'making darkness visible' by revealing who wins from unsustainability are thus the tactics to use (Lebel *et al.* 2006). The UN Global Compact is a good example. It articulates four sets of core values for doing businesses and managed to enlist 2,900 (mostly large) businesses to support it. Though implementation is complex, it sets the direction and a stage for discussing how to do it, and questions deviating behaviour.[22]

Highlighting operational misfits forms another source of legitimacy for action. Here, we have mostly environmental arguments for embarking on SCP policy: 'living within the ecological limits is the non-negotiable basis for our social and economic development', as it was put by the UK Sustainable Consumption Roundtable (SCR 2006). But the argument can also include points such as persistent traffic congestion in urban areas in the world, industrial sectors that under-perform economically and so on.

### Box 22.3 A gatekeeper compelled to comply with meta-values

Transfatty acids are known to contribute to heart disease and obesity, and a major Western food company therefore largely removed such acids from its products sold in the Western Hemisphere. A Latin American subsidiary, however, kept on using them. Complaints by local consumer organisations to the local company branch did not work, but when they contacted their European counterparts, and these in turn complained to the company's head offices, the Latin American subsidiary changed its product line almost overnight. The multinational simply could not deviate from what generally was seen as 'good'. Note: this is obviously a simple change, as one company had full control over that change.

---

22 A complexity that we will not further elaborate on is, of course, that such good forces can be (ab)used tactically. Probably since the beginning of time leaders wanting to pursue 'dark' plans have tried to frame them as being supportive of and compliant with the highest values present in a society. It will not be the first time that principles such as 'freedom', 'democracy', 'growth' and 'free markets' are hijacked in support of individual, unsustainable interests.

### 22.4.3.6 Short-term impact: act on opportunities, or start from the bottom up

Many actions supportive of sustainable consumption (and production) can be taken now, since they do not represent the impossible uphill battles of Don Quixote-like situations. They 'start where people (and business) are' and try to create momentum from there: usually taking small steps, looking for win–win situations or ensuring that production and consumption systems become more consistent with 'meta-values' and other boundary conditions. It then simply comes down to sound leadership to implement such options or to finding the right 'gatekeeper' who can stimulate change.

**Production side**[23]

On the production side, the ball is mainly in the court of business and government. The aim here is to enhance the environmental and social performance of production and products to a feasible extent, largely via the approaches highlighted in Section 22.3.2, on the business perspective. Typical tools include eco-efficiency approaches (cleaner production and pollution prevention pays programmes; ecodesign), management of social and environmental standards in supply and downstream chains (e.g. via corporate social responsibility [CSR] standards or labelling approaches such as those of the MSC and the FSC) and practising 'choice editing'. Industry self-regulation and backup action by authorities can support 'a level playing field' and hence broad implementation of such concepts. Furthermore, industry can use 'meta-factors' and problems ('cracks') in the system as an inspiration to develop novel, more sustainable, products, business models or other sustainable strategic innovations (see Box 22.4). Government can provide support for this specifically by supporting the innovation system for such sustainable innovation (e.g. by supporting lead markets and facilitating venture capital; see Chapter 18).

**Box 22.4 How business can build on sustainability trends**

> In Germany, the Bionade firm successfully latched onto those tacit desires of consumers that are grounded in sustainability values, such as health and chemical-free products. This small family business developed a 'non-alcoholic organically produced refreshment drink', and positioned it against regular soft drinks with artificial additives. It generated a massive buzz on the German pub scene. Combined with smart branding, this small firm built a whopping market success and made this all-natural soft drink the 'cool' lounge drink of choice (www.bionade.de [accessed 20 March 2007]).

**Market side**

On the market side, the ball is mainly in the court of the authorities. Here, the task is to make markets truly fair. Such 'market transformation' would ideally internalise externalities, abolish perverse subsidies (see Box 22.5), counter monopolies and oligopolies, promote consumer power and choice, and encourage transparency in social and environmental issues related to products. Furthermore, some limitations would be put on advertising: for example, with regard to advertising products to vulnerable groups (e.g.

---

23 See also Section 22.3.2, on the business perspective.

the young), advertising the promotion of unhealthy and unsustainable behaviour (such as smoking) and advertising that can be seen as unfair or dishonest.

### Box 22.5 How taxation distorts the mobility market

Already, under current circumstances, fuel makes up the main part of the costs of a discounted economy-class air ticket. But it is not taxed. At the same time, in Europe energy used for powering trains, buses and cars is heavily taxed. No wonder that low-cost carriers can easily beat the prices of railways for short-haul flights, yet it is exactly these flights that are relatively damaging to the environment because of energy used and lost at take-off and landing.

## Consumption side[24]

On the consumption side, the lessons from Section 22.3.4 on the consumer perspective apply. Consumers and civil society organisations can exercise 'green' choice and articulate expectations of environmental and social sustainability standards to be met in production and markets. As far as possible, within the boundaries consumers face, consumers can try to embark on more sustainable lifestyles. Such change could be promoted by government via 'software': motivational campaigns endorsed by appealing engagement, clear leadership and repetitive feedback (e.g. via smart electricity meters, 'green' credit cards with 'carbon-offset' schemes for any purchase, the direct possibility of buying 'carbon-offset' certificates when buying air tickets and fuel-use feedback meters in cars). Yet such attempts to influence motivation should go hand in hand with providing opportunities to behave sustainably via 'hardware': infrastructure that promotes sustainable choice (see Box 22.6), or forms 'no-need' contexts. Government and not-for-profit organisations can put 'green' public procurement schemes in place and focus these where spin-off can be expected by setting the right examples (e.g. providing high-quality catering in government canteens, in schools and in hospitals).

### Box 22.6 How a congestion tax changed London travel habits

The London congestion charge is a good example of striking the right balance between dissuasive and persuasive policy. Car travel in the city was what could be regarded as a nightmare, so it was unpopular with commuters anyway. In 2003 a penalty of £5 was introduced for cars entering Central London, but at the same time more and cheaper bus transport options were introduced. The alternative hence became the easier option, resulting in wide acceptance of the measure by citizens (see Holdsworth and Steedman 2005).

---

24 See also Section 22.3.4, on the consumer perspective.

### New niche systems[25]

Furthermore, nothing, of course, is preventing business, consumers or others from dealing proactively with the issue by using the 'mega' factors identified in Section 22.4.2 as inspiration for developing new sustainable products, business models or social practices. Even when they are radical and initially confined to niches where they do not need the blessings of society as a whole, they make sense. They can ripen in their niches, until regime and landscape changes make their breakthrough possible (see Box 22.7).

#### Box 22.7 How front-runners sow the seeds for Swiss sustainable agriculture

In Switzerland, in the 1990s, virtually overnight almost the whole agricultural sector shifted towards organic farming or semi-organic precision farming. In a very long preparatory phase, starting in the 1960s, and travelling via various routes, experience had been gained with organic and semi-organic production, certification and retail systems. But, for these niches to become mainstream, a financial upheaval had to take place. Under new rules from the General Agreement on Tariffs and Trade (GATT) and the World Trade Organisation (WTO), Switzerland was forced to abolish support for regular farmers—only support for sustainable farmers was allowed to continue. The rest is history (Belz 2004).

#### 22.4.3.7 Medium-term impact: build adaptive capacity for further change

For actions with impact in the medium term, a collaborative effort among societal groups is needed to see how more systemic changes in production–consumption structures can be fostered. For example, this concerns what the UK SCR (2006) calls the 'development of product roadmaps'; alternatively, in a broader sense, the Dutch Knowledge Network on System Innovation terms it 'transition management' (Chapter 20). It also links up with the call of UNEP, in its *Consumption Opportunities* report, to make changes to consumer behaviour possible by offering sustainable infrastructure that allows for sustainable consumer choice (UNEP 2002; see also Box 22.8). This type of action assumes that there is a reasonable sense of urgency and an agreed 'strategic intent' with regard to change in society but that the exact way how to realise this cannot yet be agreed on.

#### Box 22.8 How indicative planning gave Curitiba its famed public transport system

Megacities in, for example, Thailand, Indonesia and China are infamous for their gridlocked traffic and air pollution. In Curitiba, a major city in the south of Brazil, some city planners created a different history. They took the strategic decision to base further city development on principles such as minimisation of urban sprawl, retention of the historic district and use of a cost-effective express bus system as the backbone for mass transit. The approach was so successful that now 60% of the travel in the city takes place via the public bus system. The city itself is one of the most liveable in Brazil (see von Weizsäcker *et al.* 1997).

---

25 See also Section 22.3.3, on the design perspective.

Such 'enhanced self-organising capacity' is relevant, since real radical change is probably too difficult to predict. Such 'learning' approaches are nothing new for business: business 'gurus' such as Hamel and Prahalad and managers such as de Geus have become world famous with their books, *Competing for the Future* (Hamel and Prahalad 1994) and *The Living Company* (de Geus 1997). They each conveyed that the future will always be too complicated and unpredictable to plan a reaction on expected external change, so that learning and adaptive management will be the only realistic option left. Companies can act on this alone, enhancing their agility and flexibility to cope with (sustainability) problems in the future. Authors such as Kemp in this book (Chapter 20) go one step further: they have expanded such approaches for organising change 'management' across different actor groups, such as the different businesses, intermediaries, government and consumer groups involved in a specific sector. The idea is that such groups strategically interact, agree on the strategic direction of change and adjust tactics along the way in an approach that can be described as learning by doing, and doing by learning.

An equivalent of this, at a more local level, is to foster 'small-group community management' by enhancing the role of local production–consumption systems. This allows for more intimate interaction and mutual adjustment than is possible in impersonal and globalised systems.

### 22.4.3.8 Long-term impact: deliberate fundamental paradigms

However, to conclude, it is clear that the SCP agenda inevitably must also deal with more fundamental issues—difficult topics that perhaps at this stage go counter to existing trends, widely held beliefs and expectations in society. Such long-term challenges cannot be tackled head on but require a deliberative process in society (see UNEP 2002). Such 'capita selecta' on the SCP agenda include (in part, inspired by SCR 2006):

- A 'beyond the consumer economy': looking at how to create feedback loops in the market system that break through the need to create material growth as the only way of keeping the economic growth engine spinning, a topic that also requires novel business models

- 'Inequity': looking at how feedback systems in the market must be adapted to promote a reasonable level of equity, including in North–South relations

- 'Consuming less, but of what?': looking at the extent to which current market systems contribute to a loss of natural and social capital or, conversely, looking at which characteristics of patterns of consumption and production foster a high quality of life with low material input

- 'Social aspirations': looking at how to foster ways to meet social and psychological goals (e.g. mediating status) in less material ways, or how to curtail status races altogether

- 'Keeping powers in balance': looking at how to ensure a reasonable balance of power in the triangle of consumer-citizens, government and businesses, where now business sometimes seems to overpower consumers via advertising, and sometimes has much influence on government via lobbying, election support and so on

Most of these questions need informed deliberation before there may be a good opportunity to take broader action. For instance, if analyses such as the 'happy planet index' approach (Marks *et al*. 2006), or experience with alternative lifestyles as propagated by, for example, the New American Dream (see footnote 7, page 415), provide evidence that economic growth does not foster a higher quality of life, smaller or larger adjustments to this growth paradigm may take place. Fostering such investigations, showing working alternatives and, through these examples, appealing to people's highly held norms and values in high-profile ways are thus other avenues that could be followed to foster SCP.

## 22.5 Summary and conclusions

So, what are the final conclusions of the first half of the SCORE! project? Table 22.4, summarising the lessons from Section 22.4, shows that change to SCP is a complex business and that some clear rules can be discerned. The most important points include the following:

1. Do not put the burden for change solely on 'green' consumers and businesses: support mutual re-enforcement in the triangle of change
    a Business is good at improving the efficiency of products and processes. Such improvements can be stimulated via voluntary action, standards or by use of regulations setting minimum standards (e.g. CSR, FSC and energy performance standards for buildings). However, business has an only limited interest in changing consumption patterns or consumption levels
    b Consumers can, in theory, exercise sustainable choice. Such choice can be stimulated via informative instruments and campaigns. However, in practice, consumers are to a large extent 'locked in' in infrastructures, social norms and habits that severely limit consumer choice
    c It is hence inevitable that government must at least stimulate sustainable practice at the production side and stimulate availability of choice for consumers. However, it also has a role to play in directing societal change into an equitable and a sustainable direction, since individual actors are unable to realise this and create unwanted lock-ins. Rather than 'leaving things to the market'—thereby suggesting that the market is seen as a goal rather than (probably the still best) means to foster a high quality of life of the many—debate, analysis and action should be encouraged on how to use markets to realise equitable, sustainable, development
2. Understand the systemic forcefields and do not fight windmills: start bending trends, paradigms and beliefs from where people and businesses are. Usually, the degree of freedom for action in the short term is limited by meta-trends, meta-values, meta-structures and meta-beliefs at the macro level, and by interdependences and preferences of producers and consumers at the meso level. For instance, meta-paradigms such as beliefs in a free market, personal choice and consumer sovereignty are difficult to bend in the short term. Hence, stay away from 'rigid' policy ideas that prescribe consumer choice

3. Enlist the good forces and show urgency where credible: articulate positive meta-trends and meta-values and institutionalise them as guiding principles for development; articulate likely future crises if current trends persist

4. Short-term options: in the short term, many SCP solutions can be implemented immediately that make positive use of the systemic forcefields listed above. Certainly, they will need a leadership that makes a difference, if only to counter rearguard fights by 'dinosaurs' setting traps such as asking for 'evidence beyond reasonable doubt' before action can be taken. Many options that business and consumers can use, enhanced by government support, are listed in Point 1 above. Box 22.9 (page 440) lists specific points for attention on how to support consumers in realising SCP. Another clear mandate exists for 'market transformation': ensure markets are fair, do not distort competition through externalities and perverse subsidies and so on. Even radical SCP options can be tried out in niches. Such front-runners often sow seeds for change, and front-runners can often 'do their thing' very well in smartly chosen niches that are not too greatly affected by forces at the landscape and regime level

5. Medium-term options: 'roadmapping and transition management' are good strategies for societal parties to move forward if they agree on the rough direction of change but cannot do it alone or if the road is uncertain. It entails a process of mutual enforcing actions for change:
    a Radical change usually takes a long time, and 'command-and-control' approaches will usually not work. Indicative planning and the development of 'strategic intent' through a process of learning by doing along the way is likely to be much more successful
    b A process of 'visioning' and experimentation, particularly when it is not totally clear in which direction the change has to go, is essential
    c 'Flagship' (niche) experiments with new practices and systems should ideally act as 'stepping stones' for potential future new sociotechnical constellations

6. Long-term options: these are the more difficult issues to tackle and are the ones that require discussion of widely held beliefs and paradigms in society. Such an approach requires informed deliberation on issues such as:
    a The underlying growth engine in our markets
    b How markets contribute to fairness and equity
    c How consumption supportive of sustainability can be discerned from consumption that is destructive to institutions and non-market goods providing quality of life
    d How to develop novel and dematerialised ways of realising social aspirations, and how this relates to novel business models
    e How to maintain a fair balance of power in the triangle of business, government and consumers (e.g. by questioning the role of advertising and the media)

438  SYSTEM INNOVATION FOR SUSTAINABILITY 1

**TABLE 22.4 The production–consumption chain and leverage points for change** (continued opposite)

| Landscape | Meta-structures: infrastructure, geopolitical facts, etc.<br>Meta-values: individual sovereignty, democracy, free markets and trade, growth, fairness<br>Meta-trends: individualisation, internationalisation, intensification, informatisation<br>Meta-shocks: wars, crises, natural disasters | | | |
|---|---|---|---|---|
| **Regime** | Production ⇄ Markets ⇄ Consumption | | | |
| **Time-horizon of impact** | **Actions and leading actor** | | | **Dominant leverage point** |
| | Production | Markets | Consumption | |
| **Short-term impact**<br>Goals and direction: agreement<br>Means: fairly clear<br>Main problem: overcoming opposition of 'laggards' | *Business*<br>● Apply CP, PPP, ecodesign<br>● Manage supply and downstream chains (e.g. see CSR, FSC, MSC, etc.)<br>● Apply choice editing<br>● Promote industry self-regulation on the above<br>● Use 'meta' factors as inspiration for new sustainable products, business models, and other strategic innovations (e.g. via experience design)<br><br>*Government*<br>● Provide a level playing field to support the above (covenants, regulations, standards)<br>● Foster the greening of innovation systems and support sustainable (niche) entrepreneurs<br>● Articulate and impose sustainable meta-values | *Government*<br>● Internalise externalities<br>● Abolish perverse subsidies<br>● Counter monopolies and oligopolies and promote consumer power and choice<br>● Promote transparency on social and environmental issues related to products<br>● Set basic advertising norms: fair, not promoting damaging offerings, and not directed at vulnerable groups | *Consumers, citizens and NGOs*<br>● Exercise sustainable choice<br>● Start steps towards lifestyles of health and sustainability (LOHAS)<br>● As citizen and worker, articulate and impose sustainable meta-values<br><br>*Government (combine the below for effect!)*<br>● GPP (focus on visible examples with ripple effects, e.g. providing high-quality school meals)<br>● Provide infrastructure for sustainable choice of similar quality; create no-need contexts<br>● Motivate via appealing engagement and leadership and via repetitive feedback (e.g. smart meters)<br><br>*Business*<br>● Promote sustainable consumer feedback (e.g. smart meters, green credit cards)<br>● Apply sustainability marketing and demand-side management | Technical and incentive change |

TABLE 22.4 (from previous page)

| Time-horizon of impact | Actions and leading actor | Dominant leverage point |
|---|---|---|
| **Medium-term impact**<br>Goals and direction: agreement, at least on the sense of urgency for change<br>Means: not clear<br>Main problem: focusing direction and learning about best means | • Business: develop 'competing for the future' capabilities<br>• Government (as initiator, in conjunction with business and NGOs): start processes of product roadmapping, indicative planning, transition management and other learning and visioning approaches to overcome lock-ins and stimulate a sustainability focus for long-term change<br>• All: develop and test alternatives in niches ('lifeboats')<br>• All (with an emphasis on citizens and government): stimulate small-group management by, e.g., fostering locality and the creation of local feedbacks | Enhancing self-organising capacity and learning |
| **Long-term impact**<br>Goals and direction: controversial<br>Means: no insight in means–ends relations<br>Main problem: 'managing' a mental revolution—in a nice way! | All: foster deliberation on the more fundamental issues related to markets, governance and growth:<br>• Beyond the consumer economy: what does the sustainable growth engine look like?<br>• Inequity: how do we promote markets that foster a fair level of (in)equity?<br>• Consuming less or less material: when does it help to reach a high quality of life?<br>• Social aspirations and status: how do we reach this in an immaterial way, or stop this race altogether?<br>• Power balances: how do we restore balance in the triangle of business, government and citizens? | Adapting goals and paradigms |

CSR = corporate social responsibility; FSC = Forest Stewardship Council; MSC = Marine Stewardship Council

### Box 22.9 Sustainable consumption and production and the consumer

- Radical improvements that do not require change of consumer behaviour (e.g. a zero-energy house) are probably easier to implement than are improvements that require change
- Change in consumer behaviour is likely only if three components are addressed simultaneously: motivation and intent; ability; and opportunity
- The alternative opportunity should be at least as attractive as the existing way of doing things—not only in terms of functionality but also in terms of immaterial features such as symbolic meaning, identity creation and expression of dreams, hopes and expectations
- The motivation or intent must be addressed not only by traditional methods such as incentives and education but also via small-group community management and exemplary normative behaviour by role models. Since motivation and intent are often unconscious, special attention to how to overcome the role of habits is needed

Each of these issues poses fundamental questions about the way our market-based system works and about the institutions that have been developed to support it. Gathering credible evidence on how consumption and production systems can be organised more efficiently in providing quality of life (see Marks *et al.* 2006) and showing inspiring examples of alternative ways of doing things are also tactics to be pursued.

In sum, there is usually no 'silver bullet' that can bring about a radical change towards sustainability. The findings of SCORE! show the need to apply policy mixes that range from informative instruments, via price incentives, to regulatory pressure. Yet the complexity is not so great that the task is impossible. Action can be undertaken now, at all levels. Business can embark on providing sustainable products and services, latching on to tacit and explicit expectations of consumers. Government has already sufficient mandate to embark on actions such as market transformation, providing a level playing field for 'green' products and 'green' production on the production side and to encourage 'green' consumer choice. Consumers can already express voting power on the market. And of course all groups can already take more radical action and experiment in niches. This is probably one of the most exciting ways forward, paving the way for the more radical changes that society has to go through to meet the non-negotiable boundaries of the carrying capacity of our Earth.

# References

Ayres, R.U. (1998) 'Viewpoint: Towards a Zero Emission Economy', *Environmental Science and Technology* 32.15 (1 August 1998): 366A-67A.

Baudrillard, J. (1998) *The Consumer Society: Myths and Structures* (London/Thousand Oaks, CA/New Delhi: Sage Publications).
Belz, F.M. (2004) 'A Transition towards Sustainability in the Swiss Agri-food Chain (1970–2000)', in B. Elzen, F.W. Geels and K. Green (eds.), *System Innovation and the Transition to Sustainability: Theory, Evidence and Policy* (Cheltenham, UK: Edward Elgar): 97-113.
Brooks, D. (1997) 'Inconspicuous Consumption', *New York Times Magazine*, 13 April 1997: 25.
Collingridge, D., and C. Reeve (1986) *Science Speaks to Power: The Role of Experts in Policy-making* (London: Frances Pinter).
Danziger, P. (2006) *Shopping: Why We Love It, and How Retailers Can Create the Ultimate Customer Experience* (Chicago: Kaplan).
De Geus, A. (1997) *The Living Company: Habits for Survival in a Turbulent Environment* (Washington, DC: Longview).
Elzen, B., F.W. Geels and K. Green (eds.) (2004) *System Innovation and the Transition to Sustainability: Theory, Evidence and Policy* (Cheltenham, UK: Edward Elgar).
Etzioni, A (1998) 'Voluntary Simplicity: Characterisation, Select Psychological Implication and Societal Consequences', *Journal of Economic Psychology* 19.5: 619-43.
Fuchs, D., and S. Lorek (2005) 'Sustainable Consumption Governance: A History of Promises and Failures', *Journal of Consumer Policy* 28.3 (September 2005): 361-70.
Geels, F.W. (2005) *Technological Transitions and System Innovations: A Co-evolutionary and Sociotechnical Analysis* (Cheltenham, UK: Edward Elgar).
Gore, A. (2006) *An Inconvenient Truth: The Planetary Emergency of Global Warming and What We Can Do About It* (Emmaus, PA: Rodale Press).
Guba, E.G., and Y.S. Lincoln (1989) *Fourth Generation Evaluation* (Newbury Park, CA/London/New Delhi: Sage Publications)
Hajer, M.A. (1995) *The Politics of Environmental Discourse: Ecological Modernisation and the Policy Process* (Oxford, UK: Clarendon Press).
Hamel, G., and C.K. Prahalad (1994) *Competing for the Future* (Boston, MA: Harvard Business Review Press).
Hardin, G. (1968) 'The Tragedy of the Commons', *Science* 162: 1,243-48.
Hart, S., and M.B. Milstein (1999) 'Global Sustainability and the Creative Destruction of Industries', *Sloan Management Review*, Autumn 1999: 23.
Hekkert, M.P., R.A.A. Suurs, S.O. Negro, S. Kuhlmann and R.E.H.M. Smits (2007) 'Functions of Innovation Systems: A New Approach for Analysing Technological Change', *Technological Forecasting and Social Change* 74.4 (May 2007): 413-32.
Hertwich, E. (2005) 'Life-cycle Approaches to Sustainable Consumption: A Critical Review', *Environmental Science and Technology* 39.13: 4,673-84.
Hisschemöller, M. (1993) *De democratie van problemen: De relatie tussen de inhoud van beleidsproblemen en methoden van politieke besluitvorming* (*The Democracy of Problems: The Relation between Policy Problems and Methods for Political Decision-making*) (PhD thesis; Amsterdam: Vrije Universiteit Amsterdam, Vrije Universiteit Uitgeverij).
Holdsworth, M., and P. Steedman (2005) *Sixteen Pain-Free Ways to Help Save the Planet* (London: National Consumer Council).
Holliday, C., and J. Pepper (2001) *Sustainability through the Market: Seven Keys to Success* (Geneva: World Business Council for Sustainable Development; www.wbcsd.org [accessed 25 August 2007]).
Illich, I. (1978) *Toward a History of Needs* (New York: Pantheon Books).
Jackson, T., W. Jager and S. Stagl (2004) *Beyond Insatiability: Needs Theory, Consumption and Sustainability* (Guildford, UK: University of Surrey, Centre for Environmental Strategy).
Kemp, R., and S. Van den Bosch (2006) *Transitie-experimenten: Praktijkexperimenten met de potentie om bij de dragen aan transities* (*Transition Experiments: Practical Experiments with the Potential to Contribute to Transitions*) (Delft, Netherlands: Kenniscentrum SysteemInnovaties en Transities, TNO).
Kipling, R. (1899) 'The White Man's Burden', *McClure's Magazine*, 12 February 1899.
Kuhn, T.S. (1962) *The Structure of Scientific Revolutions* (Chicago: University of Chicago Press).

Lebel, L., S. Lorek, D. Fuchs, P. Garden, D.H. Giap, J. Manoochehri, H. Shamshub, A. Tukker and U. de Zoysa (2006) *Enabling Sustainable Production and Consumption Systems* (USER Working Paper, WP-2006-08; Chiang Mai, Thailand: Unit for Social and Environmental Research).

Lomborg, B. (2001) *The Sceptical Environmentalist: Measuring the Real State of the World* (Oxford, UK: Oxford University Press).

Mandelbaum, M. (2002) *The Ideas that Conquered the World: Peace, Democracy and Free Markets in the Twenty-first Century* (New York: Public Affairs/Perseus Book Group).

Marks, N., S. Abdallah, A. Simms and S. Thompson (2006) *The (Un)Happy Planet: An Index of Human Well-being and Environmental Impact* (London: New Economics Foundation/Friends of the Earth).

Max-Neef, M. (1991) *Human Scale Development: Conception, Application and Further Reflections* (New York: Apex Press).

Meadows, D. (1999) *Leverage Points: Place to Intervene in a System* (Hartland, VT: The Sustainability Institute).

——, J. Randers and D. Meadows (2004) *Limits to Growth: The 30-year Update* (White River Junction, VT: Chelsea Green Publishing).

Meijkamp, R. (2000) *Changing Consumer Behaviour through Eco-efficient Services: An Empirical Study on Car Sharing in the Netherlands* (PhD thesis; Delft, Netherlands: Technical University of Delft).

Nelson, R. (1993) *National Systems of Innovation: A Comparative Analysis* (Oxford, UK: Oxford University Press).

Princen, T. (2005) *The Logic of Sufficiency* (Cambridge, MA: MIT Press).

Rifkin, J. (2002) *The Hydrogen Economy* (New York: Tarcher/Putnam).

Rotmans, J. (2003) *Transitiemanagement: Sleutel voor een duurzame samenleving* (*Transition Management: Key to a Sustainable Society*) (Assen, Netherlands: Van Gorcum).

Sabatier, P.A. (1987) 'Knowledge, Policy Oriented Learning and Policy Change: An Advocacy Coalition Approach', *Knowledge* 8.4: 649-92.

Schön, D.A., and M. Rein (1994) *Frame Reflection: Towards the Resolution of Intractable Policy Controversies* (New York: Basic Books).

Schumacher, E.F. (1973) *Small Is Beautiful* (New York: Harper & Row).

—— (1979) *Good Work* (New York: Harper & Row).

SCR (Sustainable Consumption Roundtable) (2006) *I Will if You Will: Towards Sustainable Consumption* (London: Sustainable Development Commission/National Consumer Council, www.sd-commission.org.uk).

Shove, E. (2003) *Comfort, Cleanliness and Convenience: The Social Organisation of Normality* (Oxford, UK: Berg).

—— (2004) 'Efficiency and Consumption: Technology and Practice', *Energy and Environment* 15.6: 1,053-65.

Simon, H.A. (1957) 'Theories of Decision Making in Economics and Behavioural Science', *American Economic Review* 49: 253-83.

Spaargaren, G. (1997) *The Ecological Modernisation of Production and Consumption: Essays in Environmental Sociology* (Wageningen, Netherlands: Landbouw Universiteit).

Stern. N. (2006) *The Economics of Climate Change* (The Stern Review; Cambridge, UK: Cambridge University Press).

Suh, S. (2004) *Materials and Energy Flows in Industry and Ecosystem Networks: Life-cycle Assessment, Input–Output Analysis, Material Flow Analysis, Ecological Network Flow Analysis, and their Combinations for Industrial Ecology* (PhD thesis; Leiden, Netherlands: CML Leiden University).

Tukker, A. (2005) 'Leapfrogging into the Future: Developing for Sustainability', *International Journal for Innovation and Sustainable Development* 1.1: 65-84.

—— (ed.) (2006) 'Environmental Impacts of Products', *Journal of Industrial Ecology* 10.3 (Special Issue).

—— (2007) *Sustainable Consumption by Certification: The Case of Coffee. A Review of Coffee Certification Schemes, their Market Penetration and Impact—and the Invisible Consumer* (working paper for SPACES; Delft, Netherlands: TNO, Draft January 2007).

—— and U. Tischner (eds.) (2006) *New Business for Old Europe: Product Service Development, Competitiveness and Sustainability* (Sheffield, UK: Greenleaf Publishing).

——, P. Eder and E. Haag (2001) *Ecodesign: European State of the Art. Part I. Comparative Analysis and Conclusions* (ESTO Report EUR 19583 EN; Seville, Spain: DG JRC IPTS; esto.jrc.es/reports_list.cfm [accessed 25 August 2007]).

UN (United Nations) (2006) *The Millennium Goal Report* (New York: UN; unstats.un.org/unsd/mdg/Resources/Static/Products/Progress2006/MDGReport2006.pdf [accessed 25 August 2007]).

UNEP (United Nations Environment Programme) (2002) *Consumption Opportunities* (Geneva: UNEP).

Vlek, C.A.J., A.J. Rooijers and E.M. Steg (1999) *Duurzamer Consumeren: Meer Kwaliteit van Leven met Minder Materiaal? (More Sustainable Consumption: More Quality of Life with less Material?)* (research report for the Dutch Ministry of Environment; Groningen, Netherlands: COV, Groningen University).

Von Weizsäcker, E.U., A. Lovins and L.H. Lovins (1997) *Factor Four: Doubling Wealth, Halving Resource Use* (London: Earthscan Publications).

Weinberg, A.M. (1972) 'Science and Transscience', *Minerva* 10.2: 209-22.

# Abbreviations

| | |
|---|---|
| ACSIS | Australian Centre for Science Innovation and Society |
| B2B | business-to-business |
| B2C | business-to-consumer |
| B2G | business-to-government |
| BMBF | Bundesministerium für Bildung und Forschung (German Federal Ministry of Education and Research) |
| BOP | base of the pyramid/bottom of the pyramid |
| BRASS | Centre for Business Relationships, Accountability, Sustainability and Society (Cardiff University) |
| CAS | Chinese Academy of Science |
| CCGT | combined-cycle gas turbine |
| CCP | Cities for Climate Protection (ICLEI) |
| CEE | Central Eastern Europe |
| CEO | chief executive officer |
| CEU | Central European University |
| CFC | chlorofluorocarbon |
| CFL | compact fluorescent light bulb |
| CHP | combined heat and power |
| CI | Consumers International |
| CJD | Creutzfeldt–Jakob disease |
| CNNIC | China Internet Network Information Center |
| $CO_2$ | carbon dioxide |
| COICOP | Classification of Individual Consumption According to Purpose |
| CSR | corporate social responsibility |
| DfS | design for sustainability |
| DSM | demand-side management |
| DVD | digital versatile disc |
| EEA | European Environment Agency |
| EEG | Erneuerbare-Energien-Gesetz (Renewable Energy Law, Germany) |
| EiC | Eco-Innovative Cities |
| EMUDE | Emerging User Demands for Sustainable Solutions |

| | |
|---|---|
| EPA | Environmental Protection Agency (USA) |
| ETAP | Environmental Technologies Action Plan |
| ETC-WRM | European Topic Centre on Waste and Resource Management |
| EU | European Union |
| FSC | Forest Stewardship Council |
| GATT | General Agreement on Tariffs and Trade |
| GDP | gross domestic product |
| GEMIS | Global Emission Model for Integrated Systems (Öko-Institut) |
| GM | General Motors |
| GMAC | GM Acceptance Corporation |
| GNP | gross national product |
| GSM | Global System for Mobile Communications |
| HANPP | human appropriation of net primary production |
| HEI | higher education institution |
| HiCS | Highly Customerised Solutions |
| ICLEI | International Council for Local Environmental Initiatives |
| ICLEI–A/NZ | ICLEI Australia and New Zealand |
| ICSID | International Council of Societies of Industrial Design |
| ICT | information and communications technology |
| ICTRANS | impact of ICT on transport and mobility |
| IEA | International Energy Agency |
| IGO | intergovernmental organisation |
| IIED | International Institute for Environment and Development |
| IÖW | Institut für ökologische Wirtschaftsforschung (Institute for Ecological Economy Research) |
| IPAT | impact = population × affluence × technology |
| IPCC | Intergovernmental Panel on Climate Change |
| IPR | intellectual property rights |
| ISOE | Institute for Social-Ecological Research |
| IUCNNR | International Union for Conservation of Nature and Natural Resources |
| JRC–IPTS | Joint Research Centre, Institute for Prospective Technological Studies |
| LCA | life-cycle assessment |
| LCP | least-cost planning |
| LCSSM | life-cycle service-system map |
| LENS | Learning Network on Sustainability |
| LETS | Local Economic Trading System |
| LG | local government |
| LOHAS | lifestyles of health and sustainability |
| MDI | Motor Development International |
| MEPSS | Method for PSS Development |
| MSC | Marine Stewardship Council |
| NAFTA | North American Free Trade Agreement |
| NEPP4 | 4th National Environmental Policy Plan (Netherlands) |
| NGO | non-governmental organisation |
| $NH_3$ | ammonia |
| NIS | national innovation system |
| NMI | Natural Marketing Institute |

| | |
|---|---|
| NOA | need–opportunity–ability |
| NO$_x$ | nitrogen oxides |
| NPD | new product development |
| NSI | National System Innovation |
| OECD | Organisation for Economic Cooperation and Development |
| OFFER | Office of Electricity Regulation (UK) |
| ÖI | Öko-Institut e.V. (Institute for Applied Ecology, Germany) |
| ÖÖI | Österreichisches Ökologie-Institut (Austrian Institute for Applied Ecology) |
| PCR | post-consumer recycled |
| PEFC | Programme for the Endorsement of Forest Certification schemes |
| PROSECCO | product and service co-design process |
| PSS | product-service system |
| R&D | research and development |
| RCD | resource-conscious design |
| ROHS | Reduction of Hazardous Substances Directive (EU) |
| RTI | research, technology and innovation |
| RTTT | Race to the Top (IIED) |
| SCORE! | Sustainable Consumption Research Exchanges |
| SCP | sustainable consumption and production |
| SCR | Sustainable Consumption Roundtable (UK) |
| sdbm | Sustainable Design of Business Management |
| SDO | sustainability design-orienting |
| SDOS | sustainability design-orienting scenarios |
| SDS | Strategic Design Scenarios |
| SDS | Sustainable Development Strategy (EU) |
| SEPA | State Environmental Protection Administration (China) |
| SIFO | National Institute for Consumer Research (Norway) |
| SIS | sectoral innovation system |
| SM | stakeholder matrix |
| SME | small or medium-sized enterprise |
| SO$_2$ | sulphur dioxide |
| SpD | system–product design |
| SPSS | sustainable product-service systems |
| SRI | socially responsible investment |
| SusProNet | Sustainable Product Development Network |
| SUV | sport utility vehicle |
| SVN | sustainable value network (Wal-Mart) |
| TNC | transnational corporation |
| TNO | Organisation for Applied Scientific Research (Netherlands) |
| UBA | Umweltbundesamt (German Federal Environment Agency) |
| UMTS | Universal Mobile Telecommunication Network |
| UN | United Nations |
| UN DESA | UN Department of Environmental and Social Affairs |
| UNCED | UN Conference on Environment and Development |
| UNEP | UN Environment Programme |
| UNEP DTIE | UNEP Division of Technology, Industry and Economics |
| UNESCAP | UN Economic and Social Commission for Asia and the Pacific |

| | |
|---|---|
| WBCSD | World Business Council for Sustainable Development |
| WLAN | Wireless Local Area Network |
| WSSD | World Summit on Sustainable Development |
| WTO | World Trade Organisation |
| WWF | formerly World Wide Fund for Nature |

# About the contributors

**Maj Munch Andersen** is a senior scientist at DTU, Denmark, the Danish Technical University. Her work focuses on innovation studies, particularly within eco-innovation processes and nanotechnology, both at the firm, inter-firm, technology and innovation system level. She has a special interest in the analysis of environmental and innovation policy and the relationship between the two. She has previously worked at the Copenhagen Business School as well as spending four years on innovation policy within the Danish Ministry of Trade and Industry and the Danish Ministry of Science where she was responsible for developing a green innovation strategy.

Univ.-Prof. Dr **Frank-Martin Belz** holds the chair of Brewery and Food Industry Management at the Technische Universität München (Germany). He received his PhD in 1995 from the University of St Gallen (Switzerland). The focus of his research team is on 'innovation and marketing of consumer goods in the context of sustainable development'. For years, Frank-Martin Belz has published extensively on sustainability marketing and has taught courses on the subject at different European universities.

Dr des. **Jasper Boehnke** has completed his PhD studies at the Institute for Economy and the Environment, University of St Gallen, Switzerland, and worked as a researcher with the project 'micropower in residential buildings' (2005–2007), funded by the Swiss National Science Foundation. His research interests concern sustainable development in general as well as sustainable management and sustainable energy. He also has a special interest in business models in energy markets. In addition to his academic research activities, he works with the European Energy Exchange in Leipzig, Germany.

**Martin Charter** is the Director and Visiting Professor of Sustainable Product Design at The Centre for Sustainable Design at the University College for the Creative Arts at Canterbury, Epsom, Farnham, Maidstone and Rochester, UK. Since 1988, he has worked at director level in 'business and environment' issues in consultancy, leisure, publishing, training, events and research. He is the author, editor and joint editor of various books and publications including *Greener Marketing* (Greenleaf Publishing, 1992 and 1999), *The Green Management Gurus* [e-book] (1996), *Managing Eco-design* (Centre for Sustainable Design, 1997) and *Sustainable Solutions* (Greenleaf Publishing, 2001). Martin has an MBA from Aston Business School in the UK, and has interests in sustainable product design, green(er) marketing, and creativity and innovation.

## ABOUT THE CONTRIBUTORS

**Tom Clark** has 25 years' management consulting experience. He has both undergraduate and postgraduate qualifications in engineering, business, and environment. Areas of work have included development of sustainability strategies and plans within the public and private sector, eco-design, environmental technology, cleaner production, environmental management systems, waste management, energy management, sustainable buildings, emissions trading schemes, sustainable transport and ecological management. His clients have included international, central and local government and government agencies, computes from, telecommunications, oil and gas, mining electronics, electrical, automotive, food, metals and plastics, engineering, building and construction, transportation (all modes), retailing and distribution, banking, insurance and professional services.

**Sophie Emmert** studied policy sciences at Twente University in The Netherlands and currently works as researcher at the Business Unit Innovation and Environment of TNO Built Environment, Delft, The Netherlands. She has worked with the International Centre for Integrated Assessment and Sustainable Development in Maastricht. Her main research interests are transition management and related policy processes. Sophie took responsibility for the day-to-day management of the SCORE! project.

**Tareq Emtairah** is a research engineer at the International Institute for Industrial Environmental Economics, Lund University, Sweden, and a frequent lecturer in Europe and the Middle East in the field of preventative environmental management and policy. He has a BS in engineering from Rutgers University in the USA, a master's in Environmental Management and Policy from Lund University, and a postgraduate research certificate in R&D management from Tokyo Institute of Technology. Current research activities and interests include strategic environmental management, environmental innovation and corporate social responsibility.

**Theo Geerken** is project leader for the Product and Technology Studies group which forms part of the expertise Centre of Integrated Environmental Studies at the Flemish research institute VITO. This group is dedicated to research for both industry and government through the application of methods and concepts such as LCA, ecodesign, substance flow analysis, cleaner production, sustainability evaluation and technology assessment. Theo is a civil engineer in physics. He has industry experience in the reprographic sector consisting of 16 years in research, product development and engineering and integrating environmental goals into business practice. He has worked at VITO since 2000.

**Casper Gray** is a Co-director of Wax RDC (www.wax-rdc.com), working with multinational and start-up companies such as Clarks and Active Disassembly Research Ltd on research, design and consultancy projects to further progress towards sustainability. He collaborates closely with The Centre for Sustainable Design and the University College for the Creative Arts (UCCA) at Farnham, UK, where he also acts as a visiting lecturer in sustainable design for a variety of course programmes. He holds a BA (Hons.) in Furniture and Product Design from Nottingham Trent University and an MA in Sustainable Product Design from UCCA. He has previously worked as a designer in the UK and Milan.

**François Jégou** is a Strategic Design Consultant with a degree in industrial design, and teaches as visiting professor at the Faculty of Design of the Politecnico in Milan and La Cambre School of Visual Art in Brussels. Since 1990 he has run the consultancy DALT based in Paris and latterly the Strategic Design Scenarios in Brussels, specialising in co-designing scenarios and new product-service system definition. François is active in various fields including: sustainable design, interaction design, cognitive ergonomics, senior-friendly design, compliance and security of pharmaceutical products, and innovation in food products. He is involved in several EU research projects, promotes the www.sustainable-everyday.net platform and the www.solutioning-design.net network.

**René Kemp** is senior researcher at UNU-MERIT (a research and training centre of United Nations University working in close collaboration with the University of Maastricht), ICIS (International Center of Integrated Assessment and Sustainable Development) and Drift (Dutch Research Institute for Transitions). He is an economist with a long-standing interest in issues of innovation and the environment. He has pioneered ideas in strategic niche management and transition management. He has provided consultancy advise to several high-level policy initiatives, including the Environment Council in July 2004 and the fourth National Environmental Policy Plan in The Netherlands which adopted the transitions approach. He is the project leader of two large research projects, one on sustainable mobility and the other on measuring eco-innovation. Further information about these projects and René Kemp can be found at kemp.unu-merit.nl.

Dr **Klaus Kubeczko** is senior researcher at the Austrian Research Centers ARC in the division of systems research in Vienna. He has a background in economics, social sciences and electrical engineering. He has also worked at IIASA and INNOFORCE, a project centre of the European Forest Institute. His main research interest lies in the fields of foresight and strategy development in technology and innovation policy, long-term transition of sociotechnical systems and systems innovation for sustainable development. He is engaged in public-funded research as well as in consultancy projects.

**Saadi Lahlou** graduated as Statistician Economist at the Ecole Nationale de la Statistique et de l'Administration Economique (ENSAE, Paris) and has a PhD is Social Psychology (Ecole des Hautes Etudes en Sciences Sociales, Paris). He also holds degrees in Human Ethology and Ecology, and in Human Biology. After studying facial mimics of newborns in a CNRS-INSERM research unit, he joined the CREDOC (Research Centre for Lifestyles and Social Policies, Paris) where he directed the Consumer Forecasting Department from 1987 to 1993. He then joined EDF R&D Division where he headed the Internal Social Forecast Group from 1993 to 1998, devoted to organisation studies. He was Head of Cognitive Studies at EDF R&D Division from 1998 to 2000. He is currently Head of the 'Laboratory of Design for Cognition'; this lab runs an industrial test-bed for futuristic hybrid (physical/digital) work environments, where actual workers perform their everyday activities under observation.

**Benny Din Leong** is Assistant Professor, Leader of the Asian Lifestyle Design Research Lab and former discipline leader of BA (Hons.) in Industrial and Product Design of the School of Design at Hong Kong Polytechnic University. Leong practised industrial design in Asia, Europe and China respectively after his graduation from the Hong Kong Polytechnic and the Royal College of Art at London in the late '80s. He has worked for, and with, companies such as Philips (Netherlands), HP (France), Alessi (Italy), Eckart+Barski (Germany), Cuckoo (Korea) and Huawei (China) on various pioneer and strategic design consultation projects. In recent years, Leong has been researching ways to revitalise traditional Chinese culture for sustainable design and contemporary design practice, and actively promoting the notion of Design for Sustainability in China. Together with Prof Ezio Manzini, he co-founded the Chinese Network on Design for Sustainability and co-written a book titled *Design Vision: A Sustainable Way of Living in China* (Ningnan Publishing House, 2006).

**Ezio Manzini** is Professor of Design at the Politecnico di Milano; there he coordinates the Doctorate in Design. He is visiting lecturer at the Tohoku University (Japan) and at the Wuxi University (China), and honorary fellow at the Australian Centre for Science, Innovation and Society at the University of Melbourne. His works are focused on strategic design, service design, design for sustainability and on social innovation in everyday life. Major recent books are *Sustainable Everyday* (with F. Jégou; Edizioni Ambiente, 2003) and *Design Vision: A Sustainable Way of Living in China* (with B.D. Leong; Ningnan Publishing House, 2006). Several papers can be found at www.sustainable-everyday.net/manzini.

Dr **Oksana Mont** is an Associate Professor at the International Institute of Industrial Environmental Economics at Lund University, Sweden. She has a PhD in Technology, an MSc in Environmental Management and Policy and an MSc in Biology and Chemistry. Her research interests lie in sustainable consumption and production, product-service systems, sustainable lifestyles, innovation studies and waste management. She works in close contact with policy-makers, business and consumer organisations and teaches undergraduate, master and PhD courses at universities in Europe and Latin America, and is involved in 'educate the educators' programmes.

**Dario Padovan** graduated in Political Sciences and has a PhD in Sociology. He is currently Senior Researcher at the University of Turin, teaches sociology and works at the Department of Social Sciences of the University of Turin. Dario has published articles on the sociology of ethnic relations, racism, the history of sociology and environmental sociology. He is presently undertaking research on urban ecology, urban unsafety, health and prejudice, and the role of experts on environmental conflicts. He has managed international conferences on eco-sustainable development and ecological crises. He is also a member of the editorial board of *Democracy and Nature*, *Theomai Journal* and *Cosmos and History*.

Dr **Harald Rohracher** is senior researcher at the Inter-University Research Centre for Technology, Work and Culture (IFZ) in Graz, Austria, and at the Faculty for Interdisciplinary Studies, University of Klagenfurt. He was director of IFZ from 1999 to 2007. His study background is in physics, sociology and science and technology policy. In his research he is interested in the social shaping of technology, the sociotechnical transformation of energy systems, and the role of end-users in technological innovation.

**Chris Ryan** is Professor and Co-director of the Australian Centre for Science Innovation and Society at the University of Melbourne and Director of the Victorian Eco-innovation Lab. He was foundation professor of Design and Sustainability at RMIT University in Melbourne, initiating the Australian EcoReDesign program (1993–97), working with 20 Australian companies to develop new greener products and a new ecodesign methodology. From 1998 to 2003 he was professor and Director of the International Institute for Industrial Environmental Economics at Lund University, Sweden. He is joint editor of *D4S: Design for Sustainability: A Global Guide*, a forthcoming UNEP guide to ecodesign and imaging sustainability (RMIT University Press, 2008).

**Gerd Scholl**, economist, is senior researcher at the Institute for Ecological Economy Research (Institut für ökologische Wirtschaftsforschung [IÖW]), Berlin. He is head of the Ecological Consumption department. His research areas are sustainable consumption and marketing, product-service systems, sustainability in retailing, integrated product policy (IPP) and consumer perception of new technologies (e.g. nanotechnologies). He has worked in projects for the European Commission (DG Environment, DG Research), UNCTAD, and a number of German public authorities (e.g. Federal Environment Ministry, Federal Research Ministry). He has co-authored, among others, *Product Policy in Europe: New Environmental Perspectives* (with F. Oosterhuis and F. Rubik; Springer, 1996) and has published recently in reviewed journals, such as the *Journal of Cleaner Production*.

**Irmgard Schultz**, doctor of political science, is senior researcher at and co-founder of the Institute of Social-Ecological Research (ISOE) in Frankfurt am Main, Germany. She heads the research department 'Everyday Life Ecology and Consumption'. Her specific profile is inter- and transdisciplinary sustainability research and 'gender and environment'; her main fields of research are: everyday life ecology and sustainable development; sustainable consumption and production patterns (sustainable consumption and lifestyles in different action fields: waste, construction, nutrition), gender equality and social aspects in corporate social responsibility; and CSR and consumers. Most recently she participated in the EU project RARE: 'Rhetoric and Realities: Analysing Corporate Social Responsibility in Europe' (www.rare-eu.net).

**Immanuel Stieß** is a social scientist and holds a PhD in planning sciences. He coordinates the research area 'Energy and Everyday Life' at the Institute for Social-Ecological Research (ISOE), Germany. His research interests include social-ecological lifestyle research on housing and energy use, energy use and social exclusion, and life-cycle management of urban neighbourhoods. Currently, he is working in a research project on social marketing strategies, promoting energy-efficient modernisation of private homes.

**Eivind Stø** is the Director of Research at the National Institute for Consumer Research (SIFO), Norway. He gained a Mag.art. in political science from the University of Oslo in 1972. He was an assistant at the Norwegian Election Programme from 1972 to 1976; and from 1976 to 1998 he worked for the Norwegian Fund for Market and Distribution Research. Since 1989, he has been a researcher at SIFO, as Head of Research from 1990, and Director of Research from 1998. He has initiated, participated and coordinated several European projects. His research interests include consumer complaints, consumer policy, sustainable consumption, energy use and nanotechnology.

**Pål Strandbakken** gained a Mag.art. in Sociology from the University of Oslo in 1987. He worked on the Alternative Future project until 1992 when he joined the National Institute for Consumer Research (SIFO), Norway, as a research fellow. In 2007 he finished his PhD on product durability and the environment at the University of Tromsø. His research interests include product durability, ecological modernisation, sustainable consumption, energy use and nanotechnology.

**Harald Throne-Holst** gained his MSc in Chemical Engineering from the Norwegian Institute of Technology at the University of Trondheim in 1994. He joined the National Institute for Consumer Research (SIFO), Norway, as a researcher in 1996. His research interests include sustainable consumption, energy use, climate change, technology development and nanotechnology.

Ass. Prof. **Ursula Tischner** specialised in Eco- and Sustainable Design of products and services during her architecture and design studies. After her master's graduation, she worked at the German Wuppertal Institute for Climate, Environment and Energy, where she was involved in developing concepts such as eco-efficiency, MIPS, Factor 4 and Factor 10, Ecodesign, Eco-innovation, etc. In 1996 she founded her own company, econcept, Agency for Sustainable Design, in Cologne where she carries out research and consulting projects with small and large companies and other organisations. She runs training and educational courses and programmes, such as the Sustainable Design Program at the Design Academy Eindhoven, The Netherlands, publishes books, and is member of networks, juries, evaluation and standardisation bodies around Eco- and Sustainable Design.

Dr **Arnold Tukker** joined TNO in 1990 after some time working for the Dutch Environment Ministry. Over time, his focus shifted from life-cycle assessment, material flow analysis and risk assessment to interactive policy-making and sustainable system innovation and transition management. In 1998 he published a book on societal disputes on toxic substances, for which he was awarded a PhD from Tilburg University. He has published about 40 peer-reviewed papers, 5 books, 10 book chapters and 150 other publications, and is frequently asked as invited speaker worldwide. In his career, he has been awarded over €15 million in mainly international research grants. He currently manages the research programme on Transitions and System Innovation within TNO Built Environment and Geosciences, Business Unit Innovation and Environment. This programme was evaluated as one of TNO's top-ranking programmes during the 2006 scientific assessment exercise. Arnold is the initiator and manager of the SCORE! network.

## ABOUT THE CONTRIBUTORS

**Edina Vadovics** is currently a PhD candidate at the Environmental Sciences and Policy Department of the Central European University (CEU) in Budapest, Hungary. Her research focuses on sustainable consumption and sustainable communities. Prior to her studies and research at CEU, she worked in environmental and sustainability management, and delivered training courses in the field both for companies and students in higher education. As a volunteer, she has been involved with creative communities both in the Eastern and Western part of Europe. She is also president of GreenDependent Sustainable Solutions Association.

For 15 years **Carlo Vezzoli** has been researching and teaching design scenarios, strategies, methods and tools for products, services and systems for sustainability. At the Faculty of Design of the Politecnico di Milano he is a professor of Product Design for Environmental Sustainability and of System Design for Sustainability, and director of the Research Unit Design and System Innovation for Sustainability (DIS, INDACO) department. Among other projects, he is coordinator of the international Learning Network on Sustainability (LENS, www.lens.polimi.it), of curricula development on Design for Sustainability focused on product-service system innovation, funded by the Asia Link Programme, European Commission.

**Gunnar Vittersø** holds a master's degree in Human Geography from the University of Oslo. After graduation in 1993 he worked as a research fellow at the Project for an Alternative Future. Since 1996 he has been working as a researcher at the National Institute for Consumer Research (SIFO), Norway. He is currently working on a PhD thesis on recreational consumption and sustainable rural development. His general research interest is sustainable consumption including sustainable food consumption, alternative food distribution and leisure consumption.

Dr **Marcus Wagner** is assistant professor at the Business School of the Technical University of Munich where he is currently on leave as Marie Curie Fellow at the Bureau d'Economie Théorique et Appliquée in Strasbourg. He is an associate research fellow at the Centre for Sustainability Management in Lüneburg where he also teaches on the MBA programme 'Sustainament'. His research is concerned with innovation and technology management as well as sustainability management in areas such as cooperation in innovation and sustainability related and environmentally related innovation.

Dr **Matthias Weber** is head of technology policy department at Austrian Research Centers ARC, systems research division, in Vienna, and lecturer at the Vienna University of Economics and Business Administration. He has a background in process engineering, political sciences and innovation economics. His main research interests are in the transformation of sociotechnical and innovation systems, in research and technology policy, in R&D collaboration network dynamics, and in foresight for policy strategy development.

Dr **Peter Wells** is a Reader in the Centre for Automotive Industry Research, UK, which he joined in 1990, and a lecturer in Logistics and Operations Management. He has a wide knowledge of the industry overall, particularly in an environmental context. Other research interests include the distribution, retail and marketing of cars, and the history of car design. Dr Wells has also published papers on wealth and sustainability, celebrity, and local eco-industrialism among other diverse interests. In 2002 Dr Wells became a founder member of the ESRC-funded Centre for Business Relationships, Accountability, Sustainability and Society (BRASS) with a ten-year programme to analyse the concept of Micro Factory Retailing and sustainable automobility. He is editor of *Automotive Environment Analyst*, and a member of the editorial board of the *International Journal of Innovation and Sustainable Development*.

**Tim Woolman** is Project Coordinator at The Centre for Sustainable Design, UK, supporting international projects, tool development and research in eco-innovation and sustainable consumption and production. This follows research in Eco-Product Innovation and Clean Manufacturing Technologies at Warwick Manufacturing Group, developing enablers for smaller engineering companies to help them make innovative step changes in their environmental performance. Prior to this, Tim worked for 15 years in automotive product development with GKN and Ricardo managing both detail design and client project delivery to cost, quality and performance targets; and recently as a Senior Design Engineer specifying components to comply with End-of-Life Vehicles legislation.

Prof. Dr **Rolf Wüstenhagen** is Vice Director of the Institute for Economy and the Environment, University of St Gallen, Switzerland. He teaches sustainability management and energy entrepreneurship at St Gallen and within the Community of European Management Schools (CEMS)'s Master of International Management programme. His research agenda focuses on processes for successful market introduction of new energy technologies, with a particular emphasis on sustainable consumption and production. He is a member of the Swiss Federal Energy Research Commission and has recently been a Visiting Professor at University of British Columbia and Wilfrid Laurier University (Canada). Prior to his academic career, Rolf worked with a leading European sustainability investment company with a focus on energy venture capital investments.

# Index

*Note:* page numbers in *italics* indicate figures and tables.

**Adaptive Foresight**
  conceptual foundations and experiences 345-66
**Adaptive and strategic planning**
  *see* Adaptive Foresight
**Adaptive technology**
  portfolio approach 355-6
**Affluence**
  based on consumption per capita 117-18
**Agriculture**
  transition to sustainable, in The Netherlands 385-6, *386*
**Allen, Paul** 50
**Anderson, Ray** 50
**Asset Recycling Programme** 396
**Assimilative capacity**
  waste emissions 24
**Attitudes**
  *see* Consumer behaviour
**Australia**
  ecodesign in eight local councils 197-211
  local government 203
**Australian Centre for Science Innovation and Society** 202
**Austria**
  Adaptive Foresight framework 345, 365-6
  adaptive policies project example 361, 363
  strategy development process 364
**Automotive industry**
  business models 80-96, *91*
  closed-loop product-service system 90
  economic pressures 89-90
  growing concerns over sustainability 90

**B2B**
  *see* Business-to-business
**B2C**
  *see* Business-to-consumer
**B2G**
  *see* Business-to-government
**Backcasting**
  application to scenarios 362
**Base-of-the-pyramid (BOP) economies**
  agenda for 408
  SCP challenges 4, 6
**Basic**
  organic retail chain 127
**Behaviour**
  *see* Consumer behaviour
**Behavioural innovation**
  sustainability potential 161-6, *162*, *163*
  *see also* other innovation types listed following 'Innovation'
**Behavioural intention** 238
**Ben & Jerry's** 60
**Benefits**
  consumer perceptions 122
  promotion of sustainable energy 71-2, 74-5
**Biodiversity**
  transition in The Netherlands 384-6
**Biogas power plants** 76

The Body Shop   60
Bonding social capital   272, 273, 278
BOP
  see Base-of-the-pyramid economies
Boycotts   239-40, 244, 251
BP
  support for Kyoto Protocol   131
  vision of sustainable development   125
Branson, Richard   50
BRASS
  see Centre for Business Relationships, Accountability, Sustainability and Society
Bridging social capital   272, 273, 278
Brundtland Report   15, 22, 25, 236
Business
  corporate sustainability statements   125
Business drivers and strategies   47-50
Business management
  sustainable design of business management   174-5
Business models
  in the automotive industry   86-96
  defining the concept   84-6
  key elements   74
  in management research   72-4
  needed to deliver SCP   81-4
  relevance in sustainable energy context   74-7, 77
  sustainable innovation   58-60
  and value creation   85-6
Business and SCP
  contributions to change   409-11, 420, 421
  fostering change   61-3
  limitations and obstacles to change   63-5
  the move towards sustainable production   50-2
  product certification   53-4
  products and production processes   51-2
  strategic sustainable innovation   57-61
  supply chains   52-3
Business-to-business (B2B) markets
  customer type   46, 55
  levels of systemic change   396-402
Business-to-consumer (B2C) markets
  and consumer goods marketing   118-33
  customer type   46, 55
  levels of systemic change   395-402
Business-to-government (B2G) markets
  customer type   46, 55

Capital intensity
  power of incumbents in energy sector   72, 75
Capital stocks
  concepts of sustainability   22, 24
Car-sharing systems   125, 395, 401-2
Carbon dioxide ($CO_2$) emissions
  time path in The Netherlands   385
  see also Low-carbon technologies
Cardiff University
  BRASS   80
Carpet-leasing   397
Case, Steve   50
CEE
  see Central Eastern Europe
Central Eastern Europe (CEE)
  background information   303-4
  consumption trends   304-8, 305, 307, 308
  creative communities   309, 312-15
  gross domestic product   304, 305
Centre for Business Relationships, Accountability, Sustainability and Society (BRASS)   80
Certification
  of products   53-4
CFCs
  see Chlorofluorocarbons
CFL
  see Compact fluorescent light bulb
Change
  in business environment   47-8
  contributions to   409-13, 420, 421
  fostering change towards SCP   61-3
  limitations and obstacles to   63-5
  see also Systemic change; Transformations; Transition
Checks
  sustainability checks   21
Childcare
  CEE countries   309-10
China
  development dilemma   216-17
  pursuit of sustainable solutions   214-30
  signs of change   224-5
Chlorofluorocarbons (CFCs)   48
Choice editing
  for sustainability   55
CHP
  see Micro combined heat and power
Circular casualties   352

INDEX    457

**Climate change**
  Australian attitudes and action    204-6
  cost to the UK    48-9
  significant contribution by households    100
**Closed-loop supply chains**    85-6
**Collective sharing systems**
  norms in Swedish society    401
**Communal washing centres**
  'Sunwash Laundry'    217-18, *218*
  in Sweden    395-6, 401, 402
**Communication**
  credibility and trust    130
  sustainability marketing    130
**Communities**
  elective    189
**Community creativity**
  egalitarian view    29-31
**Community housing**
  EMUDE models    188-9
**Community solutions**
  semi-public status of resources    195
**Compact fluorescent light bulb (CFL)**    56
**Competition**
  supply side of market structure    102-3
  sustainability marketing    129
**Conceptual approach**
  followed in SCORE!    3-11
  sustainability and SCP    14-38
**Confidence**
  quality of life    280
  and trust    274, 275-6
**Congestion tax, London**    433
**Consumer behaviour**
  desire for material consumption    422
  how consumers choose    414
  models of change    246-8, *247*, *248*, 414-16
  multi-dimensional model    248, *248*
  'needs and wants'    413-14
  and sustainable consumption    235-41
  sustainable marketing    122-4
**Consumer economies**
  agenda for    408
  SCP challenges    4, 6
**Consumer goods**
  property rights and duties    257
**Consumer knowledge**
  about products and processes    243-4
**Consumer organisations**    247
**Consumer perspective**
  research shift towards    298

**Consumer research**
  attitude–behaviour model    237-9
  limitations to consumer-oriented approaches    249-50, *249*
**Consumerism**
  rapid growth in China    216-17, 220
**Consumption**
  affluence and resource scarcity    117-18
  business as stakeholder    65
  'conspicuous consumption'    262
  consuming local products    313
  contributions to change    413-16, *420*, *421*
  design-led    56-7
  and the environment    9, 235, 427-9, *428*
  factors influencing    54-5
  influencing consumer motivation    433
  interaction with production    34
  material consumption    422
  optimisation    408
  organic products    282-3
  political and ethical    239-41, 242, 244-5, 247
  potential for radical sustainable changes    234-51
  radical reduction of impact per unit    35
  shifting focus of ecodesign research    199
  and social stratification    260-6
  social-ecological research    289-98
  symbolic meaning    260-6, 263-4
  *see also* Sustainable consumption
**Consumption patterns**
  consumption as a process    246-8, *247*, *248*
  demand-side management    57
  designer responsibility    147, 412
  *see also* Consumer behaviour
**Consumption pressures**
  State of Victoria    203-4
**Contracting/leasing**
  energy sector models    76
**Contractual negotiations**
  product-service systems    258
**Control**
  symbolic control in usership    265
**Convenience**
  sustainability marketing    129-30
**Corporate social responsibility (CSR)**
  promoting sustainability    52-3
**Corporate sustainability statements**    125
**Costs**
  consumer perceptions    122
  total life-cycle costs    129-30

'Creative communities'
  in Central Eastern Europe  309
  grass-roots initiatives for sustainable development  302
  group participation  312-13
CSR
  *see* Corporate social responsibility
Cultural constraints
  barriers to SCP in China  221-2
Cultural theory
  key characteristics  26
  and management of toxic substances  26, 28
Customer costs
  sustainability marketing  129-30
Customer satisfaction
  operational sustainability marketing  128-9
Customer solutions
  operational sustainability marketing  128-9
Customer value
  commercialising sustainable energy technologies  76-7

Decisions
  transition dynamics  351
Demand-side management (DSM) programmes  57, 108-11
  energy-specific  *108*
  non-specific  *109*
Design Council Research Centre  141-2
Design for sustainability
  contributions to change  411-13, *420*, *421*
  definitions and scope  139-41, *140*
  four dimensions  144, 151-3, *151*
  limitations and potential  150-3, *151*
  model of change  144-50
  office environments  174-5
  present and potential role of design  139-42, *151*
  real problems and real solutions  160-1
  resource-conscious design  211
  rules for economic processes  19
  social equity and cohesion  146-50, *151*, *152*
  sustainable design research gaps  160-1
  system innovation  145-6
  towards 'solution-oriented' design  141-2, *412*
  user-centred  56-7
Designer responsibility  147
Development
  alternative ways in CEE countries  311-12
  *see also* New product development; Sustainable development
Disposal of products  396

Disruptive technologies  86, 87
Distrust
  organic products  282-3, 285
  quality of life  280
Doerr, John  50
'Dream society'  240, 244-5
Drive-by-wire  94
DSM
  *see* Demand-side management
Durability of products
  PSS schemes  398
Dutch government
  *see* Netherlands

E-commerce
  field for business model research  73
Ecodesign  142-4, *151*, *152*
  of services in eight local councils  197-211
Eco-efficient system innovation
  design for  144-6, *151*, *152*
Eco-innovation  57-61, 320-1
  limits to role of business  64
  and the service sector  201
  *see also* other innovation types listed following 'Innovation'
Eco-Innovative Cities programme (EiC)
  aims, process and methodology  206-9
  balancing the social and the technical  199-200
  collaboration, teleconferences and Wiki  209
  design methodology  207-8
  within local councils in Melbourne  199-201
  outsourcing by councils  211
  overview  201-4
  resource-conscious design  211
  review and outcomes  210-11
  service sector, primary focus of project  201
Eco-labelling  53-4, 240-1, 247
  signalling strategy  258
Eco-marketing
  *see* Marketing
Economic divide
  technology accessibility in China  226
Economic efficiency
  and energy efficiency  104, 105
Economic growth
  and concepts of sustainability  24, 25
Economic reality
  barriers to SCP in China  220

INDEX 459

Economies
   see Base-of-the-pyramid economies;
      Consumer economies; Emerging
      economies
EEA
   see European Environment Agency
Egalitarian view
   paradigms for sustainable development   29-31
Egg production
   ethical discussions about consumption   244
EiC
   see Eco-Innovative Cities programme
Elderly, care of
   CEE countries   310
Electricity generation
   UK and Germany   103
Electricity markets
   changes in regulations   104, 110
Emerging economies
   agenda for   408
   SCP challenges   4, 6
Emerging User Demands for Sustainable Solutions (EMUDE)
   EMUDE research process   179-81, *180*
   matrix illustrating scenarios   *182-3*
   network   149
   relational quality guidelines   193-6
   research project   209, 302, 311, 312
   strategic design examples   *184-7*
   towards synthesis of quality and access   191-6
   'ways of doing' models   181, 188-90
Emissions
   energy-related   104
   greenhouse gases   216
   targets   19, 21, 22
Emotional products
   incentives for customers   399
EMUDE
   see Emerging User Demands for Sustainable Solutions
Energy
   business models for sustainability   70-7
   consumption   100-1
      State of Victoria   204
   renewable   112
   rising prices   49
   transition to sustainable in The Netherlands   384-6
Energy efficiency
   and economic efficiency   104, 105

   energy efficiency gap   103-6
   future research   112-13
   improvements
      role of the Porter Hypothesis   106-10
Energy sector
   capital intensity and power of incumbents   72, 75
Energy Star®
   eco-labelling   53-4, 56
Energy supply
   demand–response programmes   57
Energy technologies
   key target area for SCP   71
Energy transition
   themes, goals and transition paths   386
Energy-specific demand-side management (DSM) programmes   108-11, *108*, *109*
Entrepreneurs
   investing in sustainable technology   49-50
Environment
   exploitation, boundaries and interventions   407
   future relevance of marketing   116-18
   impact of human activities   117-18
Environmental challenges
   and social capital   280-4
Environmental costs
   passing on to consumers   102-3
Environmental discourse
   shift in focus and rhetoric   235
Environmental impact
   aggregate   391-2
   influencing use of materials and energy sources   142-4, 146, *151*, *152*
   of nutrition styles   296-7, *297*
Environmental implications
   of egalitarian view   31
Environmental issues
   Australian attitudes and action   204-6
Environmental labelling
   see Eco-labelling
Environmental marketing
   see Marketing
Environmental outcomes
   function innovation   394
Environmental pressures
   and climate change   48-9
   reducing environmental pressure by consumption   427-9, *428*

**Environmental regulation**
  incentive-based or command-and-control  102
  a means to achieve SCP objectives  99
  stimulating innovation  101-2, 103
**Environmental responsibility**
  turned into profit centre  397
**Equity**
  fostering of  36, 408
  principle  146
**Ethical companies**
  implications of takeover  60-1
**Ethical consumption**  239-41, 242, 244-5, 251
  egg production  244
**Ethical products**  55
**Ethical Purchasing Index**  55
**European Environment Agency (EEA)**
  Copenhagen workshop 2006  10
**European Union**
  Environment Technologies Action Plan  321
  *European Transport Policy 2010: A Time to Decide*  395
  household expenditure  308
  mandatory energy labels  240-1
  Sustainable Development Strategy  15, 19, 33, 146, 235
  objectives  17, 18
  *see also* Central Eastern Europe (CEE)
**Evergreen Lease**
  service contract  397
**Everyday life**
  consumption research  289
  nutrition  292-3
**Evolutionary transition**  370-1, 387
**'Experience economy'**  49
**Exploitation of resources**
  need for boundaries and intervention  407

**Family-like Services**
  EMUDE models  188
**Financial crises**
  in automotive industry  94
**Food networks**  30
**Ford, Henry**  86
**Forest Stewardship Council (FSC)**
  product certification  53-4
**Freitag**
  sustainable products  52
**FSC**
  *see* Forest Stewardship Council

**Fuel cell vehicles**  94
**Function innovation**
  and PSSs  392-402
  *see also* other innovation types listed following 'Innovation'
**Functional products**
  customers not emotionally attached  399
**Functional thinking**
  design concept  143
**Future transition processes**  354-6

**Gates, Bill**  50
**GEMIS life-cycle analysis**  296
**General Electric**  76
  'ecomagination' initiative  58
**General Motors**  86
  AUTOnomy  94
  General Motors Acceptance Corporation (GMAC)  86-7
**Germany**
  renewable energy levy  112
**Global market**
  socio-ethical dimension  147
**Global resources exploitation**
  need for boundaries and interventions  407
**Global warming**
  *see* Climate change
**Globalisation**  159
**GM**
  *see* General Motors
**Goals**
  for sustainability  34-8
  trade-offs and balances  125
  *see also* Transition goals
**Governance**
  objectives of  38
  for SCP agenda  33-8
**Governmental organisations**
  marketing sustainability to  130
**Green & Black's**  60
**Green consumerism**
  in electricity market  107-8
**'Green' electricity**  75
*Green Gauge Report*
  on US consumers  123, 124
**Green marketing**
  *see* Marketing
**Greenhouse gases**
  China's emissions  216
  *see also* Climate change

# INDEX    461

Greenstar e-commerce and community centres   148
Group participation
  creative communities   312-13

H&M
  textile retailer   127
The Hannover Principles   20
Hierarchic view
  paradigms of sustainable development   31-2
Hierarchical planning
  transition management   375
High-volume producers
  automotive industry   87-8
Households
  contribution to climate change   100
Housing
  community   188-9
  example of LCA impact assessment   121
  extended   189
Human activities
  impact on natural environment   117-18
Hungary
  Ecological Institute for Sustainable Development   311
  gross domestic product   305
  household spending   307, 307
  Open Garden Foundation   312
  per capita footprint   306, 306
  Sustainable Village Project   311
Hybrid vehicles   95

ICLEI
  see International Council for Local Environmental Initiatives
ICSID
  see International Council of Societies of Industrial Design
ICT
  see Information and communications technology
Identity
  consumption and social stratification   260
IIED
  see International Institute for Environment and Development
IKEA   127
Incremental innovation
  innovation types, sustainability potential   161-6, 162, 163

see also other innovation types listed following 'Innovation'
Incumbents' power
  capital intensity in energy sector   72, 75
Indicator systems
  sustainability and SCP   15-19
Individual technologies   350
Individualist view
  paradigms for sustainable development   28-9
Industrial transformation
  solutions in SCP literature   82-3
Industrialised nations
  addressing unsustainability   81-2
Information and communications technology (ICT)
  significant role   198
  technology accessibility in China   226
Innovation
  central feature of SCP analysis   83
  for energy-efficient consumption and production   99-113
  innovation types and sustainability potential   161-5, 162, 163
  in promotion of sustainable consumption   54-5
  stimulated by environmental regulation   101-2, 103
  systemic nature of   400
  see also Behavioural innovation; Eco-efficient system innovation; Eco-innovation; Function innovation; Incremental innovation; Radical innovation; Social innovation; Sustainability innovation; System innovation; User-led innovation
Innovation cycle theory   328-9
Innovation offsets   101-2
Innovation system approach   321, 322, 326-30, 392, 417
  change processes towards PSSs   400
  framework   326-7
  functional approach   329-30
  organisational strand   327-9
  policy focus   332-4
Innovation system research   346-7
Innovation system transition model
  towards SCP   333-4
Innovation systems
  functions with sustainability objectives   353
Innovative business models
  and the automotive industry   89-96

Institute for Social-Ecological Research (ISOE)  288
Institutional trust  276, 285
Institutions
  coordinating role in transition management  375
  marketing transformational sustainability to  130-1
Interface Inc.  397, 402
Intergovernmental Panel on Climate Change (IPPC)  48
International Council for Local Environmental Initiatives (ICLEI)  202
International Council of Societies of Industrial Design (ICSID)  139
International Institute for Environment and Development (IIED)
  Race to the Top (RTTT)  52
International Survey of Corporate Responsibility Reporting, 2005  205
International Technology Roadmap for Semiconductors  107
International Union for Conservation of Nature and Natural Resources (IUCNNR)  15, 16
Internet
  new communities of interest  201
  technology accessibility in China  226
Internet and e-commerce
  field for business model research  73
Interpersonal trust  275, 276
Interventions
  change in and outside the regime  424-5
  types and time-horizons  425-6, 426
Investment
  in sustainable energy  72
  in sustainable technology  49-50
IPAT formula
  environmental impact of human activities  117-18
IPPC
  see Intergovernmental Panel on Climate Change
ISOE
  see Institute for Social-Ecological Research
IUCNNR
  see International Union for Conservation of Nature and Natural Resources

Jenbacher
  biogas power plants  76
Johannesburg Plan of Action  15, 33
Johannesburg Plan of Implementation  16, 18

Khosla, Vinod  50
Kids Clothing Chain  187, 194
Kyoto Protocol  131

Labelling
  see Eco-labelling
Landscape
  sociotechnical  324, 346, 374
Laundries
  see Communal washing centres
LCA
  see Life-cycle assessment
LCD
  see Life-cycle design
'Lead markets'
  as a core initiative  333
Leadership
  crucial to establishing SCP  66
Learning Network on Sustainability (LENS)  148-9
  limits and barriers to SCP in China  217-22, 224
Leasing
  in B2B markets  396
  customer reluctance  397
  see also Usership
Leasing practices
  concept of continual lease  397
Leasing/contracting
  energy sector models  76
LENS
  see Learning Network on Sustainability
Life-cycle assessment (LCA)
  analysing impact of products  120-2
Life-cycle design (LCD)
  system innovation  144-6, 151, 152, 412
Life-cycle environmental impacts
  of consumption  9
Life-cycle orientation
  sustainable production  128
Life-cycle of products
  changing ownership structure in PSS cases  402
  PSSs providing added value  255, 397

**Lifestyle research**
  sustainable consumption   290-2
**Lifestyle resource intensity**
  State of Victoria   204
**Linking social capital**   273, 279
**'Local'**
  as a niche of eco-innovation   200
**Local products**
  creative communities   313
**Local-scale initiatives**
  quality of relationships   194
**Local visibility**
  facilitating access to initiatives   191
**'LOHAS consumers'**   124
**London**
  congestion tax   433
**Low-carbon technologies**
  call for increased deployment   49

**Mainstreaming**
  of sustainable products   126-7
**Maintenance**
  consumer's duty for owned goods   257
**Maintenance contracts**
  for biogas power plants   76
**Management**
  demand-side   57
  *see also* Transition management
**Management research**
  business models as unit of analysis   72-4
**Management support**
  for sustainability initiatives   192-3
**Manufacturing**
  products designed for remanufacturing   396
**Marine Stewardship Council (MSC)**   131
  product certification   53-4
**Market coordination**
  managing transition   375
**Market economies**
  and sustainability   35
**Market inefficiency**
  possible failure in energy markets   106
**Market structure**
  and sustainability innovation   102-3
**Marketing**
  awareness of total life-cycle costs   130
  conception of sustainability marketing   118-32, *120*
  eco-marketing   119
  multi-channel   129
  and sustainability   55-6
**Marks & Spencer**   51
**Marrakech process**   3
**Material possessions**
  *see* Possessions
**Materials**
  rising costs   49
**Materials and energy sources**
  environmental impact   142-4, 146, *151*, 152
**MDI**
  *see* Motor Development International
**Melbourne**
  ecodesign of services in eight local councils   197-211
**MEPSS**
  *see* Method for PSS development
**Method for Product-Service System Development (MEPSS)**   152-3, 161
**Micro combined heat and power (CHP)**   75
**Migros**   127
**Mobile phones**
  technology accessibility in China   226
**Mobility**
  car sharing   125, 395, 401-2
**Modulation**
  in transition management   377
**Monitoring**
  operationalising sustainable development   19, 22
  of the transition process   382-3
**Montreal Protocol**   48
**Moral questions**
  changes in consumer values   242, 248
  ethical consumption   239-41, 248
**Morgan**
  UK niche vehicle manufacturer   88
**Motor Development International (MDI)**   92
**MSC**
  *see* Marine Stewardship Council
**Multi-segment marketing**   127

**Natural food products**   127
**Natural Marketing Institute (NMI)**
  'LOHAS consumers'   124
**Natural resources**
  *see* Resources

**Natural science**
  and concepts of sustainability   24
**The Natural Step**   20, 25, 26
**Need**
  areas of need   349-50
**Needs and wants relationship**   236-7, 413-14
  changes in consumer values   242, 248
**Negre, Guy**   92
**The Netherlands**
  4th National Environmental Policy Plan   377
  Policy Plan on the Environment and the Economy   395
  target-setting for acidifying emissions   22
  transition policies   384-6, *385*, *386*
**Network of relationships**
  social capital   272-3, 274, 275-6
**Network-structured initiatives**
  for sustainable enterprise   148-50
**Networks**
  direct access   190
  exchanging clothes   194
**'New colonialism'**   159
**New product development (NPD)**
  integrating sustainability   58-60, *59-60*
**NGOs**
  *see* Non-governmental organisations
**Niche business model**
  automotive industry   88-9
**Niche markets**   60-1, 126-7, 434
  car sharing   395, 401-2
**Niches**
  for new technologies   324-5, *324*, 331-2, 346, 373-4
**NMI**
  *see* Natural Marketing Institute
**Non-governmental organisations (NGOs)**
  marketing sustainability to   131-2
**Normative sustainability marketing**
  *see* Marketing
*Nouveaux riches*
  'conspicuous consumption'   262
**NPD**
  *see* New product development
**Nutrition**
  environmental impact of styles   296-7, *296*
  in everyday life   292-3
  typology of styles   293-6, *294*

**Objectives**
  of SCP   *18*
  of sustainable development   *16-17*
**Office sector**
  *see* Sustainable Office research project
**Operational sustainability marketing**
  *see* Marketing
**Organic products**   127, 312
  consumption   282-3
**Orienting transitions**   352-3
**Oslo**
  Soria Moria Symposium on Sustainable Consumption and Production, 1994   *18*
**Oslo Declaration on Sustainable Consumption**   3, 100
**Outsourcing**
  configuring value creation in energy sector   75-6
  and the 'liability trap'   211
**Overcapacity**
  in automotive industry   87
**Over-consumption**   242, 251
**Ownership of products**
  ownership rights   256-7
  replaced by renting/leasing   255, *261*, 263-6
  symbolic meaning of   260-3
**Ownership structure**
  shift in diverse user contexts   401

**Paradigms**
  for sustainable development   26, 27, 28-32
**Patagonia**
  sportswear and outdoor retailer   130
**PCR**
  *see* Post-consumer recycled
**Pelamis Wave Power**   75-6
**PetrolCo**   93-4
**Philips**   52
  marketing CFL bulbs   56
**Planning**
  hierarchical planning for transition   375
**Poland**
  eco-friendly hamlet   311
  per capita footprint   306, *306*
**Polarisation**
  consumer goods markets   126-7
**Policy abuse**
  barriers to SCP in China   219-30

Policy instruments
    for transition management   383-4
Policy intervention
    system innovation   376-7, 377
    in the transition scheme   374
Political appropriation
    barriers to SCP in China   219, *219*
Political consumption   239-41, 244-5
Political implications
    of egalitarian view   31
Population
    world population growth   117-18, 407
The Porter Hypothesis
    relevance to SCP   99-113
Porter, Michael E.   99
Portfolios of real options   355-6
Possessions
    symbolic perspective   260-6, 265-6
    *see also* Ownership
Post-consumer recycled (PCR)
    Synchilla fleece   130
Poverty
    eradication   36, 408
Premium producers
    automotive industry   87-8
Price-sensitivity   127
Principal–agent theory
    product-service systems   258-9
Principles
    sustainability and SCP   15-19, *20-1*
Product certification   53-4
Product concept failures
    automotive industry   94-5
Product life
    socio-ecological problems   120-2
Product-service systems (PSSs)
    applicable to automotive industry   90
    in emerging and developing contexts   148-9
    symbolic perspective on usership   255-67
    system innovation   145-6
    systemic changes and SCP   391-402, *400*
Production
    business as stakeholder   65
    dematerialisation   407
    dual focus of sustainable products   128
    enhancing environmental and social
        performance   432
    improvement of sustainable products   128-9
    interaction with consumption   *34*
    rising costs   49

Production and consumption
    challenging the prevailing orthodoxy   81-4
    process model   246-7, *247*
Products and processes
    business contribution to SCP   51-2
    consumer knowledge   243-4
Products and services
    sustainability marketing   128-9, 130
    tangible elements   129
    *see also* New product development
Programmes
    for system innovation   382
Property rights theory   256-7
Provider and user
    changing relationships of PSSs   398-9
PSSs
    *see* Product-service systems
Public and political process
    transformational sustainability marketing
        130-1
Public support
    creation and maintenance   383
Public transport networks   314, 434

Quality of goods and services
    information economics   257-8
Quality of life   32, 36-8, *37*
    confidence, trust and distrust   280, 285
    improvement   147

Race to the Top (RTTT)   52
Radical innovation   60, 64-5
    sustainability potential   161-6, *162*, *163*
    *see also* other innovation types listed following
        'Innovation'
Real options
    portfolios of   355-6
Recycling
    (PCR) Synchilla fleece   130
    PSS case   397
    and re-use   313-14
    of waste
        environmental challenges   281-2
ReEntry
    carpet reclamation initiative   397
Regulations
    *see* Environmental regulation
Relational qualities
    EMUDE design guidelines   193-6
Remanufacturing   396, 398

Rent-o-box flexible office   172, *173*
Renting/leasing
   *see* Usership
Resource-conscious design   211
Resources
   exploitation   407
   natural resources in The Netherlands   384-6
   official policy concepts   15-19
   semi-public status   195
   unrestrained consumption   214, 215, 216
Retro-distribution   129
Revenue models
   sustainable energy sector   76
Rio de Janeiro
   Conference, 1992   15, 289
   Rio Declaration   16
Roddick, Anita   60
Roper Starch Worldwide
   US consumer surveys   123
Routines and habits
   consumer behaviour models   237-9, 245
RTTT
   *see* Race to the Top

Sams, Craig   60
Satisfaction
   quality of life   32, 36-8, *37*
   satisfactional thinking   143, 144-5, 412
Schmack Biogas AG   76
SCORE!
   *see* Sustainable Consumption Research Exchange
SCP
   *see* Sustainable consumption and production
SDS
   *see* European Union, Sustainable Development Strategy
Sectoral innovation system analysis   330
Sectoral system of innovation   400
Sensible sustainability   24-5
Service
   as a three-dimensional process   264
Service clubs
   EMUDE models   190
Service delivery
   re-establishing symbolic control   265
Service design
   eco-service redesign   202, 207, 209
   PSS schemes   399

Service processes
   standardisation   265
Service sector
   EiC programme   201, 202, 203, 204, 208-9, 211
Services and products
   *see* Products and services
Sharing systems
   norms in Swedish society   401
Shopping
   environmental challenges and social capital   284
Signalling strategies
   communication quality information   257-8
Skills
   inter-personal   195
Social behaviour alteration
   strategy for sustainable China   227, *227, 228, 229*
Social capital   272-7
Social displacement
   barriers to SCP in China   222, *223*
Social equity and cohesion
   design for sustainability   146-50, *151, 152*
Social innovation
   design of sustainable solutions   178-96
   *see also* other innovation types listed following 'Innovation'
Social quality
   availability of resources   194
Social science
   and concepts of sustainability   24
Social-ecological consumption research
   framework   289-92
   nutrition styles   292-7
   overall conclusions   298
Societal expectations
   driver for change   49-50
Societal functions
   dynamics and transformation   400-2
Societal transformations
   as transitions   370-5
Socio-ecological consciousness
   three levels of   123, 132
Socio-ecological dimension
   strategic sustainability marketing   126-7
Socio-ecological impact matrix
   qualitative instrument of LCA   121-2
Socio-ecological problems
   sustainability marketing   120-2

Socioeconomic implications
  of egalitarian view  31
Socio-ethical dimension
  to design for sustainability  150
Sociological approach
  'ways of doing'  181, 188-90
Sociotechnical landscapes  324, 346, 374
Sociotechnical regimes  324-6, 332, 346, 374
Sociotechnical systems
  long-term transformation processes  345
Sociotechnical transformation
  transition processes  348-51
Solar electricity  75
Solutions
  see Emerging User Demands for Sustainable Solutions (EMUDE)
Soria Moria Symposium on Sustainable Consumption and Production, Oslo, 1994  18
Specialist producers
  automotive industry  87-8
Status symbols
  car as a status symbol  395
  products as  260, 262-3
Stirling engines  75
Strategic design
  dissemination of solutions  149
  towards a synthesis of quality of access  191-6
Strategic sustainability marketing
  see Marketing
Strategic sustainable innovation  57-61
Strategies
  for changing business environment  47-8
Strong sustainability  24
'Sunwash Laundry'
  sample sustainable service solution  217-18, 218
Supply chains  52-3
  compliance with standards in  408
  impact of shift to PSS  399-400
SusProNet  161
  product-service systems screening tool  166
Sustainability
  and business opportunity  58
  changing interpretations in design research  142-4
  and the office sector
    see Sustainable Office research project
  'weak', 'strong' and 'sensible'  24-5

Sustainability goals
  pursued through environmental policy  376
  pursued through system innovation  376-7
Sustainability innovation
  entering the market  126
  facade modules  170, 172
  and the Porter Hypothesis  99-113
  strategic  57-61
  see also other innovation types listed following 'Innovation'
Sustainability marketing
  see Marketing
Sustainability triangle  22, 23, 427
Sustainability Victoria  202
Sustainability visions  377-81, 378
Sustainable agriculture
  transition in The Netherlands  354-6, 386
Sustainable community
  interpersonal and systemic trust  275-6
Sustainable consumption  54-7, 65-6
  consumer behaviour  122
  emerging patterns in CEE  314-15
  a multi-dimensional approach  234-51
  social-ecological consumption research  288-98
  umbrella term  387
Sustainable consumption and production (SCP)
  automotive industry  81-4
  business role in  46-66
  challenges
    by type of country  4, 6, 6
    long-term  435-6
  changing consumer behaviour  241-6
  China
    limits and barriers to SCP  218-25
    search for an appropriate strategy  225-30
  design for problem solving and radical change  159-77
  developing an SCP policy  429-35
  energy sector a key target area  70-1
  example definitions and objectives  18
  goals  407-8
  marketing as a business function in SCP  116-33
  need for radical change  391-3
  need to stimulate progress  66
  as part of sustainable development  33
  policy and action required  376, 427
  and the Porter Hypothesis  99-113
  priority areas  8
  production–consumption regime in a landscape context  423

PSS dynamics and transformation required 400-2
relevance 2-3
research exchanges 4-8
research still required 338-9
stakeholder approach required 251
a system approach 138-53, *151*, 423-4, *423*
system transition processes 320-39
systems perspective, integration via 418-36, *420*, *421*
transition management 369-87
unsustainable production and consumption 242, 422

**Sustainable Consumption Research Exchange (SCORE!)**
conceptual approach 3-11
final conclusions of first half of project 437-40, *438*, *439*
hypothesis 7, 418-19
knowledge communities involved 7, *7*
structure and timetable 10

**Sustainable design**
*see* Design for sustainability

**Sustainable development** 118-19
Australian local government support 205-6
and SCP 14-38
some definitions and objectives *16-17*

**Sustainable energy**
transition in The Netherlands 384-6

**Sustainable energy technologies**
challenges in commercialisation 71-2

**Sustainable Marketing Knowledge Network** 56

**Sustainable mobility**
transition in The Netherlands 385-6, *386*

**Sustainable Office research project** 160, 161, 164, 167-75
method for sustainability assessment 165
methodology and tools 170
overview of development phases *171*
Rent-o-box flexible office 172, *173*
Revital facade 170, *172*
sustainability evaluation 170, 172-5, *172*, *173*
sustainable design of business management 174-5

**Sustainable solutions**
*see* Emerging User Demands for Sustainable Solutions (EMUDE)

**Sustainable urban living relationship model** 277-85

**Sustainable use of natural resources**
transition in The Netherlands 384-6

**Sweden**
communal washing centres 395-6, 401, 402

**Switzerland**
'green' electricity 75

**Symbolic control**
in and through usership 265

**Symbolic perspective**
enhancing symbolic meaning of usership 264-5
products conveying cultural meaning 260-6, 263-4, *264*

**Synchilla fleece** 130

**System innovation**
experiments for 382
in innovation systems 345-66
innovation types *162*, 163
life-cycle design (LCD) 144-6, *151*, 152
policy focus 331-2
and PSSs 394
research 345-6
to pursue sustainability goals 376-7
transition management model 331-2
*see also* other innovation types listed following 'Innovation'

**System innovation approach**
multi-level model 324-5, *324*, *325*, 416-18, 419, 423-4
oriented at environmental policy 321-6

**System innovation policy**
contributions to change 416-18, *420*, *421*

**System innovation and sustainability**
criteria 164
identifying sustainability potential 165
*see also* Sustainable Office research project

**System optimisation** 393

**System redesign** 393-4

**System transition processes**
different approaches 334-6
emphasis on policy rather than analysis 335-6
limitations 337-8
policy recommendations 331-4
representation and role of market 335
for SCP 320-39
sociotechnical regime transformation model 325-6, 332
transition management 323-5, *324*, *325*

**System-level changes** 392-3

**System–product design (SpD) process**
towards sustainable solutions for China 217-30

**Systemic change**
and PSS cases 393-7, 400-2
and SCP 391-402
*see also* Change; Transformations; Transition
**Systemic trust** 275, 276, 285
**Systems theory**
concepts of sustainability 22, 23, 24

**Target setting**
operationalising sustainable development 19, 22
**Technology**
investment in sustainable 49-50
for resource productivity and energy efficiency 117-18
**Technology accessibility**
overcoming China's economic divide 226
**TH!NK**
battery electric vehicle 93
**Timing**
marketing sustainability innovation 127
**Toyota**
environmental impact research 122
hybrid car, Prius 95, 127
Toyota Production System 87
**Trade-offs**
ecological, social and economic objectives 125
**Transaction cost theory**
product-service systems 258
**Transformation processes**
long-term, multi-level 346
**Transformational sustainability marketing**
*see* Marketing
**Transformations**
societal 370-5
*see also* Change; Systemic change; Transition
**Transition**
four phases of 370-5, 371-5, *372*, *373*
sustainable 376-7
**Transition agendas**
joint agendas 380-1
**Transition arena**
selection of participants 381
**Transition cycles** 383
**Transition dynamics** 351-2
**Transition fields** 349-51
**Transition goals**
transition management model 370-1, 377-9, *378*, 379-81

**Transition management** 354-6
model for transitions to sustainability 377-9, *378*
multi-level model of technological change 324, *325*, 331-2, 346, 373
for SCP 323-5, *324*, *325*
solutions in SCP literature 82-3, 84
for sustainable consumption 369-87
system innovations 346
**Transition paths** 385, *386*
**Transition process**
evaluation 382-3
**Transition towards sustainability**
changing role of design research 142-4
**Transport**
demand management 57
environmental challenges and social capital 284
environmental view 129
public transport networks 314, 434
**Trust**
and confidence 273, 275-6
organic products 282-3, 285
quality of life 285
trust-based relationships 195-6
**Trustinfood Project** 282

**UNCED**
*see* United Nations, Conference on Environment and Development
**Unemployment**
CEE countries 310
**UNEP**
*see* United Nations, Environment Programme
**Unilever** 131
**United Kingdom**
Ethical Purchasing Index 55
**United Nations**
Conference on Environment and Development (UNCED), Rio de Janeiro 1992 15
Conference on Environment and Development (UNCED), Rio Declaration *16*
Department of Environmental and Social Affairs (UNDESA) 3
Environment Programme (UNEP) 3, 147-8, 387
10 Year Framework on SCP 116, 235
Millennium Goals 15, 17
**Upgrading**
PSS schemes 398

**Urban life**
  sustainable   272-5
**Use value and exchange value**   236-7
**User accessibility**
  social innovation   191-3
**User and provider**
  changing relationships of PSSs   398-9
**User-centred design**   56-7
**User-led innovation**
  *see also* other innovation types listed following 'Innovation'
**User orientation**   349
**Usership**
  product-service systems   255-67
  *see also* Leasing

**Value**
  consumer perceptions   122
**Value creation**
  and business models   85-6
  outsourcing value creation in energy sector   75-6
**Value proposition**
  promotion of sustainable energy   74-5
**Values**
  changes in consumer values   242, 248
**Victoria, Australia**
  consumption pressures   203-4
**Visions and values**
  normative sustainable marketing   124-5, 132

**Wal-Mart**
  fish purchasing   53
  sustainable value networks   51
**Waste management**
  environmental challenges   281-2, 285
**Waste and material flows**
  State of Victoria   204
**Waste-stream recovery**   85-6
**Water consumption**
  State of Victoria   204

**Water utilities**
  influencing demand   57
**Wave energy converters**   75-6
**'Ways of doing'**
  everyday life models   181, 188-90
**WBCSD**
  *see* World Business Council for Sustainable Development
**WCED**
  *see* World Commission on Environment and Development
**Weak sustainability**   24
**Wealth disparity**
  barriers to SCP in China   221
**Well-being**
  need to redefine ideas   141
**WhisperGen®**   75
**Whole Foods Markets**   127
**Wholefood products**
  consumption   282-3
**Windows of opportunity**
  drivers for change to SCP   245-6, 248
**World Business Council for Sustainable Development (WBCSD)**   28
**World Commission on Environment and Development (WCED)**   16
  Brundtland Report   15, 22, 25, 236
**World Conservation Strategy**   15
**World Summit on Sustainable Development (WSSD)**   2, 3, 100, 116
  Johannesburg Plan of Action   15, 33
  Johannesburg Plan of Implementation   16, 18
**WSSD**
  *see* World Summit on Sustainable Development
**WWF**   15, 16, 131

**Xerox Corporation**
  Asset Management Programme   396-7, 402